MOLECULAR DYNAMICS FOR TRIBOLOGY

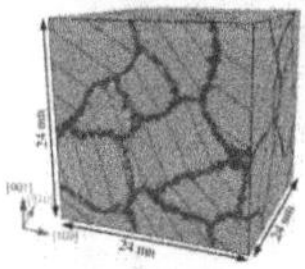

MOLECULAR DYNAMICS FOR TRIBOLOGY

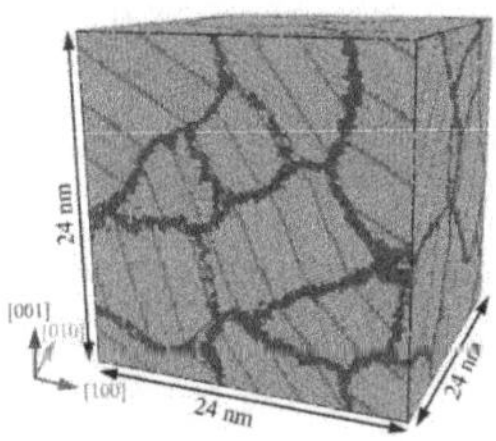

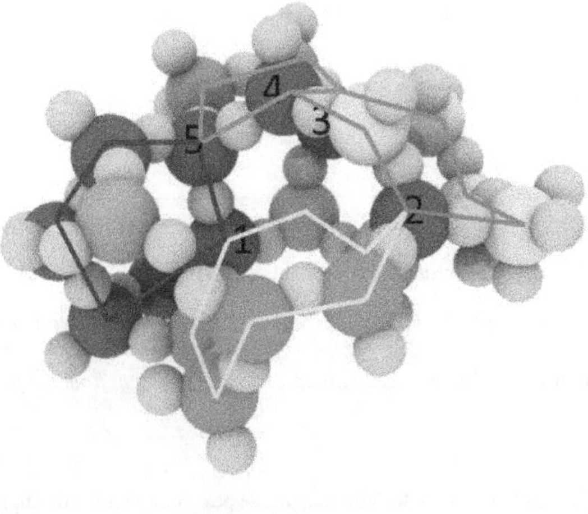

Liang Fang
Xiamen University Tan Kah Kee College, China

World Scientific

NEW JERSEY • LONDON • SINGAPORE • BEIJING • SHANGHAI • HONG KONG • TAIPEI • CHENNAI • TOKYO

Published by

World Scientific Publishing Co. Pte. Ltd.

5 Toh Tuck Link, Singapore 596224

USA office: 27 Warren Street, Suite 401-402, Hackensack, NJ 07601

UK office: 57 Shelton Street, Covent Garden, London WC2H 9HE

Library of Congress Cataloging-in-Publication Data

Names: Fang, Liang (Professor on mechanical engineering) author
Title: Molecular dynamics for tribology / Liang Fang,
Xiamen University Tan Kah Kee College, China.
Description: New Jersey : World Scientific, [2026] | Includes bibliographical references and index.
Identifiers: LCCN 2025011023 | ISBN 9789819812943 hardcover |
ISBN 9789819812950 ebook for institutions | ISBN 9789819812967 ebook for individuals
Subjects: LCSH: Microelectromechanical systems--Simulation methods |
Molecular dynamics--Simulation methods
Classification: LCC TK7875 .F36 2025 | DDC 621.38101/13--dc23/eng/20250612
LC record available at https://lccn.loc.gov/2025011023

British Library Cataloguing-in-Publication Data

A catalogue record for this book is available from the British Library.

摩擦磨损的分子动力学
Originally published in Chinese by China Machine Press Co., Ltd.

For any available supplementary material, please visit
https://www.worldscientific.com/worldscibooks/10.1142/14307#t=suppl

Desk Editors: Murali Appadurai/Steven Patt

Typeset by Stallion Press
Email: enquiries@stallionpress.com

Preface

With the increasing maturity of Micro-Electro-Mechanical Systems (MEMS) applications and the growing demand for large-scale integrated circuit manufacturing, the tribological issues that arise during manufacturing and application have garnered significant attention from scholars both domestically and internationally in recent years. The tribological research within these two domains falls under the field of nanotribology. In terms of applications, scholars have primarily focused on reducing friction and wear, whereas in manufacturing, the interest lies in the scientific utilization and control of friction and wear. A quintessential example is Chemical Mechanical Polishing (CMP) in chip manufacturing, where the goal is to maximize the material removal rate (MRR) while minimizing defects, which is contrary to the objective in applications where minimizing material wear is desired.

The rapid development of nanotribology research is closely tied to advances in modern microscopic detection technologies, such as Scanning Electron Microscopy (SEM), High-Resolution Transmission Electron Microscopy (HRTEM), X-ray Photoelectron Spectroscopy (XPS), Scanning Tunneling Microscopy (STM), Atomic Force Microscopy (AFM), and Nanoindentation Instruments (NI). Among these, the novel nanoindentation instruments, which integrate

AFM capabilities, not only facilitate nanoindentation tests but also enable nano-scratch tests, yielding precise measurements of mechanical behaviors at the nanoscale or microscale, such as indentation hardness, elastic modulus, yield strength, and fracture toughness. Some advanced nanoindentation devices can also conduct tests on micro/nano-scale materials, such as compression, tension, fracture, fatigue, friction, and wear, significantly advancing the field of nanotribology. Theoretically, the development of molecular simulation techniques, particularly molecular dynamics simulation software, has undoubtedly provided a catalyst for nanotribology research. These simulations can overcome the limitations of real nanoscale experiments by using large-scale parallel computing to simulate nanoscale friction and wear processes. The collaboration between molecular simulation theory and modern microscopic detection technologies has extended our understanding to the atomic and molecular levels, making cross-scale research truly feasible.

Like traditional tribology, nanotribology involves friction, wear, and lubrication and is an interdisciplinary field related to mathematics, physics, chemistry, mechanics, mechanical engineering, crystallography, and materials science. This book focuses on a highly specialized area within this broad field, namely, the use of molecular dynamics simulation software LAMMPS to simulate the nanoscale wear processes of various materials, aiming to derive general principles that could lay a foundation for practical applications in this domain.

Chapter 1 provides a brief overview of the development and application value of molecular dynamics. Chapter 2 outlines the principles, potential functions, and simulation tools of molecular dynamics. Chapter 3 begins with the theory and experiments of nanoindentation, followed by a discussion of the regularities observed in the nanoindentation of single-crystal copper and single-crystal silicon using molecular dynamics simulation. The final section of this chapter discusses the uniaxial tensile behavior of single-crystal nanowires, proposing the idea that material scale significantly influences material properties. Chapter 4 systematically explores the simulated experiments and findings on the nanoscale two-body and three-body abrasive wear of single-crystal copper under air lubrication, followed by a discussion on the nanoscale abrasive wear of single-crystal silicon. It highlights that one of the key differences between nanoscale and macroscopic abrasive wear is the need to account for the elastic recovery of the worn surface, leading to an improved criterion for distinguishing sliding and rolling of ellipsoidal abrasives at the nanoscale. Chapter 5 analyzes the behavior of single-crystal copper under nanoindentation, two-body, and three-body abrasive wear in the presence of a water film, extending the criteria for abrasive rolling and sliding to conditions with water lubrication. The presence of a water film reduces the friction coefficient and affects the material removal rate. Chapter 6 directly addresses the CMP process, systematically discussing the nanoindentation and wear behavior of single-crystal silicon surfaces covered with an amorphous silica film and analyzing the material removal mechanisms in the presence of a water film. It also explores the wear process using non-rigid abrasives, finding that non-rigid abrasives result in higher material removal efficiency and better surface quality of single-crystal silicon, with no defects in the substrate. In Chapter 7, the author systematically studies the deformation behavior of nanocrystalline copper using molecular dynamics and, combined with phase field modeling, constructs a more realistic polycrystalline model. This model quantitatively analyzes the movement of grain boundaries and the evolution of twinning, providing theoretical support for the rational design of crystal structures in the future.

This book is a continuation of my previous monograph on "Abrasive Wear" which has been published by China Science Publishing & Media LTD. The content presented here summarizes over ten years of the author's latest research. The main chapters and computational data were contributed by my doctoral students, including Dr. Jiapeng Sun (Associate Professor at Hohai University), Dr. Junqin Shi (Associate Professor at Northwestern Polytechnical University), Dr. Juan Chen (Associate Professor at Taiyuan University of Science and Technology), Dr. Meng Zhang (Postdoctoral Fellow at the University of Tokyo), and outstanding master's student Mr. Xiangzheng Zhu (Mindray Bio-Medical Electronics Co., Ltd.). I would like to extend my heartfelt thanks to all my students, especially Dr. Meng Zhang, who made significant contributions to the typesetting, organization, and writing of Chapters 1, 2, and 7. I also express my gratitude to the National Natural Science Foundation of China for funding the publication of this book (Project No. 51375364: Nanotribology of Three-Body Abrasive Wear in MEMS). I take this opportunity to thank my colleagues and other graduate students in my research group for their strong support during my research work. I am also grateful to all my friends, family, and relatives who have supported me. This book was completed at Xiamen University Tan Kah Kee College.

Errors in the book are inevitable, and I hope readers will understand and provide constructive feedback to help the author make corrections in future editions.

About the Author

Liang Fang is a Professor at Xiamen University Tan Kah Kee College. Before his retirement, he was a Professor at the State-Key Laboratory for Mechanical Behavior of Materials (SKL-MBM) at Xi'an Jiaotong University. He is the Editorial board member of the journal *Wear* published by Elsevier Group PLC and the journal *Lubrication Science* published by John Wiley & Sons, Inc. He is the recipient of the Outstanding Science and Technology Worker Award from the Lanzhou Institute of Chemical Physics, Chinese Academy of Sciences, and the Advanced Organizational Work Award from the Chinese Mechanical Engineering Society. His research interests are in molecular dynamics in wear process and the modeling and computer simulation of abrasive wear processes in materials. He has published about 209 papers with H-Index 34.

Contents

Chapter 1

Introduction

1.1 History of the Development of Molecular Dynamics Simulations

Molecular Dynamics (MD) is a simulation method that uses Newtonian mechanics and computer numerical solving of molecular systems. It samples different states of the molecular system to compute the configurational integrals and uses the results as a basis for further calculations of thermodynamic quantities and other macroscopic properties. Specifically, molecular dynamics studies the trajectories of atoms in a system using atomic nuclei. The initial atoms' displacement, velocity, and position change under an interaction potential function and by solving Newton's equations of motion. We can obtain the velocity, acceleration, and other relevant physical information of the atoms at different moments. The macroscopic properties of the substance are derived based on the statistical theory of thermodynamics.

The classical molecular dynamics approach was first proposed by Alder and Wainwright [1, 2] in 1957 and was initially applied to the "hard-sphere" liquid model. Their work led to the discovery of the "liquid-to-crystalline phase transition" predicted by Bethe and Kirkwood [3] in 1939 based on statistical mechanics, which later became known as the Alder phase transition. In 1963, Rahman [4] investigated the simulation of molecular dynamics of liquids using a continuous potential model. In 1980, Andersen *et al.* [5] created the constant pressure molecular dynamics method. Gillan *et al.* [6] extended the technique to non-equilibrium systems with temperature gradients in 1983, resulting in a set of molecular dynamics methods for non-equilibrium systems. Nose *et al.* [7] completed the creation of the constant temperature molecular dynamics method in 1984. For semiconductors and metals, where modeling potential functions is difficult, Car *et al.* [8] proposed a first-principles molecular dynamics method that combines electron theory and molecular dynamics methods. In 1991, Cagin *et al.* [9] further proposed a molecular dynamics method of grand canonical ensemble applied to the adsorption treatment. The rapid development of computer technology in the late 1980s, coupled with the proposal and development of multibody potential functions, led to further advancements in molecular dynamics simulation techniques.

Classical Molecular Dynamics (Classical MD) determines interatomic interaction potentials from experimental results or empirical models. These potentials are small enough to solve larger-scale problems but lack portability. Different empirical parameters may need to be determined for different problems. Before the 1980s, molecular dynamics simulations generally used the pair potential model, which only considered Coulombic forces and short-range interactions between neighboring atoms. It assumed that the system energy was the sum of the energies of the particles. Pair potentials provide a good description of almost all inorganic compounds except for metals and semiconductors. Some commonly used pair potentials are the hard-sphere potential, the Lennard–Jones (LJ) potential, the Morse potential, the Johnson potential, etc. Each of these potentials has its strengths in specific problems.

The positions of atoms in a polyatomic system affect the distribution of the electron cloud in a given space, which in turn influences how atoms interact with each other. This led to the consideration of many-body effects between particles and the development of many-body potential structures. One of the first such potentials, proposed in the early 1980s by Daw and Baskes [10], is known as the embedded-atom method (EAM). The EAM potential effectively describes interactions between metal atoms and is widely used in describing metallic systems. However, it is not applicable to organic molecules held together by covalent bonds or semiconductor materials. To address this, Daw and Baskes [11] proposed the Modified Embedded Atomic Nuclei Method (MEAM) which accounts for the non-spherical symmetry of the electron cloud, making it suitable for substances containing covalent bonding interactions. The MEAM model is semi-empirical, approaching the problem from a local electron density perspective and using experimentally obtained data for its parameters, such as lattice constants, transition energies, and elastic moduli.

During the equilibration of the system and after reaching equilibrium, a large amount of raw data, such as atomic positions, momentum, energy, and forces, are saved for each step or each N-step. From these saved data, the following parameters can be obtained through direct calculation or statistical analysis:

(1) basic energy, structural, and mechanical properties, some of which data will be used to correct the empirical potential function,
(2) coefficient of thermal expansion, melting point, and phase diagram related to pressure and volume,
(3) structural defects and diffusion, grain boundary structure, and sliding properties,
(4) thermal fusion, free energy difference between different phases, and heat conduction properties,
(5) radial distribution function and diffusion coefficient,
(6) description of processes and phenomena, e.g., spraying, vapor phase deposition, fast plastic flow, crack growth and fast fracture, nanoindentation, shock wave propagation, explosions, radioactivity, ion bombardment, cluster effects, and nano-friction.

The molecular dynamics computational method is very fast and has a larger computational system than the first-principles method. However, MD also has some limitations:

(1) The available potential functions are very limited, especially for multicomponent systems, while the accuracy of the potential functions is often questioned.

(2) The size of the simulated system still does not reach macroscopic dimensions, and the time scale is limited to the nanosecond order of magnitude.
(3) Electromagnetic properties are not available.

1.2 Application and Significance of MD Simulation

The progress of modern science relies on the synergy of experiment, theory, and simulation. Simulation, in particular, acts as a bridge between theory and experiment, playing a pivotal role in uncovering experimental phenomena and forecasting theoretical outcomes. Molecular dynamics simulation stands out as the method that most closely mirrors experimental conditions, providing a microscopic view of system evolution from its fundamental level. It reveals the mechanisms and laws governing experimental phenomena. For a system consisting of atomic nuclei and electrons, the equations of motion (e.g., Newton's equation, Hamilton's equation, or Lagrange's equation) are solved. Each nucleus is considered to be influenced by all other nuclei and electrons. By analyzing the forces on the system's particles and determining their positions and velocities using classical or quantum methods, the state of the particles' motion is established. This makes it possible to calculate the system's structure and properties. Due to progress in the semiconductor industry, computer algorithms, and potential functions, molecular dynamics research has gained extensive attention and has been applied to diverse fields, such as condensed matter physics, materials science, nanomechanics, nuclear technology, biomedicine, and reaction chemistry [12]. As a result, molecular dynamics research has garnered significant interest.

For certain physical processes with extremely rapid changes, it is challenging to carry out dynamic observation and statistical analysis using experimental methods. Molecular dynamics (MD) simulation has become widely utilized for such research. High-energy particle collision serves as a typical example [13]. Through the MD method, we are able to simulate energetic particle collisions under varying conditions. This involves individual atoms or other types of particles acquiring energies ranging from tens of electron volts (eV) to thousands or even mega-electron volts (MeV), particularly in the simulation of irradiation damage. During the simulation, particles collide with the material at extremely high speeds, enabling observation of defect creation and annihilation processes, as well as visualization of point defects and clusters. However, the time scale limitation of MD prevents accurate simulation of defect evolution. Nevertheless, advancements in computational methods are gradually addressing this limitation. Additionally, MD is capable of simulating processes such as phase transitions [14, 15], shock vibrations [14, 16, 17], and explosion mechanics, which are characterized by their extremely rapid nature.

The use of molecular dynamics simulations can replicate traditional mechanical characterization tests, such as tensile, compression, and shear tests. However, due to the high strain rates in the simulation, it is challenging to lower them [18, 19]. Additionally, obtaining interatomic forces by solving the complex Schrödinger equation in larger scale systems is computationally inefficient. As a result, the potential function needs to be simplified. Nevertheless, MD simulations can still provide insights into the deformation mechanisms of polycrystalline materials, including the release of dislocations, the formation of twins, and microstructure evolution. Furthermore, MD simulations can predict the effects of various parameters on the mechanical properties and deformation behavior of materials, which is valuable for material design. Therefore, research on MD simulations' impact on the mechanical properties of materials primarily focuses on the deformation process of material microstructures, the influence of defect evolution on mechanical properties, and the generation and expansion of cracks.

Frictional wear is a major contributor to the failure of industrial equipment and parts, resulting in significant economic losses. In the United States, statistics indicate that the annual economic loss due to friction and wear amounts to 4% of the gross national product (GNP) [20–24]. Molecular dynamics (MD) has been utilized to understand the microscopic mechanism of frictional wear. MD is used to create various contact models, such as flat-flat, rough peak-flat, and rough peak-single peak contacts, to study the impact of contact area, load, temperature, relative velocity, and crystal orientation on the mechanical properties related to friction and wear [25].

Additionally, MD is employed to simulate the cutting process, particularly the friction and wear behaviors between the tool and the material matrix. This helps investigate adhesion effects, material transfer laws, and energy dissipation mechanisms of different materials during the cutting process [26].

Surface coating involves applying a coating to a material's surface to achieve specific mechanical properties. Molecular dynamics is used in surface engineering for various applications [27]. For instance, MD can simulate the surface deposition process and its properties by showing how lower energy atoms interact, as well as the impact of atom generation at interfaces, diffusion, dislocations, lattice deformations, and relevant mechanical and microscopic friction properties. Moreover, MD can simulate surface modification processes, stress states, and surface cracks.

As information technology and ultra-precision technology advance, the integration level of large-scale integrated circuits increases while the diameter of silicon wafers grows and the chip's characteristic line width decreases [28]. Additionally, the shrinking of devices has made micro/nano-electromechanical devices (MEMs/NEMs) increasingly popular. These devices find use in military, medical, biological, environmental control, sensors, and various aspects of daily life [29, 30]. The manufacturing of large-scale integrated circuits and MEMS devices requires strict adherence to high standards. This involves processing silicon wafers with atomic-level flatness and achieving ultra-smooth surfaces with zero surface and sub-surface defects. Currently, chemical mechanical polishing is the primary technology capable of meeting these requirements. During chemical mechanical polishing, the surface of single-crystal silicon undergoes oxidation due to the chemical reagents in the polishing solution, forming a thin amorphous SiO_x film. This film alters the nature of the original single-crystal silicon wafer. Recent studies have focused on the impact of macroscopic process parameters on the surface quality and polishing efficiency of silicon wafers during chemical mechanical polishing.

Additionally, molecular dynamics is widely utilized in various natural and industrial systems, including biochemical molecules (such as proteins, enzymes, nuclei, and polysaccharides), polymers (such as safety glass, fats, pharmaceuticals, nylon, rubber, and oil molecules), and other systems that consist of a large number of atoms and molecules [33].

1.3 Trends in MD Simulation

The study of multiscale phenomena is an interdisciplinary subject in materials science, physics, and mechanics [34]. This involves examining the behavior of solid materials across various scales, from atomic structure to macroscopic level, in order to understand how materials behave. By understanding these cross-scale behaviors, we can better design material structures and accurately predict how materials will perform in real-world applications. Multiscale research is gaining increasing attention in areas such as mathematics, physics, chemistry, and materials science. For example, the Materials Genome Initiative (MGI) is a significant multiscale research effort in the field of materials. The MGI aims to establish connections between the atomic arrangement of materials, the formation of phases, the creation of microstructures, and the macroscopic properties and service life of materials. Through this process, we can accelerate materials development, reduce costs, and increase the success rate of materials design [35]. MD plays a critical role in multiscale studies. MD simulations and Monte Carlo methods, including *ab initio* molecular dynamics [8, 36], provide valuable information, such as thermodynamic properties, reaction pathways, structure, point defects, dislocation mobility, grain boundary energy, grain boundary mobility, and precipitated phase size, among others [37, 38]. Additionally, MD simulations can offer important input parameters for Monte Carlo and phase field evolution equation solving.

In 1985, Car and Parrinello [36] introduced a method to improve the accuracy of MD simulations by integrating the virtual dynamics of electrons into traditional MD. This method, known as the Car–Parrinello (CP) method, combined the density functional theory with MD, treating the degrees of freedom of electrons and nuclei as a unified system. This innovation allowed for the direct use of first-principles calculations based on local density functional theory for the simulation of statistical mechanics, significantly expanding the scope and depth of computer simulation experiments.

More recently, researchers have been exploring the combination of MD with finite elements for multiscale analysis. In this method, force and displacement are the most commonly used variables to establish a connection between the continuous medium region and the discrete atomic region. However, using Cauchy stress from continuous medium theory and virial stress from inter-particle interactions in multiscale calculations can lead to errors due to their different behaviors. To address this issue, researchers have adopted displacement as a continuous variable for the interface between the continuous medium region and the discrete atomic region. Stress transfer in both the fine and macroscopic regions is achieved through the stress intensity factor. This approach aims to provide more accurate and consistent results in multiscale calculations.

Since 1998, researchers have been striving to enhance the realism of traditional MD simulation models by experimenting with different construction methods. They introduced the Voronoi method to construct polycrystalline systems [40–42], which is widely used due to its simplicity. However, the flat faces of the grain boundaries in models constructed using this method greatly diminish the coupling of nanopolycrystals during deformation, thereby limiting certain deformation behaviors [43, 44]. Subsequently, they developed the sintering method using molecular dynamics [45], which is more reasonable. Nonetheless, the crystals created using this method contain various voids, and the quality of the sintering relies entirely on the pressure and time of sintering, making it difficult to control and introduce other variables into the simulation. In recent times, researchers have also developed methods to construct polycrystalline model systems using images taken by TEM or SEM, as well as CAD images. This demonstrates the significance placed on simulation modeling.

The study of algorithmic software is another important aspect to consider. For researchers in MD simulation, Atomeye is a widely known software developed by Mr. Ju Li. It's a small but powerful visualization tool that can handle millions of atomic systems with impressive rendering effects. However, becoming proficient in using Atomeye requires familiarity with its commands, which may be challenging for beginners. As computer computing continues to advance, more visualization software is emerging for the Windows operating system, such as OVITO and VMD. Those tools integrate features for individual defect analysis and statistics. Additionally, researchers have developed numerous defect analysis methods, including Command Neighbor Analysis (CNA) [47, 48], Centrosymmetry Parameters (CSP) [49], and Dislocation Analysis (DXA) [50, 51], which are discussed in detail later.

With advancements in Graphics Processing Unit (GPU) hardware and software technology, certain simple parallel problems in MD simulation can be computed using GPUs, such as the accelerated computation of EAM potential functions. Compared to CPUs, GPUs are not only more cost-effective, but they also have a significantly larger number of computational cores per unit, potentially reaching thousands. As a result, computational efficiency can be greatly improved. Currently, the development of GPU-based MD is progressing rapidly [52, 53]. If interested, you can explore the GPUMD recently developed by Zheyong Fan [54, 56].

In recent years, machine learning for training potential functions based on first-principles calculations has been a significant advancement in computational materials. Accurately trained potential functions demonstrate accuracy comparable to first principles while significantly reducing computational effort. Their effort is only linearly related to the number of simulated atoms, whereas typical density functional calculations are generally cubic in relation to the number of atoms. Therefore, machine learning-based empirical potentials can be utilized to simulate ultra-large-scale complex material systems with high accuracy. Accurate machine learning empirical potentials need to be simultaneously trained using both force and energy obtained from first principles. Since force is a first-order derivative of energy, this necessitates very large datasets to achieve accurate training results, posing a significant challenge to the training methodology [57].

References

[1] Alder BJ, Wainwright TE. Phase transition for a hard sphere system. *J Chem Phys*, 1957, 27: 1208–1209.

[2] Alder BJ, Wainwright TE. Studies in molecular dynamics. 1. General method. *J Chem Phys*, 1959, 31: 459–466.

[3] Bethe HA, Kirkwood JG. Critical behavior of solid solutions in the order-disorder transformation. *J Chem Phys*, 1939, 7: 578–582.

[4] Rahman A. Correlations in motion of atoms in liquid argon. *Phys Rev*, 1964, 136: A405–A411.

[5] Andersen HC. Molecular-Dynamics simulations at constant pressure and-or temperature. *J Chem Phys*, 1980, 72: 2384–2393.

[6] Gillan MJ. Diffusion in a temperature gradient. *Springer Science*, Bosten, MA, 1983.
[7] Nose S. A unified formulation of the constant temperature molecular-dynamics methods. *J Chem Phys*, 1984, 81: 511–519.
[8] Car R, Parrinello M. Structural, dynamical, and electronic-properties of amorphous-silicon—an Abinitio molecular-dynamics study. *Phys Rev Lett*, 1988, 60: 204–207.
[9] Cagin T, Pettitt BM. Molecular-Dynamics with a variable number of molecules. *Mol Phys*, 1991, 72: 169–175.
[10] Daw MS, Baskes MI. Semiempirical, quantum-mechanical calculation of hydrogen embrittlement in metals. *Phys Rev Lett*, 1983, 50: 1285–1288.
[11] Daw MS, Baskes MI. Embedded-Atom method—derivation and application to impurities, surfaces, and other defects in metals. *Phys Rev B*, 1984, 29: 6443–6453.
[12] Chen Juan. *Molecular Dynamics Simulation of Mechanical and Frictional Wear Behaviour of SiO_2/Si Bilayer Nanomaterials* (in Chinese). Xi'an: Xi'an Jiaotong University, 2019.
[13] Malerba L, Terentyev D, Olsson P, *et al.* Molecular dynamics simulation of displacement cascades in Fe-Cr alloys. *J Nucl Mater*, 2004, 329: 1156–1160.
[14] Kadau K, Germann TC, Lomdahl PS, *et al.* Microscopic view of structural phase transitions induced by shock waves. *Science*, 2002, 296: 1681–1684.
[15] Sandoval L, Urbassek HM. Solid-solid phase transitions in Fe nanowires induced by axial strain. *Nanotechnology*, 2009, 20: 325704.
[16] Chaisse F, Heuze O. Fundamental property of the solid equation of state infered from shock waves physics. *Mater Res Soc Symp Proc*, 2002, 731: 297–300.
[17] Li WH, Yao XH, Branicio PS, *et al.* Shock-induced spall in single and nanocrystalline SiC. *Acta Mater*, 2017, 140: 274–289.
[18] Sainath G, Choudhary BK. Molecular dynamics simulations on size dependent tensile deformation behaviour of [110] oriented body centred cubic iron nanowires. *Mater Sci Eng A*, 2015, 640: 98–105.
[19] Sainath G, Choudhary BK, Jayakumar T. Molecular dynamics simulation studies on the size dependent tensile deformation and fracture behaviour of body centred cubic iron nanowires. *Comput Mater Sci*, 2015, 104: 76–83.
[20] Liu Gaohang, Liu Guangming. Failure and failure analysis of engineering materials and structures (in Chinese). *Foreign Metalwork*, 2006, 1: 6–9.

[21] Qu Xiaobin, Chen Jianmin, Zhou Huidi, *et al.* Current status and development trend of research on wear failure of materials and its prevention (in Chinese). *J Tribol*, 1999, 19: 187–192.

[22] Li Shunmeng. *Mechanical Fatigue and Reliability Design* (in Chinese). Beijing: Science Press, 2006.

[23] Xue Qunji, Zhang Jun. Progress of micro tribological research (in Chinese). *J Tribol*, 1994, 14: 360–369.

[24] Luo Jianbin, He Yu, Wen Shicheng, *et al.* Tribological challenges of micro- and nano-fabrication technologies (in Chinese). *J Tribol*, 2005, 25: 283–288.

[25] Liu Pei, Han Xiuli, Sun Dongli, *et al.* Research progress in molecular dynamics simulation of material friction and wear (in Chinese). *Mater Sci Technol*, 2017, 25: 26–34.

[26] Liang Yingchun, Guo Yongbo, Chen Mingjun. New advances in the study of diamond tool wear during nanofabrication (in Chinese). *J Tribol*, 2008, 28: 282–288.

[27] Wang Ze, Li Guolu, Wang Haidou, *et al.* Current status of molecular dynamics simulation and its application in surface engineering (in Chinese). *Mater Herald*, 2014, 28: 91–100.

[28] Huang QH. International technology roadmap for semiconductors (ITRS) 2013 edition overview(3) (in Chinese). *China Integr Circuit*, 2014, 23: 14–26.

[29] Ayazi F, Najafi K. High aspect-ratio combined poly and single-crystal silicon (HARPSS) MEMS technology. *J Microelectromech Syst*, 2000, 9: 288–294.

[30] Zhai Yujing, Yang Kaiyong, Pan Yao. History, current status and prospect of gyroscope (in Chinese). *Fly Missiles*, 2018, 12: 84–88.

[31] Xu J, Luo JB, Wang LL, *et al.* The crystallographic change in sub-surface layer of the silicon single crystal polished by chemical mechanical polishing. *Tribol Int*, 2007, 40: 285–289.

[32] Estragnat E, Tang G, Liang H, *et al.* Experimental investigation on mechanisms of silicon chemical mechanical polishing. *J Electron Mater*, 2004, 33: 334–339.

[33] Chen Zhenglong, Xu Weiren, Tang Lida. *Theory and Practice of Molecular Simulation* (in Chinese). Beijing: Chemical Industry Press, 2007.

[34] Chai LH. Research progress in multiscale science (in Chinese). *Adv Chem*, 2005, 17: 186–191.

[35] Zhao Jicheng. Introduction to the Material Genome Project (in Chinese). *Nat Mag*, 2014, 36: 89–104.

[36] Car R, Parrinello M. Unified approach for molecular-dynamics and density-functional theory. *Phys Rev Lett*, 1985, 55: 2471–2474.

[37] Berendsen HJC, Postma JPM, Vangunsteren WF, *et al.* Molecular-Dynamics with coupling to an external bath. *J Chem Phys*, 1984, 81: 3684–3690.

[38] Soler JM, Artacho E, Gale JD, *et al.* The SIESTA method for ab initio order-N materials simulation. *J Phys Condens Matter*, 2002, 14: 2745–2779.

[39] Wang Lusheng. *Multiscale Simulation of Deformation and Failure of Metallic Materials Based on Molecular Dynamics and Finite Element Method* (in Chinese). Chongqing: Chongqing University of Technology, 2018.

[40] Voronoi G. Nouvelles applications des paramètres continus à la théorie des formes quadratiques. *J Reine Angew Math*, 1908, 134: 198–287.

[41] Schiotz J, Ditolla FD, Jacobsen KW. Softening of nanocrystalline metals at very small grain sizes. *Nature*, 1998, 391: 561–563.

[42] Schiotz J, Vegge T, Di Tolla FD, *et al.* Atomic-Scale simulations of the mechanical deformation of nanocrystalline metals. *Phys Rev B*, 1999, 60: 11971–11983.

[43] Zhang M, Chen J, Xu T, *et al.* Effect of grain boundary deformation on mechanical properties in nanocrystalline Cu film investigated by using phase field and molecular dynamics simulation methods. *J Appl Phys*, 2020, 127: 125303.

[44] Zhang M, Xu T, Li ME, *et al.* Constructing initial nanocrystalline configurations from phase field microstructures enables rational molecular dynamics simulation. *Comput Mater Sci*, 2019, 163: 162–166.

[45] Kadau K, Germann TC, Lomdahl PS, *et al.* Molecular-dynamics study of mechanical deformation in nano-crystal line aluminum. *Metall Mater Trans A*, 2004, 35a: 2719–2723.

[46] Prakash A, Hummel M, Schmauder S, *et al.* NANOSCULPT: A methodology for generating complex realistic configurations for atomistic simulations. *Methods X*, 2016, 3: 219–230.

[47] Stukowski A. Structure identification methods for atomistic simulations of crystalline materials. *Model Simul Mater Sci Eng*, 2012, 20: 045021.

[48] Ackland GJ, Jones AP. Applications of local crystal structure measures in experiment and simulation. *Phys Rev B*, 2006, 73: 054104.

[49] Kelchner CL, Plimpton SJ, Hamilton JC. Dislocation nucleation and defect structure during surface indentation. *Phys Rev B*, 1998, 58: 11085–11088.

[50] Stukowski A, Albe K. Extracting dislocations and non-dislocation crystal defects from atomistic simulation data. *Model Simul Mater Sci Eng*, 2010, 18: 085001.

[51] Stukowski A, Bulatov VV, Arsenlis A. Automated identification and indexing of dislocations in crystal interfaces. *Model Simul Mater Sci Eng*, 2012, 20: 085007.

[52] Fei Hui, Zhang Yunquan, Wang Ke, *et al.* Parallelisation and implementation of GPU-based molecular dynamics simulation (in Chinese). *Comput Sci*, 2011, 9: 276–278.

[53] Lin Jianghong, Lin Jinxian, Lu Toon. Multi-Core CPU and GPU acceleration for molecular dynamics simulation (in Chinese). *Comput Appl*, 2011, 3: 843–847.

[54] Fan ZY, Chen W, Vierimaa V, *et al.* Efficient molecular dynamics simulations with many-body potentials on graphics processing units. *Comput Phys Commun*, 2017, 218: 10–16.

[55] Fan ZY, Pereira LFC, Wang HQ, *et al.* Force and heat current formulas for many-body potentials in molecular dynamics simulations with applications to thermal conductivity calculations. *Phys Rev B*, 2015, 92: 094301.

[56] Fan ZY, Pereira LFC, Hirvonen P, *et al.* Thermal conductivity decomposition in two-dimensional materials: Application to graphene. *Phys Rev B*, 2017, 95: 144309.

[57] Cooper AM, Kstner J, Urban A, *et al.* Efficient Training of Accurate Neural Network Potentials by Including Atomic Forces via Taylor Expansion: Application to Water and a Transition-Metal Oxide. *npj Comput Mater*, 2020, 54: 1–14.

Chapter 2

Principles and Methods of Molecular Dynamics

2.1 Principles of Molecular Dynamics

Through continuous exploration of materials, an integrated research approach combining experimentation, simulation, and theory has been developed. Computer simulation serves as a crucial link between experimentation and theory, filling in the gaps in our understanding of the microscopic deformation mechanisms of materials and significantly reducing the design cycle for new materials. As early as the 1950s, Alder and Wainwright used molecular dynamics (MD) to simulate interactions between spheres [1, 2]. However, the simulation systems were limited by the available computer hardware at the time. Today, with rapid advancements in computer technology and hardware, simulation systems can model systems as small as a few particles or as large as millions—even hundreds of millions—of particles. These advancements enable us to explore phenomena such as the Hall–Petch effect in polycrystalline metal materials. Classical molecular dynamics utilizes Newtonian mechanics to model the motion of systems composed of atoms and molecules. The system's particle interactions are described using empirical potential functions, as the interactions between electrons are neglected. This obviates the need to solve the complex Schrödinger equation, thereby allowing for larger simulated systems and longer simulation times. The classical MD method is commonly employed to investigate the properties of polycrystalline materials at the nanoscale and to analyze interactions between different defects in the materials. Within the simulated system, the forces on the particles are examined, along with their positions and velocities at a specific moment in time, to discern the particles' motion states. Subsequently, statistical mechanics is employed to compute the system's structure and properties.

According to Newtonian classical mechanics, for a simple atomic system, the acceleration of atom $i(a_i)$ can be expressed as follows:

$$a_i = \frac{f_i}{m_i} \tag{2.1}$$

where f_i is the external force on the atom and m_i is the mass of the atom. The information about the details of the motion of the atom can be obtained by this formula, and the force on the atom can be obtained from the gradient of the potential function $Uf_i = -\nabla_i U$. The details of the motion of an atom are usually described by the position (q_i), velocity $(\dot{q}_l)$, and momentum vector information (P_i) using Lagrangian and Hamiltonian quantities.

In the generalized coordinate system, the system is in the $t = t_1$ and $t = t_2$. The equation of motion between the two positions of $S = \int_{t_1}^{t_2} L(q, \dot{q}_l, t)dt$ taking the minimum value. $L(q, \dot{q}_l, t)$ denotes the Lagrangian quantity, which indicates that there is a quantity that can describe the state of the system, and its integral over time is the evolution quantity of the state of the system. Such an evolution quantity needs to satisfy the principle of minimization, and the Lagrangian equations can be derived through the variable time integral as follows:

$$\frac{d}{dt}\left(\frac{\partial L}{\partial q'}\right) = \frac{\partial L}{\partial q} \tag{2.2}$$

In MD, for a conservative system of n atoms, the Lagrange equation s can be expressed as the difference between the kinetic and potential energies of the system:

$$L = \frac{1}{2}\sum_{i=1}^{n} m_i \dot{q}_l^2 - \sum_{i<j} U(r_{ij}) \tag{2.3}$$

And the Hamiltonian quantity can be expressed as the sum of kinetic and potential energy:

$$H = \frac{1}{2}\sum_{i=1}^{n}\frac{P_i^2}{m_i} + \sum_{i<j} U(r_{ij}) \tag{2.4}$$

where r_{ij} is the distance between atom i and atom j. Regularizing the Hamiltonian quantities, we obtain the force equation $-\frac{\partial H}{\partial q_i} = \dot{P_l} = F_i$ and the velocity equation $-\frac{\partial H}{\partial P_i} = \dot{q}_l$.

Lagrange's equations are usually used to solve the equations of motion of an atomic system, i.e., the velocity and position of the atoms, and can impose some conditions on the simulation such as constraints, external forces, and boundary conditions. However, Hamiltonian regular equations are more suitable for solving the kinetic evolution process and the thermodynamic state of the system [3].

2.1.1 *N-atom systems*

The total force on the N-atom system is as in Eq. (2.5), and at a given position, if the energy is constant ($dE/dt = 0$), the force acting on atom i is the negative value of the derivation of the energy. In a real MD, the forces acting are as in Eq. (2.6):

$$F(r_1, r_2, \ldots r_N) = \sum_i m_i a_i = \sum_i m_i \frac{d^2 r_i}{dt^2} \tag{2.5}$$

$$F_i = m_i \frac{d^2 r_i}{dt^2} = -\frac{dU(r_i)}{dr_i} \tag{2.6}$$

The forces acting on the atoms at a given moment can be determined from the interaction potential, which depends only on the positions of all the atoms. The next step is to solve the equation of motion. In theory, the first and second integrals will give the velocity and position. However, for a system with $6N$ atoms ($3N$ positions and $3N$ momentum), the analytical solution is not so straightforward. Therefore, MD simulation uses a mathematical iterative finite difference approach, replacing differentiation with a finite difference. This transforms the differential formulation into a finite difference formulation. For instance, using the Taylor expansion, the position at the next moment $(t + \Delta t)$ is obtained from the position at the previous moment (t) as follows:

$$r(t + \Delta t) = r(t) + v(t)\Delta t + \frac{1}{2!}a(t)\Delta t^2 + \frac{d^3 r(t)}{3!dt^3}\Delta t^3 + \cdots \quad (2.7)$$

where Δt is a small discrete step and all rates of change are assumed to be constant at time $t+\Delta t$. The actual formula is approximated by a polynomial. Then, the integration process of the Newtonian motion formula for the N atomic system is as follows:

(1) Calculate the force on all atoms from a given potential function.
(2) Based on the obtained force F_i, calculate the acceleration a_i according to $a_i = F_i/m$.
(3) Using the finite difference method (Eq. (2.15)), calculate r_i, v_i, and a_i at the moment $t + \Delta t$.
(4) Use the calculated data as the next sub-input value and repeat the process until equilibrium is reached.

In practical MD, the difference terms are usually considered to be applied to the Taylor expansion up to the third order, with truncation errors of $O(\Delta t^4)$ for higher-order terms. The solution of the equation is not very precise, but it is close enough to reality.

2.1.2 *Verlet algorithm*

In MD simulations, there are various methods for mathematically integrating Newton's equations of motion and calculating atomic trajectories. The primary goal of these methods is to maintain simulation stability and ensure result accuracy. In 1967, Verlet [4, 5] introduced an algorithm that utilized a third-order term to compute the positions of atoms at previous and subsequent time steps as follows:

$$r(t + \Delta t) = r(t) + v(t)\Delta t + \frac{1}{2!}a(t)\Delta t^2 + \frac{d^3 r(t)}{3!dt^3}\Delta t^3 \quad (2.8)$$

$$r(t - \Delta t) = r(t) - v(t)\Delta t + \frac{1}{2!}a(t)\Delta t^2 - \frac{d^3 r(t)}{3!dt^3}\Delta t^3 \quad (2.9)$$

Summing the two equations yields the final expression for the position of the atom at the moment $t + \Delta t$:

$$r(t + \Delta t) = 2r(t) - r(t - \Delta t) + a(t)\Delta t^2 \tag{2.10}$$

Therefore, $r(t + \Delta t)$ can be predicted using $r(t)$, $r(t - \Delta t)$, and $a(t)$. The first step requires the position of $r(t - \Delta t)$ at the previous moment, as in Eq. (2.9). However, its computation can be avoided by employing the normal Taylor expansion:

$$r(t + \Delta t) = r(t) - v(t)\Delta t + \frac{a(t)}{2}\Delta t^2 \tag{2.11}$$

After the first step of calculations, the second step of calculations, $r(t + 2\Delta t)$, is carried out using Eq. (2.10). The velocity did not evolve with time when using the Verlet algorithm, but the velocity is needed to calculate the kinetic energy, which is indirectly derived from the change in position during the $2\Delta t$ time interval:

$$v(t) = \frac{r(t + \Delta t) - r(t - \Delta t)}{2\Delta t} \tag{2.12}$$

The Verlet algorithm is simple, relatively accurate, time-reversible, and requires only one force calculation per step.

2.1.3 *Velocity–Verlet algorithm*

The Velocity–Verlet algorithm is one of the most frequently used algorithms in MD [6, 7]. In this algorithm, the position, velocity, and acceleration at the moment $t + \Delta t$ are obtained from the corresponding quantities at the moment t. Half a step is used to update the velocity v for half a step and the displacement r for one step:

$$r(t + \Delta t) = r(t) + v(t)\Delta t + \frac{1}{2!}a(t)\Delta t^2 + \frac{d^3 r(t)}{3! dt^3}\Delta t^3 + \cdots \tag{2.13}$$

$$v\left(t + \frac{\Delta t}{2}\right) = v(t) + \frac{1}{2!}a(t)\Delta t \tag{2.14}$$

$$r(t + \Delta t) = r(t) + v(t)\Delta t + \frac{1}{2!}a(t)\Delta t^2 = r(t) + v\left(t + \frac{\Delta t}{2}\right)\Delta t \tag{2.15}$$

The acceleration a is updated by the potential energy relationship, using the entire step size, i.e., the velocity is further updated as follows:

$$\begin{aligned} v(t+\Delta t) &= v(t) + \frac{a(t)+a(t+\Delta t)}{2}\Delta t \\ &= v\left(t+\frac{\Delta t}{2}\right) + \frac{1}{2}a(t+\Delta t)\Delta t \end{aligned} \tag{2.16}$$

The velocity is updated after obtaining the new position and acceleration (equivalent to the new force). This algorithm is easy to implement, time-reversible, accurate, suitable for both short and long step sizes, and stable because position and velocity are computed at each step. Although there may be energy fluctuations in this algorithm, there is no long-term energy drift.

2.1.4 *Initial position and velocity*

In order to solve Newton's equation of motion, we need to know the initial velocity and position of the atom. The atom can start at any position, but usually, it is specific to known lattice positions. The initial velocities of the particles can both be zero, but at a given temperature, the atom's velocity is usually random according to the Maxwell distribution or Gaussian distribution laws. For instance, the probability that an atom has an initial velocity of v at an absolute temperature of T is given by the following:

$$P(v) = \left(\frac{m}{2\pi k_B T}\right)^{\frac{1}{2}} \exp\left(-\frac{mv^2}{2k_B T}\right) \tag{2.17}$$

where k_B is the Boltzmann constant. The direction of the velocity is also chosen randomly in order to ensure that the total linear momentum is zero.

2.1.5 *Time step*

A ballerina can jump a distance of two meters in one step and stay in the air for a second. To track her movement, a camera capable of capturing 5–10 movements in a second is necessary. Similarly, when atoms in a solid lattice vibrate in 10–14 s, a unit of 10–15 s is required. This small interval of time is called a time step (Δt), and it is assumed that the velocities, accelerations, and forces of the atoms are constant over this time step so that simple algebraic operations can be performed. According to the laws of solving Newton's equations of motion, the positions of the atoms are updated at each time step. Therefore, the time step should be as large as possible to accelerate the simulation while ensuring energy storage and operational stability. A practical rule is that the motion of the atoms within a time step should not exceed 1/30th of the most critical distance, so the usual time step ranges from 0.1 to several femtoseconds.

The computational load and accuracy are conflicting issues in the computational process when determining the time step. Smaller time steps increase the computational load but improve accuracy, while larger time steps decrease the computational load but reduce accuracy. If the total energy is unstable, it means that the time step is too large. For relatively fast atomic motions caused by high temperatures, light masses, or high undulated potential energy curves, shorter time steps are preferred.

2.1.6 *Total simulation time*

A typical MD run may involve 10^3–10^6 time steps, with a total simulation time of several hundred nanoseconds (10^{-9} s). This duration is usually adequate for observing both the static and dynamic properties of the material. Running an MD calculation for too long can lead to accumulated errors and less efficient data generation. However, it is important for the total simulation time to exceed the energy relaxation time of the system in order to obtain reliable data, particularly for phenomena such as phase transitions, vapor phase deposition, crystal growth, and radiation damage quenching, where the system's equilibrium is very slow.

2.1.7 *Type of ensemble*

In the MD simulation box, each atom moves and behaves differently, leading to constant updates of the microstates of the system. Once the system is appropriately simulated and equilibrated, the culmination of all microstates in the system constitutes an ensemble, possessing the same macroscopic state and thermodynamic properties.

1. **Microcanonical ensemble (NVE):** The NVE system maintains as constant the total energy, particle number, and volume of the system, creating an isolated conservative system. In MD simulation, the particle number and volume are kept constant, and various mathematical methods are used to control the energy and keep it constant, typically through calibrated velocity methods.
2. **Canonical ensemble (NVT):** The NVT system ensures constant temperature, number of particles, and volume of the system. This simulates molecular behavior close to reality by controlling the temperature, albeit with the system's energy constantly changing. When using the NVT system for simulations, temperature control is paramount. Typically, adjusting particle velocity directly to manipulate kinetic energy is employed to regulate the temperature or maintain the system's temperature identical to that of the external heat bath in a specific system. The primary methods for temperature control include velocity calibration, Andersen's temperature control, Berendsen's temperature control, Nosé's temperature control, and Nosé–Hoover's temperature control. The NVT system is mainly utilized for simulations within this book.
3. **Constant temperature and pressure ensemble (NPT):** In the NPT system, the number of particles, temperature, and pressure of the system are kept constant. The temperature is controlled in the same way as in the NVT system, while the pressure is regulated by adjusting the volume. Various pressure control methods are used, including the constant pressure scale method, Andersen' pressure control method, Berendsen pressure control method, and Parrinello–Rahman pressure control method. These methods help maintain constant pressure by controlling the size of the protocells in the system.

2.1.8 *Temperature and pressure control*

In MD simulations under the NVT ensemble, maintaining a constant temperature is essential. This can be achieved by calibrating the atomic velocities. The average kinetic energy is associated with the average velocity and the atom's temperature, expressed by the following equation:

$$\langle E_{\text{kin}} \rangle = \left\langle \frac{1}{2} \sum_i m_i v_i^2 \right\rangle = \frac{3}{2} NkT \tag{2.18}$$

where "3" denotes the dimensional degrees of freedom in a three-dimensional system. The temperature is directly related to the mean velocity:

$$\langle v \rangle = \left(\frac{3kT}{m} \right)^{\frac{1}{2}} \propto T^{\frac{1}{2}} \tag{2.19}$$

Then, the temperature is made to increase or decrease from T to T' by multiplying each velocity component by the same factor:

$$\langle v_{\text{new}} \rangle = \langle v_{\text{old}} \rangle \left(\frac{T'}{T} \right)^{\frac{1}{2}} \tag{2.20}$$

This equation is used to gradually increase the velocity of all atoms to the desired value. Although the system is not exactly equivalent to the NVT ensemble under real, strictly thermodynamic conditions, the temperature can be guaranteed by actual velocity calibration. In MD to obtain the NVT ensemble and the corresponding thermodynamic properties (e.g., Helmholtz free energy), it is necessary to impose a heat bath interaction, i.e., coupling the system to a heat bath or considering the coordinates and velocities. In NPT systems, an extended system (pressure bath) is often used to obtain constant pressure while allowing the volume of the system to change.

2.1.9 *Energy minimization*

To find the configuration of a system with the lowest energy, a systematic approach to energy minimization is required. One method is the maximum velocity descent method, which involves calculating the initial energy U of the atom's configuration. The atom then moves in the direction of lower energy until the search line is perpendicular to the potential energy gradient. If any further update leads to higher energy U, the next search begins in the direction of the most rapid decrease (refer to Figure 2.1(a)). This process is repeated iteratively

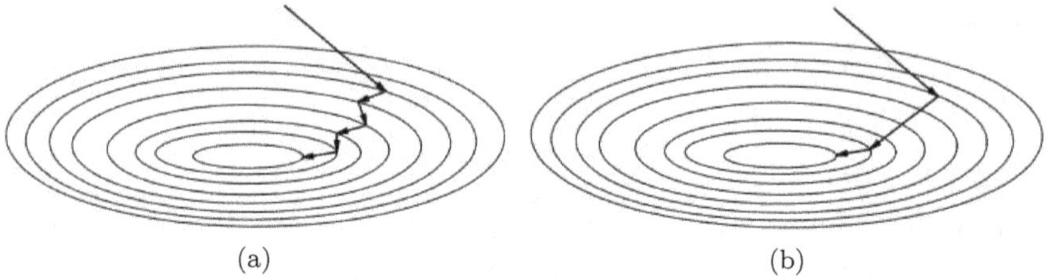

Figure 2.1. Schematic diagram of energy minimization integration method [4, 5]. (a) Fastest descent method, (b) conjugate gradient method.

until the atomic configuration's very small energy is found. However, this search path is zigzag and many orthogonal steps take too much time, resulting in a slow process.

Another method called the conjugate gradient method begins the search by moving in the direction of the steepest descent. It stops at a local minimum and then changes direction based on the new energy gradient (which is the negative force between atoms) and the previous search direction. As illustrated in Figure 2.1(b), this search path is relatively direct, with fewer steps and faster search speeds.

2.2 Intermolecular Interactions and Potential Functions

When atoms are in close proximity, they reach equilibrium through mutual attraction and repulsion, which is determined by a potential function. Atoms follow Newton's second law of motion and eventually settle at an equilibrium distance with the lowest potential energy. The force (F) acting on an atom is given by the following:

$$F = ma = m\frac{dv}{dt} = m\frac{d^2r}{dt^2} = \frac{dp}{dt} \tag{2.21}$$

where α is the acceleration/Å $\cdot$ ps^{-2}, v is the velocity/Å $\cdot$ ps^{-1}, t is the time/ps, r is the distance/Å, and p is the momentum/g $\cdot$ Å $\cdot$ ps^{-1}.

If the total energy E of the system is constant at different moments ($dE/dt = 0$), for MD, this is an isolated system and F is the negative value of the potential energy differentiated with respect to position:

$$F = -\Delta U \rightarrow F = -\Delta U(r_1, r_2, \ldots, r_N) \tag{2.22}$$

where U is the potential energy/eV. Thus, if the potential energy (a function of the distance between atoms) of a system is known, the forces acting on the atoms can be determined, and thus Eq. (2.22) can be solved for different moments.

The parameters of a potential function for a specific system are determined by adjusting experimental data or data obtained from first principles calculations. These experimental parameters encompass equilibrium lattice parameters, cohesion energy, bulk phase modulus, elastic constants, vacancy formation energy, thermal expansion coefficients, dielectric constants, vibrational frequencies, and surface energies. Each parameter is specific to the system, so before applying a particular potential function to another system or condition, its suitability must be determined.

2.2.1 *Pair potentials*

In a system containing N atoms, the interaction potential of a particular atom i with all other atoms at the same time can be described as follows:

$$U = \sum_{i<j}^{N} U_2(r_i, r_j) + \sum_{i<j<k}^{N} U_3(r_i, r_j, r_k) + \cdots \tag{2.23}$$

where the term on the right corresponds to the potential function for two atoms, three atoms, and so on. The pair potential is the simplest form of potential function, considering only the interaction of two atoms and ignoring all others.

The Lennard–Jones potential function proposed in 1924 [8] is a typical function with the following expression:

$$U_{LJ} = 4\varepsilon \left[\left(\frac{\sigma}{r}\right)^{12} - \left(\frac{\sigma}{r}\right)^{6} \right] \tag{2.24}$$

where ε is the lowest energy of the potential energy curve (related to the depth of the potential well and the cohesion energy)/eV, σ is the average mutual distance/Å between atoms when the potential energy is zero, and these two parameters are fitted for a specific material. A plot of the pair potentials is shown in Figure 2.2.

When two atoms are far apart, both gravitational and repulsive forces are zero. As the atoms get closer, they experience an attraction

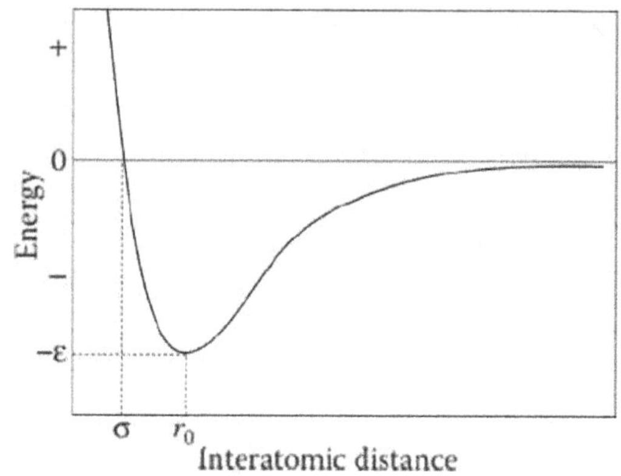

Figure 2.2. Lennard–Jones potential curve [8].

between their dipole moments, and the van der Waals interaction can be described well by the r^{-6} term. When the atoms are even closer, the Bauerian repulsive force comes into play due to electrostatic recombination, and the drastic increase in repulsive force is described well by the r^{-12} term. At the equilibrium interaction distance, the attractive and repulsive forces are balanced, resulting in the lowest energy. As the distance (r) decreases further, the energy increases rapidly from zero.

The simple pair potential effectively describes the interactions between noble gas molecules and different species of atoms. Using the Lorentz–Berthelot (L–J) mixing law, L–J potential functions for various atomic species can be derived. However, the pair potential cannot be utilized for metals, semiconductors, and other solid materials due to the significant impact of multiple atoms.

2.2.2 *Morse potential*

The Morse potential, named after physicist Philip M. Morse, is a simple analytical model for the intermolecular potential between diatomic molecules [9]. Solving the Schrödinger equation for the Morse potential has an analytical solution, making it convenient for analyzing the problem. Additionally, it is a good approximation of the microstructure of molecular vibrations because it implicitly includes bond-breaking phenomena. The Morse potential includes a property that is lacking in the harmonic oscillator model, which is the non-bonded state. In comparison to the quantum harmonic oscillator model, the Morse potential is more realistic as it can describe harmonic effects and multiplicative and combinatorial frequencies. The Morse potential can be described by the following equation:

$$U = \sum_{ij} D_0 \left[e^{-2\alpha(r-r_0)} - 2e^{-\alpha(r-r_0)} \right] \tag{2.25}$$

where D_0, α, and r_0 represent the binding energy, elastic modulus, and equilibrium distance between the two atoms, respectively, where D_0 is the depth of the potential well, which controls the "width" of the potential well.

2.2.3 *SW potential*

Covalently bonded materials, such as silicon, silicon dioxide, organic materials, and certain carbon-based materials like graphene, diamond, and diamond-like membranes, have two distinct features: high bond strength and directional bonds. When modeling these materials, it is important to consider atomic positions, bond lengths (stretching), bond angles (bending), and even dihedral angles (torsion). The first Stillinger–Weber potential functions were developed for diamond-structured silicon [9]. This potential function consists of a linear combination of two-body and three-body potentials, which are expressed as follows:

$$U = \sum_i \sum_{j>i} \varepsilon f_2(i,j) + \sum_i \sum_{j>i} \sum_{k>j} \varepsilon f_3(i,j,k) \tag{2.26}$$

where ε is the bonding energy/eV between atoms and f_2 is the interaction energy/eV between pairs of atoms, with the following equation:

$$f_2(i,j) = g_{ij} A_{ij} (B_{ij} r_{ij}^{-p_{ij}} - r_{ij}^{-q_{ij}}) \exp[(r_{ij} - a_{ij})^{-1}] r_{ij} < a_{ij} \tag{2.27}$$

Otherwise, $f_2(i,j) = 0$. g_{ij} is the softening function that controls the magnitude of the contribution of the two-body potential, A_{ij}, B_{ij}, p_{ij}, q_{ij}, a_{ij} are the parameters of the atom pairs i and j, and f_3 is the three-body potential/eV as a function of the three atoms i, j, and k, which can be expressed as follows:

$$f_3(i,j,k) = h(r_{ij}, r_{ik}, \theta_{jik}) + h(r_{ji}, r_{jk}, \theta_{ijk}) + h(r_{ki}, r_{kj}, \theta_{ikj}) \tag{2.28}$$

where r_{ij} is the interatomic distance/Å and θ_{jik} is the angle/° between the atomic distance r_{ij} and r_{ik}. When $r_{ij} < a_{jik}^{ij}$ and $r_{ik} < a_{jik}^{ik}$, the function h is expressed as follows:

$$h(r_{ij}, r_{ik}, \theta_{jik}) = \lambda_{jik} \exp\left[\frac{\gamma_{jik}^{ij}}{r_{ij} - a_{jik}^{ij}} + \frac{\gamma_{jik}^{ik}}{r_{ik} - a_{jik}^{ik}}\right] \times (\cos\theta_{jik} - \cos\theta_{jik}^{0})^2 \tag{2.29}$$

Otherwise, $h(r_{ij}, r_{ik}, \theta_{jik}) = 0$, λ_{jik}, γ_{jik}^{ij}, γ_{jik}^{ik}, a_{jik}^{ij}, a_{jik}^{ik}, θ_{jik}^{0} are the parameters of the three atoms, a_{jik}^{ij} and a_{jik}^{ik} are the truncation distance/Å of the three-body potential part, and θ_{jik}^{0} is the equilibrium angle/° between atoms j, i, and k.

The SW potential function integrates factors such as bond lengths and bond angles, which streamline the interactions between atoms and enhance the material's stability. It also addresses dislocations, point defects, certain surface structures, and liquid and mixed states in silicon. However, it has a limitation in that it can only describe a single equilibrium atomic configuration and fails to account for other stable structures. The introduction of bond order rectifies this shortcoming by assessing the strength of different bonds using a bond order parameter, enabling the attainment of steady state through the use of a potential function. A commonly cited bond sequence potential function is Tersoff [10].

2.2.4 *Tersoff potential*

In covalently bound solids, these structures are typically low-density arrangements (commonly featuring a coordination number of only 4) and exhibit strongly oriented bonds. Diamond and sphalerite structures are included in this category of solids, with bonds displaying a bond angle of 109.47° and utilizing sp^3 hybridization orbitals. Consequently, bond angles and bond orders are essential characteristics of these materials, and the strength of the bond between two atoms depends on the local environment, such as the coordination number and the bond angle. In 1988, Tersoff introduced a novel potential after recognizing the significance of this geometric factor [11]:

$$U = \frac{1}{2}\sum_{i}\sum_{j\neq i}[f_{ij}A_{ij}\exp(-\lambda_{ij}r_{ij}) - f_{ij}b_{ij}B_{ij}\exp(-\mu_{ij}r_{ij})] \quad (2.30)$$

where U is the total energy/eV of the system, the first and second terms on the right side of the equation are the attraction and repulsion terms, respectively, b_{ij} is the bond order function, which not only depends on the positions of i and j atoms but also relates to the immediate neighboring atoms around i atom, which is the main feature of the Tersoff potential function, which can be expressed as follows:

$$b_{ij} = \left[1 + \left(\beta_i \sum_{k\neq i,j} \varsigma_{ijk}\right)^{n_i}\right]^{-1/(2n_i)} \quad (2.31)$$

$$\varsigma_{ijk} = f_{ik}\exp\left[\mu_{ij}^{m_i}(r_{ij} - r_{ik})^{m_i}\right]\left\{1 + \frac{c_i^2}{d_i^2} - \frac{c_i^2}{d_i^2 + (h_i - \cos\theta_{ij_k})^2}\right\} \quad (2.32)$$

where B_i is the bond level coefficient, c_i, d_i, and h_i are the elastic constants, ς_{ijk} is the angular potential energy/eV, θ_{ij_k} is the bond angle/° between atoms i, j and atoms i, k; r_{ij} is the atomic spacing/Å and f_{ij} is the smooth truncation radius/Å, which are described as follows:

$$f_{ij} = f_c(r_{ij}, \sqrt{R_iR_j}, \sqrt{S_iS_j}) \quad (2.33)$$

$$f_c(r, R, S) = \begin{cases} 1 & (r \leq R) \\ \frac{1}{2} + \frac{1}{2}\cos\frac{\pi(r-R)}{S-R} & (R < r < S) \\ 0 & (S \leq r) \end{cases} \quad (2.34)$$

$$\lambda_{ij} = \frac{\lambda_i + \lambda_j}{2} \quad (2.35)$$

$$\mu_{ij} = \frac{\mu_i + \mu_j}{2} \tag{2.36}$$

$$A_{ij} = \sqrt{A_i A_j} \tag{2.37}$$

$$B_{ij} = \sqrt{B_i B_j} \tag{2.38}$$

where A_{ij}, B_{ij} are the pairwise potential binding energies/eV of the attraction and repulsion terms, S and R are the truncation radii/Å. The combined binding energy of the dual potential of attraction and repulsion terms λ_{ij}, and μ_{ij} are the gradient coefficients of the dyadic potential energy curves of the attractive and repulsive terms.

Although both terms of the potential function depend on distance, this type of potential function typically possesses more than six parameters that require fitting. When well parameterized, this potential effectively describes interatomic interactions and has been widely employed in various covalently bonded solids and molecules, including SiC, Si, diamond, graphene, and amorphous carbon. The silicon–oxygen Tersoff potential function has been documented in the literature [12].

2.2.5 *EAM potential*

Metallic materials are typically characterized by their excellent electrical conductivity and rapid heat conduction, which is a result of the valence electrons in metals being free. The interaction between the atom and the surrounding free electrons is a crucial component of the bonding energy. The embedded atom method (EAM), which was originally proposed by Daw and Baskes in 1984, is used to describe this property. The EAM potential divides the total potential energy of a crystal into two parts [13, 14]: One part is the interaction pair potential between atomic nuclei located on the lattice array of dots and another part is the embedding energy resulting from the interactions between the atomic nucleus and the electron cloud in which it is embedded, representing many-body interactions. The total potential energy of the system is expressed in the embedded atom method as follows:

$$U = \sum_i F_i(\rho_i) + \frac{1}{2}\sum_{j \neq i} \varphi_{ij}(r_{ij}) \tag{2.39}$$

where F is the embedding energy/eV, the second term is the pair potential term, which can take different forms as needed, φ_{ij} is the interaction potential/eV between the ith and jth atoms, and ρ_i can be expressed as follows:

$$\rho_i = \sum_{j \neq i} \rho_{ij}(r_{ij}) \tag{2.40}$$

where $\rho_{ij}(r_{ij})$ is the charge density contributed by the extra-nuclear electrons of the jth atom at the ith atom, and r_{ij} is the distance/Å between the ith atom and the jth atom. For different metals, the embedding energy function and pair potential function parameters need to be determined by fitting the macroscopic parameters of the metal.

The aforementioned potential function effectively illustrates the interactions between a majority of metals and transition metals, particularly those with face-centered cubic structures. Thermodynamic functions, plastic deformation, melting points, defects, grain boundary structures, and tribological behavior of metallic materials can be effectively described using this approach [15]. However, EAM has one drawback: It assumes that metal bonds are not directional, which is not always the case. For instance, some metals, such as magnesium and cobalt, have spherically symmetric bonds. To account for this symmetry and directional bonding, a modified EAM potential was proposed, which has since been developed to describe the interactions of semiconductor materials [16].

2.3 Periodic Boundary Conditions

In simulations of molecules, the number of particles is often limited. As a result, the calculations may be affected by the surface and finite size of the system. To minimize this impact, the simulation system must be designed with a relatively large size, which can increase the computational cost significantly or even make simulation impossible. Therefore, the selection of boundary conditions is critical for computational simulations. Commonly used boundary conditions are generally categorized into four types:

1. Free boundary conditions are commonly used for simulating large, free molecules.
2. Fixed Boundary Conditions: In this setup, all particle cells designated for computation are encircled by multiple tiers of particles that exhibit uniform structure and unchanging positions. The encircling layers must possess a thickness exceeding the interaction force range between particles. This peripheral component symbolizes the segment of the macroscopic crystal that engages with the mobile particles.
3. Flexible Boundary Conditions: These conditions more closely mimic real-world scenarios by permitting minor displacements of boundary particles, thereby reflecting the forces exerted on the interior particles.
4. Periodic Boundary Conditions: Commonly utilized in simulations of expansive systems to mitigate surface or boundary effects, these conditions facilitate interactions of the protocell with the resolved around atoms. To simulate a genuine large-scale system within the confines of a molecular dynamics protocell, periodic boundary conditions are frequently implemented. By leveraging these conditions, surface effects impacting the protocell are negated, and a quasi-infinite volume is established, providing a more precise representation of the macroscopic system. Specifically, when a particle traverses the boundary of the protocell's cubic structure, it is permitted to reenter the protocell through the opposing surface at an equivalent velocity, effectively traversing the boundary, as depicted in Figure 2.3.

During simulations that incorporate periodic boundary conditions, employing the truncated radius method for the computation of forces in non-bonded long-range interactions is crucial to avoid the erroneous outcomes, specifically due to the duplication calculations of forces acting on particles. Specifically, when the distance r exceeds the critical threshold r_c, the van der Waals potential approaches zero, signifying that the associated duplicated force becomes insignificant. This critical distance r_c is defined as the truncation radius. It is imperative that the truncation radius does not surpass half the dimensions of the simulation box to ensure the accuracy of the simulation.

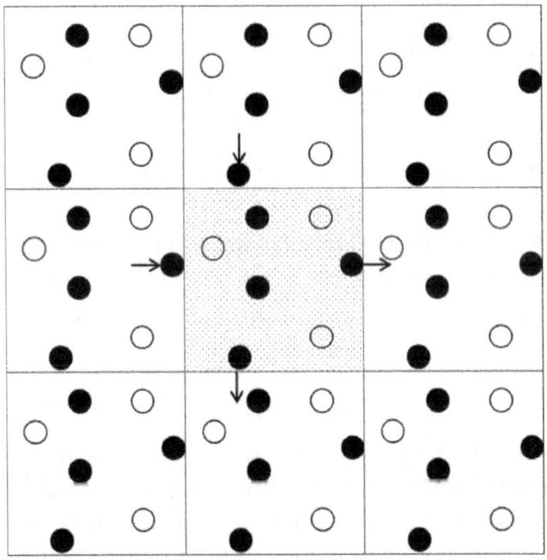

Figure 2.3. Schematic diagram of periodic boundary conditions.

2.4 Molecular Dynamics Simulation Tools

The major software packages for molecular dynamics simulations are GROMACS, AMBER, NAMD, and LAMMPS. GROMACS [17] is particularly designed for studying biomolecular systems. It is capable of simulating various molecules in solution or crystals using molecular dynamics, stochastic dynamics, or path integral methods. Additionally, GROMACS can perform molecular energy minimization, analyze conformations, and more. The software comes with GROMACS force fields for proteins, nucleotides, sugars, etc. Studies using GROMACS can span a wide range, from glass and liquid crystals to polymers, crystals, and solutions of biomolecules. It offers the great advantages in simulating the Newtonian motion of many molecular systems and has unique features that set it apart from similar software:

1. GROMACS has been optimized with various algorithms to enhance its computational capabilities. It features a user-friendly interface, and all topology and parameter files are well documented. The program does not require entering a script comment language during runtime as all operations are carried out

through simple command line operations. Additionally, the running process is step by step, allowing for the ongoing validation of the simulation's accuracy and feasibility, thereby minimizing time wastage.
2. GROMACS is straightforward, packed with features, and beginner-friendly.
3. The package includes topologies for various common proteins and nucleic acids, encompassing 20 standard amino acids and their variants, 4 nucleosides, and 4 deoxyribosides, as well as sugars and lipids.

AMBER [18] is a widely recognized molecular dynamics software designed for computational simulations of biological macromolecules, including proteins, nucleic acids, and sugars. In addition to providing empirical force fields, AMBER offers two key components. The first is a collection of molecular mechanics force fields for simulating biomolecules, which are copyright free and utilized in various simulation programs. The second component is a molecular simulation program package, including source code and demos, which is copyrighted and available for purchase. AMBER also includes programs such as *xleap* (X-windows version of leap with a graphical user interface) and *tleap* (text interface leap) for managing molecular system coordinates and parameter files. Additionally, the software features *Antechamber* for generating rare small molecule mechanics parameter files, *Sander* for MD data generation, and *pretraj* for MD simulation trajectory analysis.

NAMD (Nanoscale Molecular Dynamics) [19] is a parallel molecular dynamics code designed for fast simulation of macromolecular systems on massively parallel computers. NAMD computes atomic trajectories by numerically solving the equations of motion using empirical force fields such as AMBER, CHARMM, and DREIDING. It is known for the following features:

1. The software is highly scalable and can simulate systems at the microscopic, mesoscopic, or cross-scale microscopic levels. It is one of the best parallel processing software among many molecular dynamics programs, capable of supporting thousands of CPU operations. It also performs efficiently on a single CPU machine. NAMD can simulate systems ranging from 10,000 to 1,000,000 atoms.

2. NAMD supports various force fields such as CHARMM, X-PLOR, and AMBER, making it suitable for simulating a wide range of systems including proteins, nucleic acids, cell membranes, clusters, and CNT systems.
3. The software consists of two main parts: NAMD, which is the computational component responsible for computing trajectories, and VMD (Visual Molecular Dynamics), and other statistical analysis software for post-simulation data analysis.

LAMMPS [20], short for Large-scale Atomic/Molecular Massively Parallel Simulator, is a classical molecular dynamics code with the capability to simulate particles in various states such as liquids, solids, and gases. It supports different force fields and boundary conditions, allowing for the simulation of a wide range of systems including all-atom, polymer, biological, metallic, granular, and coarse-grained systems. LAMMPS is scalable and can handle simulations ranging from just a few particles to millions or even billions of particles. It is highly efficient when run on single-processor desktops and laptops, although it is specifically designed for parallel computers. LAMMPS can be executed on any platform with a C++ compiler and MPI, making it suitable for distributed and shared parallel machines as well as Beowulf-type clusters. Additionally, it is highly customizable and can be extended to include new force fields, atomic models, boundary conditions, and diagnostics.

In essence, LAMMPS integrates the Newtonian equations of motion for a group of molecules, atoms, and macroscopic particles while taking into account short- and long-range forces under different boundary and initial conditions. To efficiently handle interactions, LAMMPS utilizes optimized neighborhood lists to keep track of neighboring particles based on the magnitude of short-range mutual repulsive forces, thereby aiming to prevent excessive local particle densities.

LAMMPS employs a spatial discretization method to partition the simulation space into smaller three-dimensional segments on parallel machines. Each segment is assigned to a single processor, with inter-processor communication and storage of information about "ghost" atoms at the segment boundaries. In the parallel case, LAMMPS is most efficient when simulating a 3-dimensional rectangular box with a nearly uniform density. Relative to other software,

LAMMPS offers several advantages, including free open-source code, the ability to modify and extend the program as necessary, support for simulating solids, liquids, and gases, and the capability to model diverse systems such as atoms, polymers, organic molecules, and particulate materials. Additionally, LAMMPS can simulate systems with a large number of millions of particles using computational resources and allows for convenient parallel calculations.

When performing molecular dynamics simulations, it is important to select the appropriate software and force field. Each software has its own strengths and biases. GROMACS, AMBER, and NAMD are often chosen for simulating protein systems, while AMBER is preferred for DNA and RNA systems. For interface systems, DL_POLY is a powerful option. When working with material systems, LAMMPS is considered the most powerful software. For this book, LAMMPS was chosen as the molecular dynamics simulation software.

2.5 Analysis and Visualization of Defects

2.5.1 *Methods of analyzing and identifying defects*

After obtaining atomic configurations and related data from MD simulations, further analyses of the output data are necessary to obtain the covariates of interest. When it comes to the plastic deformation of metallic materials during mechanical action or frictional wear, it involves dislocations, stacking faults, and twins. There are several widely used methods for defect identification such as centrosymmetric parameters, slip vectors, coordination numbers, co-neighborhood analysis, and bond angle analysis. However, each method has its imperfections and advantages as stated in the following, so a combination of different methods is needed for practical use.

1. Coordination number: The coordination number method classifies atoms based on the difference in their coordination number. It involves counting the number of other atoms around a given atom within a specific truncation radius, typically set to the midpoint between the first and second nearest neighbor domains of the crystal. For instance, in single-crystal copper, the intact crystalline structure is face-centered cubic with a coordination number of 12. Atoms compressed near dislocation nuclei have a coordination number greater

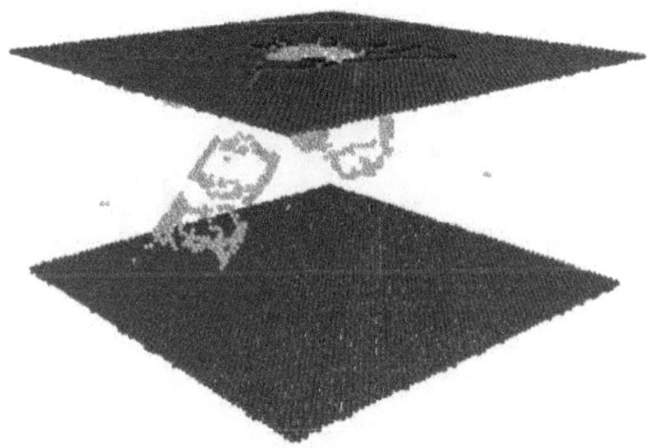

Figure 2.4. Defect morphology by coordination number method.

than 12, while atoms stretched near dislocation nuclei have a coordination number less than 12. The coordination number of atoms on the surfaces is also less than 12 and varies based on the crystal faces. For example, on the (111) face of single-crystal copper, the coordination number is 9. Interestingly, the coordination number of atoms on the lattice twins and stacking faults is the same as that of intact crystals, making these defects indistinguishable using the coordination number method. In dislocation morphology diagrams obtained using this method, the atom with a coordination number of 12 is not visible, whereas other defects such as twins and stacking faults are not identifiable as shown in Figure 2.4.

2. Centrosymmetric parameters: It is widely understood that for centrally symmetric crystal structures, such as body-centered cubic (BCC), face-centered cubic (FCC), and hexagonally close-packed (HCP) structures, the sum of the vectors from the centroid to the nearest neighboring atoms is always zero. However, when defects occur in the crystal structure, this symmetry is disrupted, and the sum of the vectors is no longer zero. As a result, it becomes possible to identify and analyze defects in the crystals. The defects can be detected and analyzed by using the centrosymmetric parameter (CSP) of a face-centered cubic metal, which can be expressed as follows:

$$\mathrm{CSP} = \sum_{i=1,6} \left| \vec{R_i} + \overrightarrow{R_{i+6}} \right|^2 \tag{2.41}$$

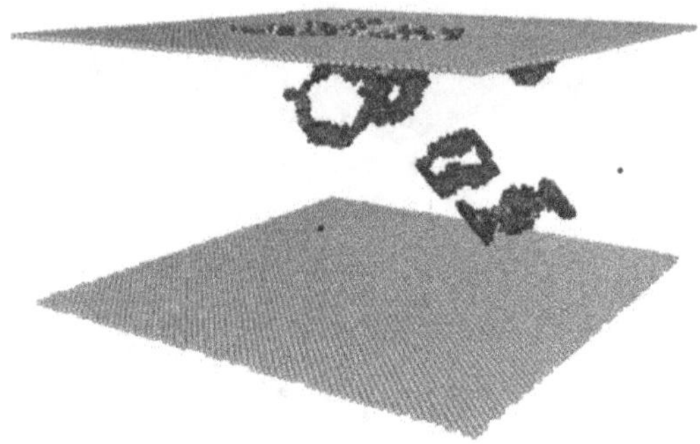

Figure 2.5. Defect morphology by centrosymmetric parameter method.

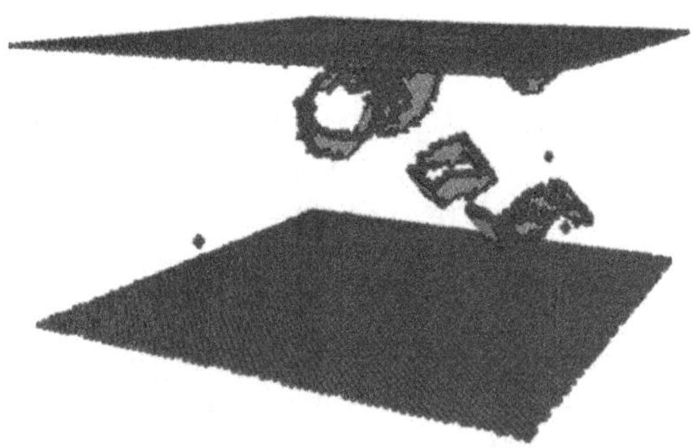

Figure 2.6. Defect morphology by co-neighborhood method.

where R_i is the nearest neighbor atom pair and R_{i+6} is the atom pair that is centrosymmetric with R_i. The CSPs of the (111) and (100) faces of the copper crystals are 1.95 nm and 2.62 nm, respectively. For example, Figure 2.5 shows the dislocation configurations of the single-crystalline copper nanoindentations obtained by the centrosymmetric parametric method.

3. Co-neighborhood analysis: The results of the centrosymmetric parameter are affected by both the thermal motion of atoms and the irregular elastic deformation of the lattice. This makes it difficult to distinguish between different defect morphologies. However, this issue is effectively addressed by using co-neighborhood analysis (CNA). Figure 2.6 displays the defect configurations of single-crystal copper nanoindentations obtained using the CNA method, where atoms with CNA equal to 1 are hidden.

CNA employs a set of four indices—denoted as i, j, k, and l—to delineate distinct crystallographic configurations. Given any pair of adjacent atoms, labeled a and b, the index i is assigned a value of 1 if they are identified as the closest neighbors; otherwise, i is designated as 2. The index j denotes the count of nearest neighbors shared between the atomic pair. Index k quantifies the number of bonds among the shared nearest neighbors, while index l serves to distinguish structures not uniquely identified by the indices i, j, and k alone. Contrasting with the centrosymmetric parameter and coordination number methodologies, CNA is well suited for the identification of regular crystal structures, offering a nuanced perspective on the atomic arrangements that constitute them. CNA takes into account not only the closest neighboring atoms but also the next-nearest neighbor atoms in some complex structures, so as to be particularly attuned to structural deformations.

4. Dislocation extraction algorithm: Dislocation extraction algorithm (DXA) is predominantly utilized for the identification of dislocation types during deformation processes and for the statistical analysis of dislocation lengths. This approach was initially proposed and subsequently refined by Stukowski and Albe [21, 22]. DXA, despite its relatively recent inception, plays an indispensable role in deciphering the deformation mechanisms within MD simulations. The technique extends its analytical capabilities to the quantification of internal crystal structures; however, it faces limitations when applied to systems comprising hundreds of thousands or even millions of atoms due to the computational intensity of the analysis. Consequently, dislocation analysis is typically reserved for critical instances of deformation. For instance, DXA is employed at the onset of plastic deformation to monitor the dislocation emission process or to examine dislocation interactions during the rheological phase. In the context of face-centered cubic (FCC) polycrystals, common dislocation types encompass perfect dislocations ($1/2\langle 110\rangle$), Shockley partial dislocations ($1/6\langle 112\rangle$), ladder-rod dislocations ($1/6\langle 110\rangle$), etc. The methodology has been seamlessly integrated into the OVITO software suite, which facilitates the comprehensive analysis and visualization of dislocations, thereby enhancing the study of material deformation at the atomic scale.

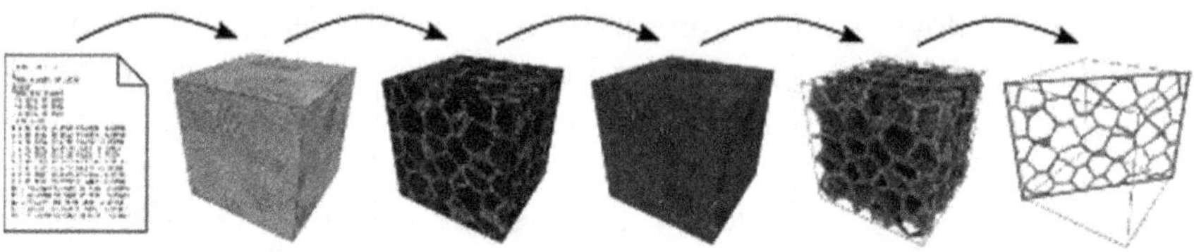

Figure 2.7. OVITO data processing, analysis, and visualization process [23].

2.5.2 *Visualization methods*

Atomic simulations, including MD, molecular mechanics, and Monte Carlo methods, generate three-dimensional atomic configurations or trajectories. These datasets necessitate subsequent analysis to extract novel scientific insights. However, the discovery, interpretation, and utilization of this pivotal information and the derivation of meaningful results are contingent upon the availability of suitable software tools. The visualization of MD simulation outcomes can be achieved through various approaches. The present authors of this publication primarily employ the standalone visualization software, OVITO [23]. OVITO is a freely available, open-source software package developed by Alexander Stukowski at the Darmstadt University of Technology in Germany. It is designed for the scientific visualization and analysis of atomic and particle simulation data. This tool aids researchers in more readily accessing material phenomena and physical processes.

In the data processing and analysis phase, as well as during the visualization of results using the OVITO software, users construct a sequence of data manipulation steps, as depicted in Figure 2.7. Each step in this sequence alters the data in a specific manner. Ultimately, these modifications are rendered visually on-screen, culminating in the generation of an output graph.

References

[1] Alder BJ, Wainwright TE. Phase transition for a hard sphere system. *J Chem Phys*, 1957, 27: 1208–1209.

[2] Alder BJ, Wainwright TE. Studies in molecular dynamics, I-general method. *J Chem Phys*, 1959, 31: 459–466.

[3] Ma Wen. Molecular Dynamics Study on the Plasticity and Phase Transformation Mechanism of Nano Polycrystalline Metals under Shock Compression (in Chinese). Changsha: PhD Dissertation, National University of Defense Technology, 2011.
[4] Verlet L. Computer "experiments" on classical fluids, I-thermodynamical properties of Lennard-Jones moleculars. *Phys Rev*, 1967, 159: 98–103.
[5] Verlet L. Computer "experiments" on classical fluids, II-equilibrium correlation functions. *Phys Rev*, 1968, 165: 201–214.
[6] Voter AF. A method for accelerating the molecular dynamics simulation of Infrequent events. *J Chem Phys*, 1997, 106: 4665–4677.
[7] Voter AF, Chen SP. Accurate interatomic potentials for Ni, Al and Ni_3Al. *MRS Online Proc Libr Arch*, 1986, 82: 175–180.
[8] Jones JE. On the determination of molecular fields, II-from the equation of state of a gas. *Proc R Soc Lond Ser A-Contain Pap Math Phys Character*, 1924, 106: 463–477.
[9] Girifalco LA, Weizer VG. Application of the Morse potential function to cubic metals. *Phys Rev*, 1959, 114: 687–690.
[10] Stillinger FH, Weber TA. Computer-simulation of local order in condensed phases of silicon. *Phys Rev B*, 1985, 31: 5262–5271.
[11] Tersoff J. New empirical-approach for the structure and energy of covalent systems. *Phys Rev B*, 1988, 37: 6991–7000.
[12] Munetoh S, Motooka T, Moriguchi K, Shintani A. Interatomic potential for Si-O systems using Tersoff parameterization. *Comput Mater Sci*, 2007, 39: 334–339.
[13] Daw MS, Baskes MI. Embedded-atom method-derivation and application to impurities, surfaces, and other defects in metals. *Phys Rev B*, 1984, 29: 6443–6453.
[14] Foiles SM, Baskes MI, Daw MS. Embedded-atom-method functions for the FCC metals Cu, Ag, Au, Ni, Pd, Pt, and their alloys. *Phys Rev B*, 1986, 33: 7983–7991.
[15] Daw MS, Foiles SM, Baskes MI. The embedded-atom method—are view of theory and applications. *Mater Sci Rep*, 1993, 9: 251–310.
[16] Baskes MI, Nelson JS, Wright AF. Semiempirical modified embedded-atom potentials for silicon and germanium. *Phys Rev B*, 1989, 40: 6085–6100.
[17] Berendsen HJC, Vanderspoel D, Vandrunen R, GROMACS—A message-passing parallel molecular-dynamics implementation. *Comput Phys Commun*, 1995, 91: 43–56.
[18] Salomon-Ferrer R, Case DA, Walker RC. An overview of the AMBER biomolecular simulation package. *Wires Comput Mol Sci*, 2013, 3: 198–210.

[19] Phillips JC, Hardy DJ, Maia JDC, Stone JE, Ribeiro JV, Bernardi RC, Buch R, Fiorin G, Henin J, Jiang W, Mcgreevy R, Melo MCR, Radak BK, Skeel RD, Singharoy A, Wang Y, Roux B, Aksimentiev A, Luthey-Schulten Z, Kale LV, Schulten K, Chipot C, Tajkhorshid E. Scalable molecular dynamics on CPU and GPU architectures with NAMD. *J Chem Phys*, 2020, 153(4): 044130.

[20] Plimpton S. Computational limits of classical molecular-dynamics simulations. *Comput Mater Sci*, 1995, 4: 361–364.

[21] Stukowski A, Albe K. Extracting dislocations and non-dislocation crystal defects from atomistic simulation data. *Model Simul Mater Sci Eng*, 2010, 18: 085001.

[22] Stukowski A, Bulatov VV, Arsenlis A. Automated identification and indexing of dislocations in crystal interfaces. *Model Simul Mater Sci Eng*, 2012, 20: 085007.

[23] Stukowski A. Visualization and analysis of atomistic simulation data with OVITO–the open visualization. *Model Simul Mater Sci Eng*, 2009, 18: 015012.

Chapter 3

Nanoindentation Process Simulation

3.1 General

Nanoindentation is a modern material testing technique based on hardness testing. It involves applying a small load to a tiny area on the material's surface while continuously measuring load, displacement, and other data during loading and unloading. This allows for obtaining mechanical property indexes such as indentation hardness, Young's modulus, indentation creep, indentation relaxation, and fracture toughness at the micro-scale. These properties help establish a connection between the material's microscopic structure and its macro-mechanical properties, guiding material design and research. The method offers advantages, such as easy specimen preparation, short testing time, control of test conditions, and highly accurate data recording.

The development of nanoindentation test methods started in the mid-1980s. In 1997, the International Organization for Standardization (ISO) began standardizing instrumented nanoindentation tests. By 2002, five international standards (ISO 14577-1-5) had been officially issued, with two more still in the process of being drafted. China has also released relevant test standards (GB/T 21838.1-4).

This demonstrates that the nanoindentation test method has gained acceptance in the science, technology, and engineering communities and has increasingly become a standard material testing method.

Nanoindentation is based on the Hertz contact theory. Oliver and Pharr [1, 2] proposed a method to calculate the elastic modulus and hardness of a material from the nanoindentation load-indentation depth curve, establishing the foundation of nano measurement techniques. In recent years, nanoindentation and test technology continue to advance. Some atomic force microscopes (AFMs) now come equipped with nanoindentation test capabilities, and corresponding finite element and molecular dynamics methods have also been used for theoretical calculations of nanoindentation.

3.2 Nanoindentation Principles and Tests

3.2.1 *Nanoindentation principle*

The choice of indenter for nanoindentation is crucial, and a popular option is the Berkovich diamond indenter, also named as the Berkovich diamond indenter. This indenter in Figure 3.1 has a triangular tip with three prismatic surfaces and a cone angle of 65.27°. The angle between the three surfaces is 120°, and it needs to meet the international standard of having a radius of the obtuse circle at the tip that is less than 200 nm. The Berkovich indenter's projected area to depth of indentation ratio is equivalent to that of the micro-Vickers hardened rectangular pyramid indenter with a cone angle of 68°.

The use of Berkovich indenters for measuring the surface properties of materials is preferred over Vickers indenters for several reasons. One key advantage is that Berkovich indenters can be made very accurately and any errors in their face-to-face angle from different suppliers typically have a negligible effect on measurement results. Additionally, Berkovich indenters can be easily ground to a sharper, single-point tip, which is not the case with Vickers indenters as they tend to have a linear tip. Although it is nearly impossible to obtain an ideal tip for an indenter in practice, it is theoretically possible to approximate the contact mechanical response of an indenter by using a spherical tip with a very small radius. To approximate the

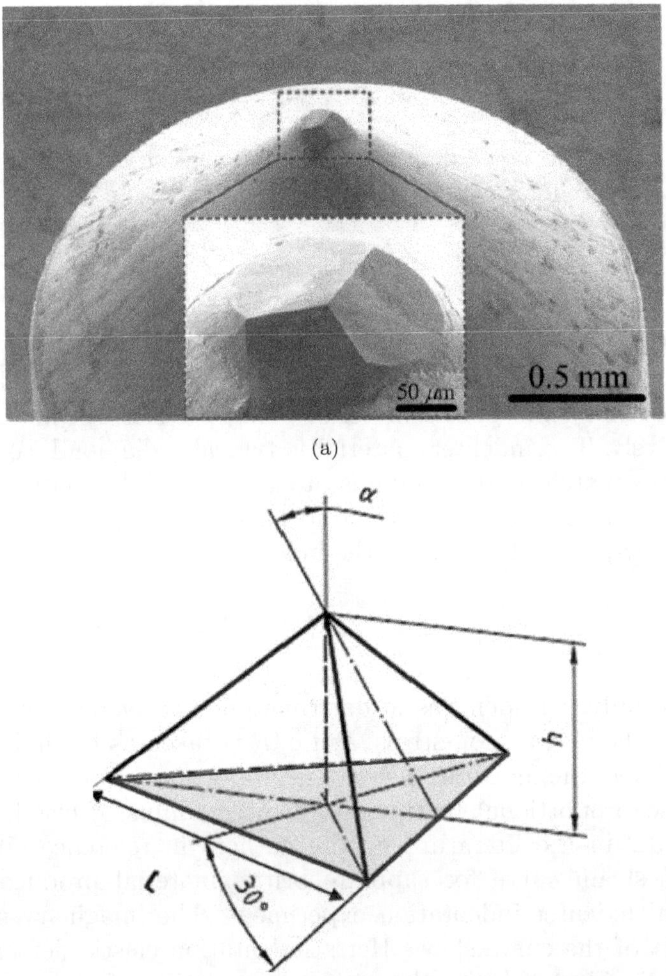

Figure 3.1. Berkovich indenter. (a) SEM image [3], (b) geometric parameters [4].

loading and unloading processes, the following conditions for spherical indenter geometry are used:

(1) **Loading process:** When a spherical indenter with radius R is pressed into a surface, if the contact radius is a and the indentation load is P, if R is significantly greater than a, based on Hertz contact

theory, certain conditions apply:

$$a^3 = \frac{3}{4}\frac{PR}{E^*} \tag{3.1}$$

where E^* is the equivalent modulus of elasticity (E^*). It can be expressed as follows:

$$\frac{1}{E^*} = \frac{1-\upsilon_1^2}{E_1} + \frac{1-\upsilon_2^2}{E_2} \tag{3.2}$$

where E_1 and ν_1 are the modulus of elasticity and Poisson's ratio of the indenter material, and E_2 and ν_2 are the modulus of elasticity and Poisson's ratio of the material being pressed by the indenter, respectively. The indenter material is typically diamond, so E^* can be approximated as the modulus of elasticity of the material being pressed into (E_2). Using equation (3.1), we can express the average pressure (p_m) as a function of the indentation stress and strain:

$$p_m = \frac{P}{\pi a^2} = \left(\frac{4}{3}\frac{E^*}{\pi}\right)\frac{a}{R} \tag{3.3}$$

If we only consider elastic deformation, the average pressure is taken as the indentation stress, and a/R is taken as the indentation strain. Then, the indentation stress is proportional to the strain, and it is also proportional to the equivalent modulus of elasticity E^*. Figure 3.2 in the literature by the authors in reference [5] shows a stress–strain curve for sapphire brittle material produced by a spherical indenter indentation experiment. The imaginary straight line part of the curve shows Hertz indentation elastic deformation, which coincides with the theoretical formulation of equation (3.3). For fully elastic deformation, the maximum principal shear stress is located at a distance of approximately $0.5a$ directly below the indenter, with a value of approximately 0.47 p_m. Plastic yielding will occur at $\tau \approx 0.5\sigma_y$ if the Tresca or von Mises shear stress criterion is applied. Expressing this in terms of the average compression, we have $p_m \approx 1.1\sigma_y$ [6, 7]. After full yielding has occurred, p_m represents what is often referred to as hardness (H).

When the indentation depth exceeds the spherical tip portion, the prismatic cone becomes the dominant geometry causing more significant plastic deformation under the surface. The size of the

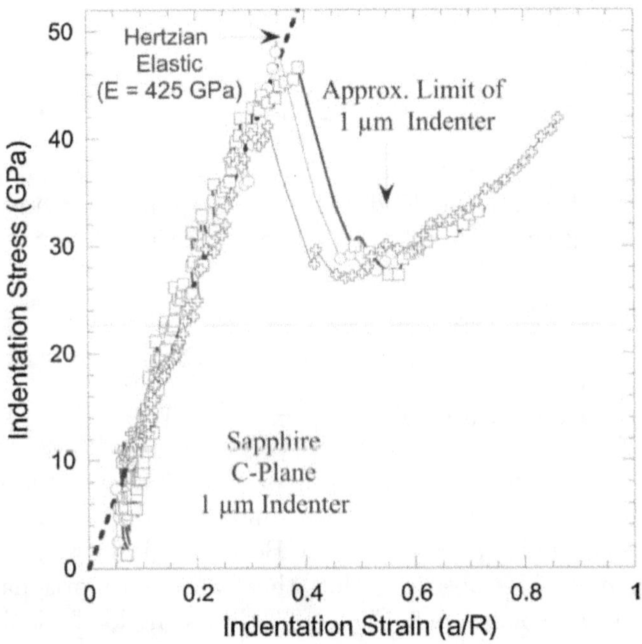

Figure 3.2. Stress–strain curves of a spherical indenter pressed into a sapphire single-crystal C-plane direction loading plate [5].

contact area at which the maximum indentation load is achieved can be used to convert the prismatic cone into an equivalent cone, simplifying the problem for easier analysis. By finding a suitable cone angle, it is possible to make the area and indentation-depth ratios of the cone equal to those of the prismatic cone. This angle can be determined from the area of the Berkovich indenter pressing into the prism face at an angle α, as shown in Figure 3.1. The area can be shown as follows:

$$A = 3\sqrt{3}h_C^2 \tan^2 \alpha \tag{3.4}$$

For cones with a half-cone angle of θ, the following equation can be used:

$$A = \pi h_C^2 \tan^2 \theta \tag{3.5}$$

where h_c is the indentation depth. The angle α is approximately 65.27° and θ is approximately 70.30° between the prismatic plane and the axis of the cone. Substituting these angle values, equations (3.4) and (3.5) become

$$A \approx 24.5h_C^2 \tag{3.6}$$

Nanoindentation is primarily used to obtain material property parameters through the load–displacement curves during testing. Assuming that the material to be indented has a well-defined elastic–plastic transition point by Fischer–Cripps [8], the load–displacement relationship is given by

$$P = E^* \left\{ \frac{1}{\sqrt{\pi}\tan\theta}\sqrt{\frac{E^*}{H}} + \left[\frac{2(\pi-2)}{\pi}\sqrt{\frac{\pi}{4}}\sqrt{\frac{H}{E^*}}\right]\right\}^{-2} h^2 \tag{3.7}$$

where E^* is the equivalent modulus of elasticity, H is the hardness of the material being pressed into, and θ is the equivalent conical half-cone angle ($\theta \approx 70.30°$ for Berkovich indenter). From the equation, it can be observed that the load is proportional to the square of the displacement. Figure 3.3 shows the load–displacement curve of Berkovich indentation of quartz glass under a maximum load of 50 mN.

(2) **Unloading process:** When unloading after the loading has reached its maximum value or has been maintained at the maximum value for a certain period, the depth of indentation decreases. Since plastic deformation is permanent and energy (heat) is dissipated during the loading process, the material does not change shape due to unloading. Consequently, pure elastic deformation occurs during unloading. The slope of the unloaded load–displacement curve, referred to as dP/dh, can be expressed as the stiffness of the material being pressed. If the indenter is assumed to be a rigid cone with a half-cone angle of θ pressed into a semi-infinite elastic plane, this can be described as follows [9, 10]:

$$P = \frac{\pi a^2}{2} E^* \cot\theta \tag{3.8}$$

The depth under the pressed-in free surface can be expressed as follows:

$$h = \left(\frac{\pi}{2} - \frac{r}{a}\right) a \cot\theta \quad r \leq a \tag{3.9}$$

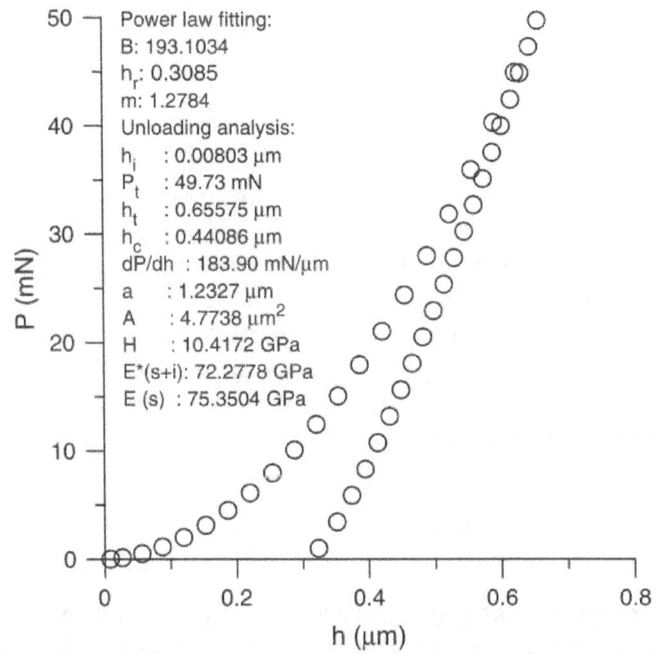

Figure 3.3. Load–displacement curves for Berkovich pressing into quartz glass [9].

When $r = 0$, equations (3.8) and (3.9) can be combined and we can get

$$P = \left(\frac{2E^*}{\pi}\tan\theta\right)h^2 \quad (3.10)$$

In the process described in equation (3.10), the load is still directly proportional to the square of the indentation depth, as indicated by equation (3.7).

The derivative of equation (3.10) yields the material's stiffness when it is pressed into

$$\frac{dP}{dh} = \left(\frac{4E^*}{\pi}\tan\theta\right)h \quad (3.11)$$

Employing Eq. (3.9) once more, and setting $r = 0$ then substituting into the previous equation (3.11), we obtain

$$\frac{dP}{dh} = 2E^*a \quad (3.12)$$

The cone is vertically pressed into the material's surface, which can be likened to a circle with area A and radius a. As a result, equation (3.12) can be expressed as follows:

$$E^* = \frac{1}{2}\frac{dP}{dh}\frac{\sqrt{\pi}}{\sqrt{A}} \tag{3.13}$$

The above equation represents the equivalent modulus of elasticity. In other words, the equivalent modulus of elasticity can be articulated in terms of the slope of the unloaded portion of the curve when the nanoindenter is pressed to its maximum load.

When the material being tested is pressed for its hardness, we can substitute in equation (3.6) to derive

$$H_{IT} = \frac{P_{\max}}{A} \approx \frac{P}{24.5h_C^2} \tag{3.14}$$

where A represents the projected area, and $P_{\max}$ and h_C values can be sourced from indentation experimental data. This indentation hardness is commonly known as instrumented indentation hardness and shows a strong correlation with Vickers hardness [11]:

$$H_V = 0.0945H_{IT} \tag{3.15}$$

In the 1950s, Tabor introduced a method to estimate the yield strength of a material based on its hardness, suggesting that the yield strength is approximately three times the hardness of most metallic materials with large E/H values. Later, Johnson proposed a more precise equation for calculating yield strength [11], taking into account the concept of a "static nucleus." For brittle materials, the Berkovich indentation test can also be used to measure fracture toughness, with the equation involving the modulus of elasticity (E) and other geometric parameters, as illustrated in Figure 3.4. The equation is as follows:

$$K_I = 0.016\left(\frac{a}{l}\right)^{\frac{1}{2}}\left(\frac{E}{H_{IT}}\right)^{\frac{2}{3}}\frac{P}{c^{\frac{3}{2}}} \tag{3.16}$$

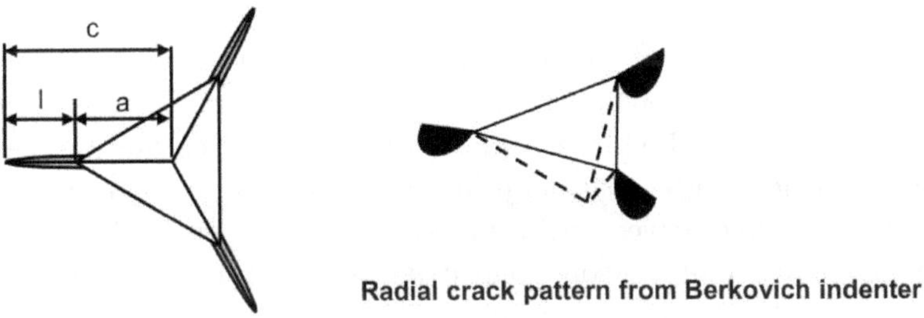

Figure 3.4. Geometrical parameters of radial cracks produced by Berkovich's indentation into brittle materials [11].

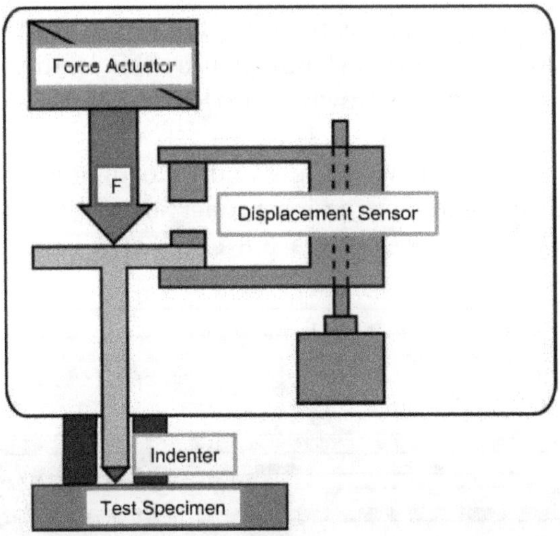

Figure 3.5. Schematic of the mechanism of the nanoindentation device [13].

3.2.2 *Nanoindentation test*

Nanoindentation is commonly performed using specialized instruments. The nanoindentation instrument typically includes a rigid frame with an indenter, a micro loading mechanism, and an indenter displacement sensor. Figure 3.5 illustrates the schematic of the nanoindentation device mechanism. Currently, nanoindentation devices are most commonly used in higher education institutions, government research institutes, and quality control laboratories.

Nanoindentation devices require high accuracy in measuring displacement (less than 1/10 nm) and loading load (within a few nN). When measuring thin film materials, the indentation depth should not exceed 1/10 of the film thickness. Piezoelectric ceramics are commonly used for loading through inductive coils or static electromagnetic field change drive, with a maximum load of generally less than 1 μN. Displacement sensors utilize inductive or capacitive changes for detection. Following is a brief description of how each component works:

(1) A commercially adopted loading method involves using a set of parallel discs (usually between two to three) to achieve precise load control through the electrostatic force. For example, the U.S. Hysitron nanoindentation instrument uses this loading method (see Figures 3.6 and 3.7). One drawback is that it can only load a vertical load of 30 μN or less.

Another way to achieve loading and unloading is by utilizing the volume expansion and contraction properties of piezoelectric ceramics when energized (see Figures 3.8 and 3.9). Typically, the load

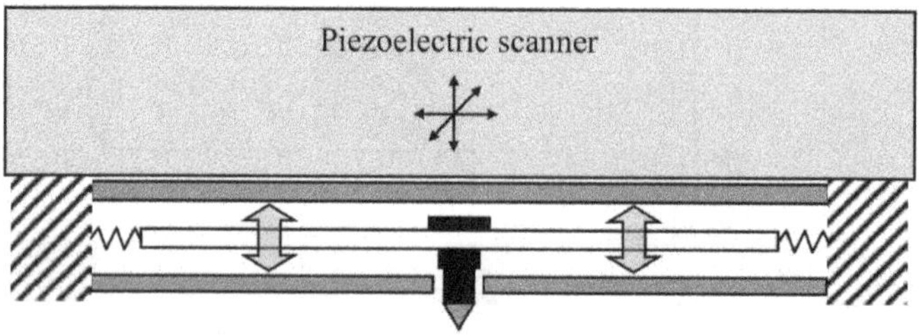

Figure 3.6. Schematic of Bruker's Hysitron nanoindentation instrument using the principle of electrostatic gravitational loading [14].

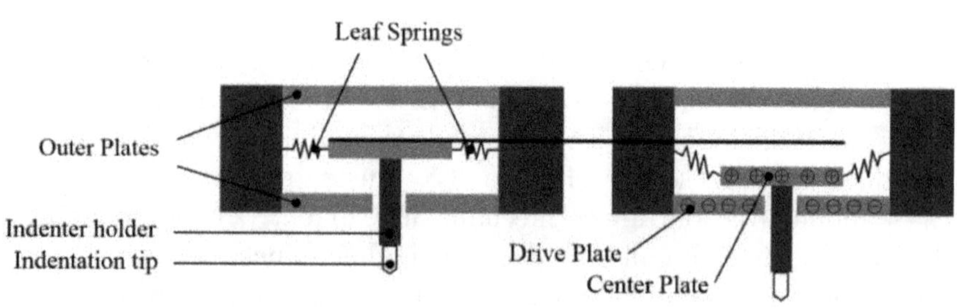

Figure 3.7. Schematic principle of electrostatic loading [15].

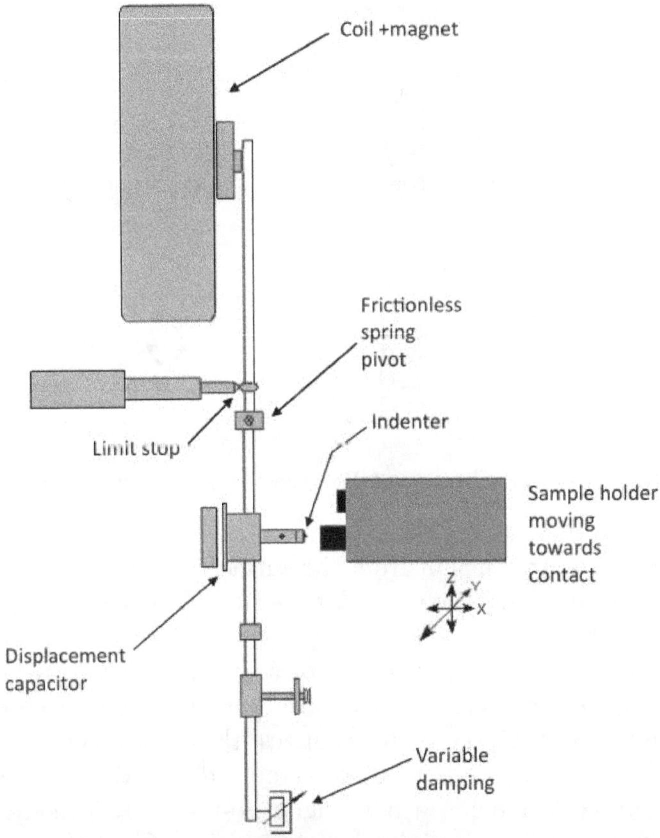

Figure 3.8. Schematic diagram of the nanoindentation instrument from micro materials using piezoelectric ceramic loading principle [14].

is applied to the indenter using the bending deformation of a disc enclosed in a metal box. Even though ambient temperature and humidity can impact the dimensions of the piezoelectric ceramics, this error can be mitigated through closed-loop control. However, this loading device is disadvantageous due to its high cost.

(2) To measure the depth of indentation or displacement, displacement sensors are used. In order to accurately measure the depth of the indentation, it's important to establish the starting point of the indentation, which is the position where the indentation depth on the surface is zero. However, determining this position is often challenging in practice. Typically, the indenter is initially set to a working

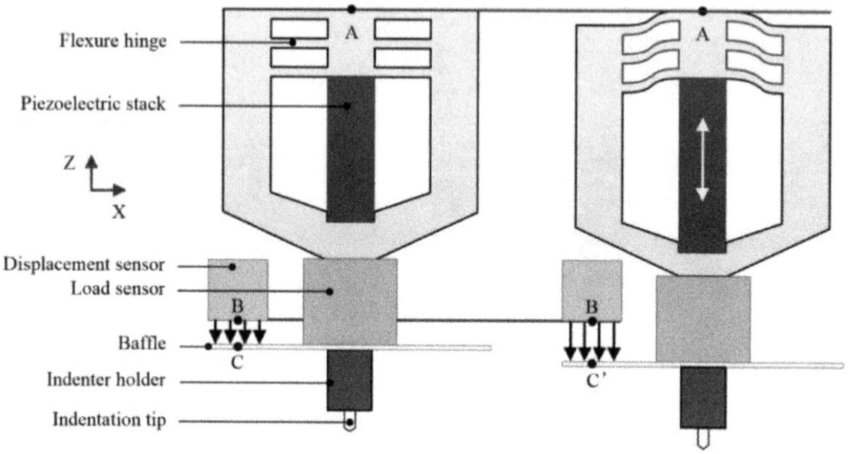

Figure 3.9. Schematic diagram of the piezoelectric ceramic drive principle [15].

distance of a certain height from the surface (usually 100 μm) before starting the test, and then it is lowered to the surface to begin the measurement.

Displacement sensing devices typically utilize two to three parallel plate capacitance arrangements and an AC bridge circuit. When the distance between the plates changes, the circuit's current can be adjusted to easily obtain the displacement data. Although the principle is simple, achieving a high signal-to-noise ratio requires large plate requirements and minimal plate spacing, which can affect the measurement range. Additionally, ensuring the plates are sufficiently parallel is important, and the use of coated glass can make them prone to damage during handling.

Currently, the linear variable differential transformer (LVDT) is also used in the market to measure displacement. This type of sensor is a linear displacement sensor that operates based on the principle of an iron core movable transformer. It consists of a primary coil, two secondary coils, an iron core, a coil skeleton, a shell, and other components. When driven by an AC excitation wave, the change in the output coil current is used to calculate displacement, replacing capacitance change with inductance change to measure distance change.

In addition to measuring in the millimeter range, this sensor can achieve measurement accuracy at the level of 0.1 μm and does not require periodic re-calibration of the zero point for a longer period. To accurately obtain the start point of the indentation of the test indent, nanoindenters are generally equipped with a reference probe to assist in determining the "zero point" of the indentation.

(3) To ensure accurate measurements, the frame needs to be rigid. Typically, granite or cast iron is used for this purpose, although aluminum alloy can also be used. However, aluminum alloy has poor vibration-damping properties. Design considerations can include the use of honeycomb or pillars to enhance the instrument's stiffness in the loading direction.

In Figure 3.10, the schematic diagram of the nanoindentation test is shown. The instrument begins by finding the "zero point" of the tested surface and then gradually increasing the indentation load. From the load–time trapezoidal curve (Figure 3.10(a)), it can be observed that the load goes from point "A" to "B," reaching the maximum value, and then starts unloading directly from point "B" (C).

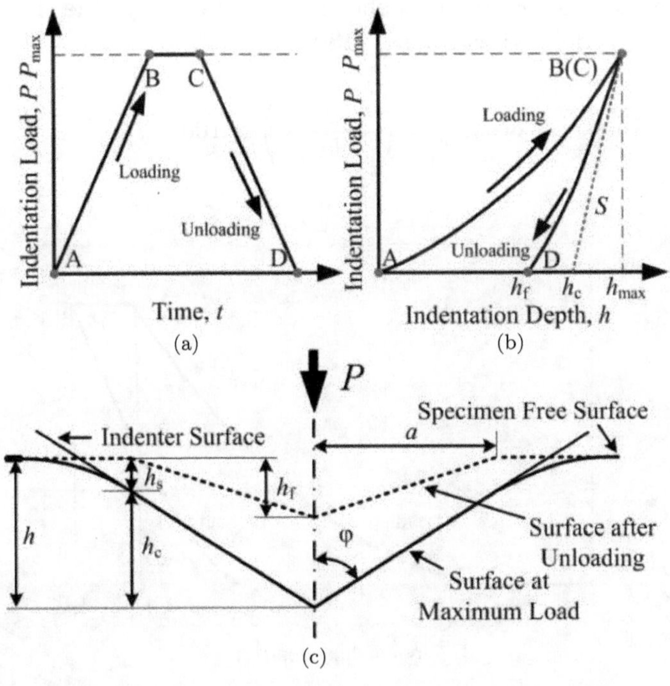

Figure 3.10. Nanoindentation test [13]. (a) Load–time trapezoidal curve, (b) Load–displacement curve, (c) Physical meaning of the calculated parameters.

Subsequently, the load returns to zero as it moves along the CD line to point "D." Corresponding to the load–time trapezoidal line, the load-indentation depth (displacement) curve (Figure 3.10(b)) forms an exponential curve rising (AC section) and falling (BCD section). By calculating this curve in Figure 3.10(b), we can determine some common mechanical properties of the surface material. During the press-in process, if the load reaches the maximum value and remains unchanged for some time (from point "B" to "C"), the depth of the press-in will continue to increase, as shown in Figure 3.11. This phenomenon is classified as material creep, representing the delayed stress–strain effect due to the viscoelasticity of the material.

To ensure the accuracy of the measurement, the frame needs to have high rigidity. Currently, most frames are made of granite or cast iron, although some use aluminum alloy, which has poor vibration-damping properties. Additionally, designers can consider using honeycomb or pillars to improve the stiffness of the instrument in the loading direction.

Based on the parameters shown in Figure 3.11, the indentation creep can be defined as follows [13]:

$$C_{creep} = \left(\frac{h_2 - h_1}{h_1}\right) \times 100\% \tag{3.17}$$

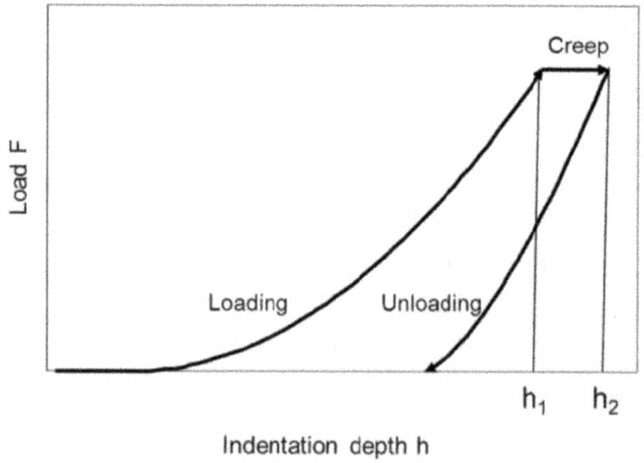

Figure 3.11. Load–displacement curves for nanoindentation tests, displacement produced by holding the load constant at its maximum value [13].

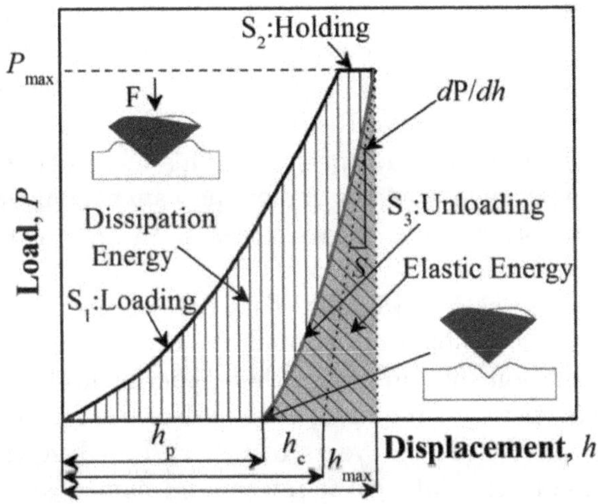

Figure 3.12. Load–displacement curves for nanoindentation interpreted from an energy dissipation point of view [16].

where h_1 represents the depth of indentation when loaded to the maximum load and h_2 represents the depth of indentation during the load holding phase.

Figure 3.10 depicts the schematic diagram of the nanoindentation test. Initially, the instrument locates the "zero point" of the tested surface. Based on the parameters in Figure 3.11, indentation creep can be defined as [13].

In Figure 3.12, there is a diagram illustrating the load–displacement curves for nanoindentation in relation to energy dissipation. The area between the loading, holding, and unloading curves represents the dissipated energy of the material, specifically the plastic deformation energy. The area under the unloading curve indicates the elastic deformation energy, which is part of the conserved energy, i.e., the recoverable energy.

3.3 Molecular Dynamics Simulation of the Nanoindentation Process

There are two main methods to study the mechanical properties of materials on the nanoscale. One method is uniaxial tensile or compression testing on the nanoscale. The other method is the nanoindentation method. When the size of the specimen is reduced to the nanoscale, the plastic deformation mode of the material under the uniaxial experiments will change. On the nanoscale, the activation of the internal Frank–Read source of the material occurs on the surface of the nanoscale specimen. This means that the results of uniaxial experiments at the nanoscale can no longer be simply generalized to other more complex cases, as they can be at the bulk scale, since the activation of surface sources is specific to one-dimensional nanoscale specimens. To study the mechanical properties of 2D or 3D nanomaterials, and to investigate other possible plastic deformation mechanisms, other experimental methods such as nanoindentation are required. Nanoindentation is an important method developed in recent years to test and analyze the mechanical properties of materials at the micro- and nano-scale. Unlike uniaxial tensile or compression experiments, nanoindentation tends to activate internal dislocation sources within the material. For example, molecular dynamics simulations show that the initial plasticity of single-crystal metallic material (100) facets in nanoindentation is activated by isotropic nucleation of incomplete dislocations in the subsurface layer of the specimen. This provides an effective tool for the study of the mechanical properties of nanomaterials.

Nanotwinning has been discovered in recent years as an effective method to strengthen small stacking fault energy face-centered cubic metals. Both *in situ* experiments and molecular dynamics simulations have shown that the strength of nano-twinned metallic materials strongly depends on the thickness of the twin layer [17–22]. The introduction of the twin layer may have a strengthening effect, a weakening effect, or no effect on the metallic material, depending on the thickness of the twin layer and the characteristic length of the specimen exterior (22). However, the mechanism of the role of the $\Sigma3$ coherent twin boundaries in the plastic deformation of metallic materials is still an urgent problem to be solved.

The present authors used molecular dynamics to simulate the nanoindentation process of single-crystal and nano-twinned copper films. We investigated the mechanical behaviors and properties of single-crystal and nano-twinned copper films and analyzed the plastic deformation mechanisms of these materials. The influence of the thickness of the twinned wafer layer on the elastic and plastic mechanical properties of single-crystal and nano-twinned copper films is revealed. The twin-assisted dislocation nucleation process is discovered, and the "intrinsic scale effect," which depends on the thickness of the twinned layer, is elucidated.

3.3.1 *Simulation methods*

Figure 3.13 shows the atomic model of the established nano-twinned copper film. The model contains a series of equally spaced Σ3 coherent twin boundaries. The thickness of the twin layer ranges from 1.25 to 5.64 nm. The size of the model is 23.5 × 23.5 × 23.5 nm and contains 1, 121, 480 atoms. x, y, and z axes are along the [−110], [11−2], and [111] directions, respectively. The indentation direction is along the z-axis direction, which is perpendicular to the direction of the Σ3 coherent twin boundary. The simulations were carried out using the Mishin [23]. The embedded atom potential proposed by Mishin is used to describe the interaction between copper atoms. The unstable layer error energy of copper predicted by Mishin embedded atom potential is 161.0 mJ · m^{-2}, the unstable twin energy is 182.4 mJ · m^{-2}, the twin energy is 18.2 mJ · m^{-2}, and the endowment layer error energy is 43.5 mJ · m^{-2}. The intrinsic stacking fault

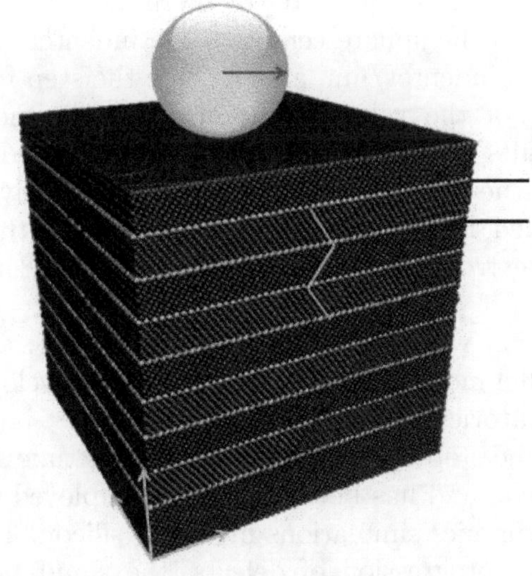

Figure 3.13. Atomic model of nano-twinned copper film.

energy follows the experimental result of $45\,\mathrm{mJ} \cdot \mathrm{m}^{-2}$ [24], therefore the Mishin embedded atom potential has been widely used in molecular dynamics simulations of copper mechanical properties.

Initially, the atoms were positioned according to the face-centered cubic lattice structure of copper, with a lattice constant of 0.3615 nm. To reach an equilibrium atomic configuration, the specimens were first equilibrated at 0 pressure and 0.01 K using the Nosé–Hoover method [25, 26]. Following this, a spherical indenter was pressed into the specimen at a speed of 10 m/s. During the pressing process, a 1 nm atomic layer at the bottom part of the specimen was frozen at a temperature of 0.01 K to prevent translational motion of the pressed specimen, while the remaining atoms were subjected to the canonical ensemble (NVT) to maintain a constant temperature of 0.01 K. In the lateral direction, periodic boundary conditions were applied to simulate an infinite periodic array of indentations, while free boundary conditions were utilized in the indentation direction. The indenter was modeled using virtual repulsive forces, that is,

$$
\begin{aligned}
F &= AH(r)(R-r)^2 \\
H(r) &= \begin{cases} 1 & r < R \\ 0 & r \geq R \end{cases}
\end{aligned} \tag{3.18}
$$

where A is the stiffness parameter of the indenter, r is the distance from the atom to the sphere center of the indenter/angstrom, R is the radius of the indenter/nm, and $H(r)$ is the step function.

The impact of the value of the stiffness parameter A on the simulation results has been thoroughly investigated. It has been observed that the simulation results remain relatively consistent when A is varied from 3 to 30 eV/angstrom^2. In this study, A is set at 10 eV/angstrom^2. Additionally, R is fixed at 8 nm throughout the simulation.

The simulation was conducted using LAMMPS, an open-source massively parallel molecular dynamics software developed by Sandia National Laboratories in the USA [27]. The Verlet integration algorithm with a time step of 1 fs was utilized. This time step is deemed suitable for silicon and has been commonly employed in a multitude of molecular dynamics simulations involving silicon. To examine the morphology and progression of defects, the simulation results are visualized in this chapter using centrosymmetric parameters.

3.3.2 *Nanomechanical behavior and properties*

3.3.2.1 *Elastic deformation*

According to the classical Hertzian contact theory, the contact load F of a rigid spherical indenter in contact with an infinitely elastic plane can be expressed as follows:

$$F = \frac{4}{3} M d^{3/2} \sqrt{R} \tag{3.19}$$

where d is the indentation depth/nm, M is the indentation modulus, which expresses the stiffness/GPa of the contact system, and in isotropic materials, M can be expressed as follows:

$$M = \frac{E}{1 - \nu^2} \tag{3.20}$$

where E is the modulus of elasticity/GPa of the material and ν is Poisson's ratio.

The load–displacement curves for single crystals and twinned copper with different twin-layer thicknesses are shown in Figure 3.14. In this analysis, the single crystal is considered as twinned copper with an infinitely thick twin wafer layer (λ). It is observed that all the curves almost overlap during the elastic deformation stage and fit well with the power function relationship. Expression (3.19) was

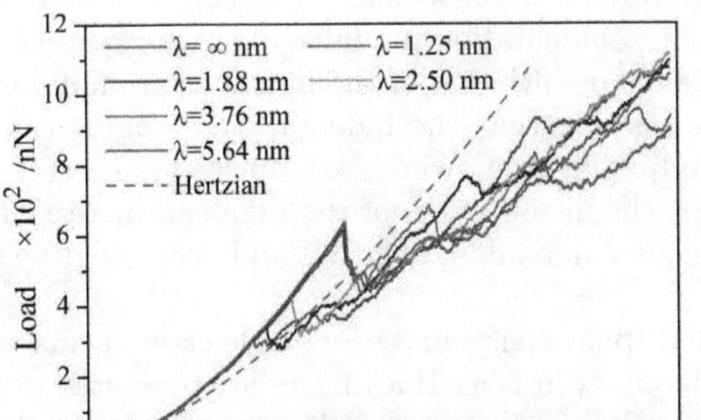

Figure 3.14. Load–displacement curves for thin single-crystal and nano-twinned copper.

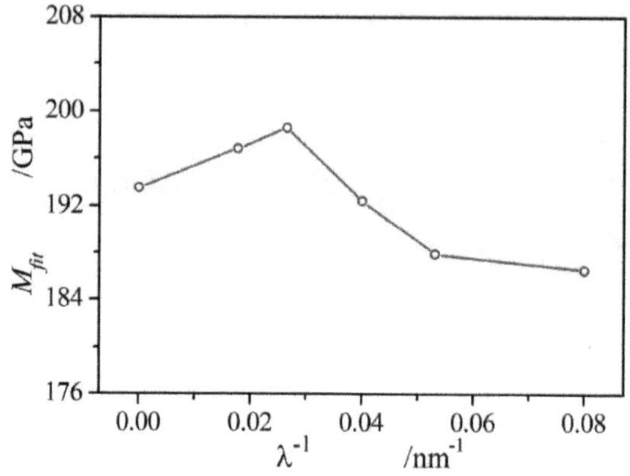

Figure 3.15. Indentation modulus versus inverse of twinned wafer layer thickness.

used to fit all the curves in order to determine the indentation modulus M_{fit}. The results are presented in Fig. 3.15. It is evident that the indentation modulus of nano-twinned copper ranges from 186.5 to 198.6 GPa. It increases and then decreases with the thickness of the twin wafer layer, reaching the maximum at 3.76 nm. However, the variation is not substantial, with only a 6.09% difference between the maximum and minimum values. The indentation modulus of single-crystal copper film is 193.5 GPa, which differs significantly from the theoretical value of 151.9 GPa. This disparity is mainly attributed to the anisotropy of the deformation of single-crystal copper films. Importantly, the indentation modulus of single-crystal Cu films falls within the range of variation of the indentation modulus of twinned Cu films, suggesting that the indentation modulus can be slightly altered by adjusting the twin layer thickness. In this study, it is observed that the introduction of the twin can increase the indentation modulus by a maximum of 2.63% and decrease it by a maximum of 3.62%.

The load–displacement curves for single crystals and twinned copper with different twin-layer thicknesses are presented in Figure 3.14. For simplicity, the single crystal is being treated as twinned copper with an infinite twin. In conclusion, the introduction of nanoconfined lattice twin boundaries does not significantly affect the elastic deformation behavior of single-crystal copper, which aligns with previous studies [21].

3.3.2.2 *Plastic deformation*

Based on Figure 3.14, all load–displacement curves exhibit a sudden decrease in load following elastic deformation. Further analysis indicates that this change corresponds to the initial dislocation nucleation, marking the onset of plasticity. Subsequently, the load shows jagged increases with increasing indentation depth, attributed to discrete dislocation events. This outcome is reminiscent of the stress discontinuity observed in stress–strain curves (under displacement loading) in micro- and nano-scale uniaxial tensile/compression tests. The discontinuity in the stress–strain curve is a fundamental characteristic of nano-sized material deformation, and its mechanism is related to the discrete nucleation and slip of finite dislocations, in contrast to the simultaneous opening and slip of numerous pre-existing dislocations on the macro-scale.

In order to further explore the plastic deformation of nano-twinned and single-crystal copper, pressure–depth curves were computed during nanoindentation. The pressure is determined by dividing the load by the horizontal projected area of the contact surface. The definition of the contact area at the atomic scale is still a matter of debate. There are two general definitions. One defines the contact area as the area enclosed by the edge contacting atoms, assuming that all atoms within the geometry enclosed by the edge atoms remain in contact. The other defines it as the sum of the single-atom areas of the actual contacting atoms. The second definition is more accurate, although defining the single-atom area presents a challenge. The difference between these two definitions is illustrated in Figure 3.16. In practical terms, the difference between the two definitions is very small in the plastic contact range. For convenience, the first calculation is used here. In the simulation, the coordinates of the contacting atoms are output in real time. Subsequently, the Delaunay triangular mesh is divided into the horizontal projection points of all the atom coordinates, and the sum of the areas of all the triangular meshes represents the required contact area. In the final calculation, Figure 3.17 illustrates the pressure–depth curve. The graph shows an initial increase in pressure with increasing indentation depth, followed by a sudden drop and then a gradual decrease. This indicates significant softening of the material after the initial plastic deformation, which is caused by dislocation slip. The sudden pressure

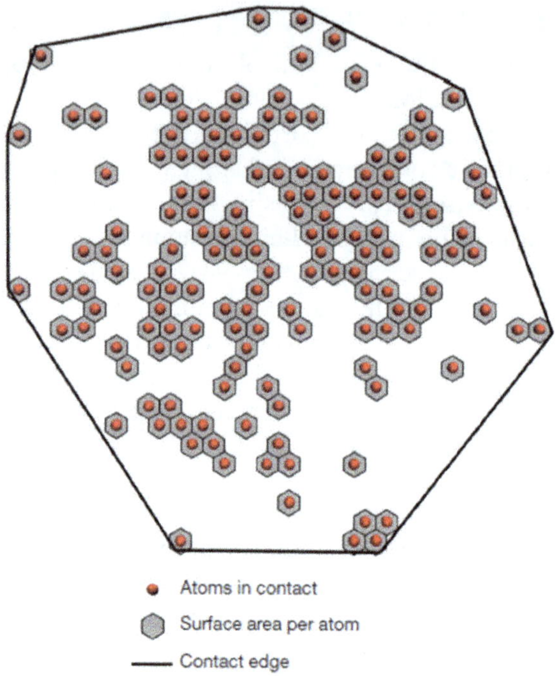

Figure 3.16. Contact area schematic [28].

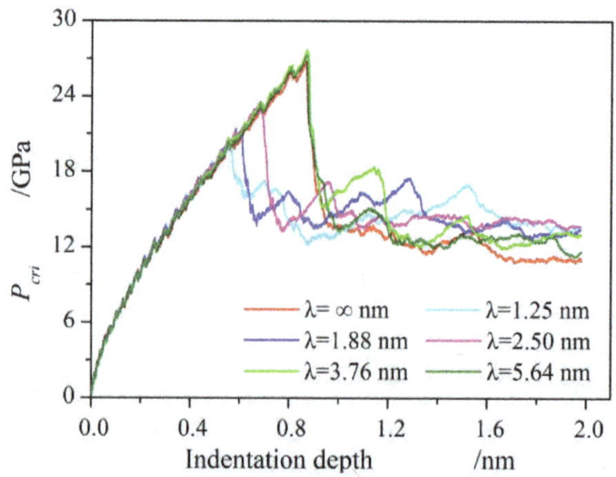

Figure 3.17. Pressure–depth curves for thin single-crystal and nano-twinned copper.

drop corresponds to a sudden change in load. The maximum pressure achievable by nanoindentation of twinned copper is seen before the drop, representing the initial pressure of plasticity. After plasticity, the indentation pressure represents the material's hardness on a macroscopic scale. As shown in Figure 3.17, both twinned and single-crystal copper exhibit a "scaling effect," with nano hardness decreasing as indentation depth increases. Twinned copper exhibits slightly greater hardness than single-crystal copper, which aligns with the commonly reported enhancement of eutectic twin boundaries.

The initial pressure of single-crystal copper plasticity (P_{cri}) is 26.79 GPa, corresponding to a critical pressure depth (D_{cri}) of 0.86 nm, which aligns with the findings in the literature [29]. Figure 3.18 illustrates the relationship between the plastic initial pressure of nano-twinned copper and the inverse of the thickness of the twinned wafer layer. It is notable that the plastic initial pressure does not consistently increase as the thickness of the twin layer decreases. Below 3.76 nm, there is a sudden decrease in the plastic initial pressure, followed by a rapid decrease as the thickness continues to reduce. This suggests that the nanoindentation process of copper nano twins exhibits similarities to the "intrinsic scale effect" caused by the grain size of polycrystalline metallic materials.

The plastic initial pressure of copper nano twins increases with the increase of the twin layer thickness until the critical point

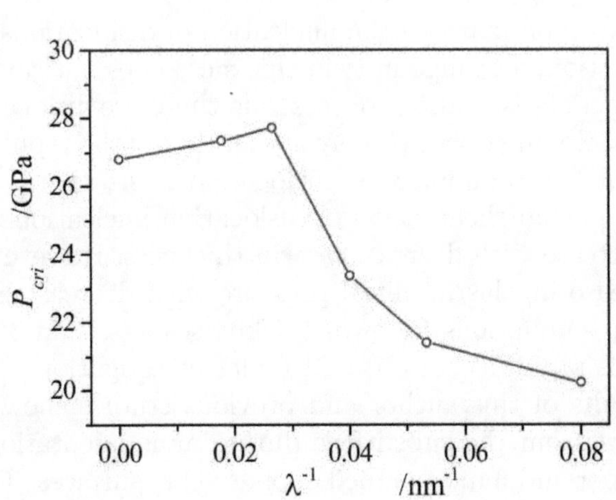

Figure 3.18. Initial pressure versus the inverse of the thickness of the twinned wafer layer.

of 3.76 nm (corresponding to the Hall–Petch relationship in polycrystalline metallic materials). However, the opposite trend occurs beyond the critical point (corresponding to the reverse Hall–Petch relationship in polycrystalline metallic materials). Nevertheless, further studies have shown that the twin-scale effect is not the same as the grain-scale effect [30, 31]. Researchers also found that the tensile strength of polycrystalline nano-twinned copper has a maximum value at a twin thickness of 15 nm. When the thickness of the twin layer is reduced below 15 nm, the tensile strength continues to decrease due to the transformation of the plastic deformation mechanism. A similar trend occurs in twinned nanowires [32].

3.3.2.3 *Mechanism of the effect of twin wafer layer thickness on initial plasticity*

Previous studies have indicated that the plastic initial pressure of nano-twinned copper increases and then decreases as the thickness of the twinned layer decreases. This plastic initial pressure corresponds to the critical shear stress of dislocation nucleation. There are two mechanisms for the effect of the coherent twin boundary on the initial plasticity of the material. One is the softening of the material due to an increase in the number of dislocation sources, while another is the strengthening of the material due to the repulsive force of the coherent twin boundary on the nucleation of dislocations [32, 33]. In nanoindentation, a change in twin thickness does not cause a change in the number of dislocation sources, and therefore dislocation source-controlled softening does not play a role. If only the repulsive force of the coherent twin boundary on dislocation nucleation is considered, the critical partial shear stress of dislocation nucleation should continue to increase with decreasing twin thickness. However, there is a rapid decrease in plastic initial pressure with decreasing twin layer thickness in simulations for twin thicknesses less than 3.76 nm. The cause of this rapid decrease is still under investigation.

The results of this author and previous studies show that dislocations start from the subsurface during nanoindentation of single-crystal copper and nano-twinned copper (111) surfaces. The distance between the point where dislocations start and the surface is very small, about 1 nm in single-crystal copper, which makes the role of the surface in dislocation formation important. Since surface atoms have higher energy than internal atoms, the energy required for dislocation formation near the surface is slightly lower than for internal formation. This influence of the surface can be represented by a mirror force. The coherent twin boundary also has an effect on dislocation formation that can be represented by a mirror force. Therefore, taking into account the influence of the coherent twin boundary and the surface, the critical partial shear stress for dislocation initiation in nano-twinned copper can be expressed as follows:

$$\text{CRSS} = \tau_0 + \tau_{\text{CTB}} + \tau_{\text{surf}} \tag{3.21}$$

where CRSS is the critical fractional stress/GPa for dislocation nucleation, τ_0 is the theoretical shear strength/GPa, τ_{CTB} is the mirror force/GPa at the coherent twin boundarys, and τ_{surf} is the mirror force/GPa at the free surface.

The parameter τ_0 remains constant regardless of the thickness of the twin layer, while τ_{surf} and τ_{CTB} vary based on the distances of the dislocation nucleation point from the free surface and the coherent twin boundary, respectively. Assuming the initial nucleation dislocation is an infinitely long linear pure screw dislocation, according to the theories of linear elasticity and mirror force, τ_{surf} can be expressed as follows:

$$\tau_{\text{surf}} = -\frac{\alpha\mu b}{4\pi r_F} \tag{3.22}$$

where μ is the shear modulus/GPa of single-crystal copper, $\boldsymbol{b}$ is the Burgers vector of dislocations, r_F is the distance/nm from the surface of the dislocation nucleation point, and β is a scale-free quantity characterizing the strength of the free surface.

Based on Chen's two-term model [34], τ_{CTB} can be expressed as follows:

$$\tau_{\text{CTB}} = \frac{\beta\mu b}{4\pi r_T} \tag{3.23}$$

where τ_T is the distance/nm from the dislocation nucleation point to the coherent twin boundary and β is a scale-free quantity characterizing the strength of the coherent twin boundary.

According to the Hertzian contact theory, the plastic initial pressure P_{cri} can be expressed as a function of the critical shear stress, i.e.,

$$P_{\text{cri}} = \frac{1}{0.48}\text{CRSS} \approx 2.083\text{CRSS} \tag{3.24}$$

Substituting expressions (3.21)–(3.23) into expression (3.24), the final expression for the plastic initial pressure is obtained as follows:

$$P_{\text{cri}} \approx 2.083\left(\tau_0 + \frac{\alpha\mu b}{4\pi r_T} - \frac{\beta\mu b}{4\pi r_F}\right) \tag{3.25}$$

When the thickness of the twin layer is less than the critical value of 3.76 nm, dislocation nucleation occurs at the coherent twin boundary. As a result, the value of τ_{CTB} becomes 0, and the plastic initial pressure decreases as the distance from the dislocation nucleation point to the surface decreases. On the other hand, when the twin layer thickness is greater than the critical value of 3.76 nm, the dislocation nucleation point occurs inside the specimen. In this case, the effect of τ_{surf} is small and almost constant with the thickness of the twin layer, as the position of the nucleation point remains the same. The plastic initial pressure increases as the distance from the dislocation nucleation point to the coherent twin boundary decreases.

It is important to note that the plastic initial pressure is only related to the distance of the dislocation nucleation point from the coherent twin boundary and the free surface, and not to the number of twin layers. The results obtained by adding only a single layer of coherent twin boundaries to single-crystal copper and then varying the distance of the coherent twin boundaries from the pressed-in surface are in line with the present results.

3.3.3 *Plastic deformation mechanism*

In Figure 3.19, the evolution of dislocations during the nanoindentation of single-crystal copper is illustrated. The atoms in the figure are color-coded based on the centrosymmetric parameter, and atoms with a centrosymmetric parameter of less than 0.1 (intact atoms) are hidden, revealing only defective and surface atoms. When the indentation depth reaches the plastic initial depth of 0.86 nm, there is isotropic incomplete dislocation nucleation on the subsurface of

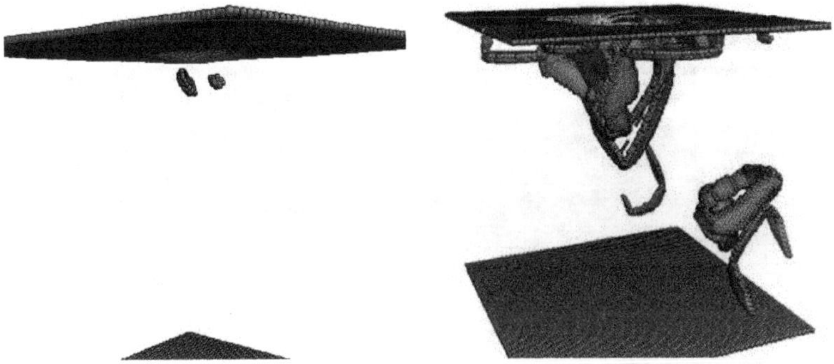

(a) (b)

Figure 3.19. Single-crystal copper nanoindentation dislocation evolution. (a) Pressure depth 0.86 n, initial nucleation of dislocations, (b) pressure depth 1.947 nm.

the specimen. This means that the three possible equivalent slip systems are activated almost simultaneously, as shown in Figure 3.19(a). Although full isotropic nucleation should result in three dislocation cores, in general, only two dislocation cores are found initially, and another dislocation will be nucleated with further loading. As the indentation depth increases, the three initially nucleated dislocations expand toward the surface and inside the specimen, forming a wide range of stacking faults. These stacking faults react with each other to create the typical prismatic stacking fault structure in the FCC single-crystal metal nanoindentation, as shown in Figure 3.19(b).

In Figure 3.20, the evolution of dislocations during the nanoindentation of copper nanotwin with a twin layer thickness of 2.50 nm is explained. The atom display rule is consistent with Figure 3.19. It is observed that the initial plasticity transforms from isotropic nucleation of incomplete dislocations in single-crystal copper at the subsurface to isotropic nucleation of twin dislocations at the twin boundaries, as illustrated in Figure 3.20(a). This is also referred to as twin-assisted partial dislocation nucleation. It is further discovered that the initial plasticity of nano-twinned copper is predominantly activated by the isotropic nucleation of twin dislocations when the thickness of the twin wafer layer is less than 3.76 nm (excluding 3.76 nm). This corresponds precisely to the sudden decrease in the plastic initial pressure once the twin wafer layer thickness falls below 3.76 nm. Accordingly, it is believed that the decrease in the plastic

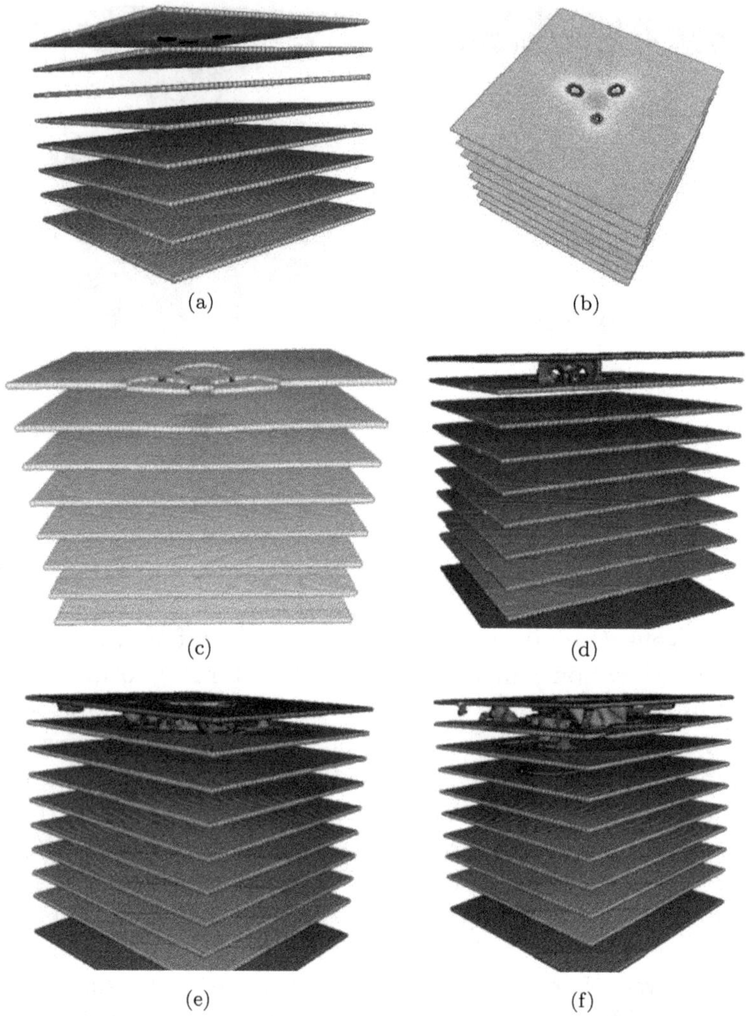

Figure 3.20. Dislocation evolution in nanoindentation of nano-twinned copper with a twinned layer sheet thickness of 2.50 nm. (a) 0.675 nm, (b) 0.685 nm, (c) 0.693 nm, (d) 0.747 nm, (e) 1.117 nm, (f) 1.697 nm.

initial pressure of nano-twinned copper is attributed to twin-assisted partial dislocation nucleation. In the case of intact crystals, the nucleation of partial dislocations necessitates the traversal of an energy barrier, specifically the unstable stacking fault energy. As for calculations by Mishin embedded in the atomic potential, the unstable stacking fault energy for copper is 161.0 mJ · m^{-2}. On the other hand, the nucleation of twinned dislocations denotes the nucleation of partial dislocations on the coherent twin boundary, which possesses a higher energy, i.e., the twin energy is 18.2 mJ · m^{-2}. The additional energy at the coherent twin boundary helps in the initiation of the partial dislocation. The nucleation of the twin occurs when the unstable stacking fault energy minus the twin energy equals 142.8 J · m^{-2}. Due to the lower energy barrier, the nucleation of the twin is favored over the nucleation of the partial dislocation, resulting in a low plastic initial pressure for nano-twinned copper.

The influence of twin dislocation nucleation on the nanoindentation properties of nano-twinned copper is significant. When the nanoindentation depth reaches 0.675 nm, the stress causes atoms in the three equivalent slip directions on the coherent twin boundary to deviate from their initial lattice position, disrupting the lattice's symmetry locally. This leads to the formation of "messy atoms" below the surface of the specimen, which are the precursor of twin dislocation nucleation. These "messy atoms" have a centrosymmetric parameter of 0.1–0.5, much smaller than that of actual dislocation atoms, which is 2.0–4.0. As the indentation depth increases, the number of "messy atoms" as well as their asymmetry also increases, accumulating large strains in the local area. Twin dislocations eventually nucleate at an indentation depth of 0.685 nm, forming three partial dislocation rings on the twin faces. The three symmetric twin dislocations have Burgers vectors of 1/6 a0 [2], 1/6 a0 [2], and 1/6 a0 [2]. This nucleation of twin dislocations occurs through the motion of "messy atoms" close to the lower surface of the coherent twin boundary. This study reveals that the coherent twins can also act as dislocation sources under suitable stress conditions, emitting dislocations, and that the energy required for twin dislocation nucleation is much lower, resulting in the softening of the material.

After the nucleation of twin dislocations, as the indentation depth increases, the twin dislocation ring will slip and expand along the coherent twin boundary. When the twin dislocations slip out of the coherent twin boundary, the coherent twin interface will move by one atomic layer distance (1/3a) into the specimen, as illustrated in Figure 3.20(c). This process is also referred to as untwinning or twin generation. When the three dislocation rings meet, the slip of the twin dislocations is impeded. New partial dislocations nucleate from the vicinity of the intersecting region of the twin dislocations and rapidly slip toward the surface, eventually intersecting the surface, as depicted in Figure 3.20(d). As the indentation depth continues to increase, the dislocations rapidly multiplicate. When the newly formed partial dislocations encounter the coherent twin boundary, they are either absorbed by the coherent-twin boundary or pass through it. At shallower depths of indentation, most of the dislocations are absorbed by the coherent twin boundary, thereby forming new twin dislocations on the coherent twin boundary, leading to untwinning or twin growth. This process causes the dislocations to be confined within the first twin region below the indenter, as shown in Figure 3.20(e). With further load increase, dislocations rapidly multiplicate and cross multiple the coherent twin boundaries, as shown in Figure 3.20(f). This twin–dislocation interaction obstructs the movement of dislocations and strengthens the material, resulting in the hardness of the nano-twinned copper being slightly greater than that of the single-crystal copper. Additionally, this interaction causes large undulations in the pressure–depth curve.

Another important factor is that a large number of dislocations are confined by the twin layer to a certain narrow superficial extent, while most of the area inside the specimen is dislocation-free. Figure 3.21 illustrates the dislocation configurations with different twin layer thicknesses at an indentation depth of 1.974 nm. It is evident that dislocations are confined to a few limited twinned regions for all twinned wafer layer thicknesses. This may result in some useful physical properties, although not as good as in the regions where the dislocations are free with highly conductive channels.

3.4 Elastic Recovery Behavior of Single-Crystal Copper Nanoindentation

3.4.1 *Introduction*

In the previous section, we briefly discussed the nanoindentation behavior of single-crystal copper and nano-twin crystals. Nanomaterials behave very differently from bulk materials after plastic deformation due to the scale effect. In the macroscopic state,

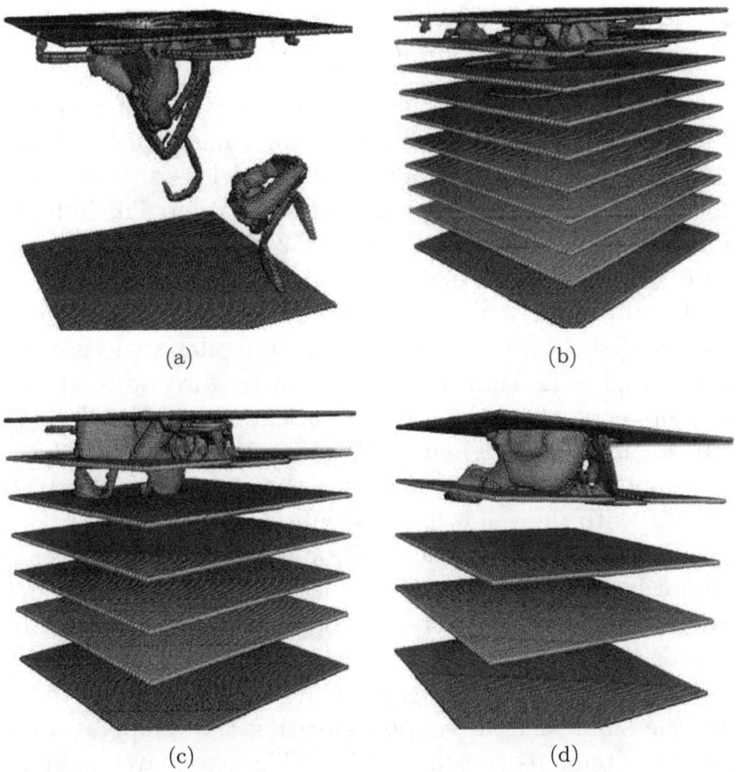

Figure 3.21. Dislocation morphology of nano-twinned copper with different twin layer sheet thicknesses at an indentation depth of 1.947 nm. (a) Single crystal, (b) 2.5 nm, (c) 3.76 nm, (d) 5.64 nm.

the amount of elastic recovery of the material after plastic deformation is usually minimal and often negligible. However, at the nanoscale, due to the scale effect and surface effect of the nano-materials, the amount of elastic recovery of the material is very large. When studying the kinematic properties of abrasive wear, it's important to consider the impact of elastic recovery at the nanoscale on the movement of abrasive material, as this factor will not be negligible. The authors will discuss the elastic recovery behavior of single-crystal copper after plastic deformation using molecular dynamics calculations of nanoindentation with constant velocity and constant load rate loading, respectively.

3.4.2 *Molecular dynamics modeling and mechanical parameters for single-crystal copper*

The three-dimensional molecular dynamics model of nanoindentation focuses on the indentation process of a nano-spherical indenter on a single-crystal copper matrix. The model uses diamond as the indenter, with a rigid definition during simulation. The indenter has a radius of 2.6 nm and consists of 11,252 carbon atoms. Its crystallographic orientations in the x, y, and z directions are [100], [010], and [001], respectively, as shown by the red atoms in Figure 3.22. The single-crystal copper matrix is a rectangular structure measuring $30.1 \times 20.4 \times 12.5$ nm in the x, y, and z directions. It contains 651,000 copper atoms and has crystal orientations of [100], [010], and [001]. In the molecular dynamics simulation experiments, the single-crystal copper matrix is divided into three parts, as illustrated in Figure 3.22, with the yellow layer serving as the fixed layer, the light blue layer as the thermostatic layer, and the dark blue layer as the Newtonian layer. The Newton layer is the region of action between the indenter and the single-crystal copper, with atoms described by the Newton equation. The thermostatic layer is responsible for transferring the heat generated during the interaction between the indenter and the single-crystal copper, ensuring the temperature of the Newton layer remains constant at 0 K. The fixed layer remains stationary throughout the calculation process to prevent shifting of the model and does not participate in the actual calculation process.

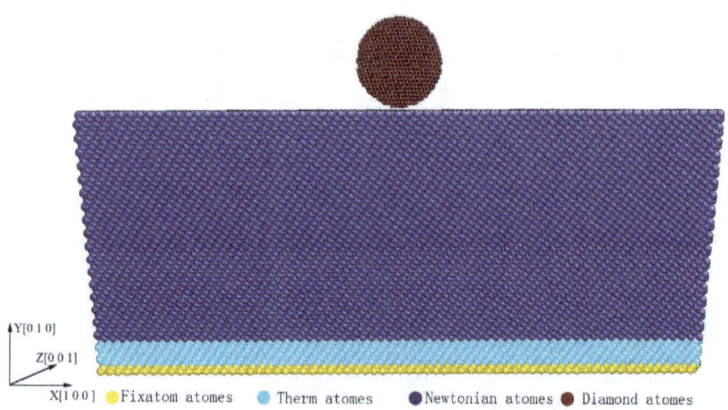

Figure 3.22. Molecular dynamics modeling of a single-crystal copper matrix.

Table 3.1. Experimental model parameters.

Modeling	Copper matrix
Potential	EAM embedding potential
Sizes	Rectangular $30.1 \times 20.4 \times 12.5\,\text{nm}$
Pacemaker	0.001 ps
Initial temperature	0 K
Relaxation time	50 ps
Press-in speed	$10\,\text{m}\cdot\text{s}^{-1}$ 10, $50\,\text{m}\cdot\text{s}^{-1}$ $100.\,\text{m}\cdot\text{s}^{-1}$
Depth of indentation at constant velocity	1 nm, 1.5 nm, 2 nm, 2.5 nm
Constant load rate	$0.1\,\text{nN}\cdot\text{ps}^{-1}$, $0.5\,\text{nN}\cdot\text{ps}^{-1}$ $1.0\,\text{nN}\cdot\text{ps}^{-1}$

The EAM potential is used to describe the potential function between Cu–Cu. This potential is widely matured in its application. The Morse mixing potential is used between Cu–C to describe the action potential between the indenter and the single-crystal copper. The experimental model parameters are detailed in Table 3.1.

The indentation hardness of single-crystal copper, as calculated with the model we created, is 3.26 GPa (Figure 3.23). This result is consistent with the value of 2.62 GPa found in the existing literature [35]. Additionally, the elastic modulus of single-crystal copper, calculated to be 121.58 GPa, aligns closely with the literature value of 124.9 GPa [36]. Therefore, it is evident that the molecular dynamics model established in this chapter is reliable.

3.4.3 *Elastic recovery under constant velocity loading conditions*

The load–displacement curve in Figure 3.23 demonstrates that as the indenter is pressed to a certain depth, the load suddenly decreases with displacement, marking the critical depth for single-crystal copper where plastic deformation begins. When the indentation depth is shallower than this critical depth, only elastic deformation occurs, and the single-crystal copper fully recovers after the load is removed. If the indentation depth exceeds this critical value, the material experiences both elastic and plastic deformation. Wang [37] and others observed that the elastic recovery coefficient of NiAl alloy varies with the content of Ni after plastic deformation. Wang proposed two methods for calculating the elastic recovery coefficient,

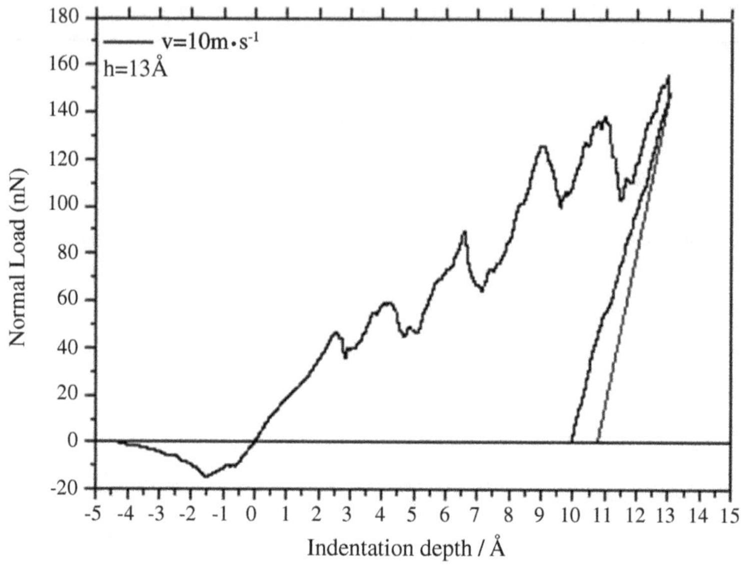

Figure 3.23. Load–displacement curves for nanoindentation into single-crystal copper.

as depicted in Figure 3.24. The elastic recovery coefficients, R_L and R_W, are defined as follows:

$$R_L = \frac{D_0 - D_1}{D_0} \tag{3.26}$$

$$R_W = \frac{W_0 - W_1}{W_0} \tag{3.27}$$

At the nanoscale, three-body wear is influenced by the scale effect and surface effect of nanomaterials. The elastic recovery of nanomaterials after plastic deformation plays a significant role in the movement mode of abrasive materials and must be taken into consideration. Nanoindentation experiments revealed that the indentation depth of a single-crystal copper substrate decreased after the indenter was pressed to a certain depth and then released. Over time, the formation of stacking faults and dislocations in the single-crystal copper gradually reduced, as depicted in Figure 3.25. Upon unloading, the indenter force reduction on the single-crystal copper caused internal defects to move toward the substrate's surface, forming steps. The figure also indicates a gradual reduction and eventual disappearance

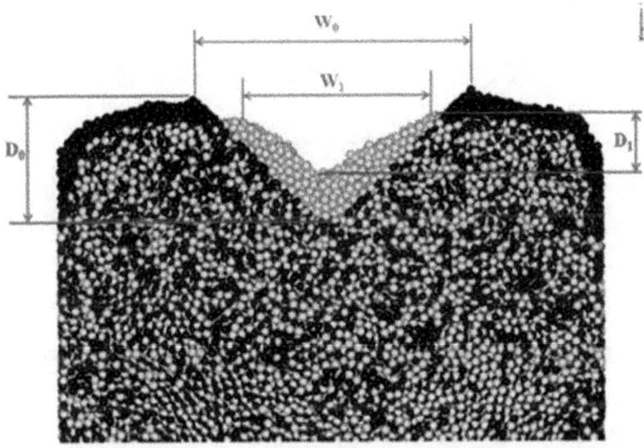

Figure 3.24. Wang's model for calculating the coefficient of elastic recovery [37].

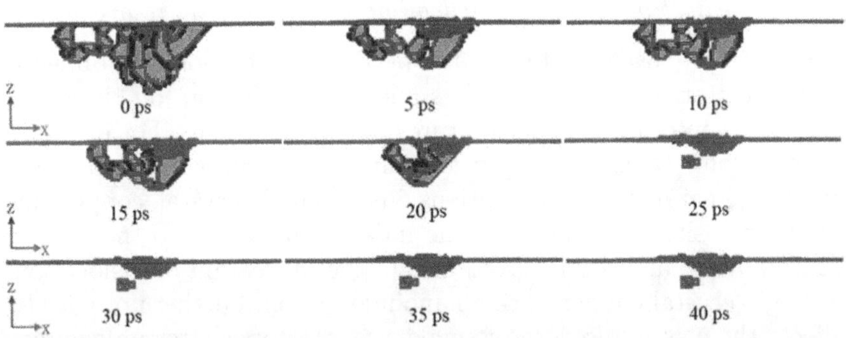

Figure 3.25. Configuration of internal defects in single-crystal copper after unloading as a function of time.

of stacking faults in the matrix post-unloading. The elastic recovery of single-crystal copper is closely related to the movement of its internal defects, and factors affecting the formation of these defects are likely to impact the elastic recovery of the material. As nanomaterial processing entails plastic deformation, the material's elastic recovery will influence the surface condition of nanomaterials after processing. Therefore, calculating the elastic recovery of nanomaterials will provide valuable guidance for their processing.

The author of this book uses a spherical indenter, and equation (3.26) is used to solve the elastic recovery coefficient. The coefficient

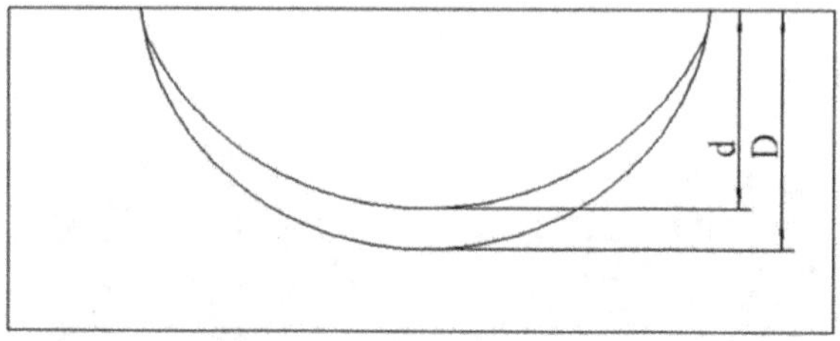

Figure 3.26. Method of calculating the coefficient of elastic recovery.

is expressed in terms of α, as shown in equation (3.28), and the calculation method is illustrated in Figure 3.26:

$$\alpha = \frac{D - d}{D} \tag{3.28}$$

3.4.3.1 *Effect of displacement loading rate on elastic recovery*

Scholars have discovered that the speed at which a load is applied in nanoindentation experiments has a significant impact on the development of internal defect structures in nanomaterials. For example, Cheng Dong *et al.* [38] found that higher loading speeds lead to a greater generation of dislocations inside single-crystal copper and result in a shallower depth of the indenter pressed into the copper. This shows that loading speed affects the nucleation of dislocations in single-crystal copper during nanoindentation. Furthermore, it also affects the elastic recovery of single-crystal copper after unloading.

In the molecular dynamics simulations, we used different indenter speeds—10 m/s, 50 m/s, and 100 m/s—with the same size and depth of indentation on single-crystal copper. After reaching an indentation depth of 15 Å, we maintained the displacement and allowed a relaxation interval of 100 ps before unloading at the same speeds. We obtained the load–displacement diagram (Fig. 3.27) and the load–time curve (Fig. 3.28) for both the indenter and the single-crystal copper substrate. From the graphs, we observed that higher loading speeds result in greater load at the specified depth and more significant load relaxation during the initial stage of relaxation. After sufficient time, the load decreases. This behavior can be explained by the elasticity and viscoelasticity of nanomaterials. In bulk materials, the speed of elastic deformation propagation is much larger than the

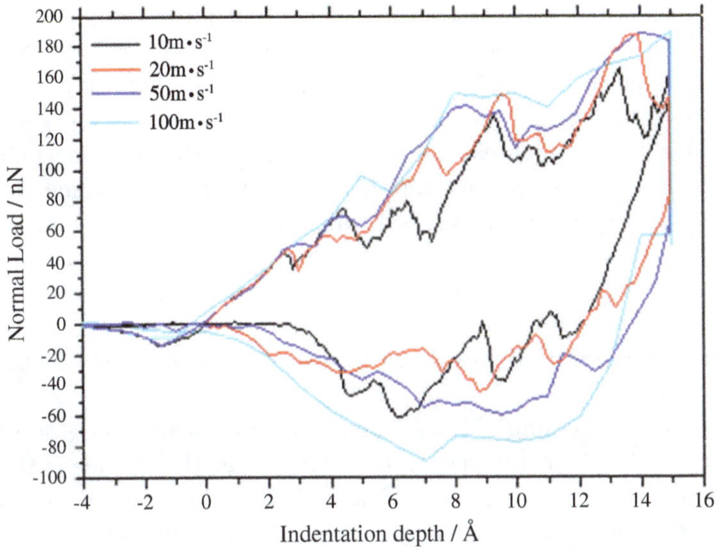

Figure 3.27. Load–displacement curves for nanoindentation at different indentation speeds.

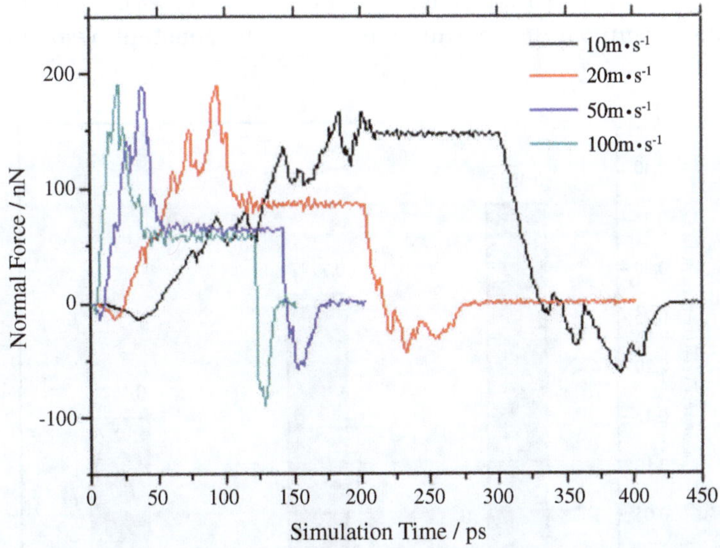

Figure 3.28. Load–time curve for nanoindentation at different indentation speeds.

speed of loading and does not affect defect formation. However, in nanomaterials, the speed of elastic and plastic deformation is related to the indentation speed of the indenter. A higher indenter speed leads to more serious plastic deformation of the single-crystal copper. The matrix creates numerous prismatic dislocation rings, and the density of dislocation pile-up is also higher when the indenter speed is high, which also has a large impact on the single-crystal copper, causing greater energy stored in the matrix. In the initial relaxation stage, the more the energy stored in the matrix is transferred to the surface, the more the force between the indenter and the matrix is reduced, resulting in a smaller load after relaxation is stabilized.

The rate of loading of the atoms at the nanoscale affects the rate of increase of the load between the indenter and the single-crystal copper substrate from Figures 3.27 and 3.28. This, in turn, affects the plastic deformation of the single-crystal copper and its elastic recovery. The uniform unloading process after relaxation also affects the elastic recovery. The amount of displacement of the atoms directly below the indenter after unloading provides the results shown in Figure 3.29, which represents the variation of elastic recovery coefficients for unloading at different speeds at constant velocity after

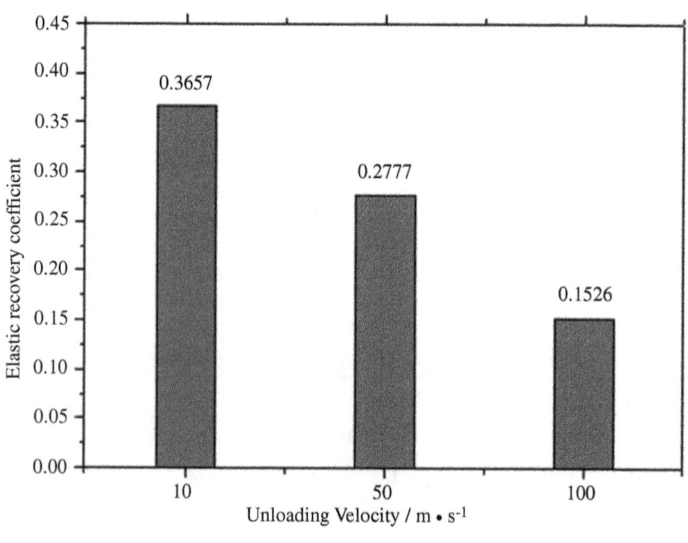

Figure 3.29. Elastic recovery of single-crystal copper at different unloading speeds after relaxation.

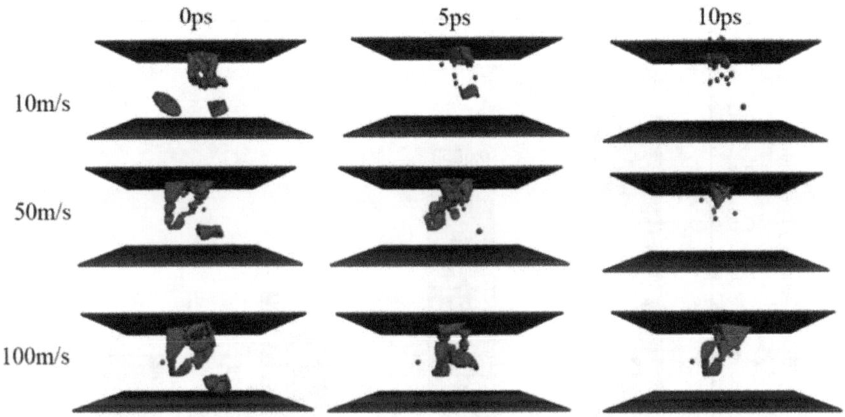

Figure 3.30. Atomic configurations of single-crystal copper pressed into the same depth with different loading speeds.

relaxation. From the figure, it can be seen that the elastic recovery coefficient decreases as the unloading speed increases. This is because the greater the speed of the indenter, the greater the impact on the single-crystal copper matrix, leading to the formation of a large number of stable stacking fault structures in the single-crystal copper, as shown in Figure 3.30.

At small speeds, single-crystal copper mainly forms prismatic stacking fault rings, which make the structure unstable when unloaded to the substrate surface. This results in increased elastic recovery and a decrease in the elastic recovery coefficient. As the load increases, the prismatic stacking fault ring in the matrix gradually increases, leading to the formation of a stable stacking fault structure. The interaction of this structure prevents defects from moving toward the matrix surface, resulting in increased elastic recovery and a decreased elastic recovery coefficient.

3.4.3.2 *Effect of indentation depth on elastic recovery*

In the study of the elastic recovery of single-crystal copper, it was found that this property is influenced not only by the loading rate but also by the depth to which the indenter is pressed into the substrate. To investigate this, a molecular dynamics simulation was conducted. The indenter was moved toward the substrate at a speed of 10 Å per unit of time. Upon reaching depths of 10 Å, 15 Å, 20 Å, and 25 Å,

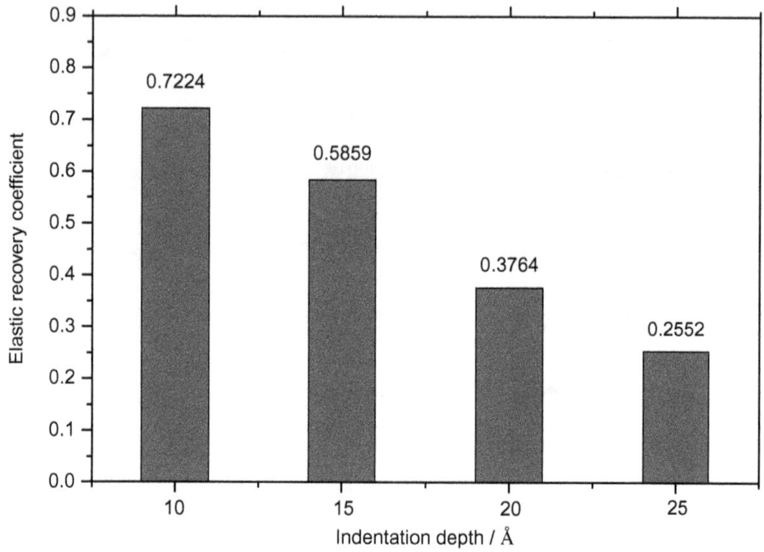

Figure 3.31. Elastic recovery coefficients for sudden unloading at different indentation depths.

it was suddenly unloaded in order to determine the elastic recovery coefficients at these different depths. These results are presented in Figure 3.31.

In Figure 3.31, it is evident that under constant indentation speed, the elastic recovery coefficient of single-crystal copper gradually decreases with the increase of the indentation depth. When the indentation depth is small, plastic deformation occurs in the single-crystal copper matrix, leading to the main formation of prismatic stacking fault ring (as shown in Figure 3.32). This structure makes it easy to move to the surface of the substrate after unload, resulting in an increase in elastic recovery and the elastic recovery coefficient. If the indentation depth is larger, a significant number of intertwined prismatic stacking faults are generated within the substrate, resulting in the formation of a complex and stable defective structure. Upon unloading, these inter-stacking fault structures hinder each other's movement to the surface of the substrate, causing a decrease in the elastic recovering and the elastic recovery coefficient consequently.

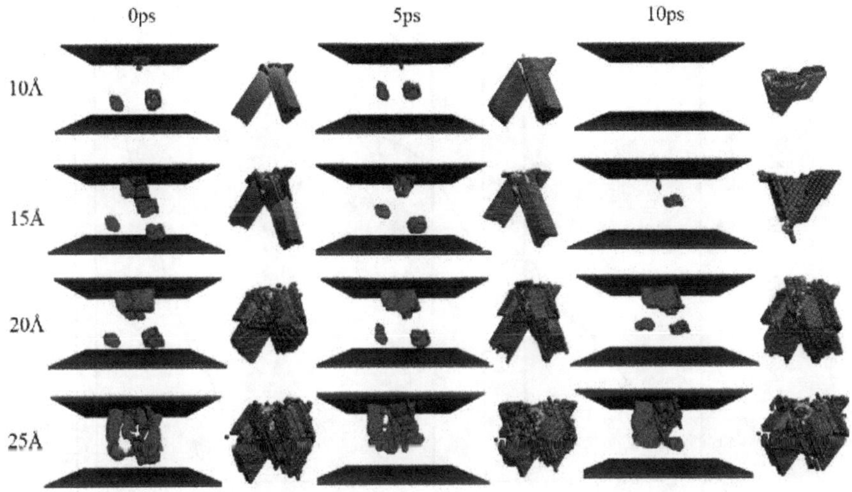

Figure 3.32. Atomic configurations of single-crystal copper at different indentation depths after unloading.

3.4.4 *Elastic recovery under constant load rate loading conditions*

In nanoindentation tests, there are primarily two loading methods: one involves constant velocity loading with controlled displacement and the other involves constant load rate loading with controlled load. The choice of loading method can influence the formation of defects within single-crystal copper and, consequently, the elastic recovery after unloading. For instance, when applying a constant load rate of 0.50 to conditions such as loading to 100 nN with a holding pressure with a time of 500 ps and then unloading at the same loading rate, a load–displacement curve can be obtained, as shown in Fig. 3.33. The curve illustrates that during the initial stage of the process, the load increases linearly with the displacement, indicating elastic deformation rather than plastic deformation within the single-crystal copper. This elasticity coefficient represents the elastic deformation of single-crystal copper (such as Figure 3.34 stage I).

In stage II, as the indentation depth gradually increases, the load decreases slowly. This happens because when the load reaches a certain value, the stacking faults form within the single-crystal copper matrix, hindering the rapid movement of the indenter into the matrix

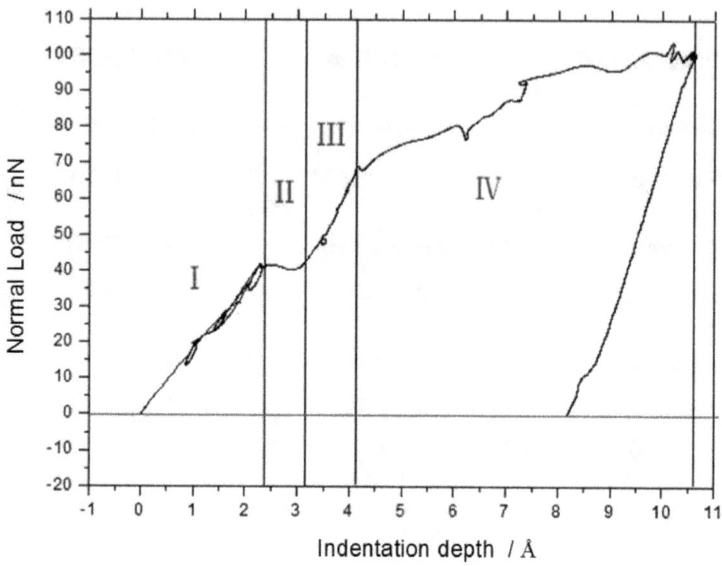

Figure 3.33. Load–displacement curve with constant load rate loading.

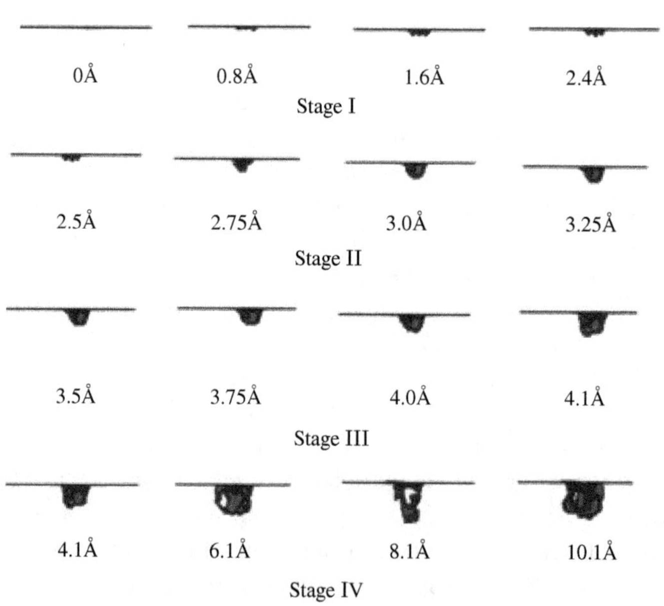

Figure 3.34. Layer fault configurations corresponding to each loading stage.

(Figure 3.34, stage II). This leads to a stage of plastic deformation, causing stress release in the region of the indenter and single-crystal copper. Consequently, the force on the indenter and single-crystal copper matrix decreases, allowing the indenter to press further into the material. In the third stage, as the indentation depth increases, the load gradually rises, and the curve becomes almost parallel to the stage I curve, with an equal slope. This stage can be approximated as part of the single-crystal copper's elastic deformation process. No new stacking fault formation or dislocation expansion occurs inside the single-crystal copper (Figure 3.34, stage III). In stage IV, the load gradually increases with the indentation depth, but the curve slope is smaller than that of stages I and III, indicating both elastic and plastic deformation. This means that the internal defects of the single-crystal copper matrix in this stage move further into the material as the indenter is pressed in. The general diagram showing the configuration of each stage corresponding to the generation and expansion of stacking faults in single-crystal copper during the loading process at a constant load rate is depicted in Figure 3.34.

The method of applying a constant load to single-crystal copper affects its internal structure and elastic recovery after unloading. It's important to study the factors influencing elastic recovery under constant load conditions.

3.4.4.1 *Effect of loading rate on elastic recovery*

On the atomic scale, when a material is subjected to an external force, the force is transmitted through the interaction and movement of atoms. The speed at which the force propagates depends on the movement of the atoms. Under certain temperature conditions, differences in external loading speed can affect the movement of the atoms, consequently impacting the formation and change of the material's internal structure as well as the size of its elastic response.

To conduct constant rate loading, the material is loaded at rates of $0.2\,\mathrm{nN \cdot ps^{-1}}$, $0.5\,\mathrm{nN \cdot ps^{-1}}$, and $1.0\,\mathrm{nN \cdot ps^{-1}}$, reaching a maximum load of 100 nN. The material is held at the maximum load for 500 ps and then unloaded at the same loading rate. Load versus time curves were obtained and are shown in Figure 3.35.

In Figure 3.36, the elastic recovery coefficient is shown under different loading rate conditions. It's evident from the figure that

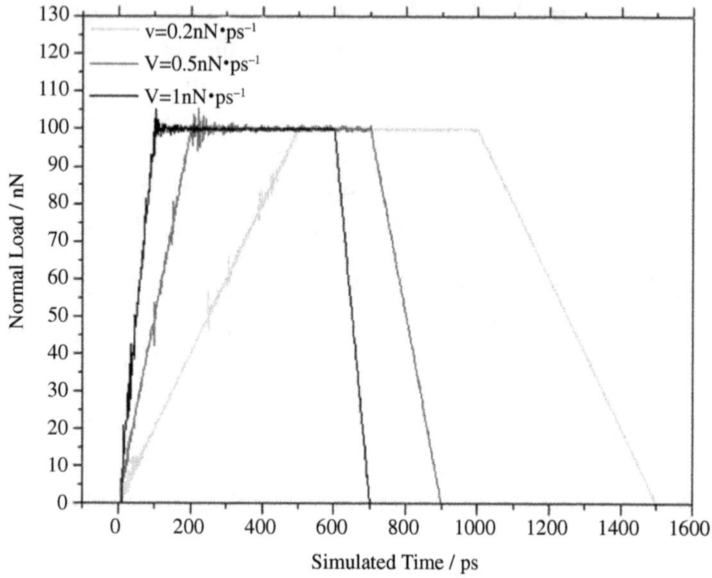

Figure 3.35. Load versus time curve.

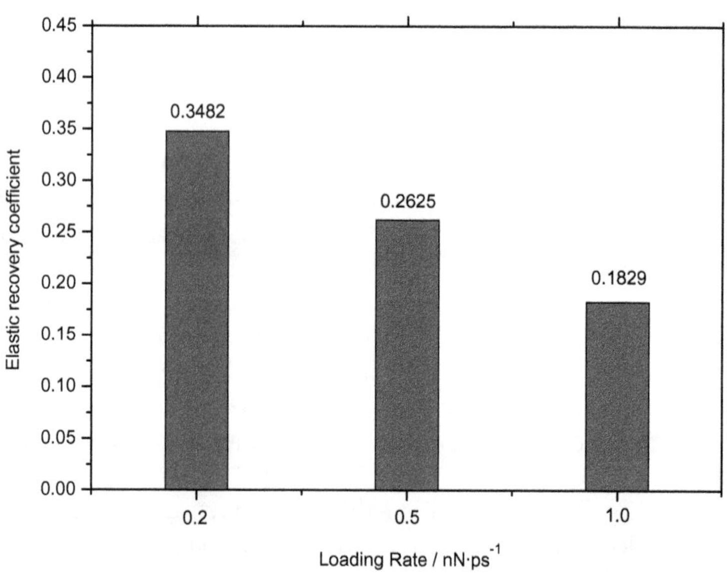

Figure 3.36. Elastic recovery coefficients at different loading rates.

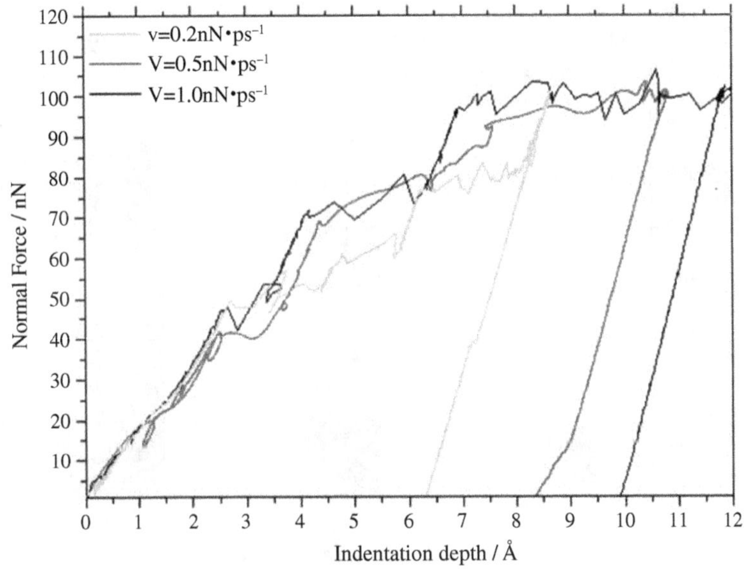

Figure 3.37. Load–displacement curves under different loading rates.

the elastic recovery coefficient of single-crystal copper after unloading is influenced by the loading rate. Specifically, the larger the load rate, the smaller the elastic recovery coefficient of single-crystal copper after unloading. Additionally, Figure 3.37 illustrates load–displacement curves under different loading rates. It can be observed that at a load rate of $1.0\,\text{nN} \cdot \text{ps}^{-1}$, the depth of indentation is $12\,\text{Å}$ at the maximum load, while at a load rate of $0.2\,\text{nN} \cdot \text{ps}^{-1}$, the maximum indentation depth is only $8\,\text{Å}$. This suggests that the indentation depth increases with greater loading rates at the same maximum load. Furthermore, at larger loading rates, a significant number of the stacking faults and dislocations are generated within the single-crystal copper matrix, as evidenced by the trend of the total length of the dislocation lines, as shown in Figure 3.38. The length of dislocations formed changes with the load loading rate. At a loading rate of $0.2\,\text{nN} \cdot \text{ps}^{-1}$, the length of dislocations is very short, and only a small number of the stacking faults are formed. However, at a loading rate of $1.0\,\text{nN} \cdot \text{ps}^{-1}$, when the loading reaches its maximum value, the dislocation length is the largest. At this point, dislocation interaction becomes more serious, and a large number

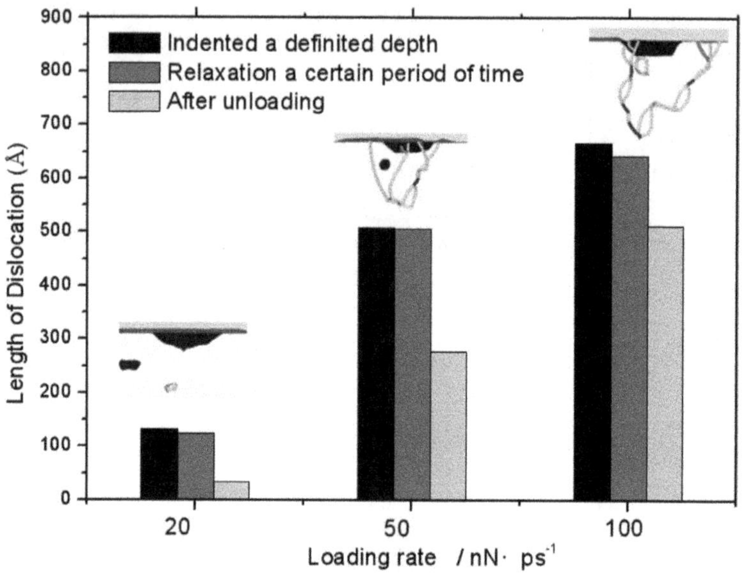

Figure 3.38. Variation of dislocation length at different loading rates.

of stacking faults are generated in the matrix. Due to the interaction of dislocations, stable stacking faults are formed in the single-crystal copper matrix. The existence of these stable defects leads the atoms to the stable state with the lowest energy. After unloading the load, a large number of defective atoms exist in the interior of the single-crystal copper matrix and do not move to the surface of the single-crystal copper, reducing the amount of elastic recovery of the single-crystal copper after unloading.

The elastic recovery coefficient of single-crystal copper decreases as the loading rate of the load increases. The fundamental reason for this is the formation of a large number of dislocations and stacking faults during the loading process. The existence of these defects reduces the amount of movement of the matrix to the surface of the indentation after unloading, thereby decreasing the elastic recovery coefficient.

3.4.4.2 *Effect of load magnitude on elastic recovery*

The recovery properties of nanomaterials can be investigated using the method of constant load loading rate, which makes it easy to control the magnitude of the load. This method was used to study the nanoindentation and elastic recovery properties of single-crystal copper. The magnitude of the indentation load directly affects the depth of indentation of the indenter into the substrate, as well as the amount of plastic deformation of the single-crystal copper substrate.

Load–displacement diagrams were obtained for different loads (100 nN, 150 nN, and 200 nN) using the method of constant load loading rate at a loading rate of $0.2\,\text{nN} \cdot \text{ps}^{-1}$. Figure 3.39 shows the load–displacement diagrams under different loads. The deeper the loaded indenter is pressed into the substrate, the more serious the plastic deformation in the substrate. After unloading, the curve of atomic elastic recovery with time below the indentation is shown in Figure 3.40. Within 5 ps after unloading, the elastic recovery amount and recovery rate of the single-crystal copper matrix are very large. During the loading process, as the indenter is continuously pressed into the single-crystal copper matrix, the lattice of the atoms changes, and a large number of stacking faults and dislocations are generated. After unloading, the inter-atomic repulsive force causes the atoms to return to a stable distance, resulting in atomic spacing becoming larger, and the atoms underneath the indentation moving

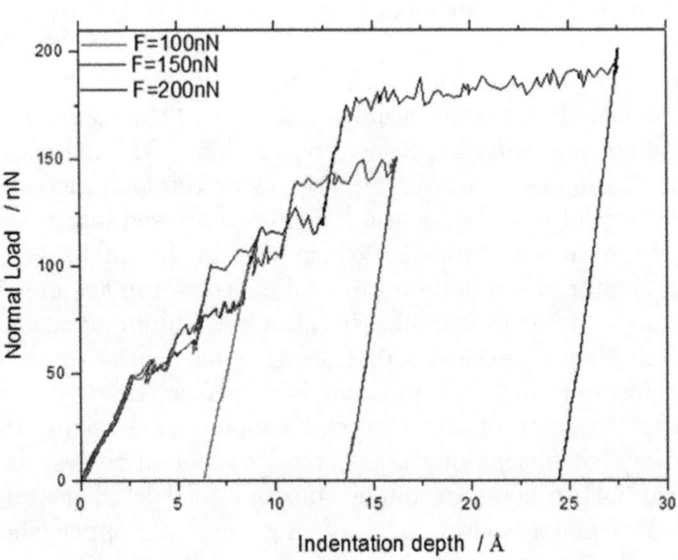

Figure 3.39. Load–displacement curve under different loads.

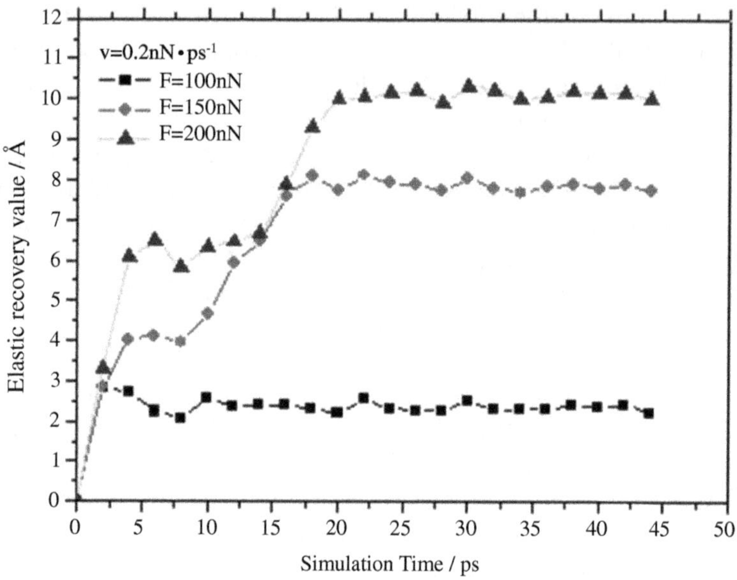

Figure 3.40. Elastic recovery curves of single-crystal copper with time under different loads.

upward so that the amount of elasticity recovery is very large in a short period.

After 5 ps of unloading, the atoms' displacement gradually increases over time and eventually stabilizes. This change in atomic displacement is caused by dislocation motion in the matrix. During this phase, the elastic recovery increases as the load increases. The greater the load, the deeper the indenter is pressed into the matrix, leading to more severe plastic deformation in the single-crystal copper and greater accumulation of residual stresses in the matrix with more energy. After unloading, the atoms at the indentation in the matrix experience residual stress, causing movement to the matrix surface, forming steps on the matrix's surface, thereby increasing the elastic recovery of single-crystal copper. As a result, the elastic recovery coefficient of single-crystal copper increases, as shown in Figure 3.41. Therefore, under constant load loading and in the condition of plastic deformation of single-crystal copper, the larger the load, the larger the elastic recovery coefficient of single-crystal copper.

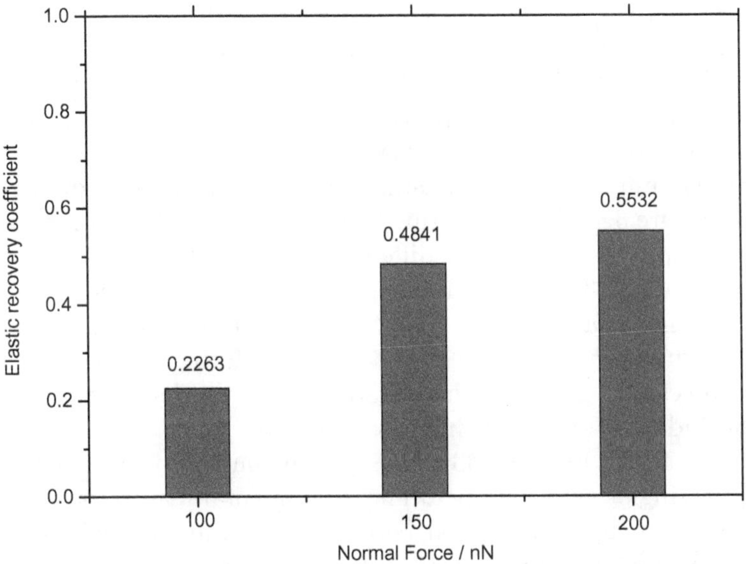

Figure 3.41. Variation of elastic recovery coefficient of single-crystal copper under different loads.

3.4.5 *Creep and stress relaxation in single-crystal copper*

The creep properties of crystals under macroscopic conditions are influenced by various factors, including the direction of force applied to the material, temperature, and crystal structure. The stress relaxation and creep properties of single-crystal copper have been studied using molecular dynamics calculations with both constant displacement rate loading and constant load rate loading. Different loading methods exhibit different types of creep in single-crystal copper. Under constant rate loading, when the indenter is pushed to a certain depth and the position of the indenter remains unchanged, the force between the indenter and the substrate decreases, resulting in varying degrees of load reduction, which is known as stress relaxation. When loaded with a constant load, reaching a specific load and maintaining it for a certain period result in an increase in the depth of the indenter, demonstrating the creep phenomenon. Despite their differences, the essence of the two phenomena is the same.

In Figure 3.37, we can see the time load curves for constant displacement under different loading rate conditions. The figure indicates that as the displacement loading rate increases, the load relaxation during the relaxation process also increases. This phenomenon, known as stress relaxation, causes a reduction in load after reaching the maximum load under constant displacement rate loading conditions. Figure 3.42 illustrates this. The amount of relaxation of single-crystal copper is affected by different loading rates.

In Figure 3.43, the displacement–load curves of single-crystal copper are shown at different load rates under constant load-loading rate conditions. The figure illustrates that during the holding pressure process, the load remains constant while the displacement depth of the indenter gradually increases (as seen in the relatively light-colored stage in Figure 3.43). This phenomenon indicates significant creep in single-crystal copper. The fundamental reason for the creep phenomenon in single-crystal copper is the time-limited nature of force transfer through the atoms in the material. When the loading rate is high, the force transfer takes more time, leading to the force

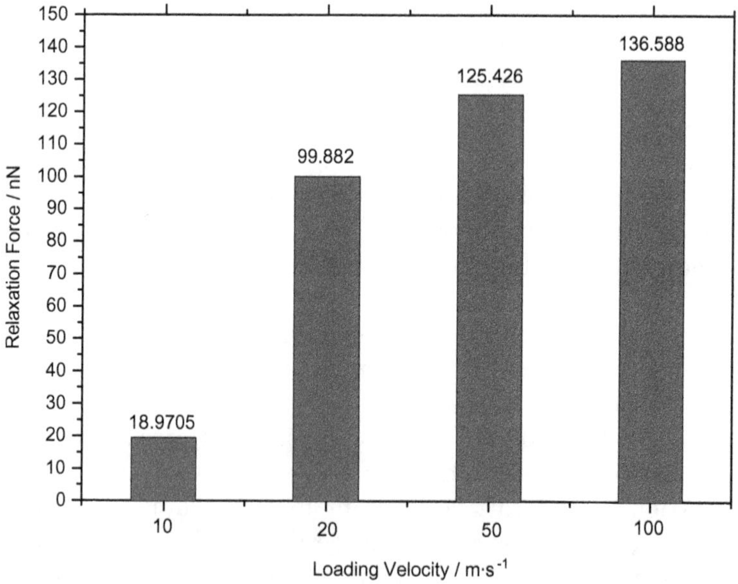

Figure 3.42. Load relaxation during relaxation at different displacement loading rates.

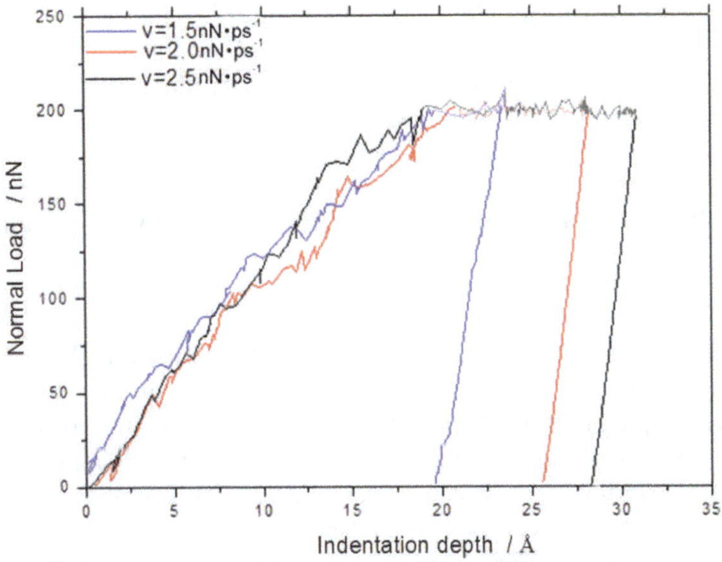

Figure 3.43. Displacement–load curves at constant loading rates for different loads.

between the matrix and the indenter being greater than the force between the atoms under the indentation region's subsurface layer. As a result, the force on the indentation's surface continues to be transmitted superficially during the relaxation process, causing the matrix to continue deforming plastically and increasing the depth of the indenter's indentation.

3.4.6 *Summary*

Through the study of the elastic recovery of single-crystal copper matrix under different loading methods, it has been observed that a larger loading rate results in a smaller elastic recovery and coefficient of elastic recovery. Additionally, increasing the indented depth leads to a subsequent change in the coefficient of elastic recovery. When a constant load rate is applied and the material is loaded to the same load, a greater load rate results in smaller elastic recovery, whereas a greater load results in greater elastic recovery and coefficient. Under both approaches, when pressure is held constant for a period, stress relaxation occurs. On the other hand, when a constant load rate is applied and the pressure is held constant, creep occurs. In both phenomena, the underlying mechanisms are the same, involving the transfer of force between atoms in nanomaterials, resulting in what is known as time-delay behavior.

3.5 Nanoindentation Behavior of Single-Crystal Silicon

The study of the nanoindentation behavior and abrasive wear behavior of single-crystal silicon is crucial for chip manufacturing as it is the main material for large-scale integrated circuits and chips. There has been extensive research on the high-pressure phase transformation of single-crystal silicon. Wu and Huang [39] investigated the phase transformation of single-crystal silicon in the nanoindentation and nanogrinding process using transmission electron microscopy and Raman spectroscopy. They found that bc8 (Si-III) and r8 (Si-XII) nanocrystals are embedded in the amorphous phase during the process. The same author subsequently found similar phase transition paths in nano-scratch tests [40]. Gassilloud [41] also found r8 (Si-XII) present in the amorphous equivalent in nano-scratch tests. Furthermore, previous experiments found that the bc8 (Si-III) and r8 (Si-XII) phases were transformed from the Si-II phase during unloading. However, there is a lack of credible evidence for the transformation of amorphous to bc8 (Si-III) and r8 (Si-XII) phases.

3.5.1 *Phase identification and characterization methods*

In the study of high-pressure phase transformations of single-crystal silicon caused by nanoindentation and nano-abrasive wear, researchers face a challenge in identifying the phase structure due to the small phase region. To address this issue, this study focuses on enlarging the phase transition region using a large indenter or abrasive diameter. Additionally, the coordination number (CN), radial distribution function (RDF), and bond angle distribution function (ADF) are used to comprehensively identify and characterize the phase structure. These methods help systematically characterize the short-range and long-range ordering of the phase structure, leading to more reliable and credible results.

Previous findings have identified five crystal phases—Si-I, Si-II, Si-III, Si-XII, and bct5—that generally occur in nanoindentation and nano-abrasive wear. To distinguish between these crystal structures, the CN is initially used to identify individual atoms and visualize their distribution characteristics in the analysis set. Subsequently, the RDF and bond ADF of the single-phase regions are calculated to further confirm the phase structure and analyze the long-range structural information characterizing the structure and its distortion.

3.5.1.1 *Phase identification method based on coordination number*

Conventional coordination number methods are commonly used in molecular dynamics simulations to identify, characterize, and visualize structures and defects, such as dislocations during the plastic deformation of metals. However, these methods are not always able to fully distinguish between the Si-I, Si-II, Si-III, Si-XII, and bct5 phases since all these structures have quadruple coordination numbers. The diamond-structured Si-I phase has four nearest neighbor atoms at atmospheric pressure, with a distance of 0.235 nm. The Si-II phase has four nearest neighbor atoms at a distance of 0.242 nm, as well as an additional two nearest neighbor atoms at a slightly larger distance of 0.257 nm. The bct5 phase has one nearest neighbor atom at a distance of 0.231 nm and four nearest neighbor atoms at a somewhat larger distance of 0.244 nm. As a result, Si-I, Si-II, and bct5 can be identified using the conventional coordination number method. For instance, at a suitable truncation radius of 0.28 nm, the phase structures with coordination numbers 4, 5, and 6 correspond to the Si-I, bct5, and Si-II phases, respectively. When identifying the Si-I phase atoms, it's important to note that they contain both Si-III and Si-XII phase atoms. This is because Si-III and Si-XII have four closest neighboring atoms at distances of 0.237 and 0.239, respectively. To differentiate between these three phase structures with the same coordination number, we need to consider the distribution of atoms at larger truncation radii. The Si-III phase is distinguished by the presence of a unique non-bonded atom at 0.34 nm when the pressure is 2 GPa. On the other hand, the Si-XII phase has non-bonded atoms at 0.323 or 0.383 nm [42]. The Si-I phase can be identified by the presence of 12 next-nearest neighbor atoms at 0.383 nm. Therefore, an atom with a coordination number of 4 has one next-neighbor atom in the suitable cut-off range (0.28–0.35 nm), characterizing it as the Si-III/Si-XII phase, while an atom with 12 next-nearest neighbor atoms is related to the Si-I phase. As previous studies have shown that Si-III/Si-XII phases are often mixed, this book does not further distinguish between these two phases. This method of phase structure identification, which considers both near-neighbor and next-nearest neighbor atoms, has been successfully used in previous research [43].

3.5.1.2 *Radial distribution function and key angle distribution function*

The radial distribution function concerns the likelihood of finding other atoms (denoted as "j atoms") at varying distances from a specific atom (referred to as the "i atom") in space. On the other hand, the radial and bond angle distribution functions pertain to the orderliness of matter for a specific particle. The expression for the radial distribution function g(r) is as follows [44]:

$$g(r) = \lim_{dr \to 0} \frac{p(r)}{4\pi (N_{\text{pairs}}/V) r^2 dr} \tag{3.29}$$

where $g(r)$ is the radial distribution function, r is the distance/nm of the computed particle pairs, $p(r)$ is the average number of particle pairs whose distances are found within the spherical shell layer of $r \sim r + dr$; V is the total volume/nm of the computed system 3, and N_{pair} is the number of unique atom pairs in the computed set 1 and set 2.

When set 1 and set 2 contain two different atoms, we have

$$N_{\text{pair}} = N_1 - N_2 \tag{3.30}$$

When set 1 and set 2 contain the same number and type of atoms, we get

$$N_{\text{pair}} = N_1 \cdot (N_1 - 1) \tag{3.31}$$

where N_1 is the number of atoms in set 1 and N_2 is the number of atoms in set 2.

The RDF provides a statistical description of the atomic arrangement across the entire computation area. It reflects the statistical distribution of atoms within a sphere centered on a single atom at a distance less than a certain cut-off radius (usually around 5 Å). If the computation area is too small (with a characteristic length less than 5 Å) and most atoms are located at the boundary, the structural information represented by the RDF, especially for long distances, may be unreliable. To obtain trustworthy results, the RDF necessitates that the size of the computation area is sufficiently large (at least greater than twice its cut-off radius, in this case, 1 nm). In molecular dynamics simulations, ensuring that the characteristic length of a single-phase region is larger than 5 Å on all three spatial scales requires a sufficiently large model. If the RDF is calculated for a large area, the result will capture the overlapping effects of multiple phases, and the irregular distribution of atoms near phase boundaries can reduce the reliability of the result. Although the angular distribution function (ADF) calculation only accounts for the nearest neighbor atoms, an increase in the proportion of mixed phases and atoms near phase boundaries can also reduce the accuracy of the calculation. To enhance the precision of RDF and ADF calculations in molecular dynamics simulations, it is essential to maximize the size of the calculation area on one hand and implement measures to minimize the impact of the boundary on the other.

In the nanoindentation simulations, we first used the CN method to obtain the distributions of Si-I, Si-II, Si-III/Si-XII, and bct5 phase atoms. Then, we calculated the RDF and ADF for each single-phase region. The dimensions of the typical Si-II, Si-III/Si-XII, and bct5 single-phase regions are as follows: $1.9 \times 1.4 \times 1.8$ nm, $2.2 \times 1.7 \times 2.5$ nm, and $5.4 \times 3.1 \times 5.4$ nm, respectively. The smallest length is three times the truncation radius, which ensures reliable results. To minimize the influence of boundary atoms, we considered a 0.5 nm region around the computational region for neighboring atoms. This means that the computational region of atom j is extended to 0.5 nm outside the computational region of atom i, as illustrated in Figure 3.44.

3.5.1.3 *Implementation of the algorithm*

The calculation of CN, RDF, and ADF is carried out by analyzing the atom trajectory data once the LAMMPS computation is complete.

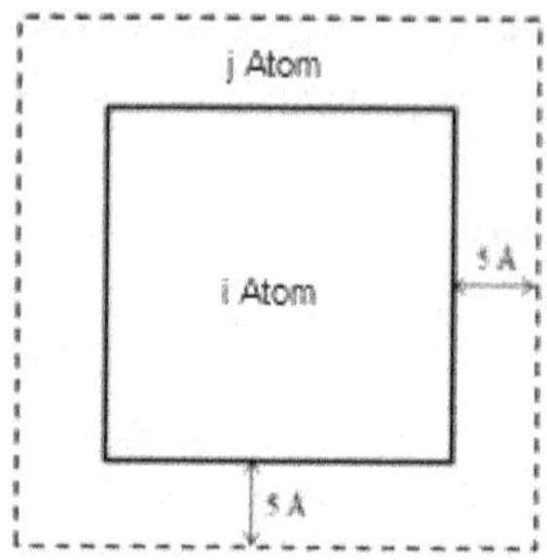

Figure 3.44. ADF and RDF calculations.

The most time-consuming part of calculating CN, RDF, and ADF is obtaining the neighborhood table for each atom. Using an ordinary search algorithm to determine the neighborhood of each atom would require traversing all other atoms once, leading to an algorithmic complexity of $O(N2)$. This becomes impractical when dealing with millions of atoms. To address this, the lattice indexing algorithm is employed to compute the neighborhood table of atoms. This algorithm divides the center cell into smaller cubic lattices with dimensions of $m \times m \times m$. With a lattice side length, r_{cell}, larger than the truncation radius (5 Å), atoms with non-zero interactions are located in either the same lattice as the current atom or the 26 lattices adjacent to it. Therefore, the simulation only requires calculating interactions between atoms in these lattices. Each lattice contains approximately $N_c = N/m^3$ particles, and calculating interactions involves around $27N_c$ pairs of atoms for a given atom. The lattice indexing algorithm is effective for saving computation if $27N_c < N$. As long as $m > 3$, $N_c = N/m^3$ ensures that $27N/m^3 < N$, reducing the computational load of the lattice indexing algorithm (Figure 3.45).

3.5.2 *Phase transition of single-crystal silicon in nanoindentation*

The computational model for nanoindentation of single-crystal silicon is similar to the one shown in Figure 3.32, but instead of single-crystal copper, we use single-crystal silicon. Figure 3.46 illustrates the phase distribution on the cross-section of the specimen after nanoindentation on the (100) surface of single-crystal silicon at an

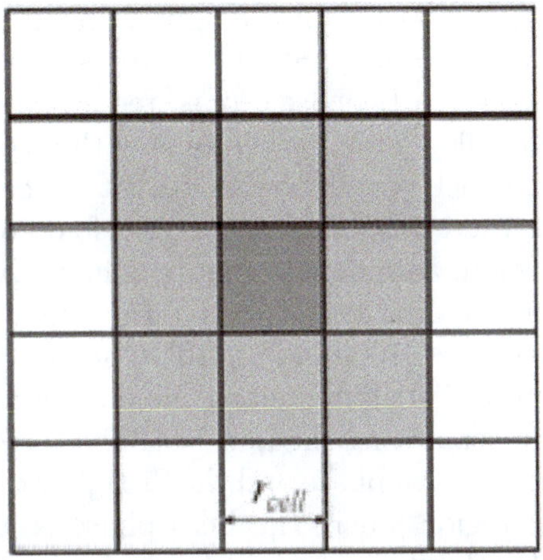

Figure 3.45. Lattice indexing algorithm.

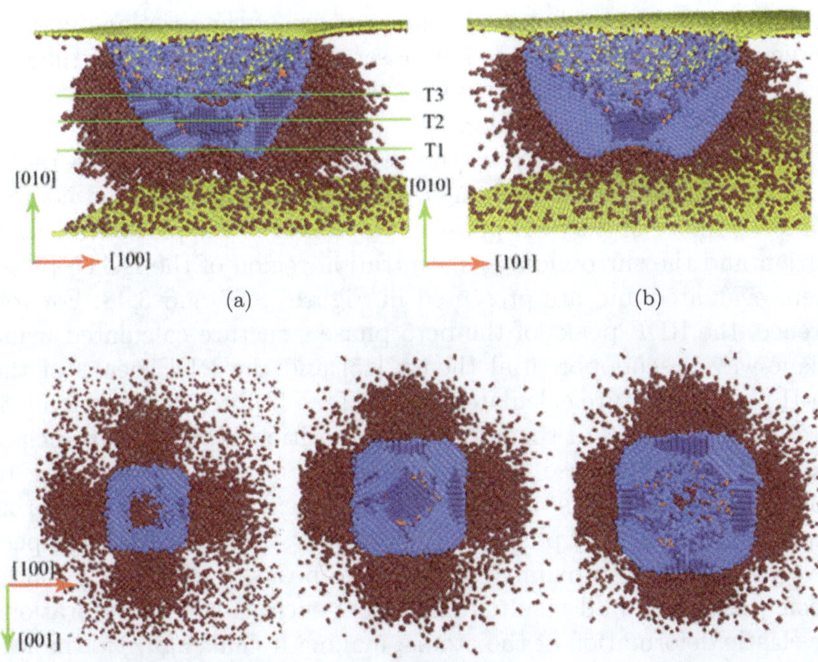

Figure 3.46. High-pressure phase distribution of single-crystal silicon due to nanoindentation. (a) (001) section, (b) (−101) section, (c) T1 section, (d) T2 section, (e) T3 section.

indentation depth of 4.50 nm. In the figure, atoms are colored based on their coordination number. Sky blue atoms represent bct5 phase atoms, blue atoms are Si-II phase atoms, red atoms are Si-III/Si-XII phase atoms, and yellow atoms are atoms with a coordination number of less than 3, which are surface atoms. At the center of the phase transition region, there's a relatively large 6-fold coordination number Si-II phase region, as shown in Figure 3.46(a). Additionally, four Si-II phase regions are distributed around the phase transition region along the directions of $\pm[100]$ and $\pm[010]$. The bct5 phase surrounds the Si-II phase and is distributed along the directions of [101], $[\bar{1}\,0\,\bar{1}]$, $[1\,0\,\bar{1}]$, and $[\bar{1}\,0\,1]$, separating the four Si-II phase regions. The phase interface between the bct5 phase and the Si-II phase is irregular, with a wide transition region where the bct5 phase is interspersed with the Si-II phase. Furthermore, several layers of atoms directly below the indenter remain as a mixture of the original diamond structure and Si-III/Si-XII phase atoms. At small indentation depths, the layers of atoms directly below the indenter are of the pristine diamond structure (not shown here). The observed characteristics of the distribution of Si-II and bct5 phases are consistent with the results reported in the literature [43].

To verify the presence of the Si-II and bct5 phases, the radial distribution function, bond angle distribution function, and coordination number of the Si-II phase at the center of the phase transition region and the surrounding concentration region of the BCT5 phase were evaluated and are presented in Figures 3.47 and 3.48. For reference, the RDF peaks of the bct5 phase structure calculated using plane-wave pseudopotential theory [45] and the RDF peaks of the Si-II phase structure calculated by the Tersoff potential function [46] are also illustrated in the figures. The position of the first principal peak of the Si-II phase was determined to be 2.51 Å. According to theoretical predictions, the first principal peak and the second principal peak of the Si-II phase are located at 2.42 Å and 2.57 Å, respectively. Due to the proximity of these two peaks, a slight variation in their positions could result from factors such as thermal vibrations or elastic deformation of the atoms, making it challenging for molecular dynamics simulations to distinguish between them. The bct5 phase exhibited a primary peak at 2.39 Å, which fell between the theoretical initial peak at 2.31 Å and the subsequent primary peak at 2.44 Å. The Si-II phase had two primary peaks in the bond angle distribution function at 75° and 104°, as well as two small peaks at

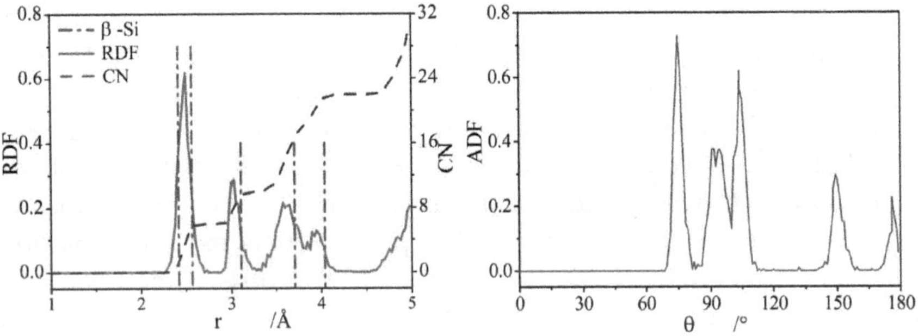

(a) (b)

Figure 3.47. Structural features of the S-II phase. (a) Radial distribution function, (b) key angle distribution function.

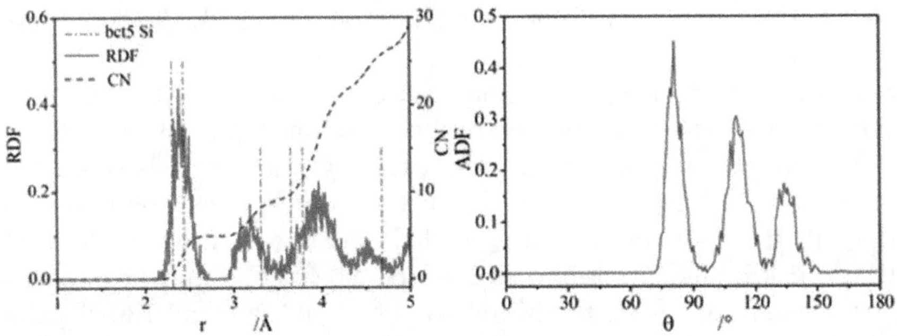

(a) (b)

Figure 3.48. Structural features of the bct5 phase. (a) Radial distribution function, (b) key angle distribution function.

149° and 176°. The bct5 phase had three peaks at 81.9°, 110°, and 136°. The broadened peaks in the radial distribution function and bond angle distribution function indicate that the lattice underwent substantial elastic deformation. Both the radial distribution function and bond angle distribution function peaks confirmed the presence of Si-II and bct5 phases.

A considerable quantity of mesostable phase atoms were discovered in the phase transition zone and directly beneath the indenter along the ±[100] and ±[010] axes, as depicted in Figure 3.47. According to their coordination number, these atoms were determined to have a fourfold coordination number, indicating that they are Si-III/Si-XII phase atoms. Previous experimental studies have demonstrated that Si-III/Si-XII phases only emerge during the unloading process and have not been observed during the loading process. The molecular dynamics investigation carried out by Kim *et al.* [43] also revealed the existence of these two phases, suggesting that they serve as the transition phases for the conversion of the Si-I phase to the Si-II and bct5 phases. A careful examination of the figure reveals that these mesostable phase atoms exist in isolation and are cross-aligned with the diamond structure atoms, with no evidence of coherent mesostable atoms in the style of large sheets. In other words, these mesostable atoms do not form a stable phase region. Furthermore, after the load is released, these mesostable atoms revert to the diamond structure. This contradicts the findings that the Si-III/Si-XII phase exists only after unloading in real tests. Furthermore, the Si-III/Si-XII phase appeared prior to the Si-II and bct5 phases during indentation, and the Si-III/Si-XII phase in the unloading process was reversible before the formation of the Si-II and bct5 phases. The radial and bond angle distribution functions of the mesostable phase are illustrated in Figure 3.49, and the RDF and ADF peaks of diamond-structured silicon are also depicted in the figures for comparison. From the figures, it can be observed that all the RDF peaks are extended and the position of the first RDF peak is slightly shifted to the left. The second and third peaks of the curve are slightly shifted to the right. A small dwarf peak is present at 3.5 nm, where the coordination number identification

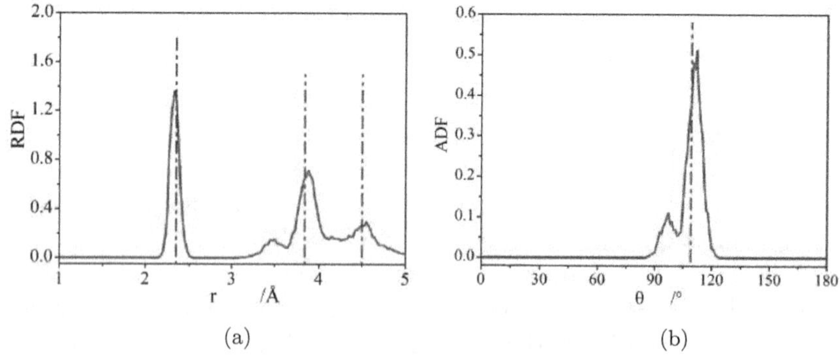

Figure 3.49. Characterization of mesostable phase structure. (a) Radial distribution function, (b) key angle distribution function.

suggests that these atoms are Si-III/Si-XII phase atoms. The intact diamond-structured atoms have an orthotetrahedral structure, and their ADF has only one peak at 109°. The mesostable phase atom possessed a major peak at 112° and a minor peak at 97.2°, which differs from the bond angles of 118° and 99° for the Si-III phase and 137° for the Si-XII phase [42, 47]. Therefore, this mesostable phase atom is not a Si-III/Si-XII phase atom but a distorted diamond cubic structure (DDS) atom.

To examine the pathways of phase transitions, the alteration of atom types within a small indentation space during nanoindentation was monitored, and the resulting statistical data are illustrated in Figure 3.50. As depicted in the figure, it is evident that the DDS atoms are directly transformed from Si-I phase atoms, and in further deformation, some of the DDS atoms revert back to the original Si-I atoms, while the rest are retained. No transformation of DDS atoms into bct5 phase atoms or Si-II phase atoms was detected. The majority of the newly formed bct5-phase atoms are directly transformed from Si-I-phase atoms, with only a tiny fraction originating from Si-II-phase atoms, which may be attributed to a change in the coordination number of atoms near the interface of the bct5 phase and Si-II phase. The formation of Si-II phase atoms is partly due to the transformation of Si-I-phase atoms and partly due to the absorption of bct5-phase atoms. This occurs because the Si-II phase is enveloped by the bct5 phase, and its expansion necessitates the consumption of bct5 phase atoms. Additionally, from Figure 3.50(a), it can be

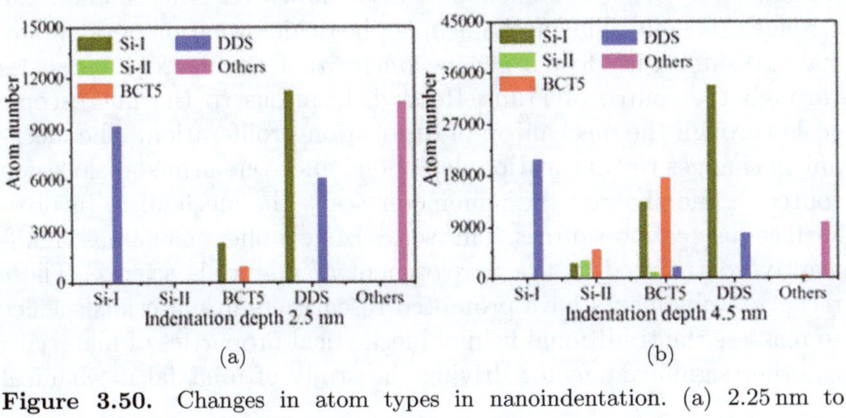

Figure 3.50. Changes in atom types in nanoindentation. (a) 2.25 nm to 2.50 nm, (b) 3.75 nm to 4.50 nm.

observed that the emergence of the bct5 phase precedes that of the Si-II phase.

The subsequent sections of this work delve into the molecular dynamics response of nanoindentation for water-containing film single-crystal silicon and single-crystal copper, as these are of great significance to the nano-abrasive wear law. Additionally, the following section focuses on the tensile behavior of single-crystal nanowires as a separate topic.

3.6 Tensile Behavior of Single-Crystal Nanowires

3.6.1 *Introduction*

Uniaxial tensile or compression testing is a widely used experimental method for investigating the mechanical properties of materials on a macroscopic scale. The stress–strain curves generated from this testing method can be utilized to construct material constitutive models, which can then be employed to predict the mechanical properties of materials and their performance under various complex stress states, shapes, and sizes. For over half a century, materials scientists have been striving to reduce the size of uniaxial tensile/compression specimens, typically referred to as the diameter of the specimen. Researchers have observed a strong "scale effect" wherein the tensile strength of the material increases as the size of the specimen decreases, from millimeters to micrometers to nanometres. This phenomenon is also known as "the smaller, the stronger" [48, 49]. The mechanism of plastic deformation of the material also shifts with decreasing specimen size, from macroscopic scales through the source of Frank–Read dislocations to the microscopic scale through the mechanism of dislocation proliferation. The mechanism changes to "dislocation depletion" and "one-armed dislocation sources", and then on the nanometer scale, the mechanism involves surface dislocation sources. This series of new phenomena has significantly contributed to the development of materials science. These recent developments have prompted researchers in materials science to reassess the traditional field of mechanical properties of materials. Another significant factor driving the study of uniaxial mechanical properties of small-scale materials is their potential applications, such as using small-scale nanowires as the basic building blocks in future NEMES devices and gold nanowires below 20 nm as the interconnecting materials in future ultra-large-scale integrated circuits. With the advancement of experimental equipment, such as *in situ* transmission electron microscopy for tensile and compression tests, and simulation calculations, such as molecular dynamics and first-principles calculations, the mechanical properties of low-dimensional materials, especially the uniaxial tensile/compressive mechanical properties, have rapidly become one of the most active areas of research both domestically and internationally in recent years.

At present, the majority of research concentrates primarily on the influence of specimen diameter on the scaling effect, while often disregarding the role of specimen length. This has led to inconsistent and contradictory results from both simulations and experiments, which do not align with the mechanical properties of nanowires studied in recent years. Previous molecular dynamics simulations have demonstrated that single-crystal face-centered cubic metallic nanowires exhibit exceptional tensile strength in uniaxial stretching (up to GPa, which is 1–2 orders of magnitude higher than that of macroscopic specimens), significant strain softening, and ductile fracture [52]. Experimental studies have also confirmed this deformation behavior [53, 54]. However, other experimental investigations have shown that single-crystal face-centered cubic metals exhibit sudden brittle fracture without significant ductile necking but instead rapidly undergo shear localization [55]. Recently, Wu [56] revealed that the brittle or ductile fracture mode is influenced by differences in specimen length and that the nanowires experience a tough-brittle transition as the length of the specimen increases.

In this chapter, we use large-scale molecular dynamics to simulate the process of uniaxial stretching of nanoscale specimens in order to investigate the mechanical behavior and properties of one-dimensional nanowires, specifically single-crystal gold nanowires. The main focus is on the "extrinsic scale effect" in the length direction. We examine how the length of the nanowires affects their mechanical properties, plastic deformation, and failure mode, covering lengths from 3.53 nm to approximately 1 μm. The results not only support the tough-brittle transition found by Wu *et al.* [56] but also reveal the variation law of length-dependent tensile strength. Our study finds an excellent balance of high-strength and high-toughness properties in ultrashort nanowires.

3.6.2 *Modeling and methodology*

3.6.2.1 *Atomic model*

In this section, a large-scale molecular dynamics approach was used to simulate the stretching of gold nanowires oriented along the [111] direction. The atomic model and coordinates of the gold nanowire are shown in Figure 3.51. The model includes an internal deformation region, a fixed layer, and a loading layer. The fixed and loading layer atoms are treated as rigid atoms and have a thickness of 0.82 nm. The nanowires are cylindrical with an initial diameter of 6.66 nm, and their initial length ranges from 3.53 to 1060.02 nm. To simulate finite-length nanowires, free boundary conditions were applied in all three directions.

3.6.2.2 *Simulation methodology*

The gold atoms are initially placed in a face-centered cubic lattice position with a lattice constant of 0.408 nm. To achieve equilibrium configurations of nanowires at 300 K, a carefully designed equilibrium method is necessary, especially for long nanowires. Due to the free surface, a low-frequency, long-period oscillation tends to form in the

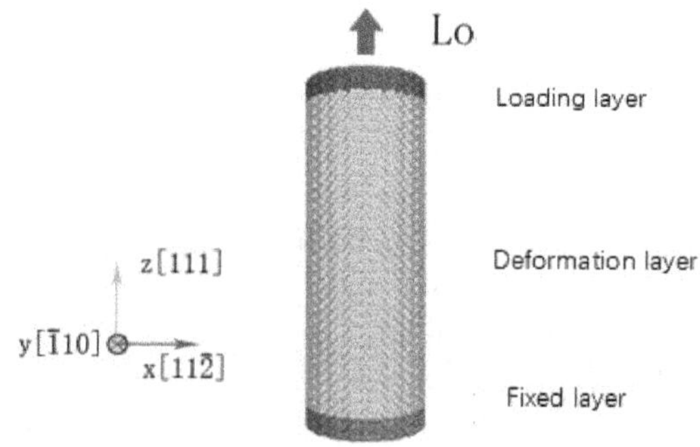

Figure 3.51. Atomic configuration of gold nanowires.

length direction of the nanowire during thermal equilibration, causing oscillations in the axial stress. To suppress this oscillation, the author utilizes the Langevin kinetic method to carefully equilibrate the nanowires at a temperature of 300 K. Subsequently, a constant velocity is applied through the carrier layer, stretching the nanowire uniaxially at a strain rate of 5×10^7, while the atoms in the fixed layer remain immobile. During the stretching process, a Nosé–Hoover heat bath is used to maintain the temperature of the nanowires at a controlled level. The interactions between gold atoms are represented by the embedded atomic potentials. Dislocations, twins, and other defects are identified and displayed using the common neighbor analysis (CNA) technique.

3.6.3 *Simulation results*

3.6.3.1 *Nanomechanical properties of gold nanowires*

Figures 3.52(a–d) display the engineering stress–strain curves in uniaxial tension of gold nanowires with different lengths: 3.53 nm, 7.07 nm, 106.00 nm, and 1060.02 nm. The gold nanowire measuring 3.53 nm exhibits a remarkable strength of 4.98 GPa and a high elongation of 275.18%. Similarly, the 7.07 nm nanowire shows a high strength of 4.19 GPa and an elongation of 174.03%. In contrast, the 1060.02 nm nanowire has a strength of 2.70 GPa and an elongation of only 3.33%. Figure 3.53 demonstrates the variation of yield strength, elongation, and modulus of elasticity of gold nanowires with different lengths. It seems that as the length of the nanowire decreases, the yield strength and elongation increase. This effect is more pronounced when the length falls below 21.2 nm, suggesting a possible change in the plastic deformation mechanism. The graph in Figure 3.53 inset shows how the elastic modulus changes with nanowire length. It reveals that the variation in elastic modulus with nanowire length is minimal. These findings indicate that the yield strength and elongation of gold nanowires are notably affected by length, but the elastic modulus is not significantly affected. Additionally, Figure 3.52 displays typical characteristics of the tensile mechanical properties of nanowires as reported in the literature, such as sudden stress reduction after elastic deformation, strain softening, and jagged stress–strain curves.

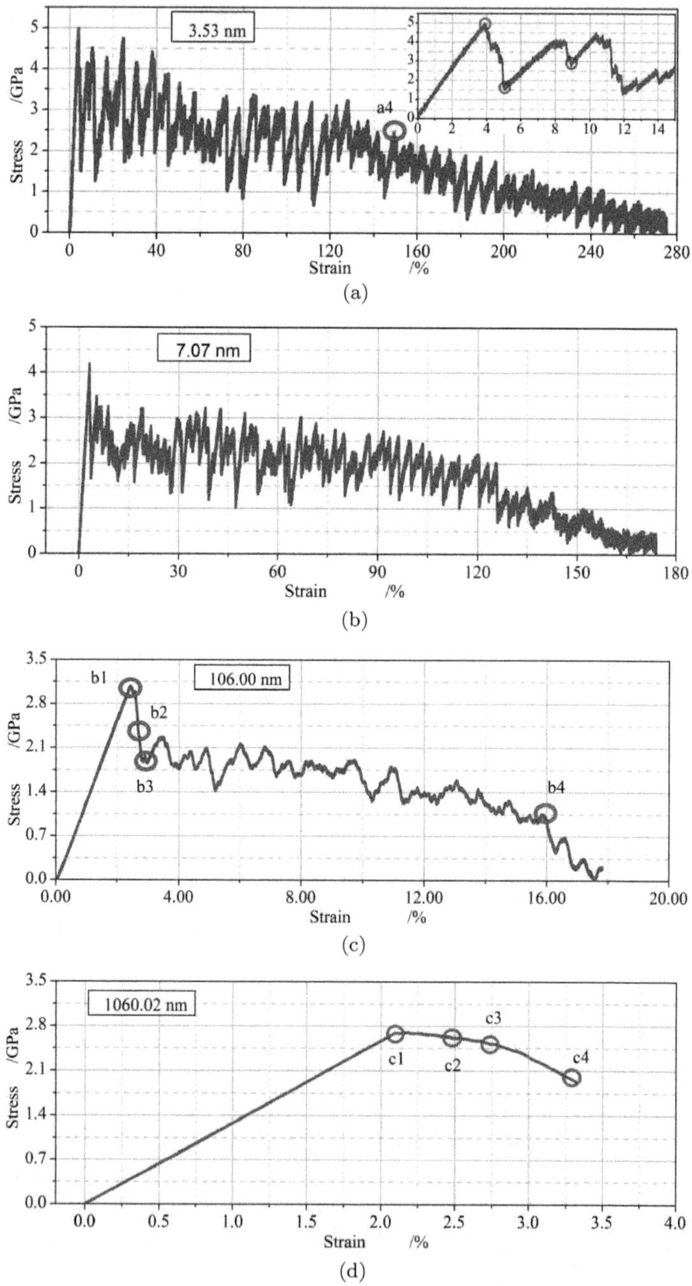

Figure 3.52. Stress–strain curve for gold nanowire. (a) 3.53 nm, (b) 7.07 nm, (c) 106.0 nm, (d) 1060.02 nm.

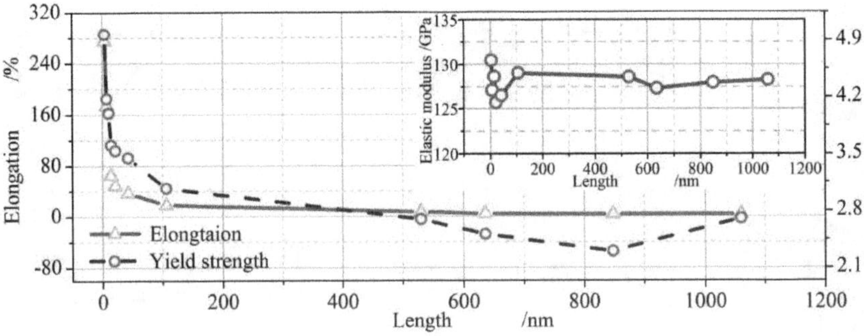

Figure 3.53. Variation of elongation, yield strength, and modulus of elasticity with nanowire length.

3.6.3.2 *Plastic deformation mechanism*

In order to investigate the origin of the unique combination of high strength and high ductility in short gold (Au) nanowires, the atomic configurations and defects that occur during the stretching of Au nanowires of different lengths were carefully analyzed. The results are presented in Figure 3.54, which includes the analysis for Au nanowires with lengths of 3.53 nm, 106.00 nm, and 1060.02 nm. The figure shows the dislocation configurations in Figures 3.52(a–i) and 3.52(l), representing the evolution of atomic configurations during the stretching process. Figures 3.52(j) and 3.52(k) show the morphology of necking. Additionally, the fractures of three nanowires are shown in Figures 3.52(m) and 3.52(o). The color scheme in the figure represents different atomic structures. Blue denotes atoms with hexagonal close-packed (hcp) structures, while red represents defective atoms instead of body-centered cubic (bcc), face-centered cubic (fcc), and hcp structures. The analysis reveals that the initial plasticity events in gold nanowires are caused by the nucleation of partial dislocations at the surface, leading to the formation of stacking faults in their slip regions, as illustrated by the blue atoms in Figures 3.54(a–c). Due to the nanowire scale constraints, the partial dislocations quickly slip out of the specimen once they nucleate. To achieve continuous plastic deformation, a large number of dislocations need to nucleate from the specimen's surface, forming numerous stacking faults inside the specimen, as shown in Figures 3.54(d–f). As the force required for dislocation slip is less than the force needed for dislocation nucleation,

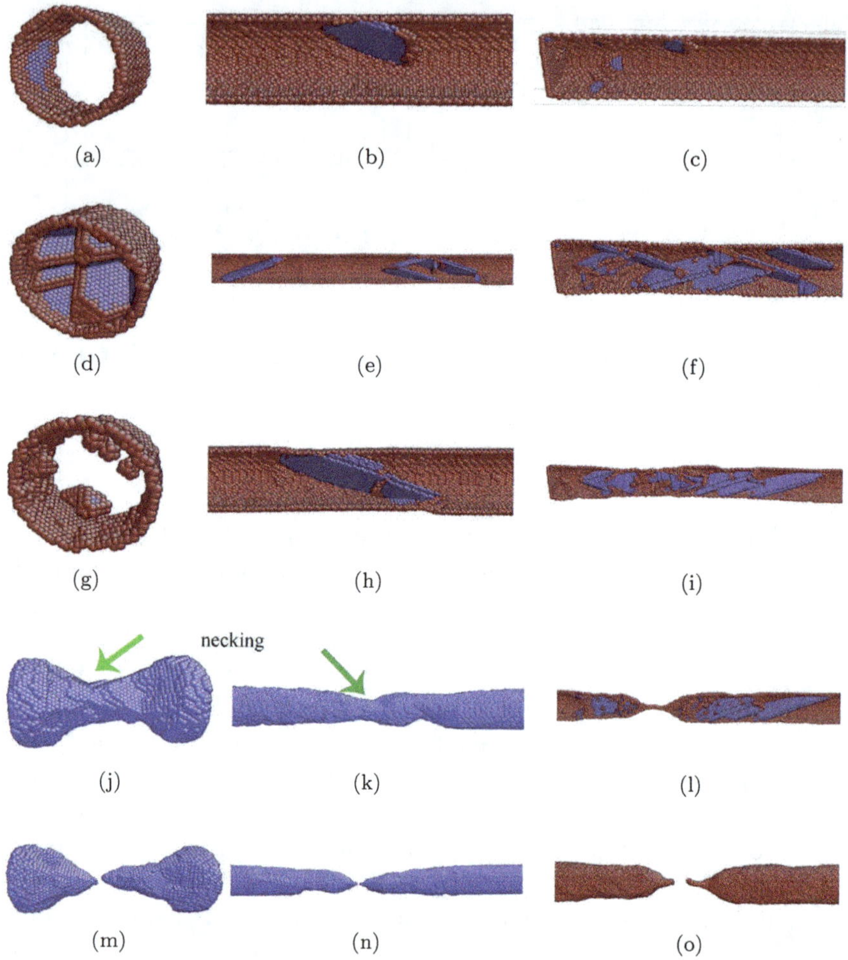

Figure 3.54. Atomic configuration evolution (corresponds to the labeled points in Figure 3.52). (a) a1, (b) b1, (c) c1, (d) a2, (e) b2, (f) c2, (g) a3, (h) b3, (i) c3, (j) a4, (k) b4, (l) c4, (m) a5, (n) b5, (o) c5.

dislocation slip leads to a rapid decrease in stress until the dislocation slips out of the specimen. Subsequently, stress re-accumulates until the next dislocation is activated. The discrete activation of dislocations results in a jagged stress–strain curve, typical in the plastic deformation of nano-sized metallic materials. Each peak on the jagged stress–strain curve corresponds to the force required to activate the dislocation, while the lowest point corresponds to the cessation of dislocation slip and the initiation of stress accumulation. After the first dislocation's nucleation, the nanowire successively experiences multiple dislocation-activated bursts on different slip systems, leading to the formation of multiple shear facets on the surface instead of shear bands on the macroscopic scale. These multiple, discrete dislocation activations consume the strain energy accumulated by stretching and sustain plastic deformation. However, all subsequent dislocation activations occur near the first dislocation nucleation point, restricting plastic deformation to a limited region.

The plastic deformation behavior of nanowires with different lengths was studied. An interesting finding was that 3.53 nm gold nanowires exhibited different dislocation mechanisms compared to longer nanowires (106.00 nm and 1060.02 nm). When subjected to tensile stress, the partial dislocation nucleation occurred at the surface of the 3.53 nm nanowires. However, the subsequent movement of dislocations was restricted, unlike longer nanowires where dislocations could move freely after nucleation. This restriction led to the rapid nucleation of more dislocations on other slip systems, forming a complex network of dislocations in the nanowire. This network reduced the stress to a minimum, as shown in Figure 3.54(d). As strain increased, the stress gradually built up to a maximum, leading to the nucleation of delayed new partial dislocations on the surface, and subsequent slip eliminates the previous leading dislocations to form the stacking faults, as shown in Figure 3.54(g). Afterward, new delayed partial dislocations nucleate on the surface and the above process is repeated, causing continuous plastic deformation of the nanowire. Similar observations of dislocation behavior were also found in long nanowires in Figures 3.54(h), 3.54(i), and 3.54(l). This research provides valuable insights into the dislocation mechanisms in nanowires of different lengths.

With further increases in tensile strain, the nanowire, which has already developed multiple slip facets due to the immovable stacking faults, is no longer able to sustain continuous plastic deformation through the activation of discrete dislocations analyzed earlier. Consequently, rapid necking forms in the localized region where the nanowire is severely deformed, as depicted in Figures 3.54(j) and 3.54(k). However, for the 1060.02 nm nanowire, no necking was observed. Instead, rapid shear localized fracture occurred, as shown in Figures 3.54(l) and 3.54(o). Prior to necking, the nanowire undergoes uniform deformation over a large area by discrete dislocation nucleation and slip mode, with stress decreasing gradually in a jagged pattern. The 3.53 nm and 7.07 nm nanowires experience uniform deformation in the length direction before necking occurs, attributable to the size limitation in the length direction. Even after necking, a substantial strain is required to activate numerous dislocation processes until failure transpires. However, other specimens of nanowire lengths fracture rapidly following necking. Consequently, the large elongation of the shorter nanowires results from the activation of numerous dislocations on different slip systems, leading to uniform deformation of the nanowires.

3.6.3.3 *Theoretical modeling of high strength of short nanowires*

In a study by Rinaldi *et al.* [59], ultra-short nanopillars made of single-crystal nickel were found to have extremely high compressive strength. The present authors suggest that the additional load and multiaxial stresses from the substrate contribute to the high strength observed in these nanopillars. Our research also indicated high strength with short nanowires. To investigate the source of this strength, the authors used the finite element method to calculate the stress state during compression of slightly longer 6.66 nm nanowires. They chose a modulus of elasticity of 130.0 GPa and a Poisson's ratio of 0.3 for gold for their calculations. They applied a displacement constraint to the bottom of the specimen and a displacement load to the top to simulate the loading conditions. Stress clouds in the axial and radial directions of the nanowire at a strain of 1.60% are presented in Figure 3.55. The figure shows large axial and radial stress gradients within the entire specimen, indicating that Saint-Venant's law no longer holds due to the specimen's limited length. The maximum axial stress is 4.49 GPa, and the maximum radial stress reaches 1.11 GPa. The authors infer that dislocation nucleation should occur at the top/bottom corner stress concentration points, but simulation results do not support this. Instead, the first dislocation nucleation tends to occur in the middle of the specimen. As a result, while multiaxial stresses and stress gradients are present, they are not the primary factors contributing to the high strength of short nanowires.

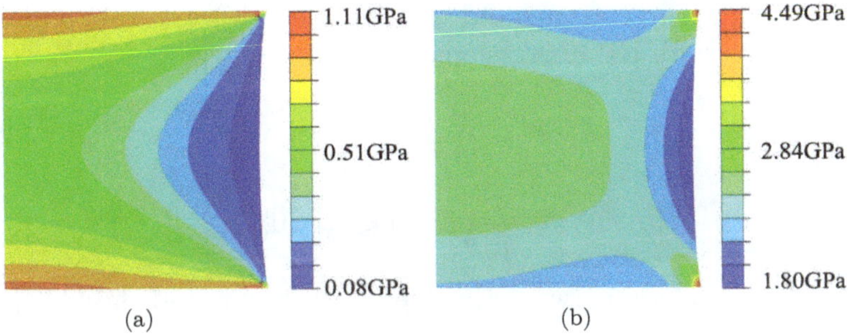

Figure 3.55. Stress cloud. (a) Radial stresses, (b) axial stresses.

Based on the fact that the fixed and loaded layer atoms at both ends of the nanowire are constrained to be rigid, it is likely that the action of the rigid atomic layers at both ends increases the nucleation stress for the first dislocation event, which in turn increases the strength. Therefore, we are attempting to estimate the impact of rigid atoms on the critical shear stress for dislocation nucleation. Similar to the concept of mirror force in classical materials science, the role of the rigid atomic layer is represented as a virtual mirror force τ_f. Using the two-phase model employed for the study of surface forces, the nanowire substrate and the fixed layer atoms are treated as two materials with different properties. When considering an infinitely long linear screw dislocation, τ_f can be expressed as

$$\tau_f = \frac{b\mu_1}{4\pi r}\left(\frac{2\mu_2}{\mu_1+\mu_2} - 1\right) \tag{3.32}$$

where b is the Burgers vector of the dislocation, μ_1 is the shear modulus/GPa of the nanowire, μ_2 is the shear modulus/GPa of the rigid atomic layer, and r is the distance/nm from the dislocation nucleation point to the fixed layer atom.

Considering that the shear modulus of the rigid atomic layer is infinite, τ_f is

$$\tau_f = \frac{b\mu_1}{4\pi r} \tag{3.33}$$

Assuming that the dislocation nucleation location is at a distance $h \cdot L$ from the fixed layer atoms, where L is the length of the nanowire and h is the scaling factor, the distances from the nucleation point to the loading layer atoms and the fixed layer atoms are r_u and r_b, respectively, with the expressions

$$r_u \approx \frac{L(1-h)}{\sin\theta} \tag{3.34}$$

$$r_b \approx \frac{h}{\sin\theta}L \quad 0 < h \leq 0.5 \tag{3.35}$$

where θ is the angle between the fixed layer atoms and the $\{111\}$ slip surface.

Substituting Eqs. (3.34) and (3.35) into (3.33) yields the mirror force of rigid layer atoms on dislocation nucleation as follows:

$$\tau_f' = \kappa\frac{b\mu_1\sin\theta}{4\pi L} \tag{3.36}$$

$$\kappa = \frac{1}{h(1-h)} \tag{3.37}$$

The factor κ represents the effect of the number of dislocation nucleation sites. Considering the role of the rigid atomic layer, the critical shear stress τ_{CRSS} for anisotropic nucleation of dislocations on the nanowire surface is

$$\tau_{\text{CRSS}} = \tau_0 + \kappa\frac{b\mu_1\sin\theta}{4\pi L} \tag{3.38}$$

where τ_0 is the critical shear stress/GPa for nucleation of infinitely long nanowire dislocations.

In equation (3.38), it is observed that dislocation nucleation tends to occur at the center of the nanowire, where h is 0.5 and the mirror force reaches a minimum. The molecular dynamics simulation indicates that for the 3.53 nm nanowire, the initial dislocation nucleation point occurs at $h \approx 0.40$, as predicted by the aforementioned model.

To quantitatively estimate the critical shear stress τ_{CRSS} of the dislocation, we can use $\mu_1 = 23.667\,\text{GPa}$, $b = 0.16657\,\text{nm}$, $\sin\theta = 2\sqrt{2}/3$, and $h = 0.5$. We can then calculate the minimum mirror force:

$$\tau_{f\,\max}' = 11.83/L \tag{3.39}$$

A nanowire's length greater than 24 nm results in a negligible minimum mirror force of 0.05 GPa. Thus, the maximum length of action

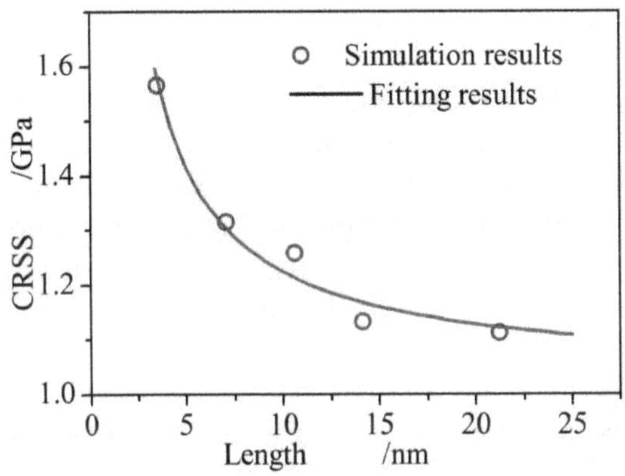

Figure 3.56. Variation of critical shear stress with length for dislocation nucleation.

for the rigid atomic layer is considered to be 24 nm. The impact of the rigid atomic layer becomes negligible when the nanowire exceeds 24 nm, and the opposite is true for shorter nanowires. Equation (3.38) shows that the critical shear stress for dislocation nucleation is inversely proportional to the nanowire length. The variation of critical shear stress for dislocation nucleation in nanowires of different lengths, simulated using molecular dynamics, is illustrated in Figure 3.56. The critical shear stress was determined by multiplying the nanowire's yield strength by the Schmid factor 0.314. The simulation results were fitted using equation (3.38), and the fitted values for κ and τ_0 are 6.60 and 1.03 GPa, respectively.

3.6.3.4 *Length-dependent tough-brittle transition*

In Figure 3.52, it is evident that the short nanowire (whose length is 3.53 nm, 7.07 nm, and 106.0 nm, respectively) experienced the jagged stress curve before ultimately fracturing. Previous studies have attributed these phenomena to discrete dislocation activation. As the nanowire length increases and the elongation decreases, the number of jagged stress changes reduces before fracture. However, the 1060.02 nm nanowire fractured quickly after the first stress decreased, indicating a brittle failure after reaching the yield strength. The material only endured minimal strain before fracturing, with an elongation of merely 3.33%. Comparing Figures 3.54(n) and 3.54(o), it is apparent that the plastic deformation of the 1060.02 nm specimen was limited to a small zone before fracturing, with numerous dislocations activated along one slip surface. These findings demonstrate a significant transition from tough to brittle behavior in uniaxial tensile testing as the nanowire length increases. Therefore, it is evident that the mechanical properties of nanowires not only exhibit a significant "size effect" at the scale of their diameter, but also demonstrate a similar "size effect" in the length direction.

Let's examine the tough-brittle transition length of nanowires. Wu *et al.* [56] proposed an analytical model to determine this length. In that model, the nanowire's cross-sectional shape is assumed to be a square with side length d, the nanowire length is L_0, and Young's modulus is E. Additionally, the total strain is represented by ε, and the angle between the slip direction and the axial direction is denoted as α. It is also assumed, based on prior studies on plastic deformation mechanisms, that all dislocations are on the same slip surface and that the slip length of the dislocations is s, as depicted in Figure 3.57. Consequently, the plastic strain along the axial direction can be expressed as $s \cdot \cos\alpha / L_0$. As slip occurs, the effective bearing area is reduced by $s \cdot d \cdot \sin\alpha$. During the slip process, the stress σ_m on the slip region changes due to the reduction of the cross-sectional area and the decrease in tensile load as follows:

$$\sigma_m = \frac{Ed}{L_0} \frac{L_0\varepsilon - s\cos\alpha}{d - s\sin\alpha} \tag{3.40}$$

σ_m will continue to increase with slip length as long as

$$\frac{L_0}{d} > \frac{\cot\alpha}{\varepsilon} \tag{3.41}$$

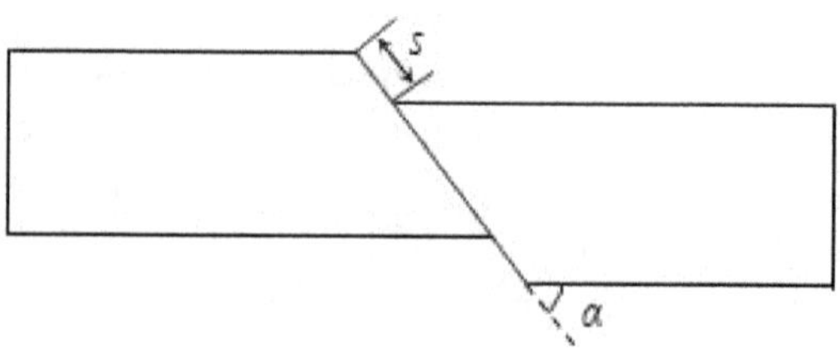

Figure 3.57. Tough-brittle transition model.

Therefore, the critical length L_c for nanowires to undergo tough-brittle transition is

$$L_c = d \cot \alpha / \varepsilon \tag{3.42}$$

If the nanowire breaks due to the activation and slip of partial dislocations, then $\cot \alpha = 2\sqrt{2}$. Assuming a molecular dynamics simulation result of $\varepsilon_y \approx 0.024$ and a nanowire diameter of 6.66 nm, substituting into expression (3.42) gives a critical length of 784.89 nm. If the breakage is due to the activation and slip of perfect dislocations, then $\cot \alpha = \sqrt{2}$, resulting in a critical length of 392.44 nm. In reality, both partial and perfect dislocations are present, as evidenced by the molecular dynamics simulation results by this author. Therefore, the critical length should fall between 392.44 nm and 784.89 nm. The molecular dynamics simulations yielded a critical length of 706.68 nm, validating the model's accuracy and indicating that the plastic deformation process of the nanowire is mainly driven by the nucleation of partial dislocations, with a relatively small number of perfect dislocations.

3.6.3.5 *Discussion and experimental design*

The studies mentioned above have shown that ultrashort nanowires with a diameter of less than 24 nm possess a great combination of high strength and high ductility. These properties are attributed to the effect of a rigidly fixed layer and the loading layer atoms on dislocation nucleation, as revealed by calculations using molecular dynamics. However, it's important to note that the artificially introduced high properties and the rigid layer may not exist in practical applications. The introduction of a rigid atomic layer is equivalent to assuming that the lattice structure at the loaded clamped-end of a real experiment is similar to the simulation conditions but with an infinite shear modulus. It's suggested that the high strength of the short nanowire is actually due to the large shear modulus of the rigid layer and is not directly related to the crystal structure of the rigid atomic layer. Therefore, it's proposed that a hard/soft/hard sandwich structure of nanowires, with a short layer of nanowires sandwiched between two ends of hard material as shown in Figure 3.58, can provide similar high strength and high ductility properties. According to this proposal, the hard-end material can constrain

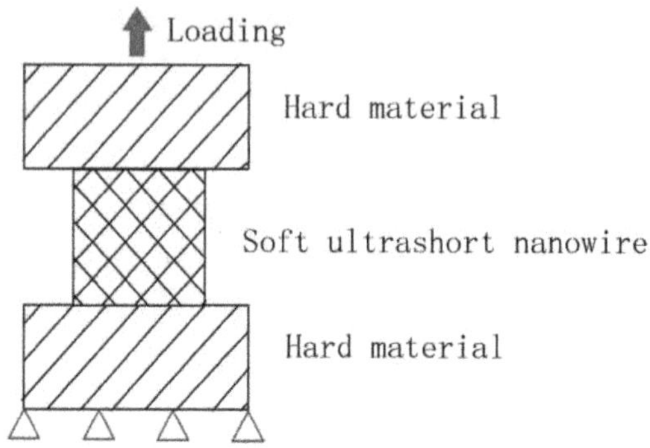

Figure 3.58. Sandwich structure figure.

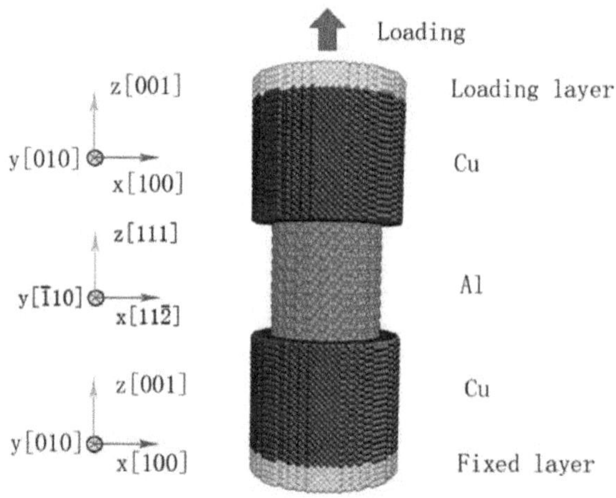

Figure 3.59. Atomic model of sandwich structure.

dislocation nucleation and movement, much like a rigid atomic layer, and is more capable of providing high strength and ductility. An experimental study program for the excellent mechanical properties of the proposed ultrashort nanowires is also suggested, wherein the hard/soft/hard sandwich structure of the nanowires is prepared by cold welding and multilayer film engraving, followed by tensile testing using an *in situ* tensile device.

In order to verify the speculations mentioned above, an atomic model of the Cu/Al/Cu sandwich structure was developed, as shown in Figure 3.59. The Cu and Al atoms and the interactions between

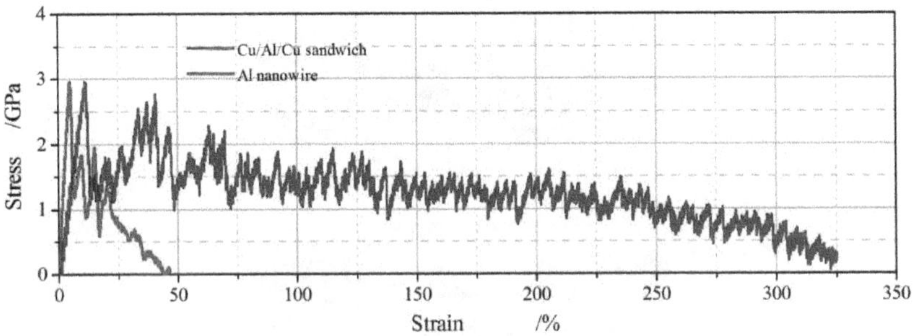

Figure 3.60. Stress–strain curve of sandwich structure.

them were expressed using embedded atomic potential functions. The initial Al nanowire has a diameter of 6.61 nm and a length of 7.01 nm, while the Cu ends have a diameter of 9.26 nm and a length of 7.01 nm. The axial direction of the Cu ends aligns with the [100] direction, and the axial direction of the Al nanowires is in the [111] direction. The simulation method used is the same as for the Au nanowires. The stress–strain curve of this sandwich structure is presented in Figure 3.60, along with the stress–strain curve of the Al nanowire with a length of 42.09 nm and a diameter of 6.61 nm for comparison. It's worth noting that the elongation of the sandwich structure reaches 325.01%, which is significantly higher than that of the Al nanowire at 46.50%. However, the strength of the sandwich structure is not higher than that of the Al nanowire, possibly because of the atomic structure of the Cu/Al interface. The evolution of dislocations during the stretching process of the sandwich structure is illustrated in Figure 3.61. It's observed that the Cu/Al interface restricts the movement of dislocations, causing dislocations to occur only in the interior of the Al and not in the Cu. Due to this limitation on dislocation movement, subsequent dislocations nucleate uniformly on the Al surface, resulting in uniform deformation of the specimen until necking occurs. This leads to a sandwich structure with a very high elongation at break.

3.6.3.6 *Summary*

In a study utilizing the molecular dynamics method, the uniaxial stretching process of finite-length gold nanowires was simulated to

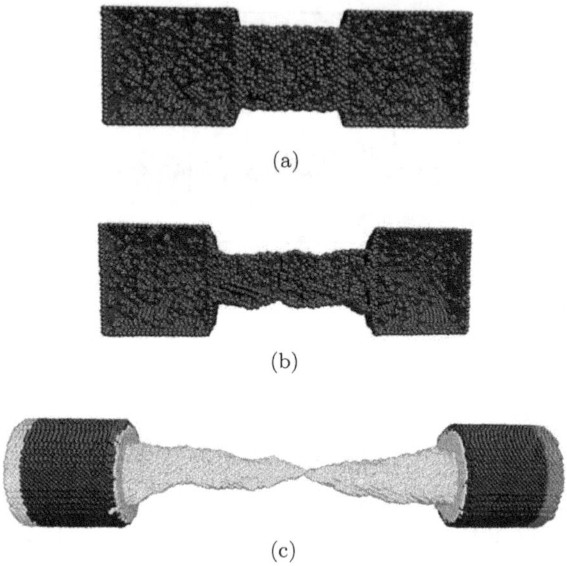

Figure 3.61. Sandwich structure dislocation evolution. (a) 11.5%, (b) 100%, (c) 325%.

analyze their mechanical properties, plastic deformation mechanism, and failure mode. The findings are as follows:

(1) During the uniaxial tensile test, it was observed that the mechanical properties of the material not only exhibit a strong scale effect in the diameter direction but also demonstrate a significant "extrinsic scale effect" in the length direction of the specimen, known as "the shorter the stronger." Furthermore, the nanowires also display the property of "the shorter, the longer."
(2) Gold nanowires, when reduced to a length below 24 nm, show an impressive combination of high strength (3–5 GPa) and high toughness (elongation >60%) properties. The excellent properties stem from constrained dislocation nucleation and slip. The author also introduces a predictive model for the yield strength of the nanowires based on the mirror force theory, with the predictions aligning with the results of molecular dynamics simulations.
(3) In this study, we propose a sandwich-structured nanowire model that aims to achieve a balance of high strength and ductility.

Through molecular dynamics simulation, we validate the feasibility of this model and provide an experimental solution for further exploration.

(4) Furthermore, we observed two distinct transitions in the mechanical properties and deformation behavior of the nanowire as its length decreases. When the length falls below 706.68 nm, the nanowire transitions from brittle to ductile fracture. Subsequently, as the length reduces to below 24 nm, the nanowire demonstrates a remarkable combination of high strength and ductility. Our analysis indicates that brittle fracture is attributed to rapid shear localization caused by the slip of numerous dislocations along the same surface. The enhanced strength of short nanowires is predominantly due to constrained dislocation nucleation and slip.

References

[1] Oliver WC, Pharr GM. An improved technique for determining hardness and elastic modulus using load and displacement sensing indentation experiments. *J Mater Res*, 1992, 7: 1564–1583.

[2] Oliver WC, Pharr GM. Measurement of hardness and elastic modulus by instrumented indentation: Advances in understanding and refinements to methodology. *J Mater Res*, 2004, 19: 3–20.

[3] Xu LX, Kong LQ, Zhao HW, *et al.* Mechanical behavior of undoped n-type GaAs under the indentation of Berkovich and flat-tip indenters. *Materials*, 2019, 12: 1192.

[4] Karimzadeh A, Koloor SSR, Ayatollahi MR, *et al.* Assessment of nano-indentation method in mechanical characterization of heterogeneous nanocomposite materials using experimental and computational approaches. *Sci Rep*, 2019, 9: 15763.

[5] Basu S, Moseson A, Barsoum MW. On the determination of spherical nanoindentation stress–strain curves. *J Mater Res*, 2006, 21: 2628–2637.

[6] Tabor D. *The Hardness of Metals.* Oxford: Oxford University Press, 1951.

[7] Johnson KL. *Contact Mechanics.* Cambridge: Cambridge University Press, 1985.

[8] Fischer-Cripps AC. *Nanoindentation.* 2nd ed., New York: Springer-Verlag, 2004.

[9] Fischer-Cripps AC. Critical review of analysis and interpretation of nanoindentation test data. *Surf Coat Technol*, 2006, 200: 4153–4165.
[10] Sneddon IN. Boussinesq's problem for a rigid cone. *Proc Camb Philos Soc*, 1948, 44: 492–507.
[11] Fischer-Cripps AC. *The IBIS Handbook of Nanoindentation.* ISBN 0958552541. Fischer-Cripps Laboratories Pty Ltd., Forestville, NSW, 2005.
[12] Dukino RD, Swain MV. Comparative measurement of indentation fracture-toughness with Berkovich and Vickers indenters. *Am Ceram Soc*, 1992, 75: 3299–3304.
[13] Nair R, Taylor M, Fischer H. Measuring hardness and more through nanoindentation. 2014, 1–12.
[14] Fischer-Cripps AC. *Nanoindentation (Third Edition).* Mechanical Engineering Series, Springer Nature Switzerland AG, 2011.
[15] Wang SB, Zhao HW. Low temperature nanoindentation: Development and applications. *Micromachines*, 2020, 11: 407.
[16] Lei M, Xu B, Pei YT, *et al.* Micro-Mechanics of nanostructured carbon/shape memory polymer hybrid thin film. *Soft Matter*, 2016, 12: 106–114.
[17] Wang J, Sansoz F, Huang J, *et al.* Near-Ideal theoretical strength in gold nanowires containing angstrom scale twins. *Nat Commun*, 2013, 4: 1742.
[18] Jang D, Li X, Gao H, *et al.* Deformation mechanisms in nanotwinned metal nanopillars. *Nat Nanotechnol*, 2012, 7: 594–601.
[19] Afanasyev KA, Sansoz F. Strengthening in gold nanopillars with nanoscale twins. *Nano Lett*, 2007, 7: 2056–2062.
[20] Cao AJ, Wei YG, Mao SX. Deformation mechanisms of face-centered-cubic metal nanowires with twin boundaries. *Appl Phys Lett*, 2007, 90: 151909.
[21] Deng C, Sansoz F. Enabling ultrahigh plastic flow and work hardening in twinned gold nanowires. *Nano Lett*, 2009, 9: 1517–1522.
[22] Deng C, Sansoz F. Size-Dependent yield stress in twinned gold nanowires mediated by site-specific surface dislocation emission. *Appl Phys Lett*, 2009, 95: 091914.
[23] Mishin Y, Mehl MJ, Papaconstantopoulos DA, *et al.* Structural stability and lattice defects in copper: Ab initio, tight-binding, and embedded-atom calculations. *Phys Rev B*, 2001, 63: 224106.
[24] Carter CB, Ray ILF. Stacking-Fault Energies of Copper-Alloys. *Philos Mag*, 1977, 35: 189–200.
[25] Hoover WG. Constant-Pressure Equations of Motion. *Phys Rev A*, 1986, 34: 2499–2500.

[26] Nose S. A Unified Formulation of the Constant Temperature Molecular-Dynamics Methods. *J Chem Phys*, 1984, 81: 511–519.

[27] Plimpton S. Fast Parallel Algorithms for Short-Range Molecular-Dynamics. *J Comput Phys*, 1995, 117: 1–19.

[28] Mo YF, Turner KT, Szlufarska I. Friction laws at the nanoscale. *Nature*, 2009, 457: 1116–1119.

[29] Ziegenhain G, Hartmaier A, Urbassek HM. Pair vs many-body potentials: Influence on elastic and plastic behavior in nanoindentation of fcc metals. *J Mech Phys Solids*, 2009, 57: 1514–1526.

[30] Lu L, Chen X, Huang X, *et al.* Revealing the Maximum Strength in Nanotwinned Copper. *Science*, 2009, 323: 607–610.

[31] Li XY, Wei YJ, Lu L, *et al.* Dislocation nucleation governed softening and maximum strength in nano-twinned metals. *Nature*, 2010, 464: 877–880.

[32] Guo X, Xia YZ. Repulsive force vs. source number: Competing mechanisms in the yield of twinned gold nanowires of finite length. *Acta Mater*, 2011, 59: 2350–2357.

[33] Deng C, Sansoz F. Repulsive force of twin boundary on curved dislocations and its role on the yielding of twinned nanowires. *Scr Mater*, 2010, 63: 50–53.

[34] Chen ZM, Jin ZH, Gao HJ. Repulsive force between screw dislocation and coherent twin boundary in aluminum and copper. *Phys Rev B*, 2007, 75: 986–989.

[35] Li LQ, Song WP, Xu M, *et al.* Atomistic insights into the loading—unloading of an adhesive contact: A rigid sphere indenting a copper substrate. *Comput Mater Sci*, 2015, 98: 105–111.

[36] Yu YN. *Princiles of Physical Metallurgy.* Beijing: Metallurgical Industry Press, 2003.

[37] Wang CH, Fang TH, Cheng PC, *et al.* Simulation and experimental analysis of nanoindentation and mechanical properties of amorphous NiAl alloys. *J Mol Model*, 2015, 21: 161.

[38] Cheng D, Yan L, Yan ZJ. Simulation experimental study of effect of running-in oil oncylinder/piston ring of diesel in running-in process. *J Dalian Marit Univ*, 2005, 31: 72–75.

[39] Wang Y, Zou J, Huang H, *et al.* Formation mechanism of nanocrystalline high-pressure phases in silicon during nanogrinding. *Nanotechnology*, 2007, 18: 465705.

[40] Wu YQ, Huang H, Zou J, *et al.* Nanoscratch-Induced phase transformation of monocrystalline Si. *Scr Mater*, 2010, 63: 847–850.

[41] Gassilloud R, Ballif C, Gasser P, *et al.* Deformation mechanisms of silicon during nanoscratching. *Phys Status Solidi A*, 2005, 202: 2858–2869.

[42] Piltz RO, Maclean JR, Clark SJ, *et al.* Structure and properties of silicon XII: A complex tetrahedrally bonded phase. *Phys Rev B*, 1995, 52: 4072–4085.
[43] Kim DE, Oh SI. Atomistic simulation of structural phase transformations in monocrystalline silicon induced by nanoindentation. *Nanotechnology*, 2006, 17: 2259–2265.
[44] Levine BG, Stone JE, Kohlmeyer A. Fast analysis of molecular dynamics trajectories with graphics processing units—radial distribution function histogramming. *J Comput Phys*, 2011, 230: 3556–3569.
[45] Boyer LL, Kaxiras E, Feldman JL, *et al.* New low-energy crystal-structure for silicon. *Phys Rev Lett*, 1991, 67: 715–718.
[46] Balamane H, Halicioglu T, Tiller WA. Comparative study of silicon empirical interatomic potentials. *Phys Rev B*, 1992, 46: 2250–2279.
[47] Crain J, Ackland GJ, Maclean JR, *et al.* Reversible pressure-induced structural transitions between metastable phases of silicon. *Phys Rev B*, 1994, 50: 13043–13046.
[48] Brinckmann S, Kim JY, Greer JR. Fundamental differences in mechanical behavior between two types of crystals at the nanoscale. *Phys Rev Lett*, 2008, 100: 155502.
[49] Uchic MD, Dimiduk DM, Florando JN, *et al.* Sample dimensions influence strength and crystal plasticity. *Science*, 2004, 305: 986–989.
[50] Hirth JPL. *Theory of dislocations.* New York: John Wiley & Sons, 1982.
[51] Zhu T, Li J, Samanta A, *et al.* Temperature and strain-rate dependence of surface dislocation nucleation. *Phys Rev Lett*, 2008, 100: 025502.
[52] Gall K, Diao JK, Dunn ML. The strength of gold nanowires. *Nano Lett*, 2004, 4: 2431–2436.
[53] Kim JY, Greer JR. Tensile and compressive behavior of gold and molybdenum single crystals at the nano-scale. *Acta Mater*, 2009, 57: 5245–5253.
[54] Kiener D, Grosinger W, Dehm G, *et al.* A further step towards an understanding of size-dependent crystal plasticity: In situ tension experiments of miniaturized single-crystal copper samples. *Acta Mater*, 2008, 56: 580–592.
[55] Richter G, Hillerich K, Gianola DS, *et al.* Ultrahigh Strength Single Crystalline Nanowhiskers Grown by Physical Vapor Deposition. *Nano Lett*, 2009, 9: 3048–3052.
[56] Wu ZX, Zhang YW, Jhon MH, *et al.* Nanowire Failure: Long = Brittle and Short = Ductile. *Nano Lett*, 2012, 12: 910–914.

[57] Schneider T, Stoll E. Molecular-dynamics study of a three-dimensional one-component model for distortive phase transitions. *Phys Rev B*, 1978, 9: 3048–3052.

[58] Grochola G, Russo SP, Snook IK. On fitting a gold embedded atom method potential using the force matching method. *J Chem Phys*, 2005, 123: 204719.

[59] Rinaldi A, Peralta P, Friesen C, *et al.* Sample-Size effects in the yield behavior of nanocrystalline nickel. *Acta Mater*, 2008, 56: 511–517.

Chapter 4

Wear Behavior of Nano-Abrasive Wear Under Air Conditions

4.1 Basic Theory of Abrasive Wear

The study of two-body abrasive wear in nano-size provides valuable insights into material wear. This form of wear, which can be easily manipulated through scratch experiments, offers experimental simplicity but presents challenges in observing microscopic processes and mechanisms at the nanoscale. To address this, molecular dynamics simulation offers a promising approach by allowing researchers to study wear at the atomic scale. For instance, Urbassek *et al.* [1, 2] examined the nano-scratching of iron (100) surfaces and identified the generation of internal dislocations and the associated reaction process. Similarly, Zhang *et al.* [3] utilized molecular dynamics simulation to study the nano-cutting process of single-crystalline copper, while Zhu *et al.* [4] delved into the relationship between cutting conditions and behavior. The results showed that the geometry of the cutting tool, the depth of cut, and the temperature of the material all have an important influence on the material's resistance to cutting. Another notable study by Sun *et al.* [5] compared the friction behavior of nanoparticles on nano single crystals and grain boundaries, revealing the role of grain boundaries in hindering dislocation expansion and adsorption, and also as a new source of dislocation expansion. Li *et al.* [6] conducted a study on the sub-surface damage and material removal caused by high-speed friction of a diamond tool on a single-crystal copper substrate. They demonstrated that higher friction speeds, larger tool sizes, and depths of cut result in larger wear volumes and higher temperatures, while low-temperature wear produces more stacking defects. While many researchers have simulated nanoscale two-body abrasive wear or nano-scratches on materials, there is still a need to enhance the understanding of the relationship between abrasive wear at the nanoscale and macroscopic conditions. Specifically, the commonalities and differences between abrasive wear at the nanoscale and macroscopic conditions are worth exploring in-depth. To simplify the study, we begin by examining nano-abrasive wear under air conditions. Subsequently, we move on to investigate abrasive wear on water-containing films in the following chapter, and finally, we discuss abrasive wear on single-crystal silicon in the context of the chemical mechanical polishing (CMP) process for the fabrication of microelectromechanical (MEMS) and ultra-large-scale integrated circuits.

4.2 Basic Theory of Abrasive Wear

Abrasive wear falls within the realm of tribology, which encompasses a variety of disciplines, such as physics, chemistry, mechanics, materials, and machinery. It poses significant challenges to industrial and agricultural production by causing substantial material losses. Managing wear resistance and effectively using wear has therefore become a crucial priority. The wear process refers to the gradual loss of material due to the relative motion of an object's operating surface. Abrasive wear involves the generation of plastic deformation or detachment of material caused by hard particles or projections on the friction surface. This loss of material from the parent body occurs under the conditions of relative motion between the objects in contact.

The wear resulting from abrasive movement can be classified into two types: two-body abrasive wear and three-body abrasive wear.

Two-body abrasive wear occurs when a rough surface or a fixed abrasive material slides over another surface, causing material removal. On the other hand, three-body wear involves abrasive materials that are loose and able to move relative to each other, creating a form of abrasive wear where rotation is also possible as they slide over the worn surface. In three-body abrasive wear, most of the abrasive material becomes embedded in a softer surface, resulting in an increase in the proportion of abrasive wear from two-body wear. If we assume that all the abrasive material is embedded in the softer surface, three-body abrasive wear essentially becomes two-body abrasive wear. Therefore, two-body abrasive wear can be considered a special case of three-body abrasive wear.

The two-body abrasive wear occurs when a harder abrasive is pressed into a worn surface under normal load and moves under a tangential force, similar to a nano-scratch test. The nano-scratch test can be used to characterize the two-body abrasive wear resistance of a material. When the material is cut by the abrasive, the cutting mechanism dominates, and studies have shown that materials with lower hardness are easier to be cut. The abrasive action can also cause material to be "ploughed" out of the groove, leading to intense plastic deformation, which can eventually result in fatigue and material spalling. This fatigue is dominated by the plastic deformation mechanism. Figure 4.1 shows the groove morphology of two-body abrasive wear. In the cross-section of the figure, A_1 and A_2 represent the area of the groove ridges (the plastically deformed portion of the material), A_g is the area of the grooves, and S is the distance traversed by the abrasive. Figure 4.1 can be used to calculate the area of pure cutting wear, which is academically represented by the f_{ab} factor:

$$f_{ab} = \frac{A_g - (A_1 + A_2)}{A_g} \tag{4.1}$$

The "f_{ab} factor" indicates the scale factor of the material when it undergoes pure cutting wear. If the value is equal to one, it represents pure cutting wear. If the value is equal to zero, it signifies pure plastic wear. In the case of brittle materials, spalling may occur directly after being ploughed by the abrasive. If the f_{ab} factor is still used, it may be greater than 1, indicating the wear can then be classified as a fracture mechanism. However, for most widely used engineering

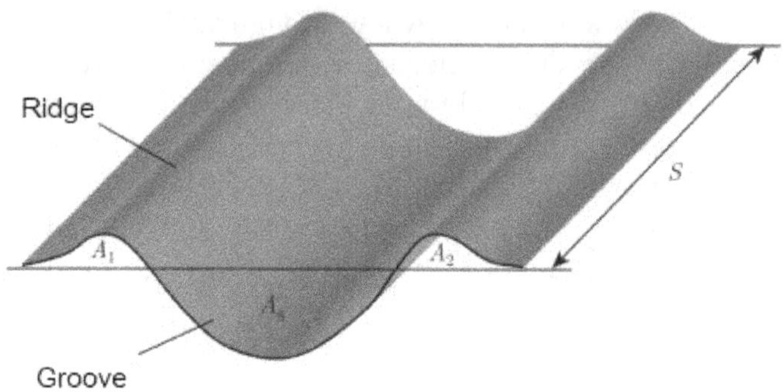

Figure 4.1. Two-body abrasive wear groove morphology.

materials, a reasonable mix of hardness and toughness is required. This means the fracture mechanism of wear should be limited, and for this reason, academic attention has often been focused on the discussion of cutting and plastic deformation mechanisms.

Three-body abrasive wear encompasses two-body abrasive wear and is more complex. In 1992, Fang *et al.* [7] created a single abrasive wear test setup to study the motion of abrasive particles in the three-body wear process, allowing for real-time dynamic observation. Based on these observations, they established the relationship (e/h) between the material characteristic load and the abrasive motion, as well as the criteria for abrasive motion modes. This was verified through experiments and numerical simulations. Figure 4.2 illustrates the force analysis of three-body abrasive particles.

In three-body abrasive wear, the abrasive in the main plane of motion comes in two states: sliding and rolling. Sliding abrasive acts as two-body abrasive wear, leading to material failure through cutting and plastic deformation wear. On the other hand, rolling abrasive primarily causes plastic deformation wear. Therefore, three-body abrasive wear mainly involves cutting and plastic deformation wear, with the former being more prominent than in two-body abrasive wear. When considering the entire three-dimensional space, there might be some rotational movement of the abrasive material about an axis of rotation that is not perpendicular to the principal plane. However, since this rotation is transient, it can be disregarded from an academic standpoint. For brittle materials, the damage caused by the rolling and sliding of the abrasive is more complex, and it's

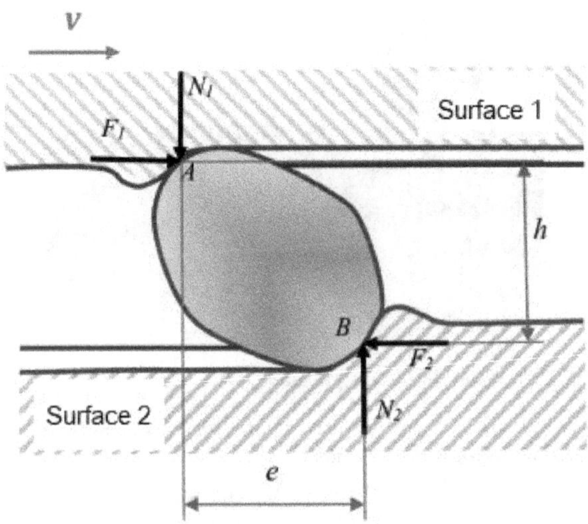

Figure 4.2. Analysis of abrasive forces in three-body abrasive wear.

essential to re-evaluate the above points. We hope that future scholars will conduct an in-depth study of three-body abrasive wear on brittle materials.

4.3 Two-Body Abrasive Wear of Nano-Scale Single-Crystal Copper Under Air Conditions

4.3.1 *Single-crystal copper two-body abrasive wear modeling*

The model illustrates the molecular dynamics of single-crystal copper two-body abrasive wear, as depicted in Figure 4.3. Under anhydrous film conditions, the model comprises of a single-crystal copper matrix with (100) crystalline surfaces and a rigid spherical abrasive. The initial velocities of the single-crystal copper atoms follow the Maxwell function. The copper matrix is divided into a fixed layer, a thermostatic layer, and a Newtonian layer. The motion of atoms in the Newtonian layer complies with Newton's second law, enabling the study of material wear mechanisms and laws. The temperature of the entire model is maintained at 300 K using the Nosé–Hoover hot-bath method in the thermostatic layer to facilitate heat transfer

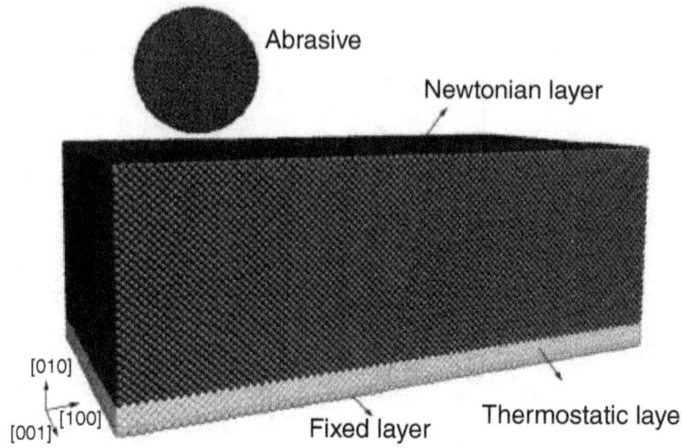

Figure 4.3. Molecular dynamics model of abrasive wear of anhydrous film single-crystal copper diameters.

between atoms. The fixed layer serves to prevent model deflection during simulation and reduce boundary effects but is not involved in the actual calculation process. To minimize computational costs and account for the effects of simulation scale, we applied Periodic Boundary Conditions (PBCs) in the X and Z directions to the substrate and fixed boundary conditions in the Y direction. The interactions between copper atoms are defined by the EAM potential, while the interactions between copper and carbon atoms in diamond are described by the Morse potential. Initially, we used the conjugate gradient method to optimize the model's energy and then conducted a simulation using the NVT system for a 100 ps relaxation period to maintain the substrate temperature at 300 K. The kinetic simulation was performed using the NVE method. For reference, you can find the molecular dynamics simulation parameters in Table 4.1 and the Morse parameters for Cu-C in Table 4.2.

4.3.2 *Molecular dynamics of abrasive wear on single-crystal copper diameters*

The Chemical Mechanical Polishing (CMP) process involves the mechanical action of abrasion from the abrasive to the substrate material. When an external load is applied, the abrasive is pressed into the substrate and then slides over the surface at a certain speed, removing material. Different process parameters, such as the external load and sliding speed, can significantly affect the substrate material

Table 4.1. Molecular dynamics parameters of single crystal copper abrasive wear.

	Specimen: Single-crystal copper	Abrasive: Diamond
Sizes	28.92 nm × 10.85 nm × 14.80 nm	Radius 3 nm
Atomic number	400140	20752
Time step	1 fs	
Initial temperature	300 K	
Sliding speed	50–200 m/s	
Indentation force	40–100 nN	
Sliding distance	0–20 nm	
Slide direction	[100] on (010) surface	

Table 4.2. Parameters of Morse potential function for Cu-C interaction.

Parameters	Numerical value
D0/eV	0.087
α/nm	51.41
γ0/nm	0.205

removal process and the quality of the polished surfaces. This study aims to investigate the effects of various process parameters on the wear process of single-crystal copper abrasives in an air environment, providing theoretical guidance for achieving higher-quality polished surfaces.

4.3.2.1 *The impact of sliding velocity on surface quality and internal defects*

The surface machining is categorized based on the sliding speed of the abrasive: general machining (abrasive speed below 45 m/s), high-speed machining (45–150 m/s), and ultra-high-speed machining (above 150 m/s). To simplify calculations, higher speeds are typically used for simulation. In this study, we consider abrasive speeds of 50 m/s, 100 m/s, and 200 m/s, sliding the same distance of 20 nm on the single-crystal copper substrate surface.

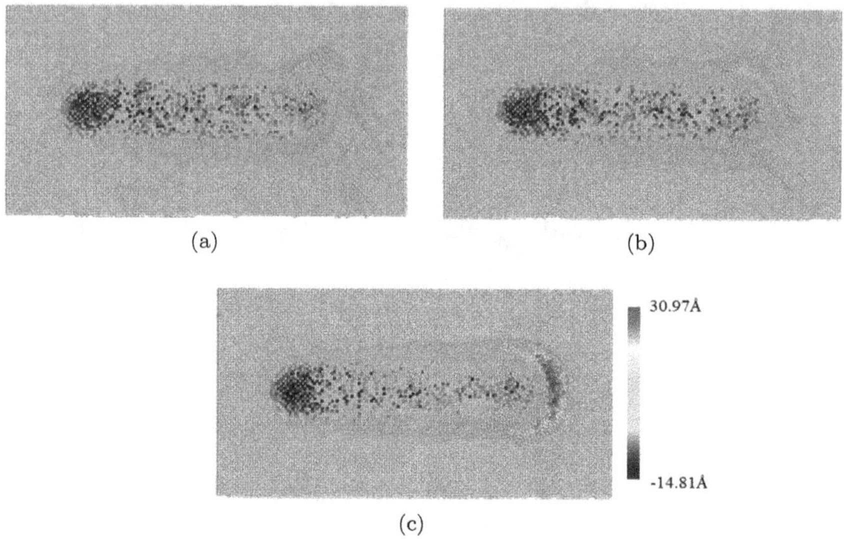

Figure 4.4. Surface morphology of single-crystal copper substrate at different sliding speeds for a load of 40 nN and a sliding length of 16 nm. (a) Sliding speed 50 m/s, (b) sliding speed 100 m/s, (c) sliding speed 200 m/s.

Figure 4.4 displays the surface topography of the single-crystal copper substrate under different sliding speeds. The sliding length was 16 nm at a load of 40 nN. Copper atoms are color-coded based on calculated displacement values. As the abrasive's sliding distance increases, the removed single-crystal copper atoms are distributed on both sides of the scratch and at the front of the abrasive. At a sliding speed of 200 m/s, a significantly larger number and volume of single-crystal copper atoms accumulate at the abrasive's front end compared to low speeds. Conversely, at low speeds, only a small number of single-crystal copper atoms are dispersed on both sides in the initial position, and this number increases with distance. At a sliding speed of 50 m/s, the copper atoms on both sides of the scratch are more evenly distributed.

The results indicate that the speed at which the abrasive moves has a significant impact on how atoms are distributed on a surface. When the abrasive moves faster under the same amount of pressure, the volume of atoms stacked at the front of the abrasive increases, and the single-crystal copper atoms that are removed are arranged more closely. This is because the movement of the abrasive takes the copper atoms away from their original positions. The higher the speed of the abrasive, the more the single-crystal copper atoms are taken away from the front of the abrasive, resulting in a greater number of copper atoms at the front. As a result, the number of copper atoms on both sides of the scratches on the single-crystal copper substrate decreases, and a small amount of single-crystal copper atoms are dispersed on the two sides of the groove at an initial stage.

During the polishing process, it is important to not only focus on the surface quality of the workpiece being polished but also pay attention to its internal quality. Defects within the material can easily become the source of dislocation nucleation and expansion of the workpiece during use, which can affect its performance and service life. A thorough examination of the formation and evolution of defects can help in obtaining workpieces of better quality.

As depicted in Figure 4.5, the atomic configurations of temporary defects within the single-crystal copper substrate are presented from left to right for sliding lengths of 4 nm, 8 nm, and 16 nm at loads of 40 nN and sliding velocities of the abrasive of 50 m/s, 100 m/s, and 200 m/s, respectively. The atoms in the figures are color-coded based on the calculated CNA values, where the red atoms represent atoms with surface or dislocation defects, the green atoms represent stacking fault atoms, and the copper atoms without distortion are hidden. The purple atoms represent diamond abrasive. During the process of material deformation, surface steps appear and atoms accumulate on the material's surface. This leads to the nucleation of dislocations within the material, causing dislocations to move in an upward or downward direction on the slip surface or to stop at the surface of the atoms, forming a dislocation ring inside the material. The internal structure of a single-crystal copper matrix contains vacancies, stacking faults, prismatic dislocation rings, V-shaped dislocation rings, and atomic clusters (refer to Figure 4.5). The internal defects of the single-crystal copper are mainly distributed below and in front of the abrasive. When the abrasive's sliding speed is 50 m/s and the sliding length is 4 nm, the formation of dislocation embryos below the abrasive occurs, which then move along the {111} sliding system and gather a large number of stacking faults and prismatic dislocation rings. These expand downward into the inside of the single-crystal copper substrate, forming a line of dislocations on the surface of the substrate. In front of the abrasive, there is a V-shaped dislocation

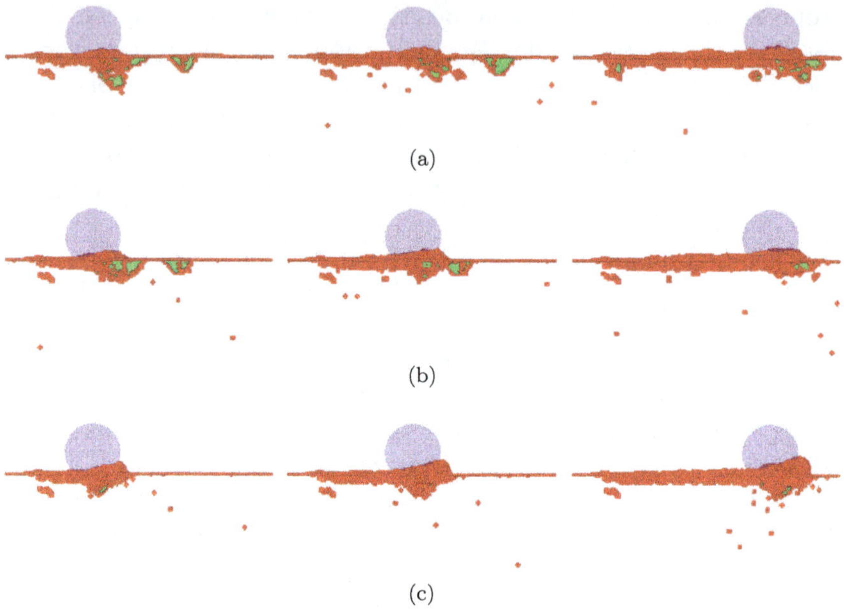

Figure 4.5. Transient defects inside single crystal copper matrix at 40 nN load and different sliding velocities. (a) sliding speed 50 m/s, (b) sliding speed 100 m/s, (c) sliding speed 200 m/s, sliding distance 16 nm.

Note: In each subplot, the left plot shows a sliding distance of 4 nm, the middle plot shows a sliding distance of 8 nm, and the right plot shows a sliding distance of 16 nm.

ring containing two Shockley dislocations $1/6\langle 112\rangle$. These two partial dislocations expand within the substrate to cross-slip internally, eventually meeting to form a compression rod dislocation $1/6\langle 110\rangle$, also known as the Lomer–Cottrell dislocation lock. This immobilized dislocation hinders the expansion of dislocations into the interior of the substrate, causing them to form a V-shaped dislocation ring inside the matrix, as shown in Figure 4.6(a). As the sliding distance increases, the prismatic dislocation ring moves upward. When the sliding speed of the abrasive is 200 m/s, the number of defects inside the single-crystal copper substrate is small and mainly located in the lower part. The defects expand a short distance into the substrate's interior, causing a large number of vacancy defects to appear inside the substrate, making them potential sources of workpiece defects during use, as shown in Figure 4.6(c).

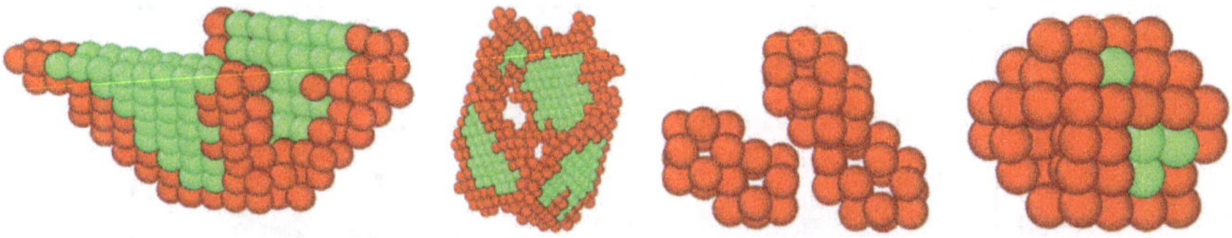

(a) (b) (c) (d)

Figure 4.6. Dislocation defects in single-crystal copper matrix. (a) V-shaped dislocations, (b) prismatic dislocations, (c) vacancy defects, (d) atomic clusters.

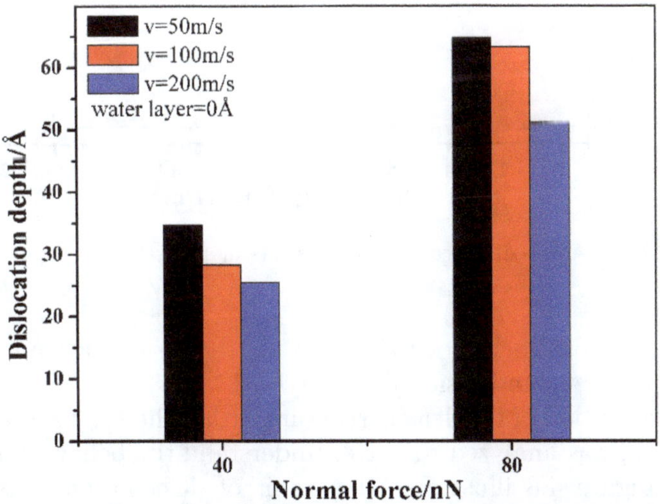

Figure 4.7. Depth of dislocation defects in single-crystal copper matrix at different sliding speeds.

In Figure 4.7, the maximum depth of internal defects in single crystal copper reaching the substrate is shown for loads of 40 nN and 80 nN at different sliding speeds. The figure illustrates that the maximum depth of dislocation defects decreases as the abrasive sliding speed increases. When the sliding speed is higher, the abrasive and substrate wear time is shorter, causing the dislocation nuclei to lose continuous support of energy and leading to annihilation. This results in fewer dislocation defects inside the substrate, with a smaller maximum depth. For instance, at a sliding speed of 200 m/s and a load of 40 N, the maximum depth of the dislocation defects is 25.52 Å. Conversely, at a lower sliding speed of 50 m/s and the same load, the

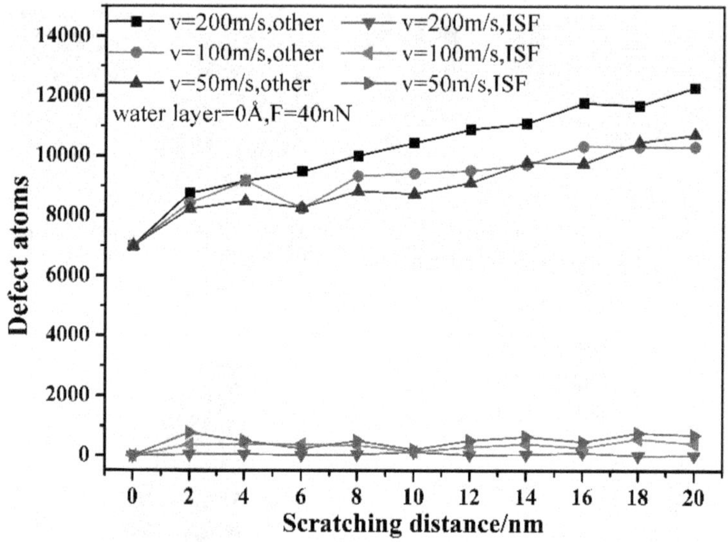

Figure 4.8. Atom–distance curves of defects in single-crystal copper matrix.

maximum depth increases to 34.61 Å, as the dislocation nuclei have more time to expand inside the substrate.

The evolution of the defect structure within the single-crystal copper matrix was analyzed to better understand the behavior of defect atoms. Figure 4.8 illustrates the count of defect atoms inside the single crystal copper body. Intrinsic Stacking Fault (ISF) refers to the stacking fault atoms, while other atoms indicate the amorphous atoms on the surface of the matrix. The increase in sliding distance results in a higher number of other atoms, as more surface defects are generated in the single-crystal copper substrate due to the movement of the abrasive. The highest number of other atoms is observed at a sliding speed of 200 m/s, while the lowest number is observed at 50 m/s. A higher sliding speed leads to a greater number of amorphous atoms on the substrate's surface. Conversely, the count of ISF atoms decreases with increasing sliding velocity. This is attributed to the short nucleation time of dislocations at higher sliding velocities, resulting in fewer dislocations and consequently a decrease in the number of ISF atoms. The fluctuating trend in the formation and movement of defective atoms inside the substrate is indicative of the fluctuating nature of ISF atoms with the movement of the abrasive.

The simulation results indicate that as an abrasive slides over a single crystal copper surface, the energy accumulated during the nanoindentation stage is gradually released due to the appearance and reaction of defects within the copper matrix. The dislocation defects are mainly concentrated below the abrasive. An increase in sliding speed reduces internal defects within the matrix, as the higher speed limits the time and energy available for dislocation formation and nucleation. This results in fewer defects within the matrix, shorter dislocation movement time, and shorter atomic reorganization time on the substrate surface with more amorphous atoms.

4.3.2.2 *Effect of sliding speed on friction and coefficient of friction*

The effect of sliding speed on friction and coefficient of friction is illustrated in Figure 4.9. It shows that the friction force increases rapidly at the beginning of abrasive sliding and then fluctuates slightly around a constant value after reaching a steady state for different sliding speeds of 50 m/s, 100 m/s, and 200 m/s under loads of 40 nN and 80 nN.

Table 4.3 provides the average friction force and the corresponding friction coefficients at three different sliding speeds. It is evident from the table that all the friction coefficients are greater than 1. At a macroscopic scale, the friction force mainly comprises the cutting force. However, at the nanoscale, the friction force consists of the adhesion force, cutting force, and resistance of the grinding chips combined. Adhesion force is a molecular force that is noticeable only when two substances are very close to each other and is not considered at the macro-scale. However, at the micro-scale, its magnitude has been found to be similar to that of the cutting force. Consequently, at the nanoscale, the friction generated during the abrasive sliding process is larger, and the coefficient of friction is also larger. This finding is consistent with the research results of Peng *et al.* [1]. The simulation results indicate that the friction coefficient increases as the sliding speed of the abrasive increases. This is because a higher sliding speed results in less dislocation nucleation and motion-dominated plastic deformation within the substrate, leading to the removal of copper atoms by the abrasive, mainly located at the front end. Consequently, the resistance and adhesion from the

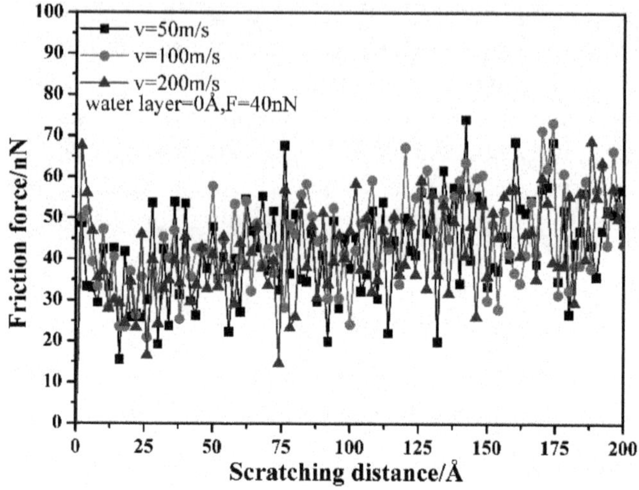

(a)

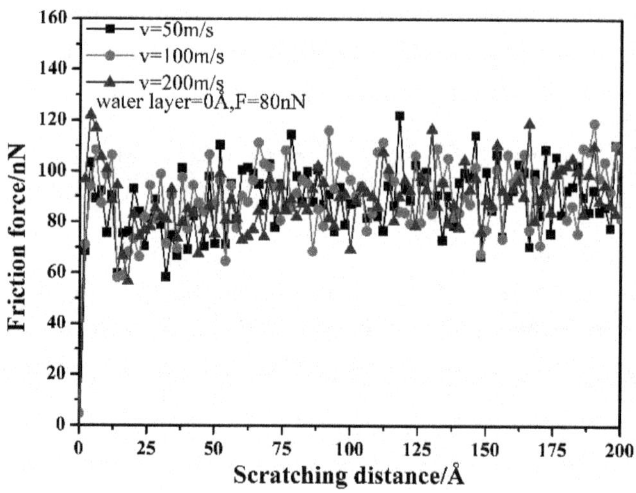

(b)

Figure 4.9. Friction–distance curves at different sliding speeds. (a) Load 40 nN, (b) load 80 nN.

abrasive chips are also increased, thereby increasing the frictional resistance during the cutting process and the coefficient of friction. This conclusion aligns with the findings of the study by Zhang Junjie *et al.* [10].

Table 4.3. Friction coefficients and average friction for different sliding speeds.

	40 nN		80 nN	
Sliding speed/m · s^{-1}	Friction/ nN	Friction coefficient	Friction/ nN	Friction coefficient
50	46.8	1.17	97.2	1.22
100	47.6	1.19	103.2	1.29
200	48.8	1.22	108	1.35

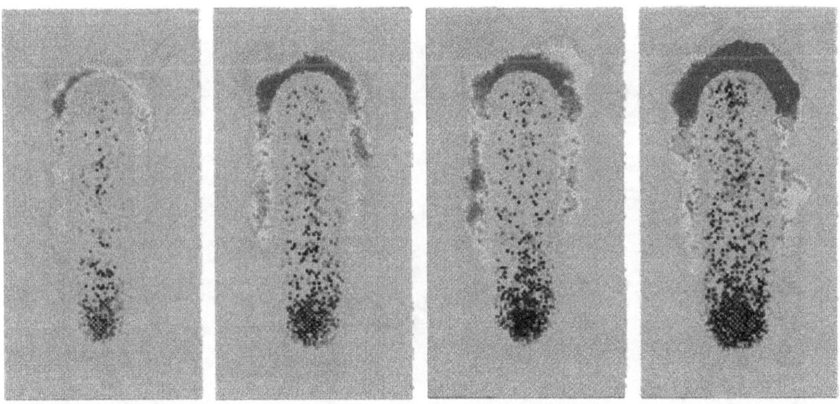

(a) (b) (c) (d)

Figure 4.10. Surface morphology of single-crystal copper at sliding distance of 16 nm and different loads. (a) load 40 nN, (b) load 60 nN, (c) load 80 nN, (d) load 100 nN.

4.3.2.3 *Effect of loading on the surface morphology and internal defects of the substrate*

The images in Figure 4.10 display the surface structure when an abrasive is moved in the direction of X at a steady speed of 100 m/s under loads of 40 nN, 60 nN, 80 nN, and 100 nN. As the abrasive shears against the single-crystal copper substrate, it creates a scratch on the surface. With increasing load, the abrasive action intensifies, leading to an increase in the width and depth of the scratch. The copper atoms pushed by the abrasive disperse sideways, while the removed single-crystal copper atoms gather in front of the abrasive

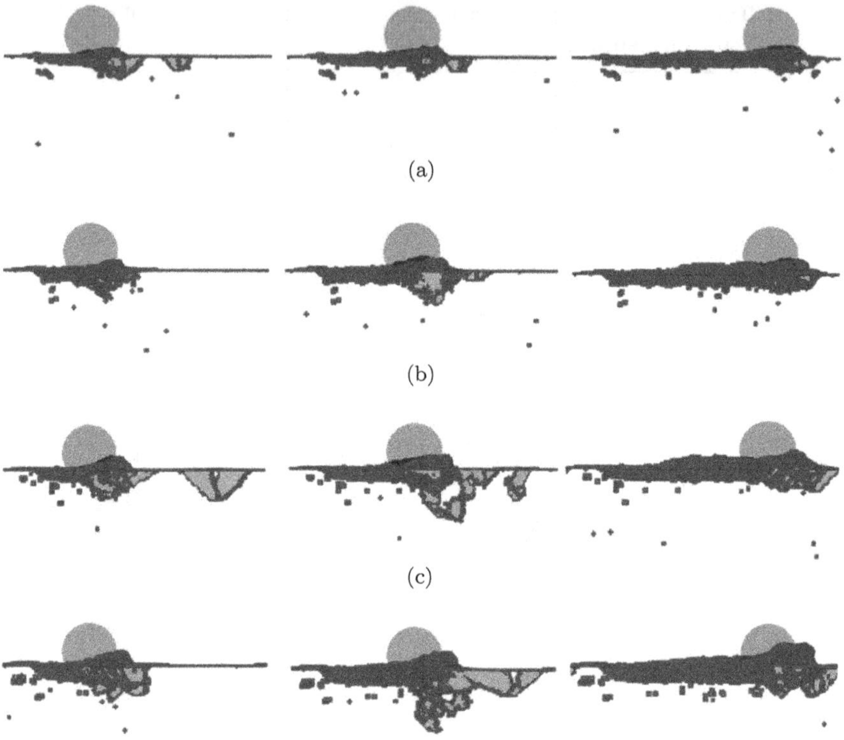

Figure 4.11. Internal transient defects of single-crystal copper under different loads. (a) Load 40 nN, (b) load 60 nN, (c) load 80 nN, (d) load 100 nN.

Note: In each subplot, the left plot shows a sliding distance of 4 nm, the middle plot shows a sliding distance of 8 nm, and the right plot shows a sliding distance of 16 nm.

and on both sides of the scratch. The higher the load, the more copper atoms gather in front of the abrasive, resulting in a higher pile height of the copper atoms on either side of the scratch.

In Figure 4.11, the dislocation transient configurations inside the single crystal copper matrix are depicted for loads of 40 nN, 60 nN, 80 nN, and 100 nN. Abrasive sliding speeds of 100 m/s and sliding distances of 4 nm, 8 nm, and 16 nm were considered. With increasing load, the interaction between the abrasive and the single-crystal copper matrix intensifies, leading to the generation of more defects inside the single-crystal copper matrix. In Figure 4.12, we can

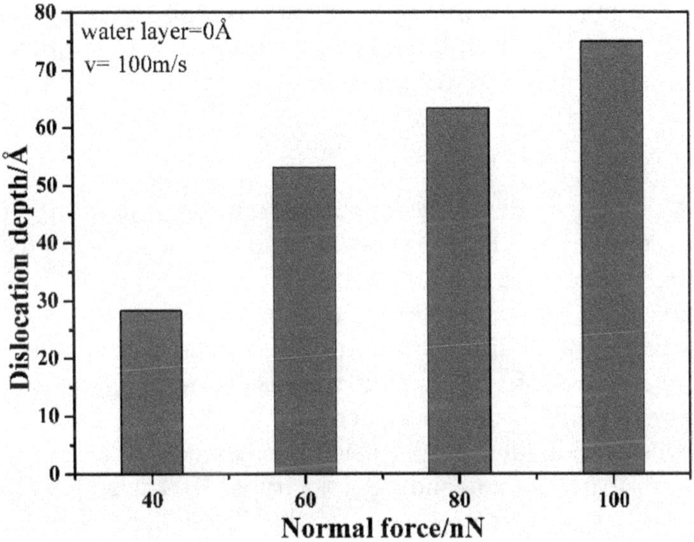

Figure 4.12. Internal dislocation depth of single-crystal copper under different loads.

see the statistical representation of the variation in the maximum depth of dislocation defects at an abrasive sliding speed of 100 m/s under anhydrous film conditions. The value of the maximum depth increases as the load increases. Compared to the transient defects inside the substrate in Figure 4.11, it is evident that higher loads cause the dislocation defects to expand further into the substrate. Specifically, the maximum depth of dislocation defects is 28.31 Å for a load of 40 nN, 53.09 Å for a load of 60 nN, 63.36 Å for a load of 80 nN, and 74.89 Å for a load of 100 nN.

The results of the simulation indicate that as the load increases, both the atoms removed from the single-crystal copper substrate and the defects inside the substrate also increase. This happens due to the increased mechanical interaction between the abrasive and the substrate. As a result, the depth of the abrasive pressed into the substrate increases, removing more material and creating more dislocations inside the substrate. These dislocations become entangled with each other and penetrate deeper into the substrate. The increase in load and the number of dislocation defects inside the substrate are not beneficial for the stability of the workpiece during use.

The number of dislocations in a crystal is represented by the dislocation density (ρ), which is defined as the total length of dislocation lines contained in a unit volume of the crystal, i.e.:

$$\rho = S/V \tag{4.2}$$

where V is the volume of the crystal/m^3 and S is the total length of the dislocation lines in the crystal/m. The unit of dislocation density ρ is m/m^3, which can also be simplified as 1/m^2. Figure 4.13 shows the dislocation density map inside the single-crystalline Cu substrate with an abrasive sliding distance of 20 nm, and it can be seen that with the increase of the loading, the dislocation density inside the single-crystalline Cu substrate increases.

The dislocation line lengths inside the matrix were analyzed under a load of 40 nN and a sliding velocity of 100 m/s, as shown in Figure 4.14. The Burgers vectors of the dislocations were analyzed using the DXA algorithm. In the figure, the colors representing the dislocations are as follows: the Shockley partial dislocation $1/6\langle 112\rangle$ is green, the perfect dislocation $1/2\langle 110\rangle$ is blue, and the ladder-rod dislocation $1/6\langle 110\rangle$ is purplish-red. Arrows on the dislocation line indicate the direction of the dislocations, and the purplish arrows

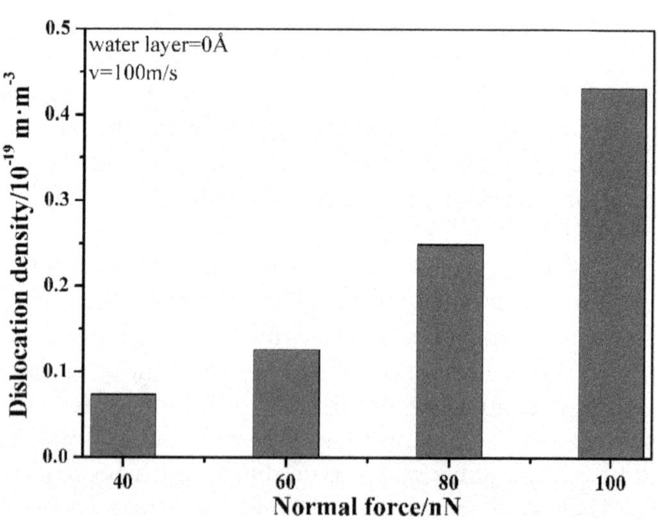

Figure 4.13. Dislocation defect density in single-crystal copper matrix at different loads.

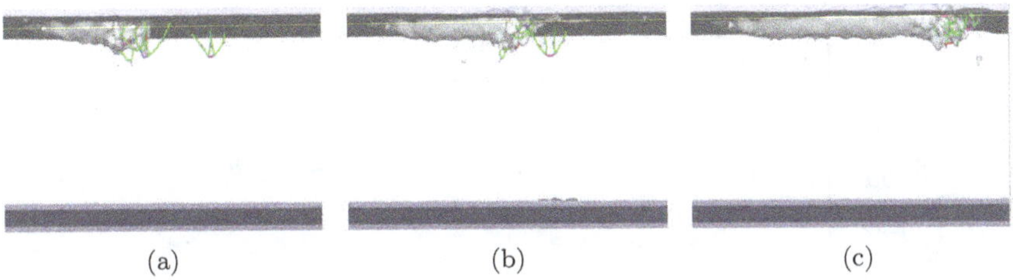

Figure 4.14. Transient dislocation lines inside a loaded 40 nN single-crystal copper substrate. (a) Sliding distance 4 nm, (b) sliding distance 8 nm, (c) sliding 16 nm.

represent the direction of the Burgers vectors. Dislocations slipping on adjacent slip planes form the ladder-rod dislocation with the Burgers vector $1/6\langle 110\rangle$. In face-centered cubic crystals, the $1/6\langle 112\rangle$ dislocation plays a significant role in plastic deformation and is often called the Schockley partial dislocation.

At a load of 40 nN and a sliding distance of 4 nm, the defects inside the single-crystal copper matrix consist of 2 ladder-rod dislocations and 15 Shockley-type dislocations. The dislocation lines intersect to form a complex dislocation network. When the sliding distance is 16 nm, the dislocation network becomes more intricate in the region in front of the abrasive action on the substrate. As the abrasive moves forward, new dislocations appear between the abrasive and the substrate. In the area where the abrasive has slid, the single-crystal copper substrate undergoes elastic recovery, leaving surface defects and a small number of vacancies. The dislocation lines that penetrate deep into the interior are annihilated. In the late stage of sliding, the dislocation lines inside the matrix become short in length and cluster densely in the shear region between the abrasive and the substrate. Under the shear action of single-crystal copper and abrasive, the effect of material removal is reached. Due to the short action time of the single-crystal copper and abrasive, the dislocation lines react with each other only briefly. The generated dislocation lines gather in the deformation region to entangle with each other. The dislocations can be released when the abrasive material leaves the action region, either moving downward independently to form a ring, getting annihilated, or remaining inside the matrix. Figure 4.15 shows the statistics of the total length of the dislocation line under different loads of single-crystal copper substrate

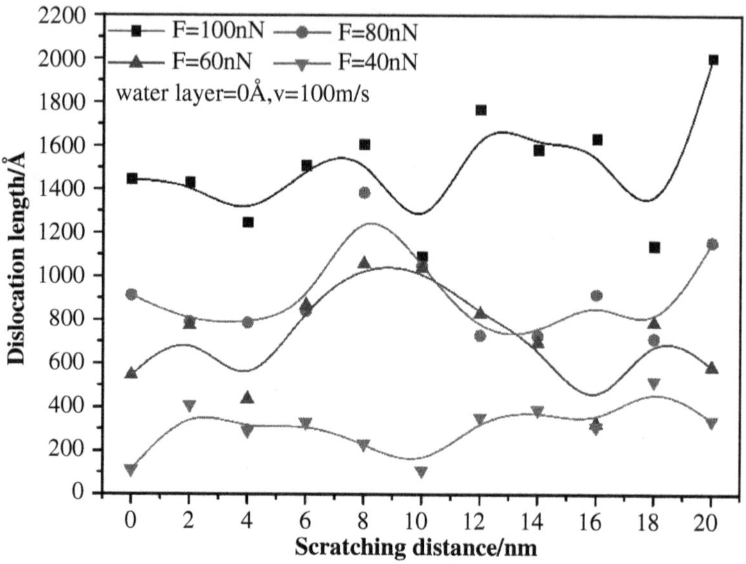

Figure 4.15. Total length–distance curves of dislocations in single-crystal copper substrate under different loads.

with increasing sliding distance. It is observed that the higher the load, the more distorted the atoms are generated and the longer the length of the dislocation line is produced.

4.3.2.4 *Effect of load on friction and coefficient of friction*

To further analyze the role of the abrasive in two-body wear that affects the substrate, we investigated the mechanical properties of the abrasive during the wear process. As shown in Figure 4.16, positive loads of 40 nN, 60 nN, 80 nN, and 100 nN were applied to the abrasive. With the increase of the sliding distance, the normal load exerted on the abrasive fluctuates up and down around a certain value under constant loading. This fluctuation is due to the vibration of the atoms themselves and the generation and release of dislocations inside the matrix. As the positive load increases, the depth of the abrasive pressed into the interior of the single-crystal copper matrix also increases, leading to greater internal deformation of the matrix. A large number of atoms gather in front of the abrasive, hindering its movement. This increased resistance in the movement

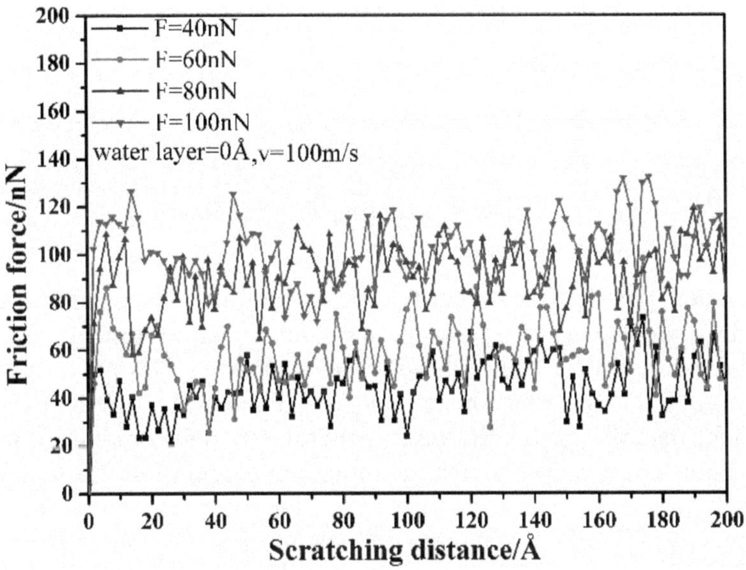

Figure 4.16. Friction–distance curves for different loads under anhydrous film.

Table 4.4. Average coefficient of friction under different normal loads.

Normal load/nN	40	60	80	100
Coefficient of friction	1.19	1.23	1.29	1.38

process results in increased friction. In the Z direction, the abrasive experiences nearly equal force on both sides of the scratch, with opposite directions that offset each other, making it negligible.

In Table 4.4, you can find the average friction coefficients under different positive loads. It's observed that the friction coefficients increase as the load increases. This happens because, under a constant load, the depth of the abrasive material pressed in increases with greater loads. As a result, more atoms accumulate at the front end of the abrasive material during movement, leading to a greater hindering force and increased friction. Consequently, the coefficient of friction also increases accordingly. It's worth noting that all the friction coefficients are greater than 1, which aligns with the conclusion of the previous simulation test.

4.4 Three-Body Abrasive Wear of Nano-Scale Single-Crystal Copper Under Air Conditions

4.4.1 *Molecular dynamics modeling of abrasive wear on single-crystal copper*

4.4.1.1 *Three-dimensional modeling of nanodiamond ellipsoidal abrasives*

When constructing molecular dynamics computational models, some regular models such as spherical and conical can be built in the self-contained module of molecular dynamics software. However, the ellipsoidal model cannot be constructed yet. The authors are focusing on the factors influencing the movement of nano-ellipsoidal abrasives, so it's necessary to construct the nano-ellipsoidal model first. Molecular dynamics can import data files that meet its format to build the model. Therefore, the nano-ellipsoidal diamond model is written in C^{++} language. The atoms within the built ellipsoidal model need to comply with the arrangement of the diamond crystal structure. The size and the axis ratio of the ellipsoidal model can be adjusted freely, and the atomic weight can be displayed. The output data file format needs to meet the requirements of the LAMMPS software input.

(1) **General idea:** The ellipsoid model is constructed from atoms, so building the model involves finding the coordinates of all the atoms. The program design should aim to output the coordinates of each atom in the required format according to the lattice law. The program design should focus on determining the relative position of each atom and developing the algorithm accordingly. Additionally, it should establish the data interface format and regulations of LAMMPS to ensure the successful importation of the model into it.

Initially, every carbon atom is filled in the diamond cell, as required. Considering that monocrystalline silicon also has a diamond cell structure, the diamond nano-ellipsoidal crystals can also be transformed into monocrystalline silicon nano-ellipsoidal crystals by simply changing the atomic types and lattice constants. Ultimately, the program input should allow the user to choose whether to generate a diamond or monocrystalline silicon structure.

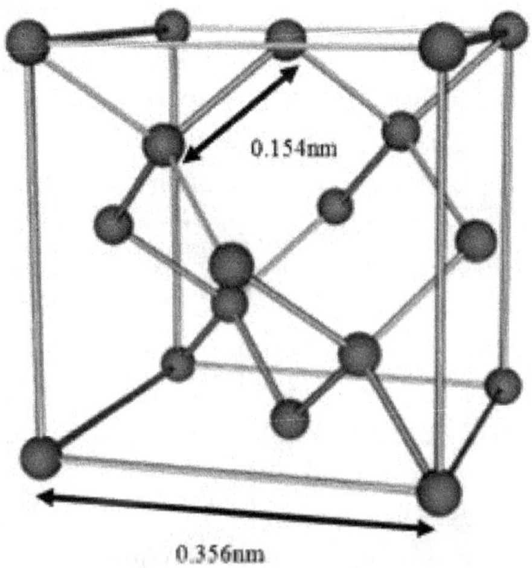

Figure 4.17. Diamond crystal structure.

The crystal structure of a diamond is illustrated in Figure 4.17. The diamond cell structure is typically in the ordinary FCC structure, containing more than four atoms. These atoms are located in the body diagonal of the four atoms at the 1/4 position and 3/4 position. Each carbon atom has a coordination number of 4, with a C–C bond length of 0.154 nm, and a lattice constant of the cell is 0.356 nm. For monocrystalline silicon, the crystal structure is the same as that of diamond, with the only difference being the lattice constant of 0.5428 nm.

In programing for crystals, the loops of the crystal cell should be the main thinking in program design. When discussing the loops, it is important to first define the principle to be followed when looping: "No more, no less". "No more" means that no duplicate atom positions should be generated during the loop, while "no less" means that no atom generation should be omitted from all the array points in the loop.

For a loop unit, its step size should be equal to the lattice constant. The next step is to determine its unit and method. When choosing the loop unit, if the cycle is based on the whole cell, it is easy to cause overlap of atomic positions. This is because some atoms will overlap at two neighboring cells as the cells continue to pile up and accumulate. If we choose to remove this part of the atom at the end, it will inevitably cause a lot of trouble and obstacles. Therefore, it is determined that the selected cyclic unit should be an atom while taking care not to miss it. In this case, the knowledge of crystallography can be used to select each cell's unique atoms. After these atoms are cycled through while accounting for the cycle step, all the atoms on the dot matrix can be obtained without duplication and without missing any.

First, we calculate the number of atoms unique to the cell. There are 8 vertices with 1/8 atom each, making a total of 1 atom. The 6 face centers have 1/2 atom each, making a total of 3 atoms. All 4 body atoms are included, giving us a total of 8 atoms labeled within the crystal cell (as shown in Figure 4.18). Cycling through these 8 atoms in turn will yield all the atoms.

For the looping approach, one of the simpler methods is to use triple nesting. Looping within the quadrant one of Cartesian coordinate system along the x, y, and z directions in limited times can generate a model of a rectangular body. By using the offset ellipsoid equation $(x-a)/a+(y-b)/b+(z-c)/c=1$ (where a, b, and c are the lengths of the three half-axes of the ellipsoid), we can constrain the coordinates to obtain the model without the outer ellipsoid atoms, as shown in Figure 4.19.

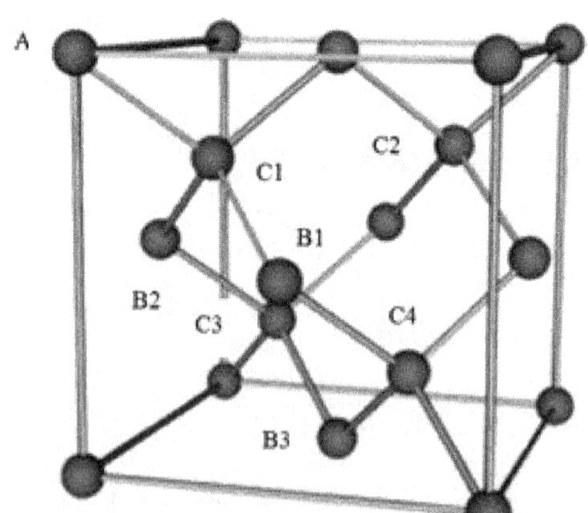

Figure 4.18. Unique atoms within a diamond cell.

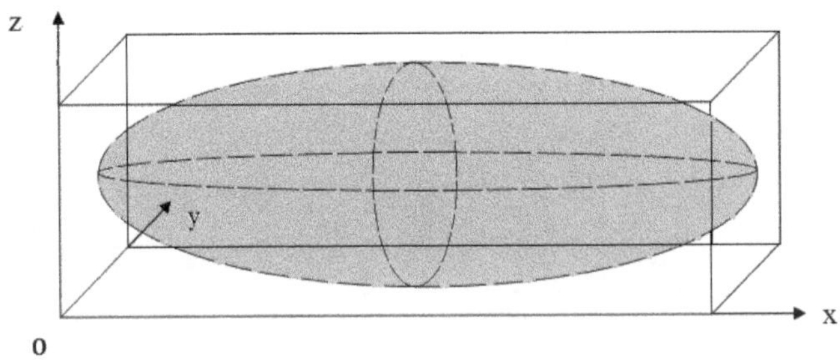

Figure 4.19. Ellipsoid generation position.

Make sure to adhere to the following requirements when preparing the output file format for the LAMMPS input interface:

1. The model's data file should be imported using the **read_data** command in LAMMPS. The data file should be in **dat** format and must include information such as the total number of atoms, atom types, atomic masses, and boundary ranges.
2. Each data point should be organized as a line of data with each line representing an atom. Atoms should be serially numbered by adding one to each line. The chosen format should be in the order of serial number-type-rectangular coordinates-polar coordinates. If rectangular coordinates are chosen, the other coordinate information can be set to zero.
3. Be aware that in LAMMPS, the default unit of length is angstroms, not nanometers.
4. Pay attention to the atom-type numbering in relation to the matrix model. The matrix atoms also have atom-type serial numbers. In cases where the matrix needs to be divided into multiple layers (e.g., thermostatic layers), the atom serial numbers of the ellipsoids need to be adjusted to the atom-type serial numbers of the matrix. When this is not considered, the numbering defaults to 1.

4.4.1.2 *Procedure flow chart*

The program flow is illustrated in Figure 4.20. When building the program, the first step is to create an output stream file called **data.dat** in the **data** format within the source file directory of the

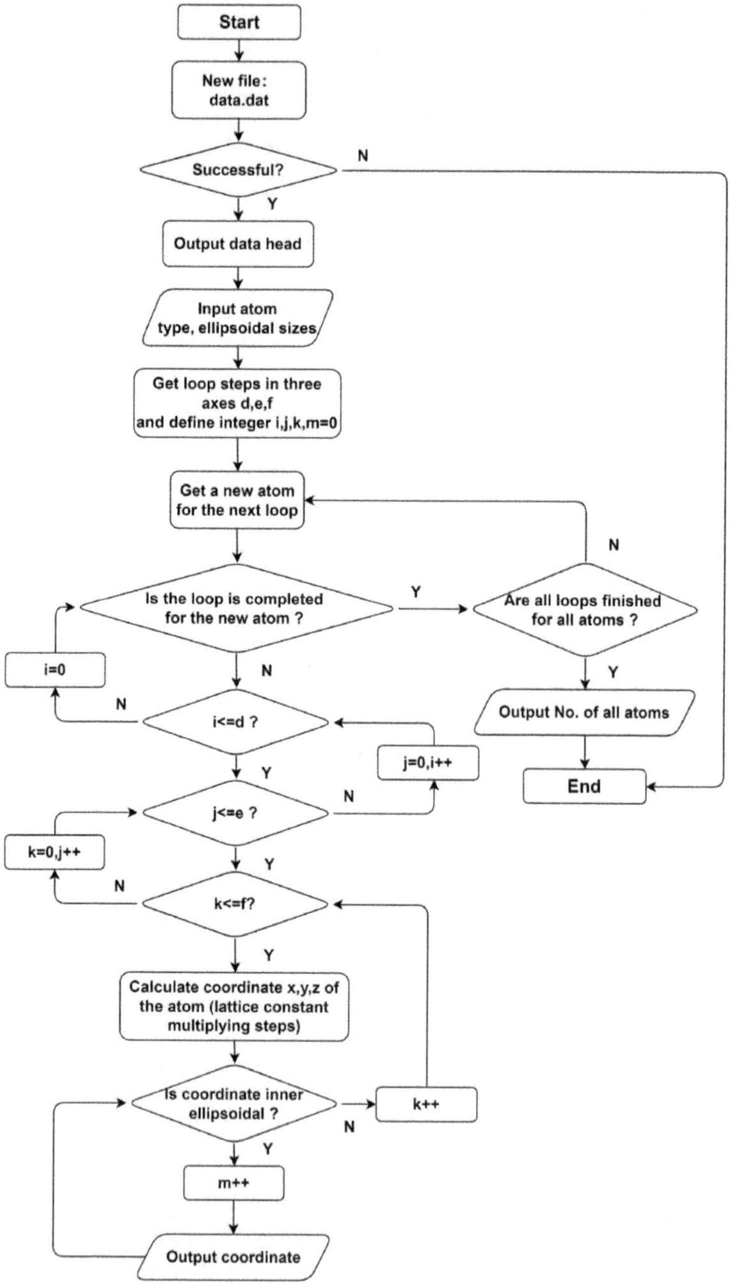

Figure 4.20. Flowchart of ellipsoid generation procedure.

C^{++} program. This makes it easier to calculate the data for export later on. Following this, the program checks if the file was created successfully. If not, the program automatically terminates and the files in the directory should be examined. If the file creation is successful, the initial data file is then exported in the format required by LAMMPS.

To define an atom type and the size of the ellipsoid, you can choose between two atom types: carbon and silicon. Carbon corresponds to the generation of diamond, while silicon corresponds to monocrystalline silicon. After selecting the atom type, the program will automatically include the atom type (carbon or silicon) in the data file. The program will treat the two atom types differently when calculating the step later. For the ellipsoid dimensions, enter the lengths of the three semi-axes (a, b, and c) in order. While the ellipsoid dimensions can be specified as desired, it's important to set the length of the rolling direction axis of wear as the longest axis during setup. This prevents rotation around the normal load loading axis during the simulation.

To start, we need to calculate the number of loop steps (d, e, f) in the three directions. We also define three zero loop integers (i, j, k) and an integer (m) with an initial value of zero as the counting value for atoms. The values for d, e, and f can be calculated by finding the maximum length on each axis and dividing it by the length of a single crystal cell. It's important to note that the calculation of d, e, and f should be based on the entire axis length (twice the half-axis length) since the cell atoms are generated within a quadrant of Cartesian coordinate system. The definition of i, j, and k is for the next three-direction loop, which is equivalent to recording the current loop position and advancing the loop by adding one each time. The integer m is used for counting atoms, starting with a value of zero. According to the LAMMPS data interface requirements, the final data file must show the total number of atoms. During the data generation process, the serial number for each atom needs to be generated, and this serial number can be used for numbering. Each time the number is incremented by one.

The next triple nested loop is the core of the entire program. As mentioned earlier, all the atoms are categorized into eight atomic loops: A, B1, B2, B3, C1, C2, C3, and C4. One of these loops needs to be operated for the above-mentioned one kind of atom.

If we assume that we are starting with the selected class A atoms, the first task is to enter the first judgment box. If the loop doesn't end, then the procedure continues. The i, j, and k are integers, respectively, with each grid point in the loop process needing to undergo an increment of one.

After handling the loop to the minimum number of bit k, the command within each loop is executed step by step. The specific content of the command is as follows:

(1) Based on the current number of loops, reduce the current coordinate value by multiplying the number of loops in each direction by the lattice constant. This is the opposite of the previous operation.
(2) Use the offset ellipsoid equation to determine if the current coordinate point is within the ellipsoid. The ellipsoid equation should be less than one after the comparison.

   - If the point is outside the ellipsoid, it will re-enter the loop system and will not proceed with the next operation.
   - If the point is inside the ellipsoid, the atomic count m is increased by one. Then, the coordinate information of the current point is outputted according to the previously described format. Finally, the loop system is resumed.

In the case where the highest bit i is greater than the total number of steps in the loop, it can be inferred that atom A has been looped completely. The program currently loops through only atom A. A new class of atoms (assumed to be B1) can be selected to restart the loop using the same process and so on. Ultimately, all eight types of atoms should be looped through to achieve the goal of finding all atom coordinates and outputting their information to the data file.

After completing the loop, the total number of atoms (m) is output to the top of the screen and then returned to the **data** file for modification. Once all the data has been output, the output stream operation should be closed. This entire process builds the model effectively.

4.4.1.3 *Correctness verification of the diamond ellipsoid model*

1. Validation of the center of mass coordinates: It's important to validate the center of mass coordinates because they are crucial for the model to operate correctly in LAMMPS. To do this, we need to compare the actual center of mass coordinates with their theoretical values. The center of mass can be calculated using a formula that averages the coordinates in all directions based on the masses. If the mass points represent the same atoms with the same mass, it's simple to average all their coordinates to find the center of mass. In Matlab, you can perform this calculation. The position of the center of mass of the ellipsoid changes with different model sizes, but the theoretical position should be the center of the form, given by (a, b, c) if the half-axis lengths are a, b, and c, respectively.

In contrast, the second algorithm has a fixed theoretical center of mass position. Because the algorithm takes into account the ellipsoid's symmetry, the actual coordinates of the ellipsoid's center of mass generated at this time should be in perfect agreement with this value, which is also tested later.

For both algorithms, three sets of data were chosen, and the difference between the theoretical center of mass coordinates and the actual center of mass coordinates was calculated. The results of the runs are shown in Tables 4.5 and 4.6. Table 4.5 shows the data for the first algorithm, and Table 4.6 shows the data for the second algorithm, with measurements on the axes in angstroms.

Based on the data from the two tables, we observed that the theoretical center of mass coordinates from the first algorithm did not match the actual coordinates exactly. However, the difference

Table 4.5. Comparison of theoretical and actual center of mass coordinates under the first algorithm.

Ellipsoid crystal type	Theoretical x value	Calculated x value	Theoretical y value	Calculated y value	Theoretical z value	Calculated z value	Error rate/%
Single-crystal silicon	70	70.016678	60	59.994355	50	50.011208	0.02
Diamond	90	89.992306	65	64.994395	40	40.003304	0.01
Diamond	20	19.999130	15	15.004060	10	10.025034	0.09

Table 4.6. Comparison of theoretical and actual center of mass coordinates under the second algorithm.

Ellipsoid crystal type	Theoretical x value	Calculated x value	Theoretical y value	Calculated y value	Theoretical z value	Calculated z value	Error rate/%
Single-crystal silicon	70	70	60	60	50	50	0
Diamond	90	90	65	65	40	40	0
Diamond	20	20	15	15	10	10	0

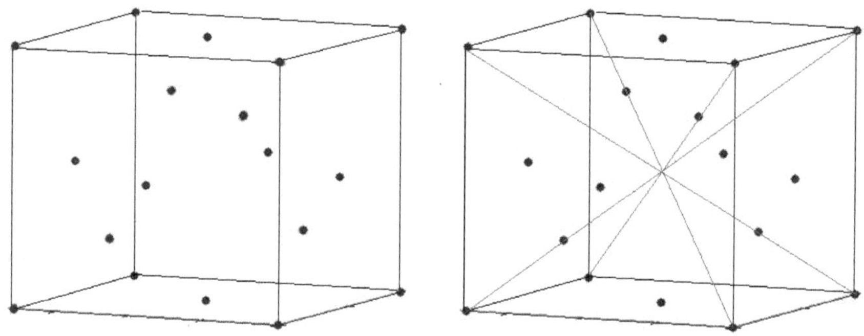

(a) (b)

Figure 4.21. Lattice structure of an atomic dot matrix shown. (a) Atomic dot matrix of the cell extracted from the crystal model, (b) schematic of the cell after making auxiliary lines.

was very small, and the error rate of the three sets of data combined did not exceed 0.1%. This meets the expected accuracy requirements. With the second algorithm, the actual center-of-mass coordinates and the theoretical values perfectly matched, demonstrating the success of the modeling using this algorithm.

2. Verification of lattice structure: The algorithm is based on the lattice structure, and we need to select a crystal cell to check if it conforms to the diamond's crystal structure. Since the crystal structures generated by both algorithms were the same, we only need to examine the model generated by one of the algorithms. The model obtained by the second algorithm is selected in the following, and the atoms of a crystal cell are obtained, as shown in Figures 4.21(a) and 4.21(b). Comparing 4.21(b) with the crystal structure image from Figure 4.17, it is evident that this is the diamond structure, confirming that this model is correct.

4.4.2 *Three-body abrasive wear modeling of single-crystal copper under constant load*

The same single-crystal copper material was utilized as upper and lower surfaces, with diamond serving as the sandwiched abrasive material held rigid. The three-body abrasive wear model of an ellipsoidal abrasive in single-crystal copper is depicted in Figure 4.22, with dimensions of 32.6 nm × 20.2 nm × 14.9 nm. The crystallographic orientations of the abrasive are aligned with the x, y, and z directions as [100], [010], and [001], respectively. The substrate consists of 575,640 atoms, and the dimensions of the abrasive vary with the ellipsoidal long-to-short axis ratio. The single-crystal copper has a face-centered cubic structure with a lattice constant of 0.35667 nm, and its crystallographic directions along the x, y, and z directions are [100], [010], and [001], respectively. The friction surface corresponds to the (010) surface. The abrasive is pressed in the y direction and slides in the x direction, moving along the [100] direction on the (010) surface. The simulation is arranged with periodic boundary conditions in the x and z directions. This means that an infinite number of periodically arranged abrasives in the lateral direction is simulated to eliminate the boundary effect. The z direction is set up as a free boundary condition. The outside atomic layers of single-crystal

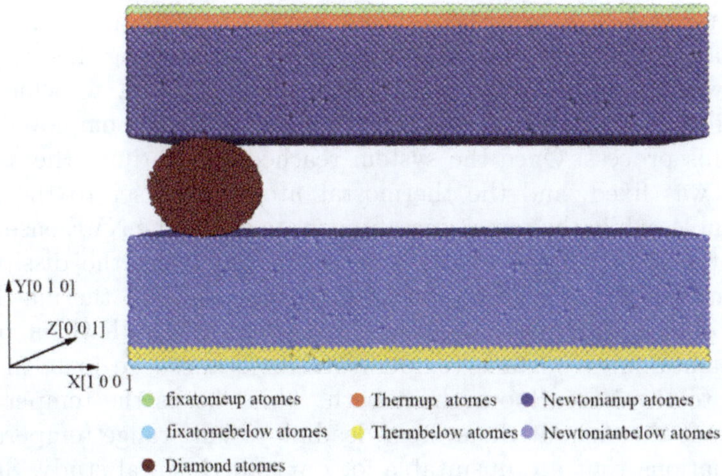

Figure 4.22. Molecular dynamics modeling of three-body single-crystal copper ellipsoidal abrasives.

copper in two blocks are fixed in place to prevent the translational movement of the copper. These fixed layers are displayed in sky blue and light green in Figure 4.22. During the related moving process, heat is generated. To dissipate the heat, constant temperature atomic layers are respectively placed above or below the fixed layers in two blocks, as shown in orange and yellow in the figure. These layers eliminate the frictional heat, ensuring that the substrate model reaches a constant temperature. The atoms in the constant temperature layers function as a heat bath, capable of absorbing or releasing heat to maintain a constant temperature. The distance between the thermostat layer and the wear surface allows the frictional heat to be gradually transferred to the heat bath and dissipated. Remember the following passage:

This process closely resembles real-world conditions. However, due to limitations in specimen size and the heat bath algorithm, it is not possible to completely avoid phonon reflection by the thermostat layer and the fixed layer in the actual calculation, leading to potential errors in the results. While some researchers have explored methods to mitigate phonon reflection from the boundaries in recent years, no fully mature method has been established thus far. In Figure 4.6, atoms represented by dark blue and blue colors are Newtonian layer atoms, which adhere to Newton's laws of motion.

In the simulation process, the entire system was initially placed in a temperature-controlled system in a heated bath to gradually increase the temperature from 0 K to 300 K. Following this, the system was allowed to relax for 50 picoseconds at 300 K to achieve an equilibrium structure. A Nosé–Hoover heat bath was employed during this process. Once the system reached equilibrium, the atomic layer was fixed, and the thermostat atoms were set to the NVT ensemble, while the remaining atoms were set to the NVE ensemble. The thermostat layer atoms were controlled using the dissipative particle dynamics method, specifically the Langevin thermal bath. This choice was made because the Langevin heat bath is better at maintaining constant temperature in localized regions, in contrast to the Nosé–Hoover heat bath, which keeps the temperature constant throughout the system, leading to long-range temperature fluctuations that are unsuitable for our experimental study. Subsequently, a specific load was applied to the simulation system to conduct molecular dynamics calculations of three-body abrasive wear

Table 4.7. Ellipsoidal three-body abrasive wear test model parameters.

Modeling	Copper matrix
Potential	EAM Embedding Potential
Sizes	Rectangular $32.6 \times 20.2 \times 14.9\,\text{nm}$
Pacemaker	0.001 ps
Initial temperature	300 K
Relaxation time	50 ps
Press-in load	40–100 nN
Movement direction	Substrate model x direction positive
Traction speed	$10\text{–}50\,\text{m}\cdot\text{s}^{-1}$
Abrasive length-to-diameter ratio	0.5–1.0

under constant load. Equal loads in opposite directions were applied to the fixed layer atoms on the upper and lower blocks, pressing the abrasive into the single-crystal copper substrate to a certain depth. Then, a specific velocity was applied to the single-crystal copper substrate on the upper plate, propelling the abrasive along the x-axis direction. Table 4.7 displays the parameters used for the molecular dynamics calculations of single-crystal copper three-body abrasive wear.

4.4.3 *Roll slip criteria for nano-ellipsoidal abrasives*

In three-body abrasive wear, the force on the abrasive is depicted in Figure 4.23. The abrasive material experiences distributed friction and normal load on the upper and lower contact surfaces, respectively. In this chapter, the distributed friction and normal load will have equivalent effects on points A and B. The friction forces on the upper and lower surfaces are denoted as F_1 and F_2, respectively, and the positive pressures are N_1 and N_2 when the abrasive material is in equilibrium:

$$F = F_1 = F_2 \tag{4.3}$$

$$N = N_1 = N_2 \tag{4.4}$$

The frictional moment $(F \cdot h)$ drives the rolling of the abrasive, while the moment of normal load $(N \cdot e)$ resists the rolling of the abrasive. When the abrasive moves in a rolling manner, the frictional

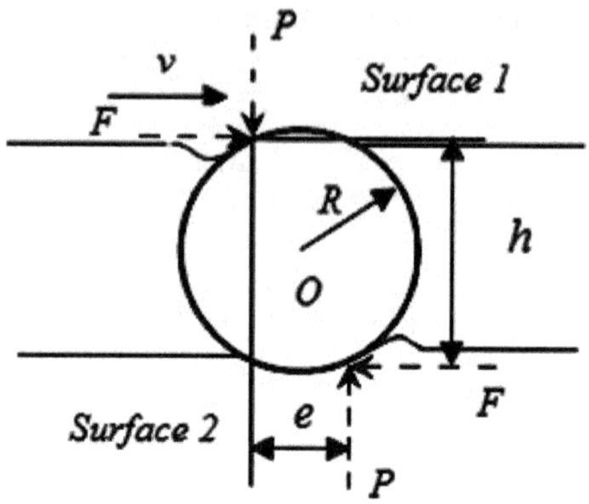

Figure 4.23. Schematic diagram of abrasive forces.

moment is greater than the rolling moment, i.e.,

$$F \cdot h > N \cdot e \tag{4.5}$$

Conversely, if the abrasive moves in a sliding motion, the frictional torque is less than the rolling torque, i.e.,

$$F \cdot h < N \cdot e \tag{4.6}$$

According to the definition of the coefficient of friction, $\mu = F/N$, the condition of abrasive rolling can be obtained as follows:

$$e/h < \mu_r \tag{4.7}$$

Similarly, the conditions for abrasive sliding are then

$$e/h > \mu_s \tag{4.8}$$

where μ_r is the coefficient of rolling friction and μ_s is the coefficient of sliding friction.

In Section 3.4, it was observed that at the nano-scale, materials exhibit significant elastic recovery during nanoindentation, unlike at the macroscopic scale where the elastic recovery of metallic materials is negligible. Therefore, to accurately predict the motion of the abrasive in nanotribological abrasion, we need to consider the role of elastic recovery. In this study, a rigid spherical diamond abrasive is placed between severely plastically deformed specimens. The radius

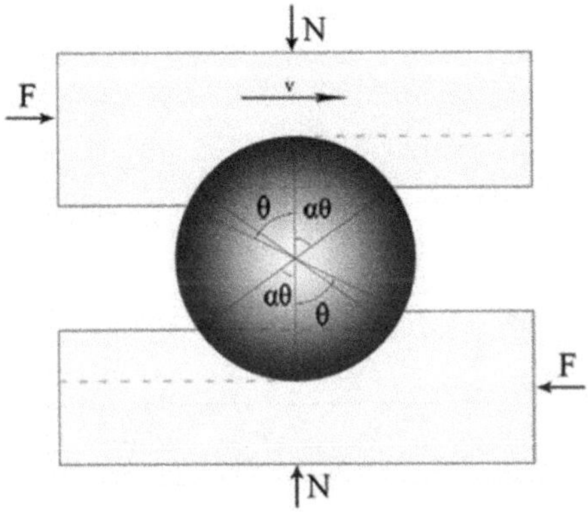

Figure 4.24. Parameters under spherical abrasives.

of the abrasive is denoted as R, the angle of indentation as θ, and the angle of elastic restitution as $\alpha - \theta$ (where α is less than 1 due to the presence of elastic recovery). The angle θ varies in the range of 0–$\pi/2$. Figure 4.24 provides a schematic diagram and parameter descriptions of the wear system in the presence of elastic recovery. The figure shows that the presence of elastic recovery introduces an additional rolling driving moment. For simplicity, the effect of this driving moment is considered as a reduction in the rolling resistance arm e.

According to the geometric conditions in Figure 4.24, the rolling resistance arm can be expressed as follows:

$$e_C = \frac{8R}{3\pi(\sin(\theta)^2 + \sin(\alpha\theta)^2)}(\sin(\theta)^3 - \sin(\alpha\theta)^3) \tag{4.9}$$

The drive force arm or friction force arm is

$$h_c = \frac{4}{3}\left(\frac{R\sin(\theta)^3}{\theta - \sin(\theta)\cos(\theta)}\right) \tag{4.10}$$

Combining equations (4.9) and (4.10) yields

$$\frac{e_C}{h_C} = \frac{2(\theta - \sin(\theta)\cos(\theta))}{\pi(\sin(\theta)^2 + \sin(\alpha\theta)^2)}\left(1 - \left(\frac{\sin(\alpha\theta)}{\sin(\theta)}\right)^3\right) \tag{4.11}$$

From equation (4.11), it can be seen that $(e/h)_C$ is a function of the indentation angle θ and the scale-free elastic recovery factor α.

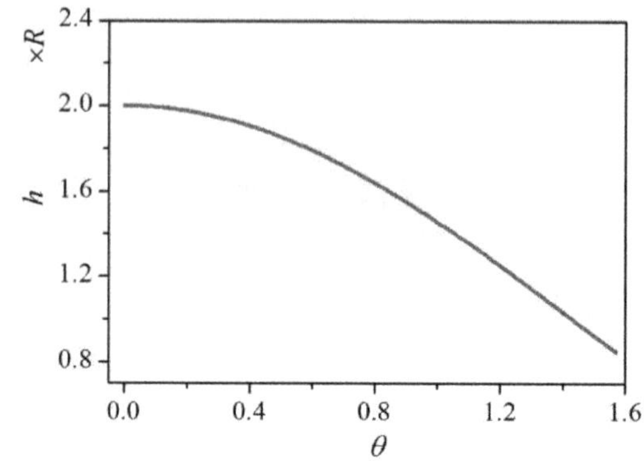

Figure 4.25. Variation of frictional moment h with press-in angle θ.

The indentation angle can be expressed in terms of the indentation depth D and the radius R of the spherical abrasive:

$$\theta = \arccos\left(\frac{R - D}{D}\right) \tag{4.12}$$

The graph in Figure 4.25 shows the relationship between the friction moment (h_C) and the indentation angle (θ). The friction moment (h_C) decreases as the indentation angle (θ) increases. When θ is close to 0 (no indentation), h_C is at its maximum value of $2R$, and when θ is at its maximum value of $\pi/2$, h_C reaches its minimum value of $8R/3\pi$. On the other hand, the normal load force arm (e_C) depends on the angle θ and the elastic unscaled recovery factor α of the material. The relationship between e_C and θ is more complex and can vary with α, so the relationship between $(e/h)_C$ and θ will not be simply increasing or decreasing in a straightforward manner.

When the scale-free elastic recovery factor α tends to 0, the model returns to that of Fang Liang [8, 13, 14]. At this point, the normal load moment is

$$e_C = \frac{8R}{3\pi}\sin(\theta)^3 \tag{4.13}$$

The frictional moment h remains constant, then $(e/h)_C$ is

$$\frac{e_C}{h_C} = \frac{2(\theta - \sin(\theta)\cos(\theta))}{\pi\sin(\theta)^2} \tag{4.14}$$

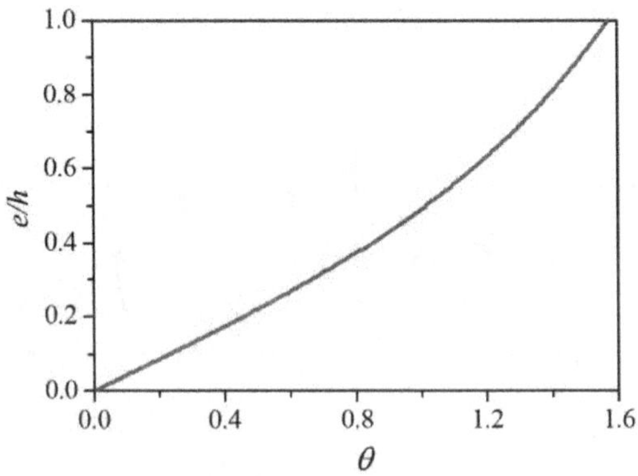

Figure 4.26. Variation of e/h with press-in angle θ in Fang Liang's model [8, 13, 14].

When analyzing equation (4.13), it is evident that e_C will be a function that steadily increases with θ. Considering h_C is a function that steadily decreases with θ, then $(e/h)_C$ will be a function that steadily increases with θ. The graph of $(e/h)_C$ concerning the indentation angle θ is shown in Figure 4.26. It is apparent from the figure that $(e/h)_C$ varies between 0 and 1 as the indentation angle ranges from 0 to $\pi/2$. This range nearly encompasses the friction coefficient of engineering materials under typical service conditions. Hence, on a macroscopic scale, there is likely to be a transition from sliding to rolling as the primary mode of motion for the abrasives. This finding aligns with Fang Liang's research [8, 13, 14].

Consider the criterion for the rolling or sliding movement of ellipsoidal abrasives in three-body abrasive wear. The ellipsoidal abrasive has two equal axis lengths, denoted as "a" and "b". When the ellipsoidal abrasive contacts the surface of a single-crystal copper substrate, the curved surface can be approximated as a spherical surface. The study shows that the error between the equivalent and actual surfaces falls within an acceptable range, indicating the feasibility of this method. For the sake of simplification, the following assumptions are made:

(a) The abrasive particles are rigid ellipsoids.
(b) The weight of the abrasive particles is not accounted for.
(c) The abrasive particles move on a main plane (which is along the x-axis) perpendicular to the material surface without rotating.

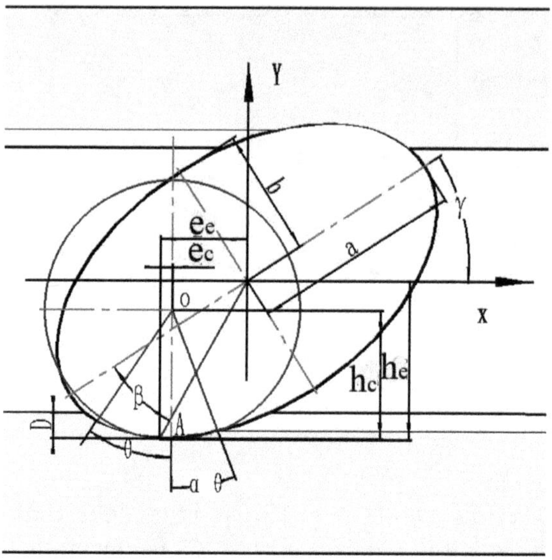

Figure 4.27. Parameters under ellipsoidal abrasives.

(d) Both abrasive force and normal pressure are evenly distributed over the contact surface.
(e) The center of the ellipsoid moves along a straight line parallel to the x-axis during rolling.

The parameters of the spherical abrasive are illustrated in Figure 4.27, with the trajectory of the ball center during rolling represented as a straight line parallel to the x-axis and the curvature of the circle denoted as (x_o, y_o).

The elliptic parametric equation is

$$\begin{cases} x = \text{acos}(\beta) \\ y = \text{bsin}(\beta) \end{cases} \tag{4.15}$$

The first- and second-order derivatives of above the elliptic equation are given by

$$y' = -\frac{b\cos(\beta)}{a\sin(\beta)} \tag{4.16}$$

$$y'' = -\frac{b}{a^2\sin(\beta)^3} \tag{4.17}$$

The radius of curvature of an ellipse at any point is

$$R = \frac{(1 + y'^2)^{\frac{3}{2}}}{|y''|} \tag{4.18}$$

Bringing Eqs. (4.15), (4.16), and (4.17) into Eq. (4.18) gives the radius of curvature as follows:

$$R = \frac{[(a^2 - b^2)\sin(\beta)^2 + b^2]^{\frac{3}{2}}}{ab} \tag{4.19}$$

The coordinates of the center of curvature are

$$\begin{cases} x_o = x - \dfrac{y'(1 + y'^2)}{y''} \\ y_o = y + \dfrac{1 + y'^2}{y''} \end{cases} \tag{4.20}$$

Bringing equations (4.15), (4.16), and (4.17) into equation (4.20) gives the coordinates of the center of the circle of curvature as follows:

$$\begin{cases} x_o = \dfrac{\cos(\beta)^3(a^2 - b^2)}{a} \\ y_o = \dfrac{\sin(\beta)^3(b^2 - a^2)}{b} \end{cases} \tag{4.21}$$

Rotating the elliptic equation by γ yields the rotated elliptic equation as

$$\begin{cases} x_1 = a\cos(\beta)\cos(\gamma) - b\sin(\beta)\sin(\gamma) \\ y_1 = a\cos(\beta)\sin(\gamma) + b\sin(\beta)\cos(\gamma) \end{cases} \tag{4.22}$$

Then, the center of curvature of the rotating ellipse (x_{o1}, y_{o1}) can be expressed as:

$$\begin{cases} x_{o1} = \dfrac{\cos(\beta)^3(a^2 - b^2)}{a}\cos(\gamma) - \dfrac{\sin(\beta)^3(b^2 - a^2)}{b}\sin(\gamma) \\ y_{o1} = \dfrac{\cos(\beta)^3(a^2 - b^2)}{a}\sin(\gamma) + \dfrac{\sin(\beta)^3(b^2 - a^2)}{b}\cos(\gamma) \end{cases} \tag{4.23}$$

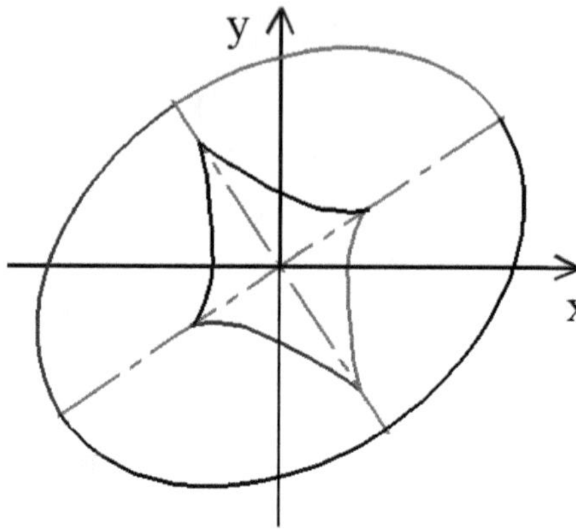

Figure 4.28. Trajectory of the curvature circle center of a rotating ellipse.

From the curvature circle trajectory image of the rotating ellipse in Figure 4.28, it can be seen that for the ellipsoidal abrasive, its resistance arm is

$$e_e = e_c + x_{o1}\mathrm{sign}(x_1) \tag{4.24}$$

The drive arm is

$$h_e = h_c + y_{o1}\mathrm{sign}(y_1) \tag{4.25}$$

Then,

$$e_e/h_e = \frac{e_c + x_{o1}\mathrm{sign}(x_1)}{h_c + y_{o1}\mathrm{sign}(y_1)} \tag{4.26}$$

Carrying equations (4.9), (4.10), (4.19), (4.22), (4.23) into (4.26) yields

$$e_e/h_e = \frac{\frac{8[(a^2-b^2)\sin(\beta)^2+b^2]^{\frac{3}{2}}(\sin(\theta)^3-\sin(\alpha\theta)^3)}{3\pi ab(\sin(\theta)^2+\sin(\alpha\theta)^2)} + \left[\frac{\cos(\beta)^3(a^2-b^2)}{a}\cos(\gamma) - \frac{\sin(\beta)^3(b^2-a^2)}{b}\sin(\gamma)\right]\mathrm{sign}(x_1)}{\frac{4[(a^2-b^2)\sin(\beta)^2+b^2]^{\frac{3}{2}}\sin(\theta)^3}{3ab[\theta-\sin(\theta)\cos(\theta)]} + \left[\frac{\cos(\beta)^3(a^2-b^2)}{a}\sin(\gamma) + \frac{\sin(\beta)^3(b^2-a^2)}{b}\cos(\gamma)\right]\mathrm{sign}(y_1)} \tag{4.27}$$

where $\theta = \arcos\left(\sqrt{1 - \frac{Fa^2b^2}{\pi H[(a^2-b^2)\sin(\beta)^2+b^2]}}\right.$, a is the length of the long axis of the ellipse/Å, b is the length of the short axis of the ellipse/Å, γ is the rotation angle of the ellipse, F is the positive load/nN, H is the material hardness/GPa, β is the parameter of the elliptic equation, and α is the coefficient of elastic recovery of the material.

From the magnitude of the $(e/h)_e$ value and the magnitude of the coefficient of friction, it can be concluded that when $(e/h)_e < \mu$, the abrasive moves in a rolling manner. When $(e/h)_e > \mu$, the abrasive moves in a sliding manner.

When the ellipsoidal abrasive has equal axis lengths on all three coordinate axes, i.e., $a = b$, the abrasive is spherical in shape, and the simplification of equation (4.27) gives

$$
\begin{aligned}
e_e/h_e &= \frac{\frac{8b(\sin(\theta)^3-\sin(\alpha\theta)^3)}{3\pi(\sin(\theta)^2+\sin(\alpha\theta)^2)}}{\frac{4b\sin(\theta)^3}{3[\theta-\sin(\theta)\cos(\theta)]}} \\
&= \frac{2(\sin(\theta)^3 - \sin(\alpha\theta)^3)[\theta - \sin(\theta)\cos(\theta)]}{\pi(\sin(\theta)^2 + \sin(\alpha\theta)^2)\sin(\theta)^3} \\
&= \frac{2[\theta - \sin(\theta)\cos(\theta)]}{\pi(\sin(\theta)^2 + \sin(\alpha\theta)^2)}\left(1 - \left(\frac{\sin(\alpha\theta)}{\sin(\theta)}\right)^3\right) \qquad (4.28)
\end{aligned}
$$

This result confirms the roll–slip criterion equation for spherical abrasives as proposed in the literature [15]. It demonstrates that the derivative of the criterion is accurate and also indicates that spherical abrasives can be considered a special case of ellipsoidal abrasives. Initially, the ellipsoidal abrasive is placed flat. As the specimen slides at a certain speed, the abrasive will be rotated to a certain angle. Therefore, under specific conditions, it is meaningful to discuss the relationship between the aspect ratio $(e/h)_e$ and the rotation angle θ. For convenience, all of the following refer back to the application of e/h instead of $(e/h)_e$.

As depicted in Figure 4.29, for a pressure angle of 60° and an elastic recovery angle of 30°, the relationship between the aspect ratio (e/h) and the rotation angle can be observed. It can be seen that when the abrasive material is rotated to a certain angle, the aspect ratio (e/h) reaches its peak value. This means that the size of

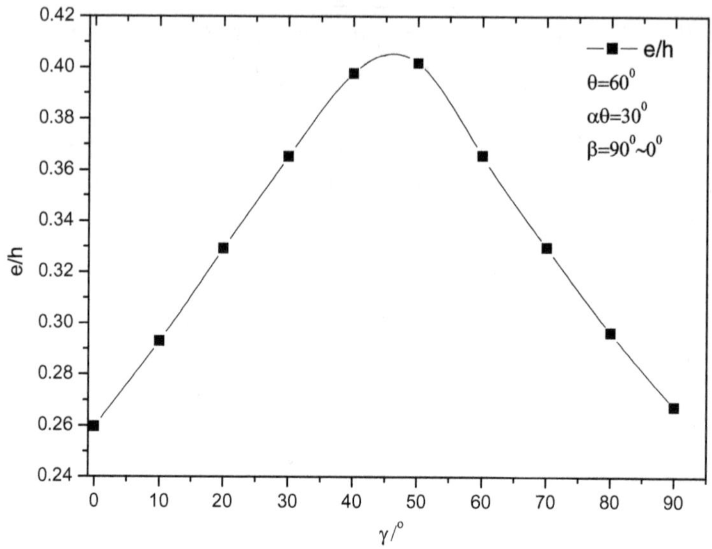

Figure 4.29. Variation curve of e/h under different rotation angles.

the friction coefficient determines the movement style of the abrasive particle. If the value exceeds the friction coefficient, the abrasive will slide. If the maximum value is lower than the friction coefficient, the abrasive will roll.

The impact of elastic recovery on the rolling of the abrasive particle at the nanoscale is illustrated in Figure 4.30 according to equation (4.28). The graph displays the variation of e/h with the elastic recovery coefficient under specific conditions: the angle of the indentation is 60°, β is 90°, the long axis length of the ellipsoidal abrasive particle is 30 Å, and the short axis length is 25 Å. Based on the figure, it is evident that as the elastic recovery coefficient increases, the value of e/h gradually decreases, causing the abrasive to tend to move in a rolling mode. Figure 4.31 depicts the change of e/h with the β angle under the conditions of a 60° indentation angle, 0.5 elastic recovery coefficient, 30 Å long axis length, and 25 Å short axis length of the ellipsoidal abrasive. The figure shows that as the β angle increases, the e/h value gradually decreases, indicating that the abrasive tends to move more in a rolling mode.

Based on contact mechanics analysis, the force in the nano-ellipsoidal abrasive is determined by examining the interaction

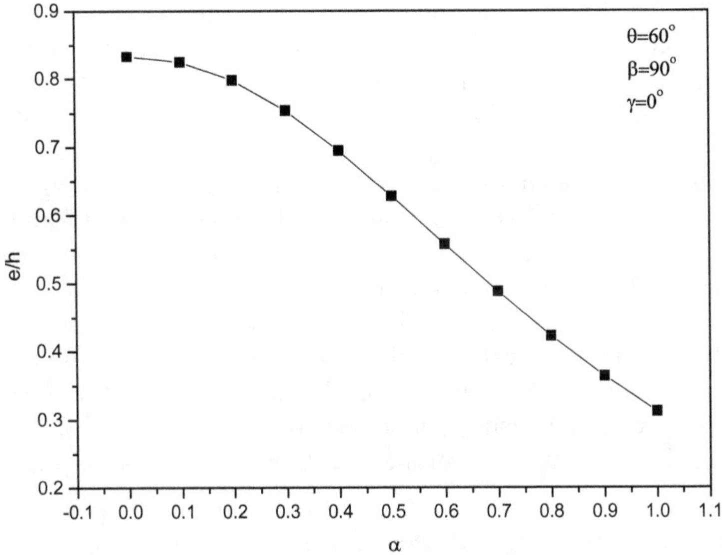

Figure 4.30. Curve of variation of e/h value with elastic recovery factor.

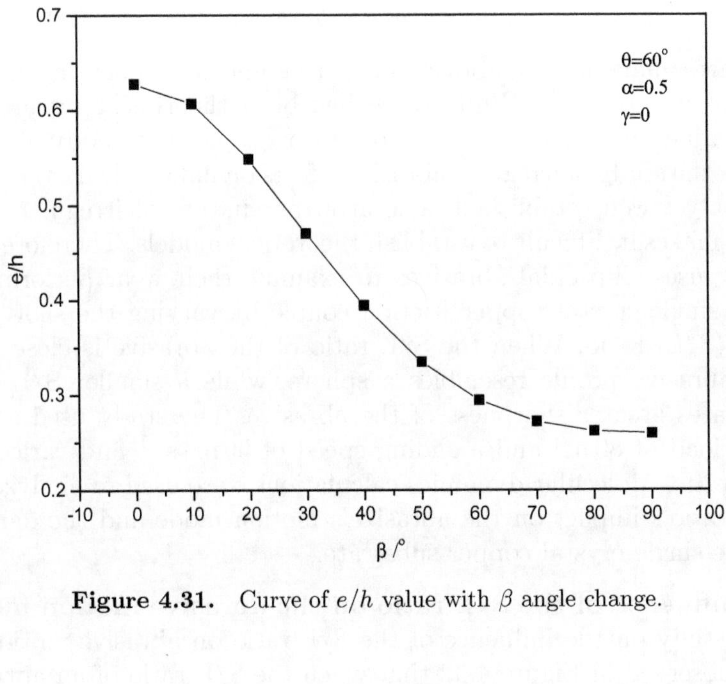

Figure 4.31. Curve of e/h value with β angle change.

between the abrasive and the surface of the friction surface. By considering the geometric relationship between the abrasive and the contacting surface, along with the geometric parameters of the abrasive shape, the mode of motion for the ellipsoidal abrasive was established using the equivalent circle approximation method. The discriminant equations were calculated and confirmed to be in close agreement with the actual abrasive motion, based on aspect ratio and traction speed conditions. Analysis of the equation parameters revealed that when the rotation angle of the abrasive material is 45°, the e/h value of the abrasive material is the largest. If the coefficient of friction exceeds the e/h value at this point, the motion of the abrasive material is rolling, otherwise, it is sliding. The e/h value decreases gradually with increasing elastic recovery coefficient and β angle, making the abrasive more prone to rolling.

4.4.4 *Nanotribological abrasive wear of single-crystal copper under air*

4.4.4.1 *The effect of short-long axis ratio on the abrasive motion and abrasive wear characteristics*

The irregular shape of abrasives has a significant impact on the wear of the friction vice. While there has been theoretical research on the influence of abrasive shape on macroscopic three-body abrasive wear, the study of irregular abrasive effects on nanoscale friction wear is not yet well established. The arbitrary nature of irregular abrasives makes it difficult to establish theoretical models. Therefore, this study uses ellipsoidal abrasives to examine their wear performance on a single-crystal copper friction couple by varying the short-long axis (S/L) ratio. When the S/L ratio of the abrasive is close to 1, the abrasive profile resembles a sphere, while a smaller S/L ratio indicates greater sharpness of the abrasive. The study used a positive load of 80 nN and a sliding speed of $50\,\mathrm{m\cdot s^{-1}}$ and varied the S/L ratio. Molecular dynamics calculations were used to analyze the S/L ratio's impact on the abrasive's motion mode and the damage to the single-crystal copper substrate.

(1) **Influence of the S/L ratio on the abrasive motion mode:** In a study on the influence of the S/L ratio on abrasive motion, it was observed in Figure 4.32 that when the S/L ratio of an abrasive

Figure 4.32. Atomic configurations of abrasive positions at different aspect ratios.

material is less than 0.83, the abrasive exhibits consistent sliding motion over time. However, when the S/L ratio is greater than or equal to 0.83, the moving angle of the abrasive changes over time, and it begins to move in a rolling manner. This was confirmed by the calculated curve of the angular velocity of rotation of the abrasive material over time, as shown in Figure 4.33, which showed that when the S/L ratio is less than 0.83, the rotational angular velocity fluctuates around zero, indicating sliding motion. On the other hand, when the S/L ratio is greater than or equal to 0.83, the angular velocity of rotation of the abrasive material is not zero, signifying rolling motion. Additionally, a negative angular velocity of rotation indicates rotation in the opposite direction.

In brief, when the load is 80 nN and the sliding speed is $50\,\mathrm{m \cdot s^{-1}}$, if the S/L ratio of the abrasive is less than or equal to 0.83, the abrasive moves in a sliding mode. If the S/L ratio of the abrasive is greater than or equal to 0.83, the abrasive moves in a rolling mode.

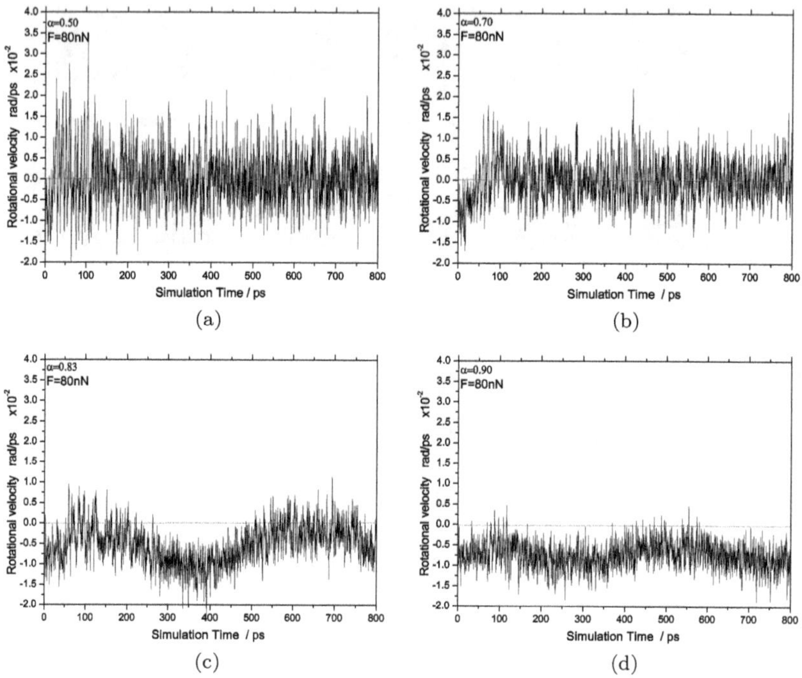

Figure 4.33. Angular velocity of rotation of abrasive at the same aspect ratio. (a) $\alpha = 0.50$, (b) $\alpha = 0.70$, (c) $\alpha = 0.83$, (d) $\alpha = 0.90$.

(2) **Effect of S/L ratio on friction force and friction coefficient:** In molecular dynamics simulations, the applied positive load is always maintained constant. Altering the S/L ratio of the abrasive material leads to a change in the friction force applied to the abrasive material and consequently, a change in the coefficient of friction of the abrasive. In three-body abrasive wear, the friction force drives the motion of the abrasive, and variations in the friction force will impact how the abrasive material moves. Figure 4.34 illustrates the relationship between positive load and friction with sliding distance for different S/L ratios. The figure indicates that the average positive load remains constant, fluctuating around 80 nN as the S/L ratio changes. With an increase in the S/L ratio, the friction curve transitions from a smooth curve to a sinusoidal curve due to the ellipsoidal shape of the abrasive material. During the rolling process, the rotation angle of the abrasive varies, causing the friction curve to exhibit a sinusoidal trend.

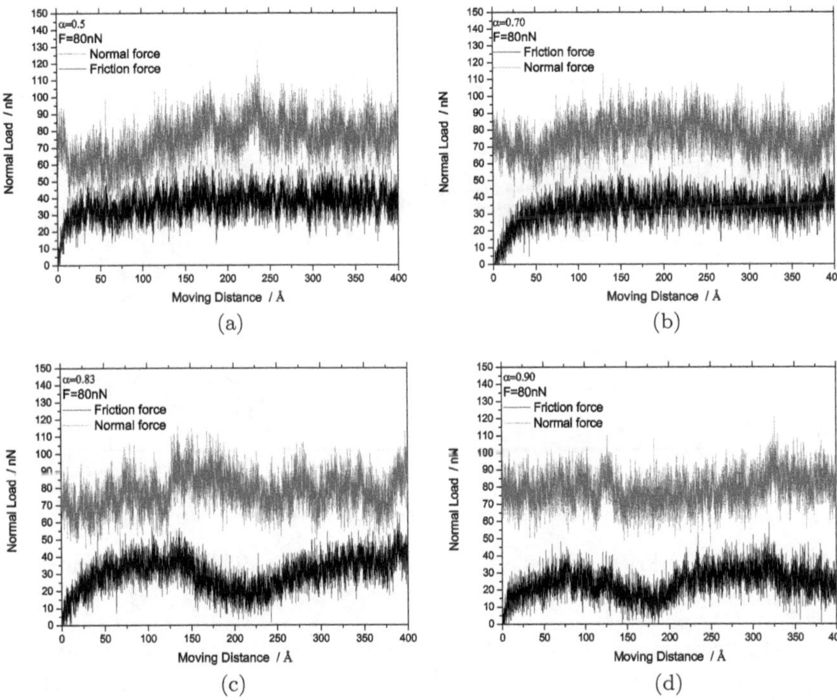

Figure 4.34. Positive load versus friction for different S/L ratios. (a) $\alpha = 0.50$, (b) $\alpha = 0.70$, (c) $\alpha = 0.83$, (d) $\alpha = 0.90$.

In Figure 4.35, the friction coefficient and the e/h curves are shown at different S/L ratios. The figure illustrates that when the S/L ratio is less than 0.83, the friction coefficient gradually increases slightly with an increase in the S/L ratio. This is because the abrasive S/L ratio increases, leading to a larger contact area and increased shear stress between the abrasive and the single-crystal copper, causing an increase in friction force and consequently, the friction coefficient. However, when the S/L ratio exceeds 0.83, the coefficient of friction decreases rapidly due to a change in the abrasive's movement mode, from sliding to rolling. This change in motion mode causes the coefficient of friction to decrease rapidly, aligning with the wear performance at a macroscopic level. The motion mode of the abrasive can be determined by comparing the relationship between e/h and the coefficient of friction. Specifically, when the abrasive moves in sliding mode (S/L ratio < 0.83), the e/h of the abrasive is greater

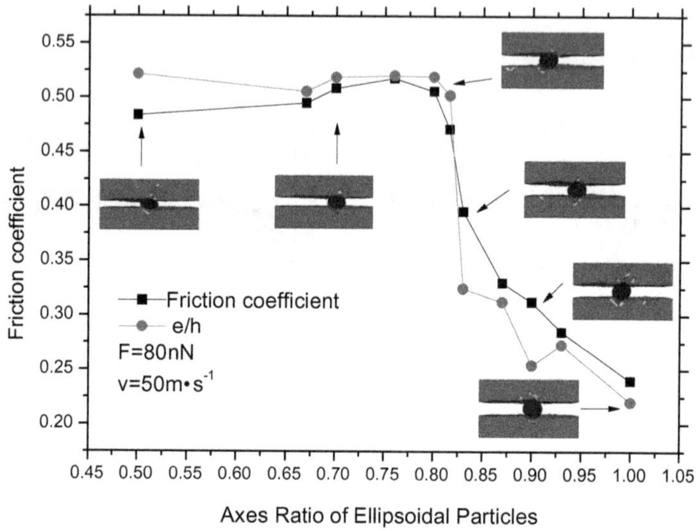

Figure 4.35. Friction coefficient and e/h curves for different aspect ratios.

than the friction coefficient. Conversely, when the abrasive moves in rolling mode (S/L ratio ≥ 0.83), the e/h of the abrasive is less than the friction coefficient. This aligns with the judgment of the abrasive motion mode in the macroscopic state, indicating that the roll–slip judgment of the abrasive at the macroscopic state is suitable for the judgment of the abrasive motion mode at the nanometer scale.

(3) **Nano-Abrasion damage for single-crystal copper by varying the S/L ratio:** Changes in the S/L ratio affect the sharpness of the abrasive, which in turn impacts the extent of damage to the single-crystal copper substrate. On a macroscopic level, an increase in abrasive sharpness results in more severe damage to the substrate. At the nanoscale, it is worth discussing the influence of abrasive sharpness on substrate damage.

Figure 4.36 illustrates the damage on the surface of a single-crystal copper substrate at different S/L ratios. The figure shows that as the S/L ratio increases, the accumulation of atoms on groove ridges on the single-crystal copper surface decreases. The quantity of atomic buildup reflects the extent of abrasive damage to the single-crystal copper substrate. A greater height of atomic buildup indicates more serious damage to the single-crystal copper. When the

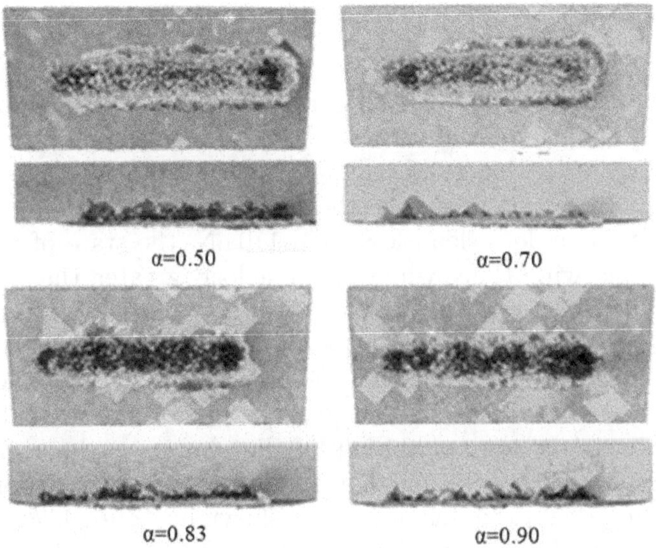

Figure 4.36. Surface damage of single-crystal copper matrix at different S/L ratios.

S/L ratio is small (≤0.83), the abrasive moves in a sliding manner, resulting in more atomic accumulation around the scratches on the substrate surface. In contrast, when the S/L ratio is larger than 0.83, the abrasive moves in a rolling manner, and there is very little atomic accumulation on the surface of the substrate. It is evident that as the S/L ratio increases, the damage of the abrasive to the single-crystal copper substrate lessens. The damage to the single-crystal copper substrate is more severe when the abrasive moves in sliding mode compared to when it moves in the rolling mode.

4.4.4.2 *Effect of load on abrasive kinematic properties and material wear performance*

In this research study, we examined the impact of consistent load and ellipsoidal abrasive with the same S/L ratio. We found that the load affects both the depth of the abrasive pressed into the substrate and the friction, including the coefficient of friction. Specifically, we maintained a fixed S/L ratio of 0.83 and a sliding speed of $50\,\mathrm{m\cdot s^{-1}}$ while varying the positive load. Our study discussed the state of motion of the abrasive and the extent of damage to single-crystal copper.

(1) **The effect of load on the motion mode of the abrasive:** When the sliding speed is $50\,\text{m}\cdot\text{s}^{-1}$, the atomic configuration of the abrasive with the change of load is as shown in Figure 4.37. Observing the state of the abrasive in the atomic configuration diagram, it is found that under a load less than 110 nN, the state of the abrasive changes with time, while under a load greater than or equal to 110 nN, the state of the abrasive does not change with time. This means that the abrasive moves in a sliding mode under a load greater than or equal to 110 nN and in a rolling mode under a load less than 110 nN. The calculated angular velocity of rotation of the abrasive further illustrates the correctness of the motion of the abrasive as judged by visual observation. The angular velocity of rotation of the abrasive material for different loads is shown in Figure 4.38. At loads less than 110 nN, the angular velocity of rotation of the abrasive is not zero, indicating that the abrasive rotates during motion and moves in a rolling manner. At loads greater than or equal to 110 nN,

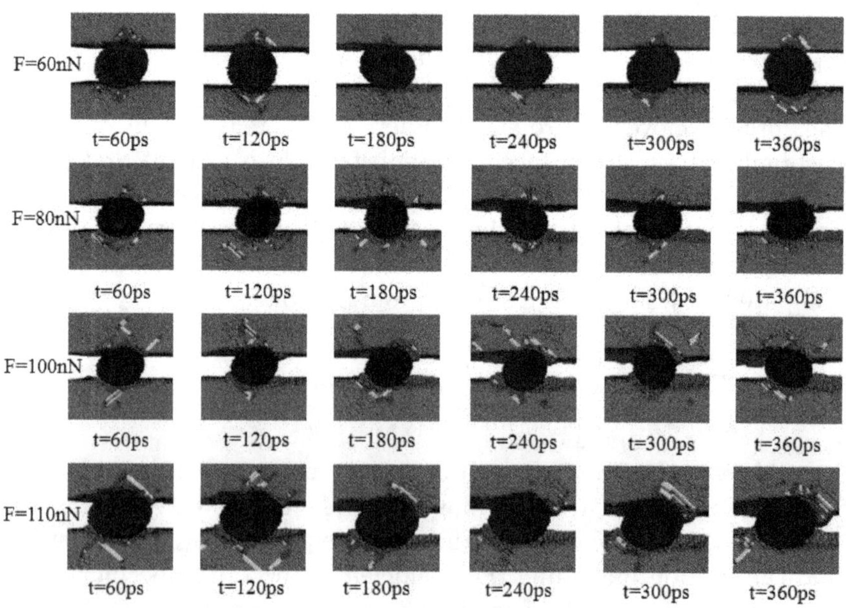

Figure 4.37. Plot of atomic configurations at different loads.

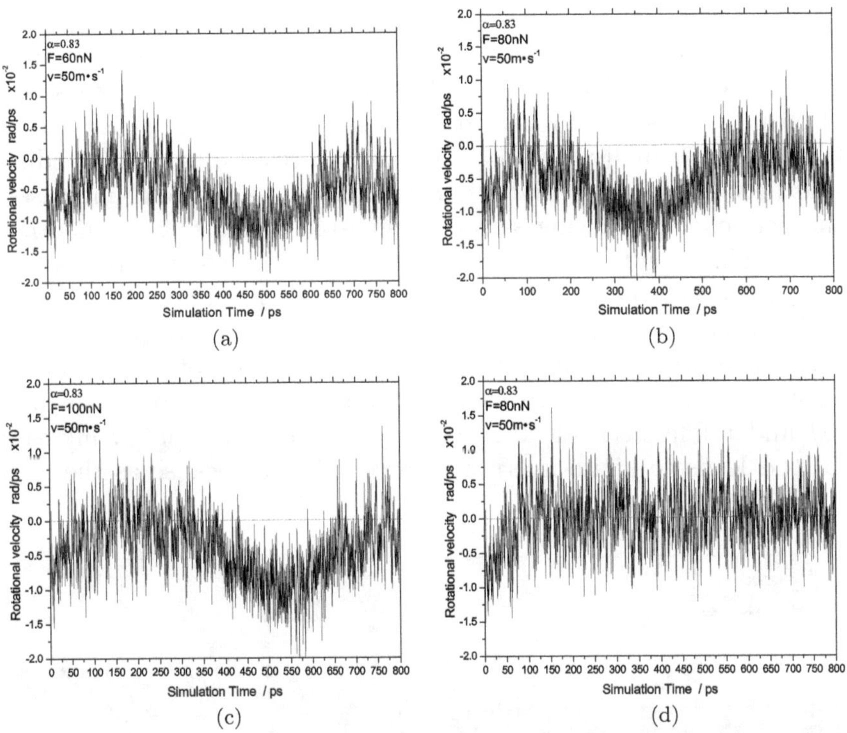

Figure 4.38. Variation curve of rotation angular velocity of abrasive with time under different loads. (a) $F = 60\,\text{nN}$, (b) $F = 80\,\text{nN}$, (c) $F = 100\,\text{nN}$, (d) $F = 110\,\text{nN}$.

the angular velocity of rotation of the abrasive material fluctuates around zero, and its average value is zero, indicating that the abrasive does not rotate during its motion and moves in a sliding manner.

In short, when the S/L ratio of the abrasive material is 0.83, and the sliding speed is at $50\,\text{m}\cdot\text{s}^{-1}$, the abrasive will be sliding if the positive load on the abrasive is greater than or equal to 110 nN. However, if the positive load is less than 110 nN, the abrasive motion will be rolling.

(2) **The effect of load on friction force and friction coefficient:** The amount of positive load placed on an abrasive affects the depth of its indentation, which in turn influences the friction coefficient and the way the material moves. For an elliptical abrasive under a constant load, a critical S/L ratio of 0.83 determines whether the

material undergoes a transition from rolling to sliding. Therefore, abrasive materials with an S/L ratio of 0.83 are specifically studied. Under a constant load at an S/L ratio of 0.83 and a sliding speed of $50\,\mathrm{m \cdot s^{-1}}$, as the applied load increases, the initial resistance to movement gradually rises, but the drop in resistance during sliding also becomes more significant. Additionally, as the load increases, the friction curve changes from a sinusoidal shape to a smoother curve. This results in a gradual increase in the amount of friction experienced by the abrasive material, as shown in Figure 4.39.

Friction is responsible for the abrasive motion, and any change in friction will also impact the roll–slip pattern of the abrasive. Figure 4.40 displays the variation curves of friction coefficient and e/h under different loads. It is evident from the figure that an increase in load leads to a rise in friction force. This is because as the load

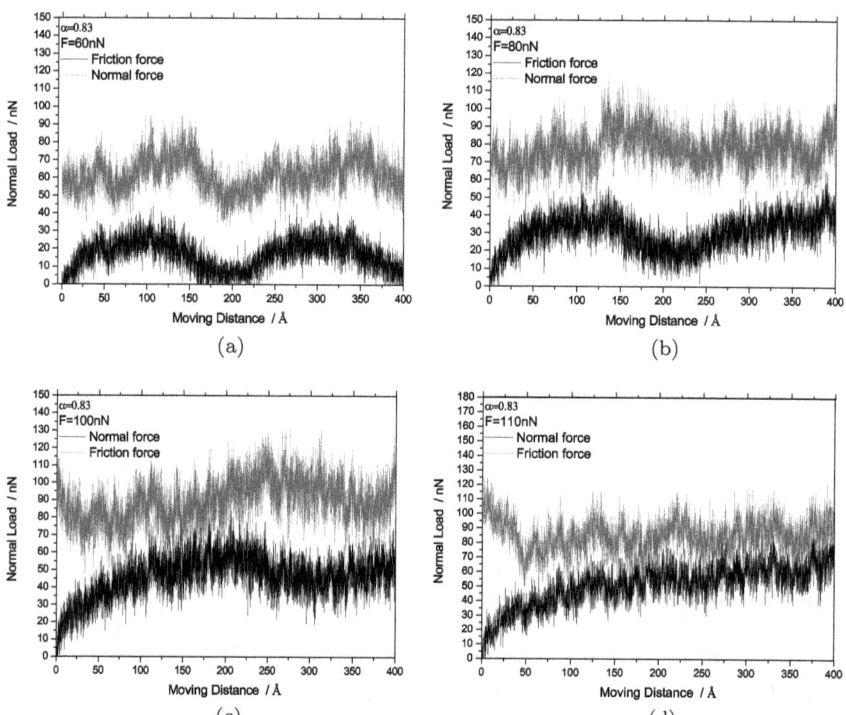

Figure 4.39. Normal load and friction under different loads. (a) $F = 60\,\mathrm{nN}$, (b) $F = 80\,\mathrm{nN}$, (c) $F = 100\,\mathrm{nN}$, (d) $F = 110\,\mathrm{nN}$.

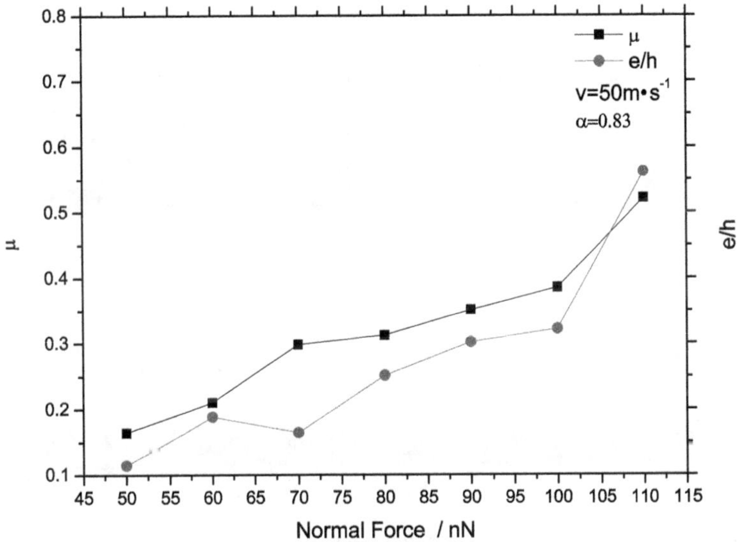

Figure 4.40. Plot of e/h versus coefficient of friction for different loads.

increases, the abrasive is pressed deeper into the substrate, resulting in greater shear force and resistance at the front of the abrasive material, causing the increase of friction coefficient. However, when the load exceeds 110 nN, the coefficient of friction experiences a sharp increase due to the transition in abrasive motion from rolling to sliding. From Figure 4.40, it can also be observed that when the abrasive moves in a sliding mode and the load is less than 110 nN, the e/h of the abrasive is greater than the coefficient of friction. On the other hand, when the abrasive moves in a rolling mode and the load is equal to or greater than 110 nN, the e/h of the abrasive is smaller than the coefficient of friction. This observation aligns with the judgment of the abrasive motion mode at the macroscale, indicating that the roll–slip judgment of the abrasive at the macroscale is applicable for judging the abrasive motion mode at the nanoscale.

When the load increases, the motion of the abrasive changes from rolling to sliding. This change in load also affects the damage of the abrasive to the single-crystal copper substrate. As the load increases, the depth at which the abrasive is pressed into the single-crystal copper substrate also increases. In the sliding process, the abrasive causes more serious damage to the substrate.

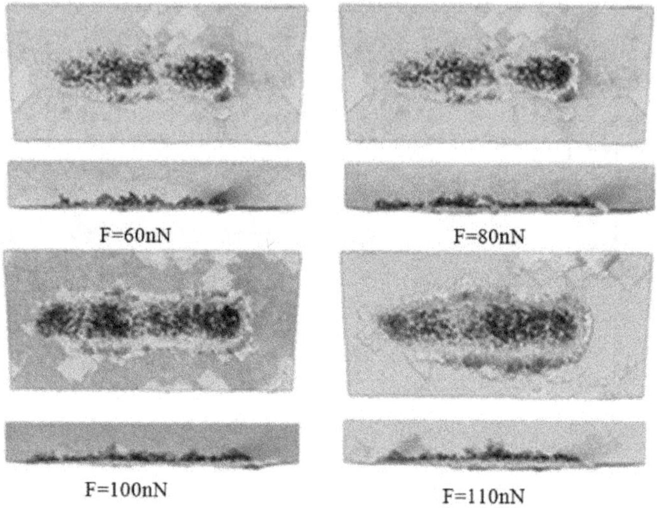

Figure 4.41. Damage on the surface of single-crystal copper matrix under different loads.

(3) **Damage under load on single-crystal copper nano three-body abrasion:** In a study on the nano-abrasive wear of single-crystal copper, Figure 4.41 shows that as the load increases, the depth of scratches on the copper surface gradually increases, as does the height of atomic accumulation on groove ridges. At a load of 110 nN, the atomic accumulation height reaches its maximum, indicating the most serious scratch damages to the copper surface. This suggests that higher loads lead to more severe damage to the single-crystal copper matrix. Additionally, when the abrasive moves by sliding, the damage to the copper substrate is more severe compared to when it moves by rolling. In industrial applications, for three-body abrasive grinding, using a high load for sliding movement of the abrasive can enhance grinding efficiency.

4.4.4.3 *Effect of sliding speed of single-crystal copper on abrasive kinematic properties and material wear performance*

The tests showed that the speed at which the surface is sliding affects how the abrasive moves the mode. Changes in sliding speed affected the interaction time between the abrasive and the single-crystal copper substrate, which in turn impacted the deformation of the substrate and the friction force and friction coefficient during wear. A load of 80 nN and an S/L ratio of 0.83 were chosen, and the sliding speed was adjusted to study the behavior of the abrasive and the damage to the copper substrate.

(1) **Impact of sliding speed on abrasive motion mode:** Under the conditions of an S/L ratio of 0.83, a positive load of 80 nN applied on the abrasive, atomic configuration diagrams at different speeds were obtained through molecular dynamics calculations and visualization methods, as shown in Figure 4.42. Based on the diagram, it is apparent that the tilt angle of the abrasive remains unchanged over time, and the particle moves by sliding when the sliding speed of the specimen is $15\,\mathrm{m\cdot s^{-1}}$ or less. However, when the specimen's sliding speed exceeds $15\,\mathrm{m\cdot s^{-1}}$, the tilt angle of the abrasive material changes over time, and it can be observed that the particle moves by rolling. Figure 4.43 displays the calculated rotational angular velocity of the abrasive material. This figure indicates that when the speed is $15\,\mathrm{m\cdot s^{-1}}$ or less, the rotational angular velocity of the abrasive is zero, implying that the abrasive does not rotate during wear

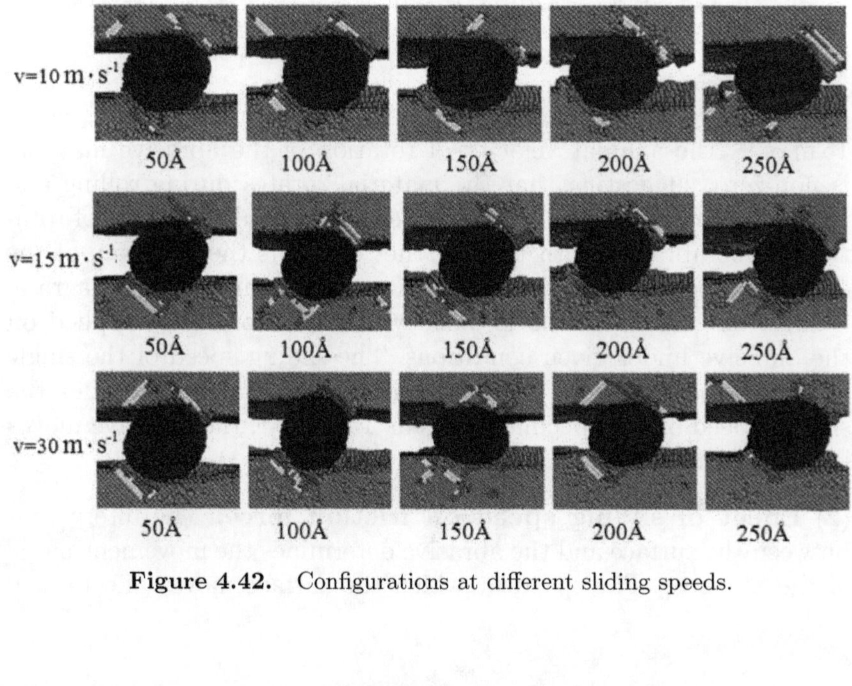

Figure 4.42. Configurations at different sliding speeds.

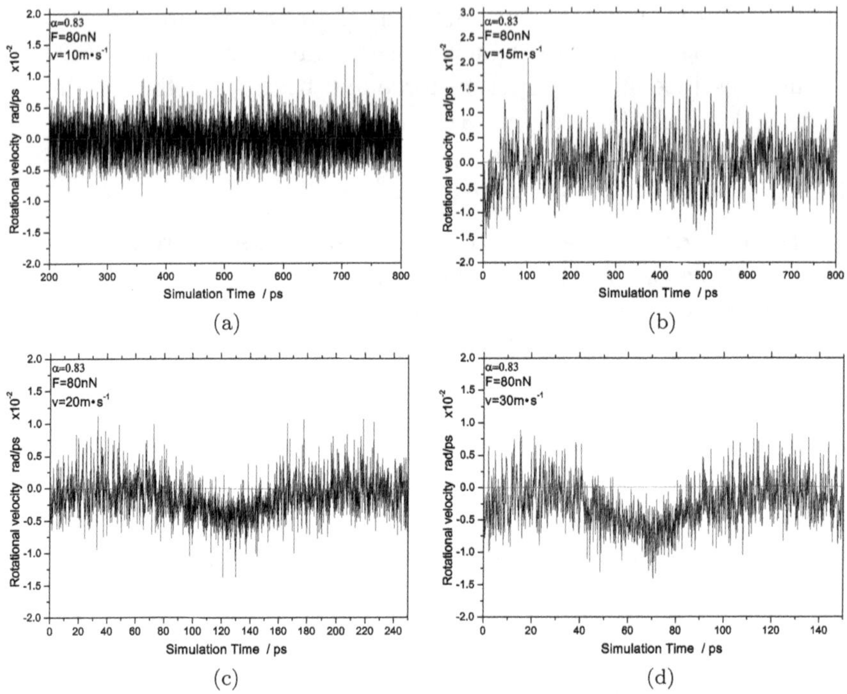

Figure 4.43. Curve of rotation angular velocity of abrasive with time for different sliding speeds. (a) $v = 10\,\mathrm{m \cdot s^{-1}}$, (b) $v = 15\,\mathrm{m \cdot s^{-1}}$, (c) $v = 20\,\mathrm{m \cdot s^{-1}}$, (d) $v = 30\,\mathrm{m \cdot s^{-1}}$.

and moves by sliding. On the other hand, when the speed exceeds $15\,\mathrm{m \cdot s^{-1}}$, the angular velocity of rotation of the abrasive material is non-zero, suggesting that the material rotates during rolling and moves in a rolling manner. The curve of the angular velocity of rotation of the abrasive with time further confirms the accuracy of the observed abrasive motion through the atomic configuration diagram.

In brief, the S/L ratio is 0.83, with a load of 80 nN applied on the abrasive under given conditions. The sliding speed of the single crystal copper specimen is less than or equal to $15\,\mathrm{m \cdot s^{-1}}$. When the sliding speed of the specimen exceeds $15\,\mathrm{m \cdot s^{-1}}$, the abrasive moves in a rolling mode.

(2) **Effect of sliding speed on friction force:** The interaction between the surface and the abrasive determines the movement mode of the abrasive. The speed at which the surface moves affects how

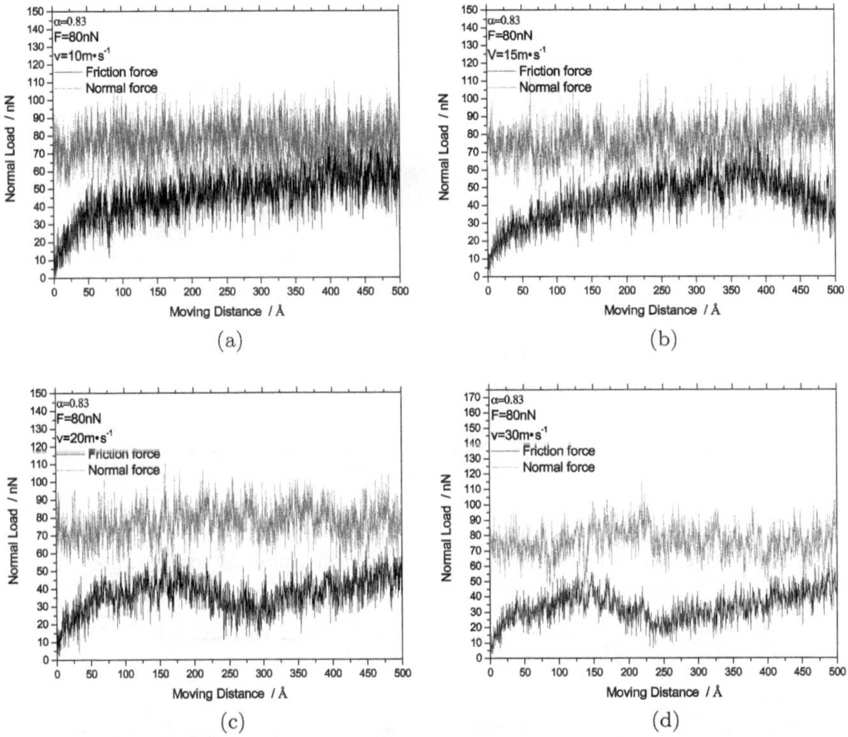

Figure 4.44. Load and friction force curves of single-crystal copper at different sliding speeds. (a) $v = 10\,\mathrm{m \cdot s^{-1}}$, (b) $v = 15\,\mathrm{m \cdot s^{-1}}$, (c) $v = 20\,\mathrm{m \cdot s^{-1}}$, (d) $v = 30\,\mathrm{m \cdot s^{-1}}$.

the abrasive moves. When the surface moves faster under a constant load, the abrasive tends to roll. When the surface moves slower, the abrasive tends to slide. Figure 4.44 illustrates the load and friction force curves at different sliding speeds. The figure shows that when the sliding speed is $15\,\mathrm{m \cdot s^{-1}}$ or less, the friction force curve follows even and smooth. However, when the sliding speed exceeds $15\,\mathrm{m \cdot s^{-1}}$, the friction force curve becomes almost sinusoidal, and the friction force decreases as the sliding speed increases.

In Figure 4.45, the graphs show the relationship between friction coefficient and e/h values at different sliding speeds. It is evident that as the sliding speed increases, the friction coefficient between the abrasive and the single-crystal copper substrate and indented depth into the substrate gradually decreases. Figure 4.46 also illustrates

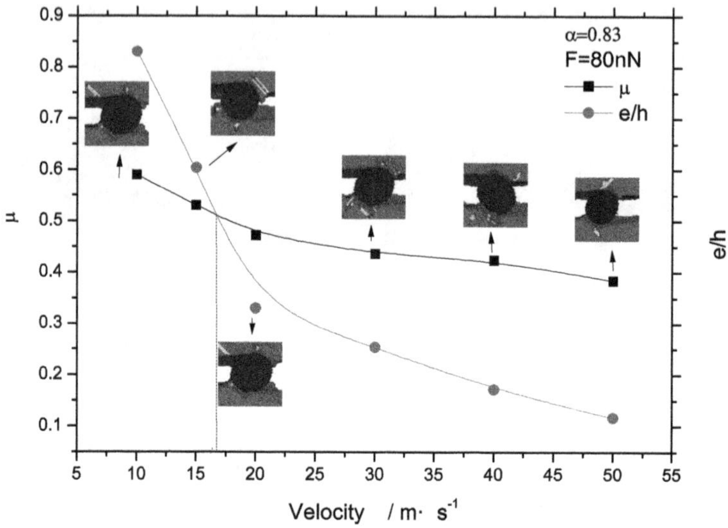

Figure 4.45. Friction coefficient curves for different towing speeds.

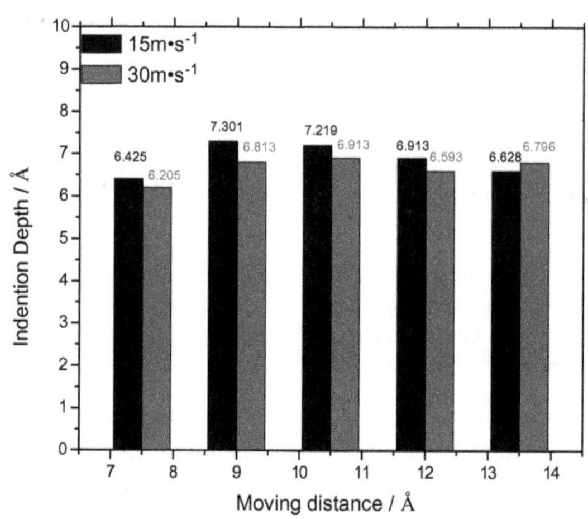

Figure 4.46. Depth of abrasive indentation into the substrate at the same distance of specimen sliding at different towing speeds.

that the depth of the abrasive pressed into the substrate decreases with increasing speed, particularly at speeds of $15\,\mathrm{m \cdot s^{-1}}$ and $30\,\mathrm{m \cdot s^{-1}}$. When the speed is greater, the depth of the abrasive pressed into the substrate decreases, resulting in a decrease in the contact area between the abrasive and the substrate, consequently reducing the friction and the coefficient of friction. Additionally, when the sliding speed exceeds $15\,\mathrm{m\cdot s^{-1}}$, the movement mode of the abrasive transitions from sliding to rolling, causing a decrease in the friction coefficient. The figures demonstrate that when the abrasive moves in a sliding mode at speeds less than or equal to $15\,\mathrm{m\cdot s^{-1}}$, and the e/h is greater than the friction coefficient, the abrasive is in sliding mode. On the contrary, the abrasive is in a rolling mode at speeds greater than $15\,\mathrm{m\cdot s^{-1}}$. The relationship between the sliding speed and the e/h value aligns with the criterion for the abrasive movement mode at the macro scale, indicating that the rolling–sliding criterion at the macro scale is suitable for determining the abrasive movement mode at the nanoscale.

The speed also determines the mode of abrasive movement. When the speed is low, the abrasive stays in contact with the substrate for a longer time, resulting in a greater depth of penetration into the substrate. Conversely, when the speed is high, the depth of penetration is lower. In single-crystal copper substrates, it was observed in Figure 4.30 that lower penetration depth leads to a higher elastic recovery coefficient. This results in the abrasive pressing less into the substrate with greater elastic recovery, leading to a smaller e/h value and a tendency for the substrate to roll. Furthermore, higher sliding speeds lead to a smaller height of the abrasive front stack, as shown in Figure 4.47. The abrasive and the point where it meets the substrate move in the direction of the short axis of the ellipse,

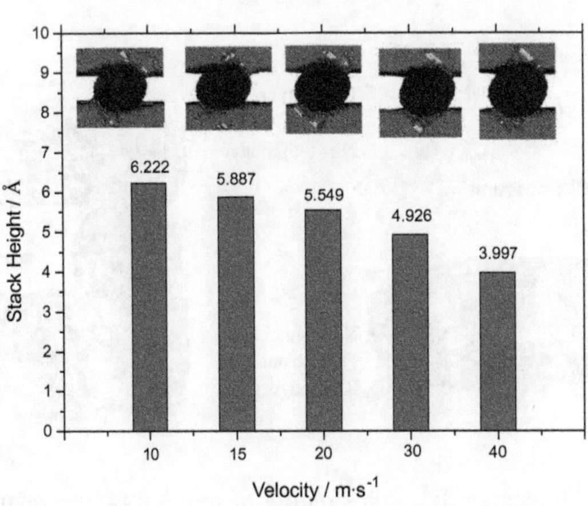

Figure 4.47. Atomic height configurations of pre-abrasive build-up at different towing speeds.

known as the smaller β, making the e/h value smaller (as shown in Figure 4.31). The combination of these factors causes the coefficient of friction to be greater than the e/h value, which in turn results in the abrasive moving in a rolling manner.

4.5 Abrasive Wear of Nano-Scale Single-Crystal Silicon Under Air Conditions

4.5.1 *Molecular dynamics simulating approach for monocrystalline silicon abrasive wear*

In Figure 4.48(a), a molecular dynamics model of two-body abrasive wear is depicted. In the simulation, the wear surface remains stationary while the abrasive moves at a constant velocity, abrading the specimen. On the other hand, in actual three-body abrasive wear, only the relative speed of motion between the upper and bottom wear surfaces and the applied load are known in advance. The mode of motion of the abrasive is determined by the interaction between the wear surface and the abrasive, which is a quantity to be found. Therefore, modeling of three-body abrasive wear should consider both the upper and bottom wear surfaces. Figure 4.48(b) presents a molecular dynamics model of three-body abrasive wear, which includes a pair of worn surfaces and an abrasive, representing a basic wear unit. The simulation applies relative motion velocity to

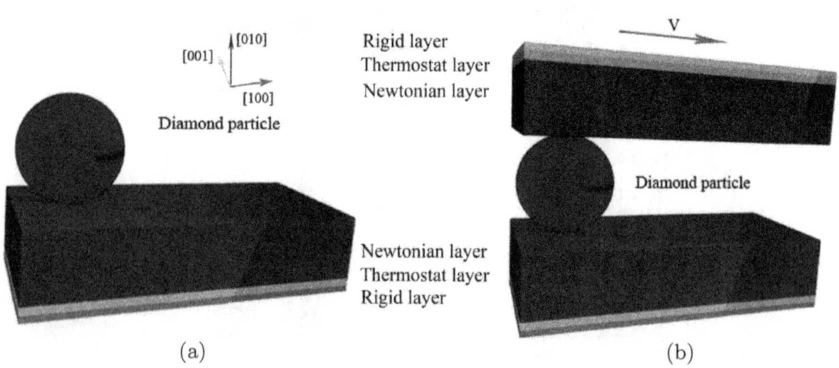

Figure 4.48. Molecular dynamics model of nano-abrasive wear. (a) Nano two-body abrasive wear, (b) nano three-body abrasive wear.

the two worn surfaces to drive the abrasive motion through the interaction between the abrasive and the worn surfaces. The translational and rotational velocities of the abrasive are captured in real time to determine how the abrasive is moving.

The nano two-body abrasive wear model, depicted in Figure 4.48, comprises a single-crystal diamond abrasive and a single-crystal silicon specimen. In contrast, the nano three-body abrasive wear model consists of a top and bottom single-crystal silicon specimen with a diamond abrasive sandwiched between them. To ensure comparability, the same sizes of diamond abrasive and specimen (first-body) were used for both nanodiamond and three-body abrasive wear tests. Two model sizes were utilized, considering accuracy and computational cost. The large model featured a diamond abrasive with a diameter of 21.72 nm and a specimen size of $70.60 \times 32.59 \times 16.29$ nm, consisting of 2,834,079 silicon atoms. On the other hand, the small model used a diamond abrasive with a diameter of 10.86 nm and a specimen size of $54.31 \times 21.72 \times 13.58$ nm, containing 808,000 silicon atoms. The initial position of the diamond abrasive was set at a distance of 0.54 nm from the specimen surface to minimize its influence on the equilibrium configuration. The model size was carefully determined through repeated computational tests to ensure minimal impact on results while keeping computational costs low. The final model was chosen to maintain a 1:1 ratio between the deformed and undeformed areas in the narrowest direction and at the maximum indentation depth. While a larger model allows for better observation of wear surface morphology, specimen deformation, and wear mechanisms, it comes with higher computational costs. Comparative studies have shown that the wear surface morphology and deformation and wear mechanisms observed in the smaller model are consistent with those in the larger model but at significantly lower computational costs. This makes it more convenient for conducting a larger number of calculations and studying the effects of parameters such as load, crystal orientation, and sliding direction on wear patterns and mechanisms.

Kindly remember the following details: The silicon atoms were initially positioned in the diamond structure with a lattice constant of 0.5431 nm. The crystallographic orientations along the x, y, and z axes were [100], [010], and [001], respectively. The (001) surface served as the abrasive surface. The abrasive was pressed in the

z direction and slid in the x direction, i.e., it moved along the [100] direction on the (001) surface. The simulations were conducted with periodic boundary conditions in the x and y directions to simulate the abrasion of an infinite number of periodically arranged abrasives in the lateral direction, eliminating boundary effects. The z direction was set as a free boundary condition. Additionally, a 1 nm thick atomic layer at the bottom of the specimen was frozen to prevent translational displacement. Furthermore, to dissipate the frictional heat from the wear process, a 1.6 nm layer of atoms was chosen as a thermostatic layer on top of the frozen atomic layer to absorb excess heat. The atoms in the thermostat layer act like a heat bath, absorbing or providing heat to nearby atoms while maintaining a constant temperature. By keeping the thermostat layer at a distance from the wear surface, friction heat can be gradually transferred to the heat bath and then dissipated. This process mimics real friction, but limitations in specimen size and heat bath algorithms can lead to errors due to phonon reflections by the thermostat and fixed layers. Researchers have attempted various methods to reduce these errors [16–19], but none have been fully successful yet. Therefore, careful selection of sliding velocity in simulations is necessary to prevent excessive frictional heat from affecting results negatively. In this study, a sliding speed of up to 200 m/s was chosen. Another common method of heat dissipation is to treat the entire specimen as an infinite heat bath, instantly dissipating excess heat. While this method is widely used for simulating processes such as nanowire stretching, bending, and nanoindentation, it is rarely used in modeling frictional and wear due to being unrealistic. Except for the frozen and thermostable atomic layers, the remaining silicon atoms move freely according to Newton's laws of motion. The same setup was applied to the upper and bottom two specimens for nanotribological abrasive wear. The diamond atoms are initially placed in diamond lattice positions with a lattice constant of 0.3567 nm. As diamond is much harder than single-crystal silicon, the diamond atoms are set to be rigid to simplify calculations, assuming that diamond deformation is negligible.

During the simulation, the entire system was initially placed in a thermostatic heat bath (NVT system) and the temperature was gradually increased from 0 K to the standard room temperature of 300 K. Subsequently, the system was equilibrated at 300 K for

50 picoseconds to establish the equilibrium structure at that temperature. This process was performed using the Nosé–Hoover hot bath method [11, 20]. After the system was equilibrated, the freezing layer atoms were immobilized, the thermostat atoms were set to the NVT ensemble, and the remaining atoms were set to the NVE ensemble. The thermostatic layer atoms were controlled using the dissipative particle dynamics method, specifically the Langevin heat bath [12]. The reason for using the Langevin heat bath method instead of the equilibrium method is that it's better at maintaining a constant temperature in a localized region, while the Nosé–Hoover heat bath method keeps the temperature constant throughout the entire system, which might lead to undesirable long-range spatial temperature fluctuations.

In two-body nano-abrasive wear, the nano-abrasive moves in the x direction at a speed of 100 m/s to abrade the specimen. In three-body nano-abrasive wear, a constant x-direction speed of 200 m/s is applied to the fixed layer atoms of one of the specimens to cause abrasion. It's important to note that when considering the translational velocity of the specimen in nanotribological abrasion, this velocity effect needs to be factored out from the calculation of the thermostatic layer temperature.

The simulation process utilized LAMMPS [21], an open-source massively parallel molecular dynamics software developed by Sandia National Laboratories in the USA. The simulations employed the Verlet integration algorithm with a time step of 2.5 fs, which is sufficiently small for silicon and has been widely used in molecular dynamics simulations of silicon.

There are two potential functions commonly used for silicon: the Stillinger–Weber potential [22] and the Tersoff potential [23, 24]. The Tersoff potential is suitable for studying single-crystal silicon phase transitions and was originally used to predict high-pressure phase transitions accurately. On the other hand, the Stillinger–Weber potential can better simulate silicon's shear behavior and dislocation behavior. It accurately predicts the stacking fault energy of silicon but overestimates the energy barrier for the transition from diamond to Si-II phase structure. Based on established single-crystal silicon nano-scratch experiments, it's observed that single-crystal silicon's deformation mechanism is phase transition under small loads, with dislocations occurring only under large loads. The Tersoff potential is more suitable for simulating the phase change plasticity of single-crystal silicon, so it is selected for the simulation of single-crystal silicon nano two-body abrasive wear and three-body abrasive wear. Additionally, the interaction between the diamond abrasive and the monocrystalline silicon specimen is expressed using a simple two-body potential function: the Morse potential.

4.5.2 *Monocrystalline silicon abrasive wear behavior*

In the context of nanoscale three-body abrasive wear, understanding the motion of the abrasive is crucial. The motion can be determined by considering the translational and rotation velocities of the abrasive's center of mass. In an ideal scenario of three-body abrasive wear, where both the specimen and the abrasive are rigid bodies, if the abrasive rolls relative to the specimen with no slip, its center velocity should be half of the sliding velocity of the specimen. This represents the minimum velocity achievable by the abrasive. Additionally, when the abrasive slides relative to the specimen in pure sliding mode, the velocity of the abrasive keeps the same with the specimen and remains constant to a maximum. In simulations, the velocity range of the abrasive was set to be 100–200 m/s. However, in real situations, due to the complex plastic deformation of the specimen, the interaction between the abrasive and the specimen becomes more intricate and challenging to predict using a suitable analytical model. Thus, it becomes necessary to simulate this behavior using molecular dynamics at an atomic scale.

The data presented in Figure 4.49 illustrates the center-of-mass velocity and rotation speed of the abrasive for both large and small models. The figure indicates that the center of mass velocity of the abrasive initially fluctuates around 100 m/s, displaying relatively high variability at the outset before gradually stabilizing as wear progresses. The fluctuation in initial velocity is a result of the abrasive's zero initial velocity, leading to sudden acceleration and subsequent large velocity oscillation. As the sliding distance or time increases, these velocity oscillations diminish and eventually plateau. Any minor velocity oscillations following stabilization are attributed to atomic steps on the specimen's surface and the surface of the abrasive. It is noteworthy that the stabilized velocity is precisely half of the specimen sliding velocity, suggesting that the abrasive

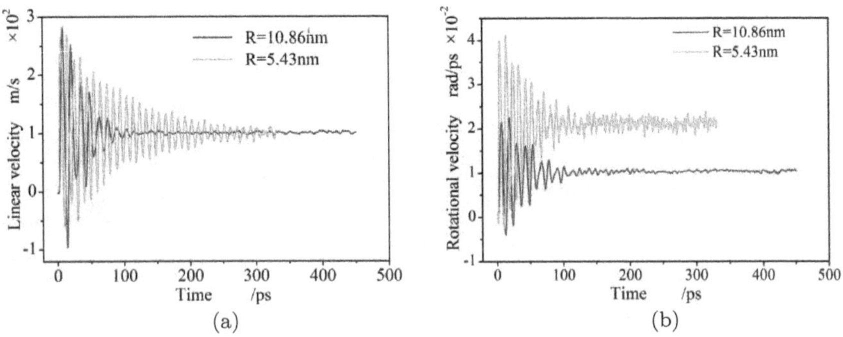

Figure 4.49. Motion characteristics of nano-abrasives. (a) Center of mass velocity, (b) rotation velocity.

tends to roll rather than slide under the conditions simulated in this study. Furthermore, the simulations indicate that smaller abrasive sizes exhibit the same behavior as larger sizes.

When a single-crystal silicon specimen undergoes significant plastic deformation, the distance between each surface contact point and the axis of rotation increases. The velocities at all contact points, except the largest contact circle, are lower than the sliding velocity of the specimen. This leads to localized atomic-scle slips at the contact points, which is also a major cause of friction in rolling friction on a macroscopic scale. Figure 4.50 illustrates the evolution process of the tangential velocity at the highest contact point on the maximum contact radius. This tangential velocity is slightly higher than the specimen's travel speed of 200 m/s, indicating the presence of atomic-scale slip. The energy dissipation due to the localized slip is generally minimal, especially in the case of large plastic deformations, and the impact of the localized slip relative to plastic shear is negligible.

The graph in Figure 4.51 illustrates the evolution of friction and normal load over time in typical nano two-body abrasive wear and three-body abrasive wear. It is evident from the figure that the friction generated by nano three-body abrasive wear is lower than that of nano two-body abrasive wear at the same normal load. Given that the abrasive motion in nano three-body abrasive wear is rolling, one might assume, based on common knowledge, that the friction generated would be smaller compared to that of nano two-body abrasive wear. However, upon closer inspection, it becomes apparent that the

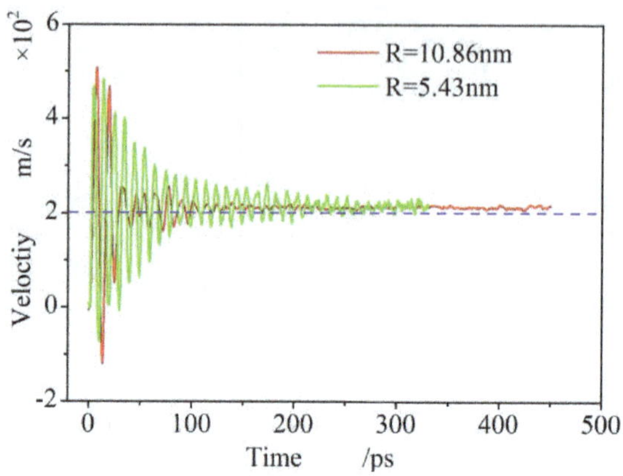

Figure 4.50. Tangential velocity at the highest contact point.

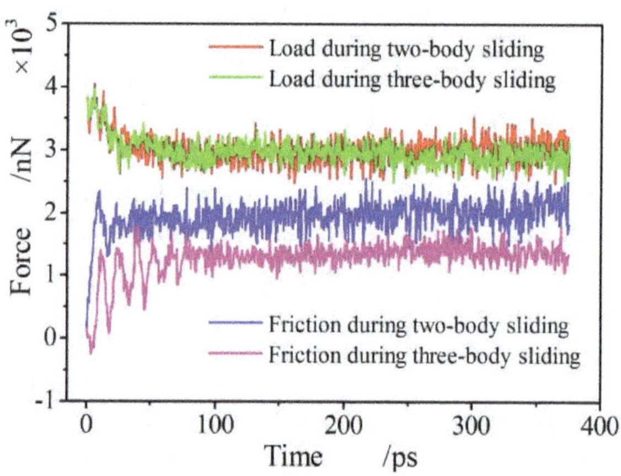

Figure 4.51. Variation of friction and normal load with time.

mechanism leading to lower friction in nano three-body abrasive wear differs from that on a macroscopic scale.

The total friction in the region of elastoplastic deformation consists of two components: adhesion force and shear force on the specimen as the abrasive slides over the surface. The adhesion force is predominantly influenced by the chemical properties of the specimen surface. Therefore, the difference in adhesion force is negligible in nano three-body abrasive wear and nano two-body abrasive wear when the same specimen model and potential function are utilized.

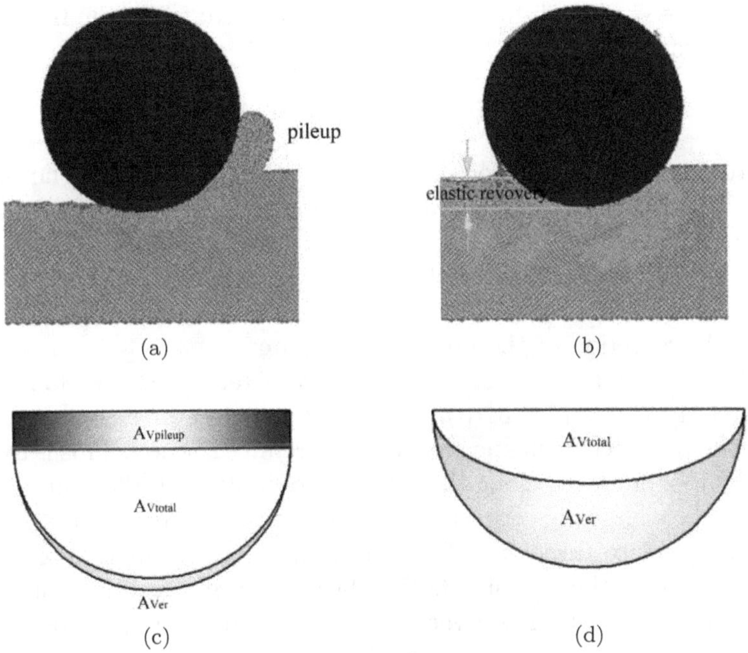

Figure 4.52. Projection of atomic configurations and contact area schematic of the wear process of nano-abrasive abrasive wear. (a) Atomic configuration in nano two-body abrasive wear, (b) atomic configuration in nano three-body abrasive wear, (c) contact area projection in nano two-body abrasive wear, (d) contact area projection in nano three-body abrasive wear.

On the other hand, the shear force is primarily linked to the plastic deformation of the specimen surface. The atomic configuration diagrams in Figure 4.52 illustrate the differences between plastic deformation on the specimen surface caused by nano two-body abrasive wear and nano three-body abrasive wear. In nano two-body abrasion, a notable feature is the substantial material build-up on the abrasive front, as shown in Figure 4.52(a), which is markedly absent in nano three-body abrasion. The plastic deformation resulting from nano three-body abrasion is characterized by significant elastic recovery of the specimen after the abrasive passes. This elastic recovery not only drives the rolling motion of the abrasive but also contributes to the low friction observed in nano three-body abrasive wear. By considering a basic contact surface calculation and assuming that the shear strength (p_m) of the specimens in nano two-body abrasive wear is the same as in nano three-body abrasive wear, the friction in nano-abrasive wear can be estimated:

$$f = p_m(A_{\text{Vtotal}} - A_{\text{Ver}} + \alpha_V A_{\text{Vpileup}}) \quad (4.29)$$

where f is the friction force, p_m is the shear strength of the material/GPa, A_{Vtotal} is the vertical projection of the contact surface/nm^2 without considering the elastic recovery and build-up, A_{Ver} is the vertical projection of the elastic recovery part/nm^2, α_V is a constant related to the distribution of the radial stresses, and A_{Vpileup} is the vertical projection of the build-up part/nm^2.

Detailed explanations of the parameters can be found in Figures 4.52(c) and 4.52(d). The symbols preceding A_{Ver} indicate that the presence of elastic recovery can reduce friction. This is due to the component force along the direction of motion on the contact surface of the elastically recovered portion of the abrasive material being in the opposite direction of the friction force. In nano three-body abrasive wear, there is significant elastic recovery and almost negligible build-up, implying a very large A_{Ver} and an A_{Vpileup} of almost 0. On the other hand, in nano two-body abrasive wear, A_{Ver} is almost 0 and A_{Vpileup} is very large. Consequently, nano three-body abrasive wear produces less friction than nano two-body abrasive wear.

Figure 4.53 illustrates the relationship between friction and normal load for nano three-body abrasive wear and nano two-body

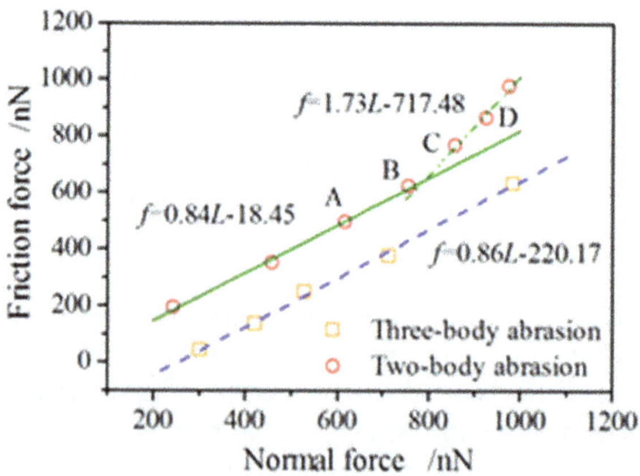

Figure 4.53. Straight lines fitted to the relationship between friction and normal load in nano two-body and three-body abrasive wear.

abrasive wear. It is evident that for nano three-body abrasive wear, the relationship between friction and normal load can be adequately described using a modified Amoton's law. On the other hand, for nano two-body abrasive wear, the friction force and normal load can be effectively described by two intersecting straight lines. For loads less than 754.34 nN, the modified friction coefficient for nano two-body abrasive wear is 0.84. In comparison, the modified friction coefficient for nano three-body abrasive wear is 0.86. The nearly identical friction coefficients for nano two-body abrasive wear and nano three-body abrasive wear suggest similar wear mechanisms. Moreover, the change in the friction law of nano two-body abrasive wear when the load exceeds 754.34 nN indicates a potential shift in the wear mechanism.

As shown in Figure 4.54, we analyzed the wear morphology of a single-crystal silicon surface after nano two-body abrasion and nano three-body abrasion. It was found that the worn morphology differed between the two abrasion methods. For instance, higher ridges were observed on both sides of the groove after nano two-body abrasive wear, and there was noticeable material accumulation at the front end of the abrasive. On the other hand, lower ridges were observed only on both sides of the groove after the wear of the nano three-body abrasive. The researchers also found that the groove of nano two-body abrasive wear was deeper than that of nano three-body abrasive wear after abrasion. After further studies, it was found that the surface morphology of the specimen after nano-abrasive wear was similar at different indentation depths, and no material build-up in the abrasive front was observed. Additionally, analysis of Figure 4.52 revealed that the grooves of nano two-body abrasive wear were deeper than those of nano three-body abrasive wear after abrasion. This difference is partly due to the material in nano two-body abrasive wear being pushed to both sides of the grooves and the front of the abrasive, and partly because the material undergoes significant elastic recovery after nano three-body abrasive wear.

Typically, we can indirectly determine the change in wear mechanism on a large scale by looking at the amount of volume loss [14]. However, this method doesn't work well for single-crystal silicon nano-abrasive wear because single-crystal silicon goes through a phase transition during wear that alters its volume, and it's difficult to measure and predict this change quantitatively. The present

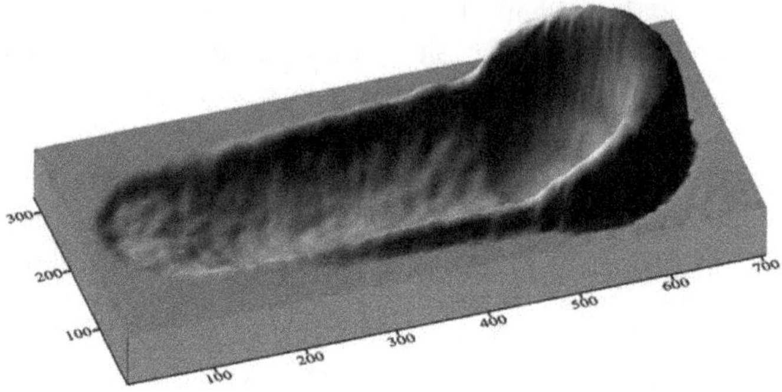

(a)

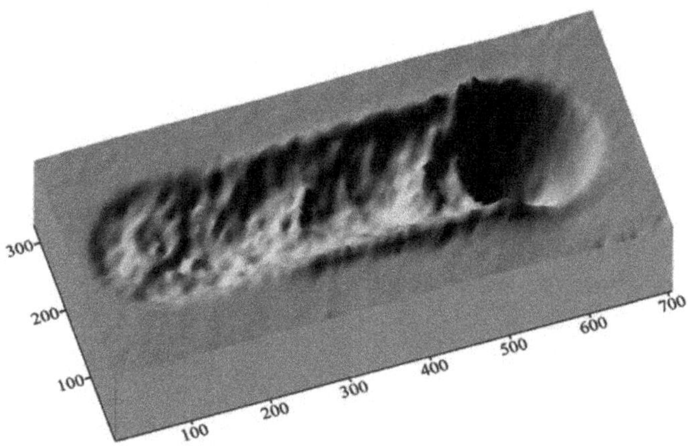

(b)

Figure 4.54. Wear groove morphology. (a) In nano two-body abrasive wear, (b) in nano three-body abrasive wear.

authors discovered that the transition of nano two-body abrasive wear mechanism aligns with the point where two fitted curves intersect on the friction-normal load curve, as shown in Figure 4.53. This point of intersection can be considered as the shift point of the wear mechanism. When the load is less than the load at this point, the wear mechanism is mainly characterized by plastic deformation grooving, and when the load is greater than the load at this point, cutting dominates the wear mechanism. To further illustrate this, a map of the atomic configuration near the transition point is provided in Figure 4.55. Figures 4.55(a–d) correspond to points A to D

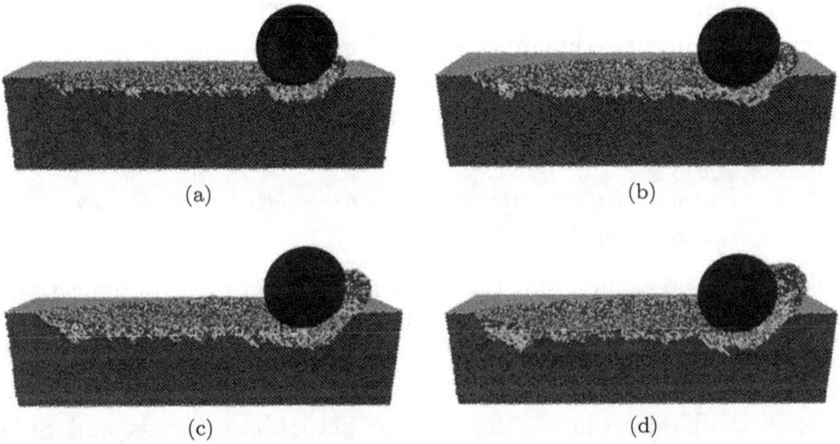

Figure 4.55. Diagram of the atomic configuration of the labeled points in Figure 4.53 for nano two-body abrasion. (a) A, (b) B, (c) C, (d) D.

in Figure 4.53, respectively. The atoms in the figure are colored based on their atomic types. Dark blue atoms represent diamond structure atoms, sky blue atoms represent bct5 phase atoms, and blue atoms represent Si-II phase atoms. It's evident that there's not much material buildup on the abrasive front at points A and B, but it's pronounced at points C and D in nano two-body abrasion. In nano three-body abrasive wear, there's no accumulation of material observed at the abrasive front at varied loads, indicating that the wear mechanism is mainly dominated by the plastic deformation grooving mechanism, which is consistent with the results from the friction-normal load curves. The results suggest that the wear mechanism of nano two-body abrasive wear changes from the plastic deformation grooving mechanism to the cutting mechanism as the indentation load increases. However, the wear mechanism of nano three-body abrasive wear is mainly dominated by the plastic deformation grooving mechanism under the experimental parameter conditions. A study by Xu Xiangjie *et al.* [25, 26] also showed that the abrasive wear of monocrystalline silicon under macroscopic conditions mainly exhibits a microscopic fracture-dominated mechanism, while under microscopic conditions (nano-scratch test), it exhibits a plastically deformed grooving mechanism. It seems that the smaller the scale of the abrasive in wear, the more the material properties and wear tend to shift toward "plastic" behavior. The inferences obtained from the molecular dynamics simulations of uniaxial tensile of gold single-crystal nanowires in Chapter 3, Section 6, are also strikingly similar to the above and warrant further research in this area.

4.5.3 *Plastic deformation of monocrystalline silicon nano-abrasives wear*

Figure 4.56 presents a plot that illustrates the atomic configurations along the mid-section following nano three-body abrasion and nano two-body abrasion. The atoms in the figure are color-coded based on their atom type. Dark blue atoms represent diamond structure atoms, while sky blue atoms denote bct5 phase atoms and Si-II phase atoms are represented by blue atoms. A detailed classification of the atoms is provided in Section 4.5.4. The figure shows that the mixed-colored phase represents the amorphous phase, indicating

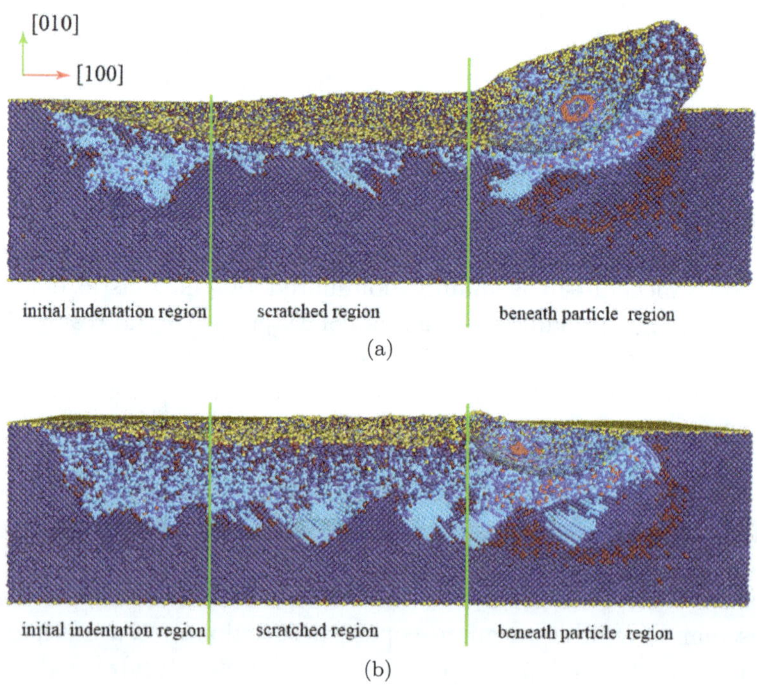

Figure 4.56. Cross-section of atomic configurations in nanowear. (a) Nano two-body abrasive wear, (b) nano three-body abrasive wear.

significant amorphization of the material both before and after abrasion. The amorphous structure appears to be slightly different in the region below the abrasive compared to after the abrasive has been scratched. Further analysis reveals that the relative atomic volume of the amorphous phase atoms directly below the abrasive is 0.76 relative to the diamond structure at 0 K in both nano two-body and three-body abrasion. After abrasion, the relative atomic volume of the amorphous phase atoms in nano three-body abrasion increases to 0.94, while in nano two-body abrasion, it is slightly smaller at 0.90. These significant changes in amorphous volume prior to and after nano-abrasive wear suggest the presence of two different amorphous network structures. This study confirms that the elastic recovery in abrasive wear stems from the expansion of the amorphous phase volume after wear. Even though the phase transitions of single-crystal silicon in nano two-body and three-body abrasive wear are similar, the phase transition layer's thickness in nano three-body abrasive wear is larger than that in nano two-body abrasive wear. This, along with the slightly larger volume of the amorphous material after abrasion, results in more significant elastic recovery in nano three-body abrasive wear. Additionally, the absence of material build-up at the abrasive front in nano three-body abrasive wear is another important reason for the significant elastic recovery. In conclusion, the elastic recovery in nanotribological abrasive wear mainly comes from the phase transition of single-crystal silicon.

As stated previously, the phase change layer in nano three-body abrasive wear is thicker than in nano two-body abrasive wear. This can be seen more clearly when the worn surface is divided into three regions: the initial indentation region, the scratched region, and the region directly beneath the abrasion, as shown in Figure 4.56. The thicknesses of the phase transformation layers in the three regions of nano three-body abrasive wear are approximately the same. However, the thickness of the phase transformation layer in the scratched region of two-body nano-abrasive wear is slightly smaller than in the other two regions. Further research revealed that the thickness of the phase transformation layer in nano three-body abrasive wear is larger than in nano two-body abrasive wear at varied indentation depths, as shown in Figure 4.57. Consequently, the deformation damage to single-crystal silicon caused by nano three-body abrasive wear is more severe than in the case of nano two-body abrasive wear.

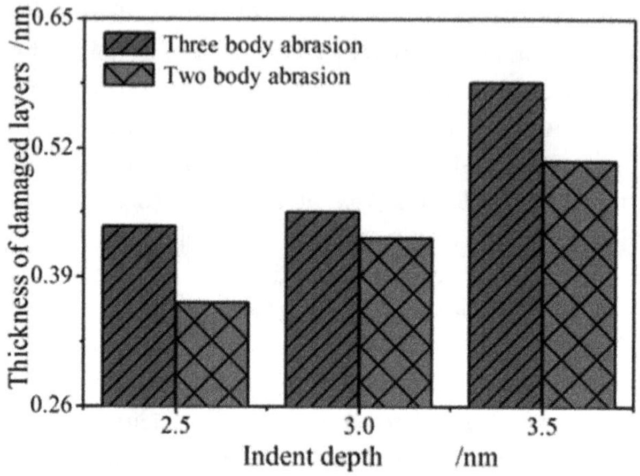

Figure 4.57. Thickness of worn phase transformation layer at different indentation depths.

4.5.4 *Phase transformation of single-crystal silicon in nano-abrasive wear*

For the purpose of clarification, the wear surface in nano two-body abrasive wear is typically divided into four distinct areas: the initial indentation region, the wear region, the abrasive action region, and the cutting region. To provide a visual representation, Figure 4.58 depicts the phase distribution characteristics of nano two-body abrasive wear in these different regions and interfaces at an indentation depth of 4.50 nm. For the sake of comparison, the phase distribution during the corresponding nanoindentation process is also illustrated in Figure 4.58(f–h). In the initial indentation region, the indentation-induced Si-II phase disappears completely after abrasion. By comparing the phase distributions in the initial indentation region before abrasive scratching (Figure 4.58(f–h)) and after scratching (Figures 4.58(a) and 4.58(c–e)), it is evident that the vast majority of the Si-II phase and bct5 generated in the nanoindentation are transformed into amorphous phases. The amorphous phase is composed of a random mixture of atoms with coordination numbers of 4, 5, 6, and so on. However, a small portion of bct5 remains and is concentrated at the bottom of the phase transition region. After the abrasive scratches, most of the indentation-generated distorted

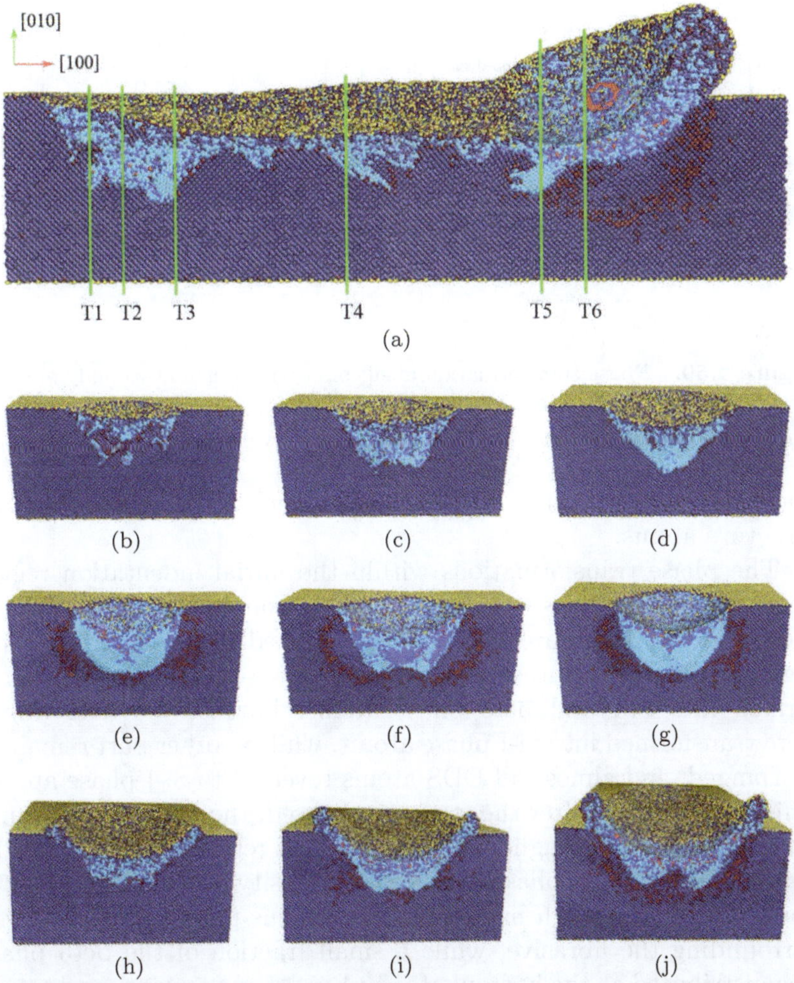

Figure 4.58. Wear phase distribution of nanodiamond abrasives. (a) Medium cross-section, (b) T1 cross-section, (c) T2 cross-section, (d) T3 cross-section, (e) nanoindentation T1 cross-section, (f) nanoindentation T2 cross-section, (g) nanoindentation T3 cross-section, (h) T4 cross-section, (i) T5 cross-section, (j) T6 cross-section. (The atoms in the figure are colored according to the coordination number: The sky blue atoms are bct5 phase atoms, the blue atoms are Si-II phase atoms, the red atoms are Si-III/Si-XII phase atoms, and the yellow atoms are atoms with a coordination number of less than 3, i.e., surface atoms.)

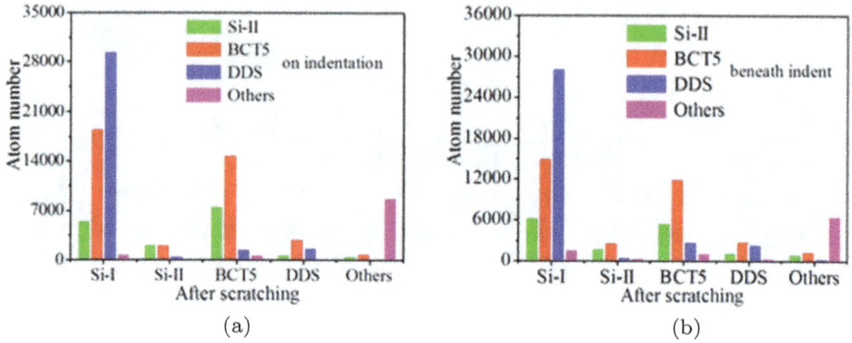

Figure 4.59. Phase transformation in silicon atoms in nano two-body abrasive wear. (a) In initial press-in zone, (b) in worn zone.

diamond cubic structure (DDS) atoms revert to the initial diamond structure atoms.

The phase transformations within the initial indentation region before and after abrasive scratching were examined, and the results are presented in Figure 4.59(a). As depicted in the figure, following the abrasive scratch, the majority of Si-II atoms were transformed into Si-I and bct5 phase atoms. Part of the bct5 atoms were transformed into Si-I phase atoms, while another part remained unchanged, and almost all DDS atoms reverted to Si-I phase atoms. This suggests that after the stress is removed, the coordination number of the atoms throughout the indentation region decreases, resulting in an amorphous phase with a lower density. Within the region of abrasive action, a high number of amorphous phases were observed surrounding the abrasive, while a small fraction of the bct5 phase was distributed at the bottom of the phase change region, and a substantial number of isolated DDS atoms were distributed around the phase change region, as shown in Figures 4.58(a), 4.58(j), and 4.58(k). It is worth noting that the presence of Si-II phase was not detected around the abrasive, which is different from the phase transformation observed during the nanoindentation process. After the abrasive scratch, the amorphous phase persisted in the amorphous region, and no experimentally observed Si-I phase of the nanocrystals was found. Part of the bct5 phase was transformed into an amorphous phase and distributed at the bottom of the phase transformation region, as illustrated in Figures 4.58(a) and 4.58(i–k). The chips formed by abrasion were entirely composed of amorphous material, as shown in Figure 4.58(a). The phase transformation in silicon atom type before and after the abrasive scratch near section T4 in Figure 4.58 is displayed in Figure 4.59(b). The phase transformation in silicon atoms is consistent with the initial nanoindentation region.

The phase transformation process in nano three-body abrasive wear is analogous to that of nano two-body abrasive wear (Figure 4.59). In the initial indentation region, the bct5 phase undergoes transformation into an amorphous phase, with a small portion of the bct5 phase retained and distributed at the bottom of the phase transformation region, as illustrated in Figures 4.60(a–e). There is no discernible difference in the atom type transition between the initial indentation region of nanotribological abrasive wear and nanodiamond abrasive wear, as shown in Figures 4.59 and 4.61(a). In the region of abrasive action, a larger area of bct5 is distributed at the bottom of the amorphous phase, as depicted in Figures 4.60(j) and 4.60(k). After the abrasive slips through, the thick amorphous phase remains on the surface of the specimen, with the bottom of the amorphous phase still retaining a significant portion of the bct5 phase, as shown in Figures 4.58(a) and 4.58(i). Compared to nano two-body abrasion, the nano three-body abrasion generates more bct5 phases, as evidenced by the atomic type transition statistics graph in Figure 4.61(b). These findings suggest that nano three-body abrasive wear causes a more substantial phase transformation than nano two-body abrasive wear, which can be attributed to the different stress states.

Upon examining the amorphous phase prior to and following the wear of the nano three-body abrasion, it is evident that the volumes of the amorphous phase beneath the abrasive and the residual amorphous phase after wear are significantly dissimilar. Are the structures of those two amorphous phases identical? To investigate this, a rectangular area was selected in the amorphous region beneath the abrasive particle, and the number of atoms contained within it was counted. Subsequently, the volumes occupied by each atom and the average number of coordination sites were calculated. The results indicate that the volume per atom in both the nano two-body abrasive wear and the nano three-body abrasive wear regions' amorphous atoms is approximately 0.76 relative to the volume per atom in the single-crystal silicon diamond structure at 0 K. This volume is

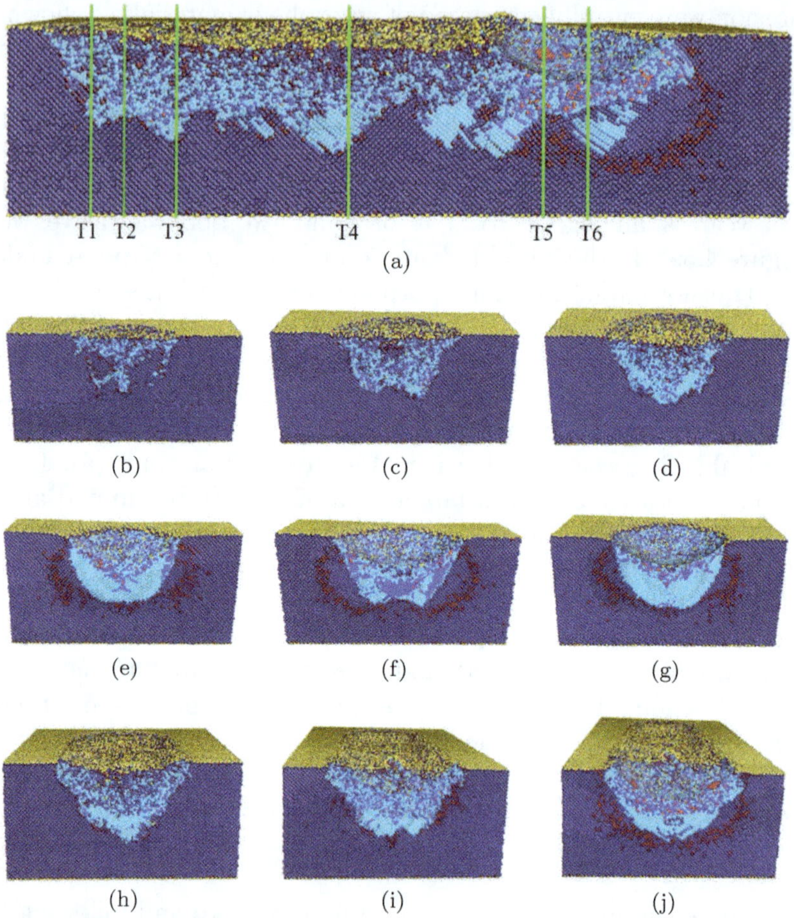

Figure 4.60. Phase transformation in silicon atom types in nano three-body abrasive wear. (a) Medium cross-section, (b) T1 cross-section, (c) T2 cross-section, (d) T3 cross-section, (e) nanoindentation T1 cross-section, (f) nanoindentation T2 cross-section, (g) nanoindentation T3 cross-section, (h) T4 cross-section, (i) T5 cross-section, (j) T6 cross-section.

consistent with the first-principles calculations of the High-Density Amorphous (HDA) [27]. Further calculations reveal that the average coordination number of atoms in the rectangular region (truncation radius 0.302 nm) is 6.27, which is consistent with the experimental value of 6.40 [28] and the CP simulation value of 6.5 [29, 30]. Consequently, it can be inferred that the amorphous phase in the

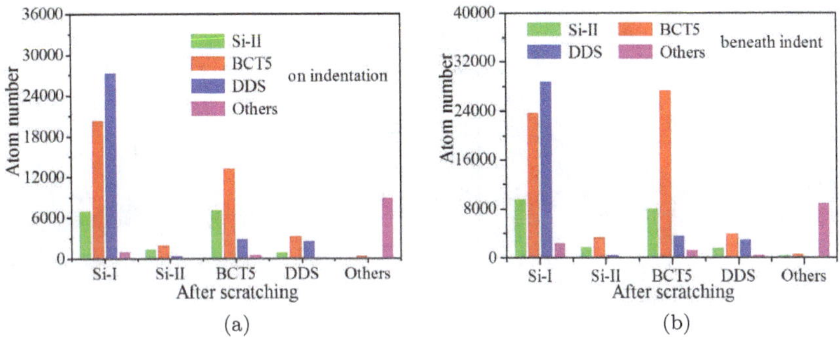

Figure 4.61. Phase transformation in silicon atom types in nano three-body abrasive wear. (a) In initial press-in zone, (b) in worn zone.

region of action below the abrasive particle is high-density amorphous. After being scratched by the abrasive, the HDA phase will be transformed into low-density metastable amorphous (LDMA) [31]. The volume per atom of LDMA on the surface of the specimen after abrasion by nano two-body abrasion is 0.94 compared to 0.90 for nano three-body abrasion, and its average coordination number is 5.02. To provide a more comprehensive comparison with existing literature, we calculated the proportion of atoms with 5-fold and 6-fold coordination numbers in HDA. The resulting value of 83.40% for HDA and 61.3% for LDMA aligns with the findings reported by Dominik [32], thereby substantiating the calibration of HDA and LDMA by the present authors. The increased volume per atom, the increased average coordination number, and the pronounced atoms with a prevalence of 5- and 6-fold coordination collectively suggest a denser amorphous network in HDA compared to LDMA.

The radial distribution function (RDF) and bond angle distribution function (ADF) for both HDA and LDMA are depicted in Figure 4.62. The absence of a third peak in the RDF corroborates the amorphous nature of both HDA and LDMA. Notably, the first peak of LDMA is more pronounced, and the second peak is shifted to a greater distance relative to HDA. The ADF of LDMA exhibits an expanded peak at the ortho-tetrahedral angle of 109°, accompanied by a shoulder at the 60° position. In contrast to LDMA, the central peak of HDA's ADF shifts to 78°, with the 60° shoulder in LDMA transforming into a minor peak. Furthermore, a minor shoulder emerges near 155°. These distinctions between the ADFs of HDA

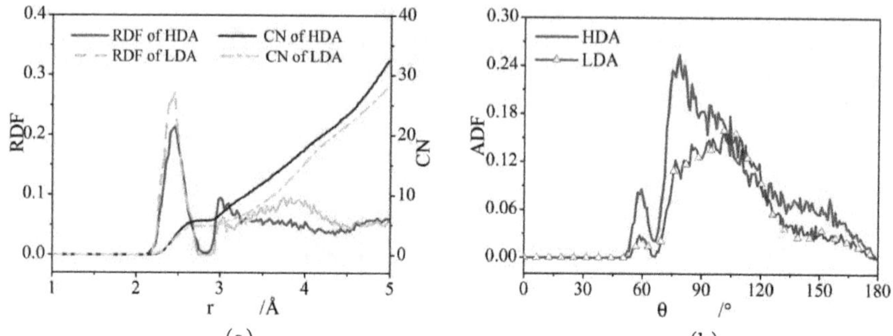

Figure 4.62. Structural features of high-density amorphous and low-density amorphous. (a) In initial press-in zone, (b) in worn zone.

and LDMA underscore that they possess two distinct amorphous network structures. It is noteworthy that LDMA does not represent a low-density amorphous silicon (LDA) in the conventional sense but rather a mesostable amorphous configuration situated between HDA and LDA. A similar LDMA structure has been identified in the decompression process following the generation of the HDA phase under hydrostatic pressure [33].

Previous studies have shown that in nano-abrasive wear under the direct action of abrasives, the diamond-structured monocrystalline silicon (Si-I) will transform into the high-density amorphous phase (HDA), and the HDA will further transform into the low-density mesostable amorphous (LDMA) phase when scratched by abrasives. The authors have not found any significant difference between nano-scratching or nano two-body abrasion in the Si-I to HDA phase transition, and no researcher has characterized the amorphous silicon network structure before and after abrasive scratching. However, it has been found that this phase transition can occur under other loading conditions, indirectly supporting the results of this research. First-principle calculations by Durandurdu [27] show that single-crystal silicon transforms to HDA at static pressures of up to 15 GPa. Deb [34] reported that crystals in porous silicon in diamond-to-top-anvil experiments partially transform to HDA and that HDA transforms back to HDA when the pressure is withdrawn. McMillan [35] studied a density-driven phase transition using a diamond-to-top-anvil experiment and found that LDA transforms to HDA during pressurization and reverts to LDA when the pressure is removed. The subsequent study provides evidence that the transformation of single-crystal silicon from diamond to high-density amorphous, and the transformation of the dense network of HDA to low-density LDMA when the pressure is removed, can occur.

Multiple studies have demonstrated that after single-crystal silicon nano-scratching, the several primary residual phases on the specimen surface are evident [36–39]. At high stresses [36] and low velocities [35], small crystalline structures, such as diamond-structured nanograins, and metastable phases bc8 and r8 grains are found sandwiched within the amorphous material. It is suggested that two potential phase transformation paths create these residual phase structures. One possibility is that the movement of the abrasive leads to direct amorphization of the single-crystal silicon. The resulting amorphous material remains on the specimen's surface after the abrasive has scratched it. Alternatively, the movement of the abrasive generates high-pressure phases, such as Si-II and bct5, which then directly transform into amorphous phases after scratching. Experimental evidence of the phase transition paths is challenging to obtain because examining the crystal structure of the specimen in real time is difficult. Molecular dynamics simulations can bridge this gap and provide accurate, real-time structural information, revealing detailed phase transition paths. Figure 4.63 presents atomic configuration diagrams at various points during the wear process of nano two-body abrasion. The coloring rules for the atoms are the same as in Figure 4.58. From the figure, it is evident that the initial diamond-structured monocrystalline silicon transforms into amorphous silicon beneath the abrasive's action. The resulting amorphous silicon remains on the specimen's surface after the abrasive has passed through. The phase transition path in nano three-body abrasion is similar to that of nano two-body abrasion. The two types of amorphous material have different network structures. In conclusion, the molecular dynamics simulations in this research not only back the experimental results [40] but also clarify that the transition from high-density amorphous to low-density metastable amorphous is the dominant phase transition path in nano two-body abrasion or nano-scratch and nano three-body abrasion.

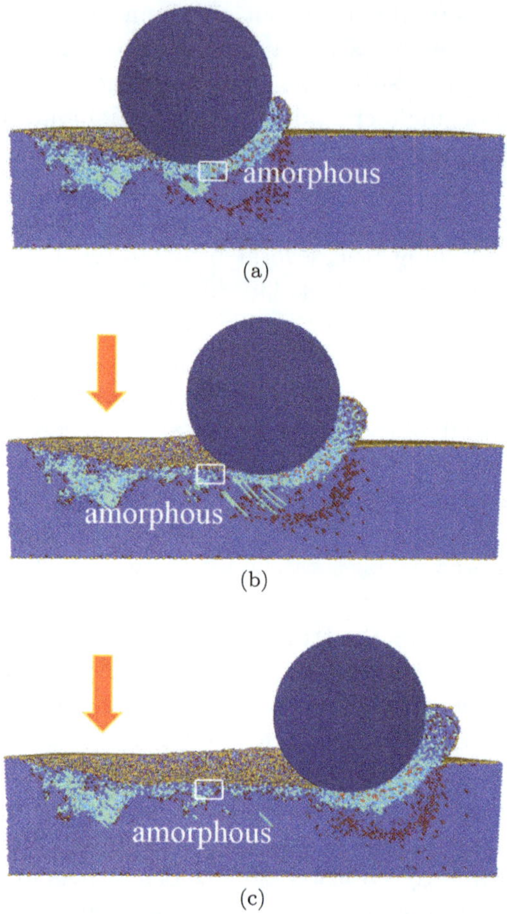

Figure 4.63. Phase transition paths in nano two-body abrasion.

4.5.5 *Stress mechanisms for phase transition in single silicon*

The sliding indenter used in nano-abrasive wear enables the direct amorphization of single-crystal silicon, without transforming it into the crystalline phases of Si-II and bct5, as seen in nanoindentation. The difference between nanoindentation and nano-abrasive wear is not likely due to loading speed, as the speed is the same in simulations. Instead, the difference is mainly because the stress states formed in single-crystal silicon specimens by nano-abrasive wear and nanoindentation are different. Previous studies have similarly attributed the different phase transitions of single-crystal silicon in nanoindentation and diamond-to-top anvil experiments to the difference in stress states.

The widely accepted belief is that applying pressure along the [100] direction of diamond-structured monocrystalline silicon leads to a flattening of the tetrahedral structure, causing a transition from diamond-structured Si-I to Si-II. This transition is primarily induced by hydrostatic pressure [41]. However, the exact stress mechanism responsible for the amorphization of single-crystal silicon remains a significant unresolved issue. Early studies linked the formation of amorphous phases in nanoindentations to hydrostatic pressure, suggesting that high bulk strain was the primary cause of single-crystal silicon amorphization [42, 43]. Subsequent research has indicated that the role of bias stress should not be overlooked. Minowa and Sumino [44] proposed that the high shear stress generated in nanoindentation contributes to the amorphization of single-crystal silicon. Zhang and Tanka [45, 46] proposed that octahedral shear stress is the main cause of the amorphization of amorphous phases in nanoindentation, nano two-body abrasive wear, and nano three-body abrasive wear. Kim [41] highlighted the significance of bias stress in the formation of amorphous phases in single-crystal silicon during nanoindentation. Additionally, Kim emphasized the role of bias stress in nanoindentation. More recently, Chrobak [47] demonstrated that a high ratio of static pressure to Mises stress encourages the amorphization of single-crystal silicon in nanoindentation, while a low ratio induces dislocation plasticity in single-crystal silicon nanoparticle nanoindentation. Numerous nanoindentation experiments have indicated that hydrostatic pressure alone does not account for several phenomena in nanoindentation, including the anisotropy of the phase transition [48], the low transition pressure [49], and the transition of the bct5 phase (that has not been observed in the experiments of diamond-to-top anvil) [50]. Those findings suggest that deformation based on shape due to bias or shear stresses plays a crucial role in the phase transition of single-crystal silicon. However, it is clear that shear stress alone is insufficient to explain the difference in phase transformation observed in nanoindentation and nano-abrasive wear. Consequently, it is reasonable to assume that neither hydrostatic pressure nor shear stress alone can accurately elucidate the differences in phase transitions under varying loading conditions. The question then arises as follows: What combination of hydrostatic pressure and shear stress promotes amorphization? And how does this combination contribute to the transformation of high-pressure phases in monocrystalline silicon?

The Viry theory is employed in this study to compute stresses at the atomic level. To address the significant stress fluctuations caused by local thermal vibrations of atoms, the stress of an atom is defined in this research as the average of the stresses of the atom and its nearest neighbor. Figures 4.64(a–f) illustrate the hydrostatic pressure and Mises stress distributions in nanoindentation, nano two-body, and three-body abrasive wear. The figures show that for all the specimens, hydrostatic pressure and Mises stresses are concentrated in a small region beneath the abrasive. The Mises stress in nanoindentation is somewhat higher than that in nano two-body and three-body abrasive wear, but there is no difference in the magnitude of hydrostatic pressure among the three cases. At first glance, it seems that the high Mises stresses are responsible for the varying phase transitions in the three cases. However, the study did not find any correlation between Mises stress or hydrostatic pressure and the distribution of phase transition regions. Therefore, neither Mises stress nor hydrostatic pressure can independently explain the phase transformation of single-crystal silicon.

In the context of nanoindentation, nano two-body abrasive wear, and nano three-body abrasive wear, the hydrostatic pressure (σ_h) and Mises stress (σ_{vos}) averaged over the stress concentration area were calculated prior to the onset of the phase transition. The calculated values for σ_h and σ_{vos} are presented in Table 4.8, along with the corresponding $\sigma_h/\sigma_{\text{vos}}$ ratios. Notably, the average hydrostatic pressure during nanoindentation is 10.15 GPa, which is comparable to the experimentally determined transition pressure of 11–12.5 GPa for the Si-I to Si-II transition. Conversely, the hydrostatic pressures of 6.93 GPa and 8.55 GPa observed in nano two-body abrasive wear and nano three-body abrasive wear are considerably lower than the transition pressure for Si-I to Si-II, suggesting the absence of a Si-I to Si-II transition in nano-abrasive wear. If the amorphization of single-crystal silicon is only attributed to shear stress, then nanoindentation may be more susceptible to amorphization than nano-abrasive wear due to its higher shear stress. However, this contradicts the

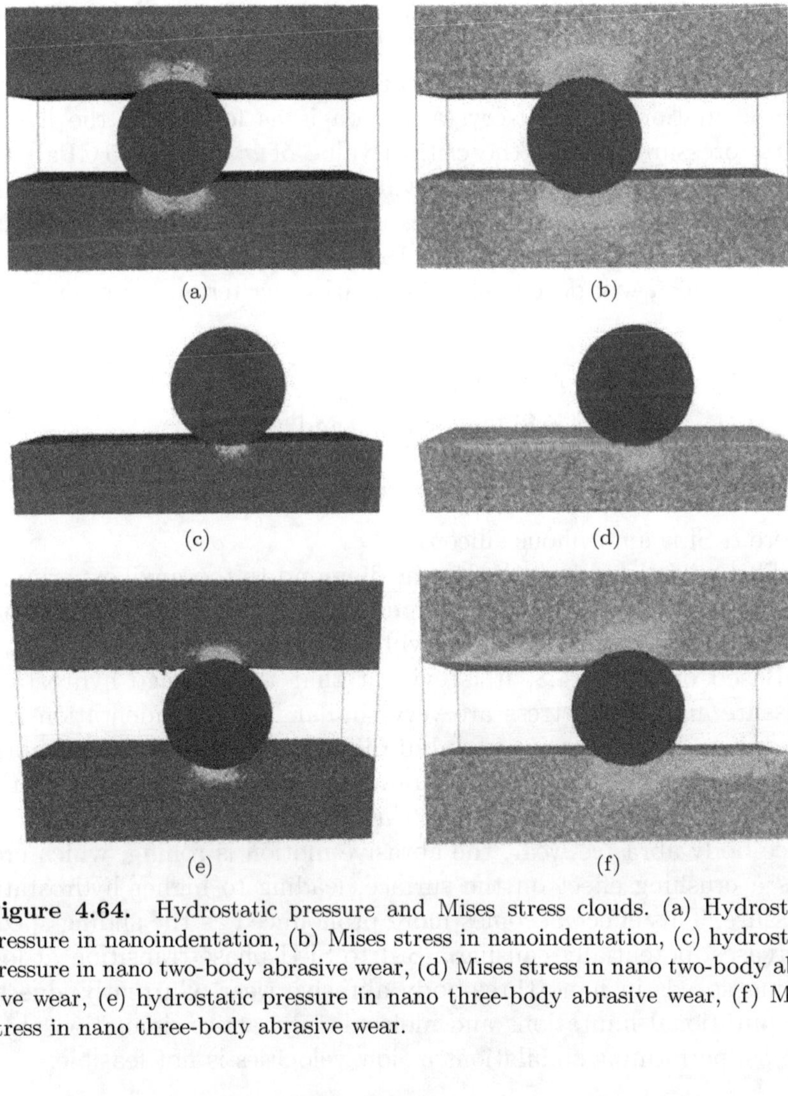

Figure 4.64. Hydrostatic pressure and Mises stress clouds. (a) Hydrostatic pressure in nanoindentation, (b) Mises stress in nanoindentation, (c) hydrostatic pressure in nano two-body abrasive wear, (d) Mises stress in nano two-body abrasive wear, (e) hydrostatic pressure in nano three-body abrasive wear, (f) Mises stress in nano three-body abrasive wear.

Table 4.8. Average hydrostatic pressure σ_h and Mises stress σ_{von}.

	Nanoindentation	Nano two-body abrasion	Nano three-body abrasion
σ_h/GPa	10.15	6.93	8.55
σ_{von}/GPa	17.85	16.89	16.08
σ_h/σ_{von}	0.57	0.41	0.53

findings of the molecular dynamics simulations and the available experimental results presented in this research. Taking into account both hydrostatic pressure and shear stress, the criterion for the phase transformation of single-crystal silicon is as follows: If the hydrostatic pressure exceeds the critical value of σ_{ch} (10–12.5 GPa), the diamond structure of Si-I phase will transition to Si-II phase. On the other hand, if the hydrostatic pressure is lower than the critical value and the Mises stress reaches σ_{cv} (16.0 GPa), amorphization of single-crystal silicon will occur. This summarized criterion is presented as follows:

$$\begin{cases} \text{Si-I} \rightarrow \text{Si-II} & \sigma_h \geq \sigma_{ch} \\ \text{Si-I} \rightarrow \alpha\text{-Si} & \sigma_h < \sigma_{ch} \quad \text{and} \quad \sigma_{von} \geq \sigma_{cvon} \end{cases}$$
$$\sigma_{ch} \approx 10\text{–}12.5\,\text{GPa} \quad \sigma_{cvon} \approx 16.0\,\text{GPa} \tag{4.30}$$

where α-Si is amorphous silicon.

This criterion also applies to the diamond-to-top-anvil experiment and the uniaxial compression experiment. Under both loading conditions, monocrystalline silicon will be transformed into Si-II phase.

Based on Table 4.8, it is evident that the average hydrostatic pressure and Mises stress are very similar for nanoindentation and nano three-body abrasive wear but different for nano two-body abrasive wear. This difference is attributed to the motion patterns of the abrasives in nano two-body and three-body abrasive wear. In nano three-body abrasive wear, the abrasive motion is rolling, which creates a crushing effect on the surface, leading to higher hydrostatic pressure. This effect becomes more pronounced as the sliding speed decreases, potentially causing a Si-I to Si-II phase transition at low sliding speeds in nano three-body abrasive wear. Currently, due to computational limitations and molecular dynamics simulation algorithms, performing simulations at low velocities is not feasible.

4.5.6 *Summary*

In-depth molecular dynamics simulations were employed to examine the phase transition mechanisms, phase distribution characteristics, phase distribution laws, and phase transition paths in nanoindentation, nano two-body abrasive wear, and nano three-body abrasive wear. The coordination number, radial distribution function, and bond angle distribution function were utilized to analyze the high-pressure phases of monocrystalline silicon Si-II and bct5, as well as two amorphous phases with distinct network structures: high-density amorphous and low-density metastable amorphous. The key findings are as follows:

(1) In nanoindentation, the primary phase change mechanism involves the transformation of diamond-structured monocrystalline silicon into β-Si and bct5 phases. On the other hand, in nano two-body and three-body abrasive wear, the amorphization of monocrystalline silicon constitutes the principal phase change mechanism.
(2) In nano two-body and three-body abrasive wear, the phase transition path entails the transformation of diamond-structured monocrystalline silicon into a high-density amorphous phase below and at the leading edge of the abrasive. After the abrasive moves through, the resulting high-density amorphous phase transforms into a low-density, metastable amorphous phase.
(3) In a comparative study of nanoindentation, nano two-body abrasive wear, and nano three-body abrasive wear, a new stress criterion for the phase transformation of single-crystal silicon is proposed. According to this criterion, if the hydrostatic pressure exceeds the critical value σ_{ch} of 10–12.5 GPa, the diamond structure of the Si-I phase will undergo a phase transition to Si-II phase. Conversely, if the hydrostatic pressure is less than the critical value, and the Mises stress reaches the critical value σ_{cv} of 16.0 GPa, amorphization of single crystal silicon will occur. This criterion highlights the significance of shear stress in the amorphization of single-crystal silicon.
(4) Additionally, in nanoindentation and nano-abrasive wear, there is evidence of isolated distribution of metastable atoms at the periphery of the phase change region under the abrasive or indenter. Previous studies referred to those as Si-III/Si-XII atoms based on the coordination number. However, it has been found that those metastable atoms are distorted diamond structure atoms that return to their initial form once the stress is removed. Further loading does not cause transformation to other phases.

References

[1] Gao Y, Ruestes CJ, Urbassek HM. Nanoindentation and nanoscratching of iron: Atomistic simulation of dislocation generation and reactions. *Comput Mater Sci*, 2014, 90: 232–240.

[2] Alhafez IA, Brodyanski A, Kopnarski M, *et al.* Influence of tip geometry on nanoscratching. *Tribol Lett*, 2017, 65: 26.

[3] Zhang J, Sun T, Yan YD, *et al.* Molecular dynamics simulation of subsurface deformed layers in AFM-based nanometric cutting process. *Appl Surf Sci*, 2008, 254: 4774–4779.

[4] Zhu PZ, Hu YZ, Ma TB, *et al.* Study of AFM-based nanometric cutting process using molecular dynamics. *Appl Surf Sci*, 2010, 256: 7160–7165.

[5] Sun K, Fang L, Yan ZY, *et al.* Atomistic scale tribological behaviors in nano-grained and single crystal copper systems. *Wear*, 2013, 303: 191–201.

[6] Li J, Fang QH, Liu YW, *et al.* A molecular dynamics investigation into the mechanisms of subsurface damage and material removal of monocrystalline copper subjected to nanoscale high speed grinding. *Appl Surf Sci*, 2014, 303: 331–343.

[7] Fang L, Kong XL, Zhou QD. A wear tester capable of monitoring and evaluating the movement pattern of abrasive particles in three-body abrasion. *Wear*, 1992, 159: 115–120.

[8] Fang L, Kong XL, Su JY, *et al.* Movement patterns of abrasive particles in three-body abrasion. *Wear*, 1993, 162–164: 782–789.

[9] Fang L, Liu WM, Du DS, *et al.* Predicting three-body abrasive wear using Monte Carlo methods. *Wear*, 2004, 256: 685–694.

[10] Zhang J. *Research on the Formation Mechanism of Crystalline Copper Nanomechanical Processing Surface Layer Based on Molecular Dynamics* (in Chinese). Harbin: Harbin Institute of Technology, 2011.

[11] Nose S. A unified formulation of the constant temperature molecular dynamics methods. *J Chem Phys*, 1984, 81: 511–519.

[12] Schneider T, Stoll E. Molecular-Dynamics study of a three-dimensional one-component model for distortive phase transitions. *Phys Rev B*, 1978, 17: 1302–1322.

[13] Fang L, Zhao J, Li B, *et al.* Movement patterns of ellipsoidal particle in abrasive flow machining. *J Mater Process Technol*, 2009, 209: 6048–6056.

[14] Fang L, Cen QH, Sun K, *et al.* FEM computation of groove ridge and Monte Carlo simulation in two-body abrasive wear. *Wear*, 2005, 258: 265–274.

[15] Sun JP, Fang L, Han J, *et al.* Abrasive wear of nanoscale single crystal silicon. *Wear*, 2013, 307: 119–126.

[16] Benassi A, Vanossi GE, Santoro GE, *et al.* Parameter-Free dissipation in simulated sliding friction. *Phys Rev B*, 2010, 82: 081401(R).

[17] Jones RE, Kimmer CJ. Efficient non-reflecting boundary condition constructed via optimization of damped layers. *Phys Rev B*, 2010, 81: 094301.

[18] Guddati MN, Thirunavukkarasu S. Phonon absorbing boundary conditions for molecular dynamics. *J Comput Phys*, 2009, 228: 8112–8134.

[19] Li XT, E WN. Variational boundary conditions for molecular dynamics simulations of crystalline solids at finite temperature: Treatment of the thermal bath. *Phys Rev B*, 2007, 76: 104107.

[20] Hoover WG. Constant-Pressure equations of motion. *Phys Rev A*, 1986, 34: 2499–2500.

[21] Plimpton S. Fast parallel algorithms for short-range molecular dynamics. *J Comput Phys*, 1995, 117: 1–19.

[22] Halicioglu T, Tiller WA, Balamane H. Comparative study of silicon empirical interatomic potentials. *Phys Rev B*, 1992, 46: 2250–2279.

[23] Tersoff J. Modeling solid-state chemistry—Interatomic potentials for multicomponent systems. *Phys Rev B*, 1989, 39: 5566–5568.

[24] Tersoff J. Empirical interatomic potential for silicon with improved elastic properties. *Phys Rev B*, 1988, 38: 9902–9905.

[25] Xu Xiangjie, Yu Bijun, Chen Lei, *et al.* Effect of sliding velocity on the wear of monocrystalline silicon at different contact scales (in Chinese). *J Mech Eng*, 2013, 49: 108–115.

[26] Li Xiaocheng, Lv Jinjun, Yang Shengrong. Study on the sliding wear properties of monocrystalline silicon and its phase transition (in Chinese). *J Tribol*, 2004, 24: 326–331.

[27] Durandurdu M, Drabold DA. High-Pressure phases of amorphous and crystalline silicon. *Phys Rev B*, 2003, 67: 212101.

[28] Waseda Y, Suzuki K. Structure of molten silicon and germanium by X-Ray-Diffraction. *Z Phys B Condens Matter*, 1975, 20: 339–343.

[29] Stich I, Car R, Parrinello M. Structural, bonding, dynamic, and electronic-properties of liquid silicon-an abinitio molecular-dynamics study. *Phys Rev B*, 1991, 44: 4262–4274.

[30] Stich I, Car R, Parrinello M. Bonding and disorder in liquid silicon. *Phys Rev Lett*, 1989, 63: 2240–2243.

[31] Demkowicz MJ, Argon AS. High-Density liquidlike component facilitates plastic flow in a model amorphous silicon system. *Phys Rev Lett*, 2004, 93: 025505.

[32] Daisenberger D, Wilson M, McMillan PF, *et al.* High-Pressure X-Ray scattering and computer simulation studies of density-induced polyamorphism in silicon. *Phys Rev B*, 2007, 75: 224118.

[33] Durandurdu M, Drabold DA. Ab initio simulation of first-order amorphous-to-amorphous phase transition of silicon. *Phys Rev B*, 2001, 64: 014101.

[34] Deb SK, Wilding M, Somayazulu M, *et al.* Pressure-Induced amorphization and an amorphous-amorphous transition in densified porous silicon. *Nature*, 2001, 414: 528–530.

[35] McMillan PF, Wilson M, Daisenberger D, *et al.* A density-driven phase transition between semiconducting and metallic polyamorphs of silicon. *Nat Mater*, 2005, 4: 680–684.

[36] Gassilloud R, Ballif C, Gasser P, *et al.* Deformation mechanisms of silicon during nanoscratching. *Phys Status Solidi A*, 2005, 202: 2858–2869.

[37] Wu YQ, Huang H, Zou J, *et al.* Nanoscratch-Induced phase transformation of monocrystalline Si. *Scr Mater*, 2010, 63: 847–850.

[38] Zhou LB, Shimizu J, Shinohara K, *et al.* Three-Dimensional kinematical analyses for surface grinding of large scale substrate. *Precis Eng*, 2003, 27: 175–184.

[39] Huang H, Wang BL, Wang Y, *et al.* Characteristics of silicon substrates fabricated using nanogrinding and chemo-mechanical-grinding. *Mater Sci Eng A*, 2008, 479: 373–379.

[40] Wang Y, Zou J, Huang H, *et al.* Formation mechanism of nanocrystalline high-pressure phases in silicon during nanogrinding. *Nanotechnology*, 2007, 18: 465705.

[41] Kim DE, Oh SI. Deformation pathway to high-pressure phases of silicon during nanoindentation. *J Appl Phys*, 2008, 104: 013502.

[42] Weppelmann ER, Field JS, Swain MV. Observation, analysis, and simulation of the hysteresis of silicon using ultra-micro-indentation with spherical indenters. *J Mater Res*, 1993, 8: 830–840.

[43] Pharr GM, Oliver WC, Harding DS. New evidence for a pressure-induced phase-transformation during the indentation of silicon. *J Mater Res*, 1991, 6: 1129–1130.

[44] Minowa K, Sumino K. Stress-Induced amorphization of a silicon crystal by mechanical scratching. *Phys Rev Lett*, 1992, 69: 320–322.

[45] Zhang LC, Tanaka H. On the mechanics and physics in the nanoindentation of silicon monocrystals. *JSME Int J Ser A*, 1999, 42: 546–559.

[46] Zhang LC, Tanaka H. Atomic scale deformation in silicon monocrystals induced by two-body and three-body contact Sliding. *Tribol Int*, 1998, 31: 425–433.

[47] Chrobak D, Tymiak N, Beaber A, *et al.* Deconfinement leads to changes in the nanoscale plasticity of silicon. *Nat Nanotechnol*, 2011, 6: 480–484.
[48] Gerbig YB, Stranick SJ, Cook RF. Direct observation of phase transformation anisotropy in indented silicon studied by confocal raman spectroscopy. *Phys Rev B*, 2011, 83: 205209.
[49] Gilman JJ. Shear-Induced metallization. *Philos Mag B*, 1993, 67: 207–214.
[50] Gerbig YB, Michaels CA, Forster AM, *et al.* In situ observation of the indentation-induced phase transformation of silicon thin films. *Phys Rev B*, 2012, 85: 104102.

Chapter 5

Wear Behavior of Single-Crystal Copper Nanofilm Materials Under Aqueous Film Conditions

5.1 General

Materials utilized in Micro-Electro-Mechanical Systems (MEMS) devices and semiconductor fabrication are subject to a multi-faceted preparation process that typically involves precision polishing via Chemical Mechanical Polishing (CMP) technology. CMP is a synergistic amalgamation of mechanical and chemical processes, which influences material removal, surface finishing, and tribological wear through a multitude of factors. Concurrently, the operational environment significantly impacts nanoscale devices. Environmental humidity and the presence of liquid water during the polishing stages—encompassing both the water within the polishing solution and that involved in the rinsing process—are conspicuous and impactful factors on material properties, both during the polishing and under operational conditions. Despite their significance, the effects of these environmental elements on material mechanics, friction, and wear are often overlooked in practical research and applications. This oversight is particularly pronounced at the nanoscale, where the scale effect magnifies even seemingly minor influences, necessitating a thorough investigation into the tribological behavior of nanomaterials in moist or aqueous environments.

To date, experimental research on the mechanical properties, elastic-plastic deformation behaviors, and underlying mechanisms of nanomaterials in liquid environments or under simple wetting conditions remains scarce. Corresponding MD simulations are seldom reported in the literature. In light of these gaps, the present authors have developed an extensive MD model, leveraging computational capabilities to delve into the intricacies of complex wetting environments. We have conducted nanoindentation simulations of single-crystalline copper under wet atmospheric conditions or within an ultra-thin water film, aiming to elucidate the transformations, deformations, and associated mechanisms of nanomaterials within these challenging environments.

5.2 Nanoindentation Behavior of Single-Crystal Copper Water-Bearing Films

5.2.1 *Introduction*

With the escalating demand for ultra-precision and miniaturization in technology, micro- and nano-scale metallic devices have become increasingly integral to modern advancements. Thin films and nanolaminates, as crucial components of nanodevices, have seen widespread application in micro/nano electromechanical systems (MEMS/NEMS) [1–3]. At the nanoscale, the introduction or presence of a third body in the friction and wear system can significantly alter the material properties through mechanisms such as pressing (embedding), rolling, or sliding, under various operating conditions. The mechanical properties of these materials are inherently contingent upon their service conditions, exerting a profound influence on the stability and lifespan of the devices. Consequently, there is a compelling need for a thorough investigation into the mechanical behavior and characteristics of materials.

The nanoindentation technique stands as a valuable tool for assessing the mechanical properties of materials, particularly in examining the plastic deformation mechanisms. While numerous nanoindentation experiments have been conducted, the microscopic operational processes often remain elusive due to the limitations inherent in experimental scale. In contrast, MD methodology offers a distinctive vantage point for probing material properties. Liang *et al.* [4] discerned the initial microstructural modifications during the elasto-plastic response through MD simulations of copper nanoindentation. Gao *et al.* [5] elucidated the dislocation initiation and reaction dynamics within iron materials. Despite the multitude of studies probing the mechanical behavior of diverse materials from multiple perspectives, these investigations predominantly focus on the materials themselves and employ simpler models, often overlooking the external environmental factors. Such factors include the wettability of material surfaces, the impact of ambient moisture, and the alterations in indentation methodologies at the nanoscale, all of which can significantly influence material deformation behavior.

Building upon those foundational studies, the present authors have employed the MD method to simulate the impact of water film adsorption on the surface of single-crystal copper films, examining the mechanical properties of single-crystal copper. The analysis of the plastic deformation behavior and dislocation evolution under varying water film thicknesses and indentation modalities has shed light on inter-atomic interactions and the initial plasticity mechanisms. This research contributes to a more profound comprehension of the deformation behavior of nanomaterials, especially those encapsulated by water films.

5.2.2 *Modeling and simulation methods*

Figure 5.1 illustrates the initial model designed for nanoindentation testing on single-crystal copper incorporating a water film. This model encompasses a spherical diamond nanoindenter, a water film, and a defect-free single-crystal copper matrix. The indenter, constructed from a diamond structural unit, has a radius of 2.53 nm and is composed of 11,954 carbon atoms. The dimensions of the single-crystal copper matrix, aligned along the x-[100], y-[010], and z-[001] crystallographic directions, are $68a \times 68a \times 35a$, with $a = 0.3615$ nm representing the lattice constant of the single-crystal copper. The matrix is structured into a boundary layer (fixed during simulation),

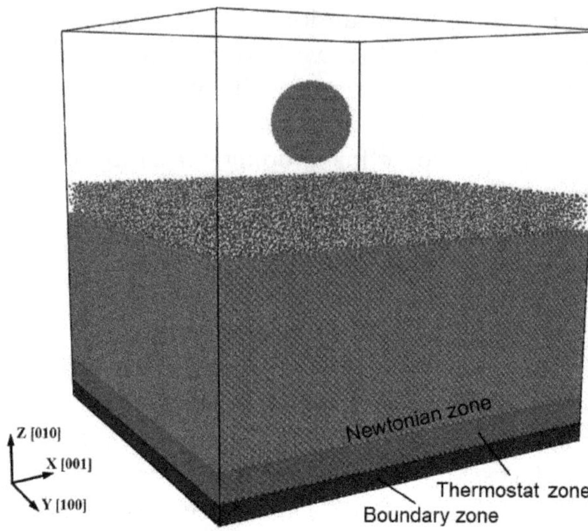

Figure 5.1. Initial model of water-bearing single-crystal copper nanoindentation.

a heat bath (thermostat) layer (serving to regulate system temperature), and a Newtonian layer, where the atoms in the latter two layers adhere to Newtonian motion laws.

The water film is a pivotal factor in examining nanoindentation and nano abrasion behaviors, necessitating a meticulous construction of its initial model. Initially, a water film with a predefined density of $1.0\,g/cm^3$ was crafted using the Amorphous Cell module within Materials Studio software. The three-dimensional configuration of this water film was customized as per the experimental requirements. Subsequently, the Forcite module was employed to endow the water film with a force field, specifically the CVFF, which assigns charges to each atom and initializes other computational parameters. The water film was then exported from Materials Studio and imported into a LAMMPS file. Utilizing the msi2lmp tool alongside pertinent commands, the water film's force field parameters were converted into an executable LAMMPS data file. Ultimately, the LAMMPS input script was scripted to read the data file, thereby finalizing the construction and integration of the water film model.

The interactive forces among copper atoms are governed by the Morse potential [6, 7]. The forces between copper and carbon atoms are delineated by the Morse potential function [8], as depicted in Equation (2.25), with parameters $D = 0.1\,\text{eV}$, $\alpha = 1.7\,\text{Å}^{-1}$, $r_0 = 2.2\,\text{Å}$, and a truncation radius of 0.4 nm. Carbon–carbon interactions were excluded from the simulation, given the diamond indenter's rigid body assumption. Water molecule interactions are characterized using the TIP4P model [9], which encompasses three fixed point charges and a Lennard–Jones center. The influence of hydrogen atoms on the system's thermodynamic properties, being minimal, led to the omission of hydrogen–hydrogen interactions; instead, Lennard–Jones potentials were applied for interactions between copper and oxygen atoms, as well as between carbon and oxygen atoms. The associated parameters are detailed in Table 5.1 [10–12]. A summary of the pertinent parameters is presented in Table 5.1.

All MD simulations were conducted utilizing the robust, large-scale MD simulation software package LAMMPS. The integration of velocities was executed via the Velocity–Verlet algorithm, employing a 1 femtosecond time step. Along the x and y axes, periodic boundary conditions were established to mimic an infinite system effectively. The temperature within the single-crystal copper thermoset layer was meticulously regulated to 300 Kelvin using the Langevin thermoset method, which facilitates the maintenance of a constant temperature by coupling the system to a thermal reservoir. Before the nanoindentation, the system underwent an equilibration procedure spanning 50 picoseconds under the NVT (constant number of particles, volume, and temperature) ensemble conditions. This relaxation process ensures that the system reaches a stable, thermodynamically consistent state. For the indentation process, both velocity-controlled and load-controlled methodologies were employed to probe the mechanical response of the material under varying experimental conditions.

Table 5.1. TIP4P and Lennard–Jones potential function parameters.

Atomic pair	ε (kcal/mol)	δ (nm)	Cut-off radius (nm)
Oxygen–oxygen	0.1554	0.31655	0.6
Copper–oxygen	0.2708	0.28877	0.5
Carbon–oxygen	0.1	0.3275	0.5

5.2.3 *Effect of water film on plastic deformation of single-crystal copper*

To elucidate the influence and underlying mechanisms of water film thickness on the plastic deformation behavior of single-crystal copper, nanoindentation simulations were conducted on single-crystal copper under conditions both devoid of an aqueous film and with water film thicknesses of 1.0 nm, 2.0 nm, and 3.0 nm. In these simulations, the indentation of the indenter was governed by displacement, specifically at a constant velocity of 10 m/s.

5.2.3.1 *Load–depth curve and dislocation evolution process*

The analysis of the elastic-plastic deformation of single-crystal copper, induced by the indentation of a spherical diamond indenter, was achieved by examining the load–indentation curve and the progression of dislocation formation. In this context, the load represents the cumulative force exerted during the interaction between the indenter and both the water film and the single-crystal copper. Figure 5.2 illustrates the load–indentation depth curves across varying water film thicknesses. In the system devoid of an aqueous film,

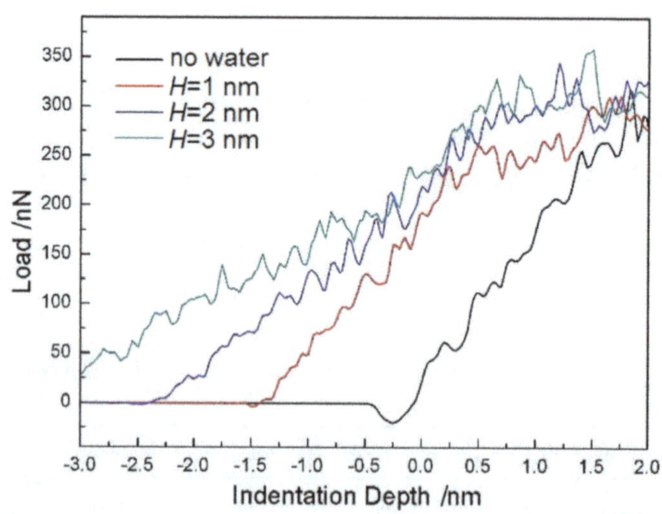

Figure 5.2. Load–pressure depth curves for different water film thicknesses.

the load–indentation curve exhibits a relatively linear trend for indentation depths less than 0.25 nm, indicative of the elastic deformation characteristics of the single-crystal copper. Despite alignment with certain [4, 13, 14] outcomes, a divergence is observed when compared with experimental data [15]. Liang *et al.* [4] have posited that this discrepancy predominantly stems from the differences in indenter dimensions and indentation depths between the MD simulations and experimental setups. Upon further increase in indentation depth, the load–indentation curve experiences a precipitous decline, signifying a stress escalation in the plastic deformation of the single-crystal copper. When a water film is present on the surface of the single-crystal copper, the load–indentation depth curve exhibits pronounced fluctuations, complicating the identification of the elastic deformation region and the point of stress inflection. Furthermore, Figure 5.2 demonstrates that the load is directly proportional to the water film thickness at low indentation depths. However, as the indentation depth increases, the impact of water film thickness on the load–pressure depth curve does not maintain a direct proportionality.

It is widely recognized that the characteristics of the load–indentation depth curve are indicative of the elastic-plastic deformation response of a matrix, which is inherently linked to the evolution of the dislocation structure within that matrix. Figure 5.3 depicts the dislocation evolution within single-crystal copper under varying thicknesses of the water film. In this visualization, red atoms signify surface atoms and those involved in dislocations, green and yellow atoms represent the initial dislocation expansion as the indentation depth (h) increases, and brown atoms denote newly nucleated dislocations. Analysis of the dislocation nucleation and expansion process reveals that post elastic deformation, small atomic clusters emerge beneath the indenter, signifying the formation of dislocation embryos. Upon further increment of the indentation depth surpassing a critical threshold (for instance, $h = 0.3\,\text{nm}$ in Figure 5.3), Shockley dislocation pairs, in conjunction with stacking faults (represented by yellow and green atoms), nucleate and evolve along the $\{111\}$ slip planes. This initial dislocation activity corresponds to the peak of stress concentration and the abrupt load drop in the load–depth curve. The Shockley pairs continue to grow, expand, and interact with the surface of the single-crystal copper, while numerous

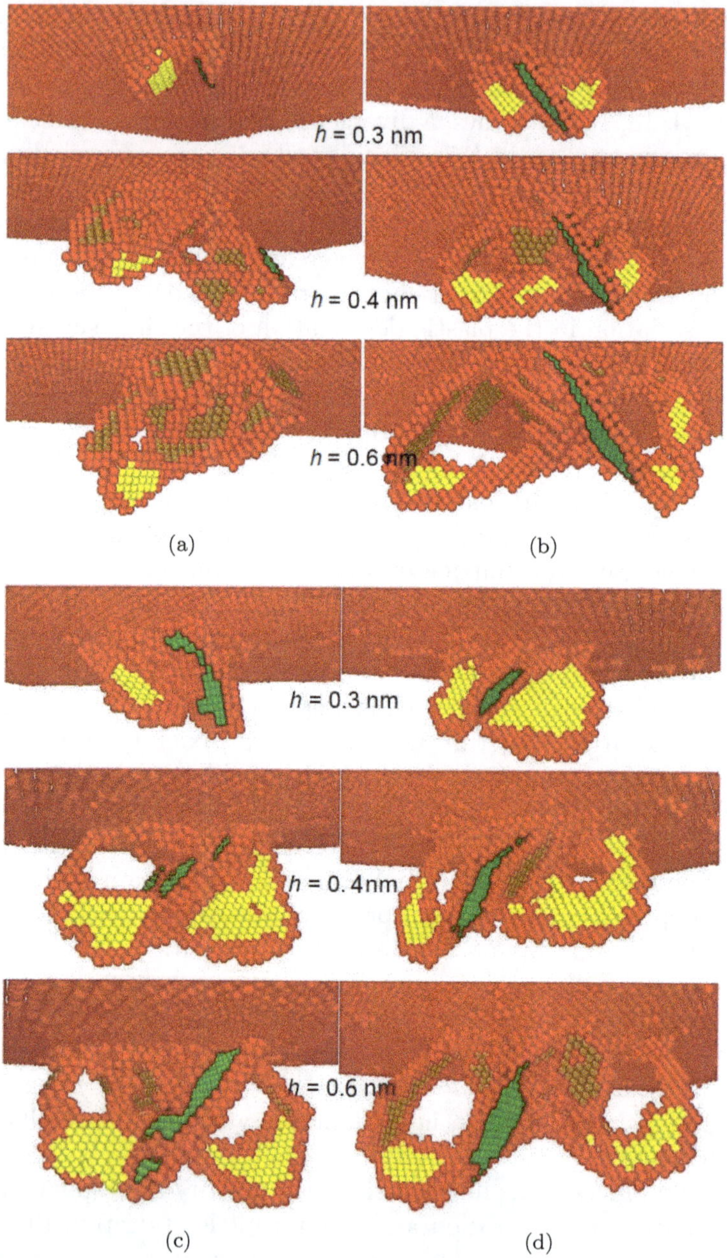

Figure 5.3. Evolution of internal plastic deformation of single-crystal copper at different water film thicknesses. (a) No water film, (b) water film thickness of 1.0 nm, (c) water film thickness of 2.0 nm, (d) water film thickness of 3.0 nm.

new dislocations nucleate, leading to the formation of a multitude of dislocation loops and Lomer–Cottrell locks (as exemplified at $h = 0.4\,\text{nm}$ and $h = 0.6\,\text{nm}$ in Figure 5.3).

Furthermore, Figure 5.3 illustrates that at a constant pressure depth, an increased thickness of the water film on the single-crystal copper surface results in a larger area and deeper expansion of the initial Shockley pairs. This suggests that the presence of the water film facilitates the complete expansion of these dislocation pairs. In contrast, in the absence of a water film, the tangles of initial dislocations become fully developed and slowly propagate into the interior of the single-crystal copper (as shown in Figure 5.3(a)). This differential in dislocation expansion implies that the water film may induce alterations in the modalities of interatomic force transfer. Figure 5.4 presents the dislocation evolution within single-crystal copper at substantial pressure depths across different water film thicknesses, with red atoms indicating surface and dislocation atoms and blue atoms signifying stacking faults.

Figure 5.2 illustrates that at an indentation depth exceeding 0.8 nm, the load–indentation curve demonstrates the phenomenon of strain bursts. These bursts trigger the nucleation of a multitude of new dislocations, which subsequently interact and become entangled with existing dislocations. Consequently, a proliferation of dislocations emanates from the subsurface of the single-crystal Cu along the $\{111\}$ slip planes, penetrating deeper into the material. The intricate interplay among dislocations fosters the emergence of dislocation rings, which form and migrate away from the indentation zone. Additionally, Figures 5.4(a–d) depict that an augmentation in the water film's thickness can instigate a more pronounced inward expansion of dislocations toward the substrate's interior. To delve deeper into the influence of the water film on the plastic deformation of single-crystal copper, a quantitative assessment was conducted by tallying dislocation line lengths within a predefined volume, as depicted in Figure 5.5. It is observable that the dislocation line length escalates with increasing indentation depth, exhibiting a proportionality to the thickness of the water film. The variation in dislocation line length is indicative of the extent of plastic deformation experienced by the single-crystal copper.

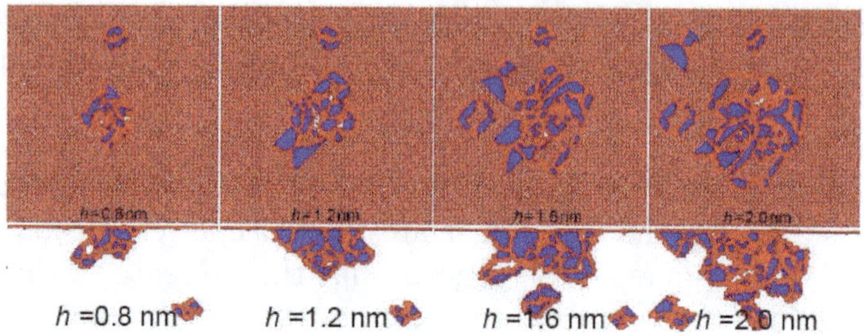

(a)

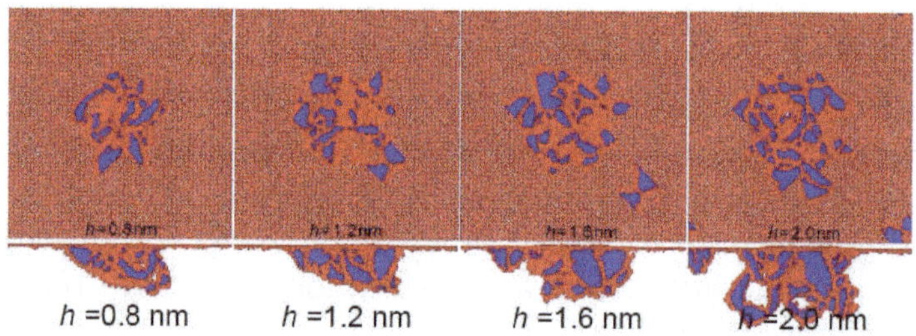

(b)

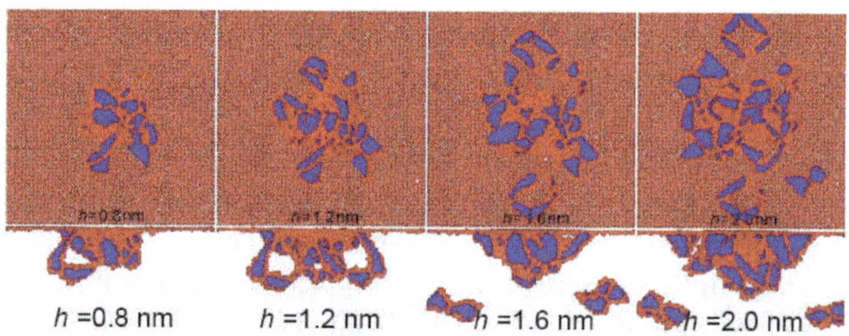

(c)

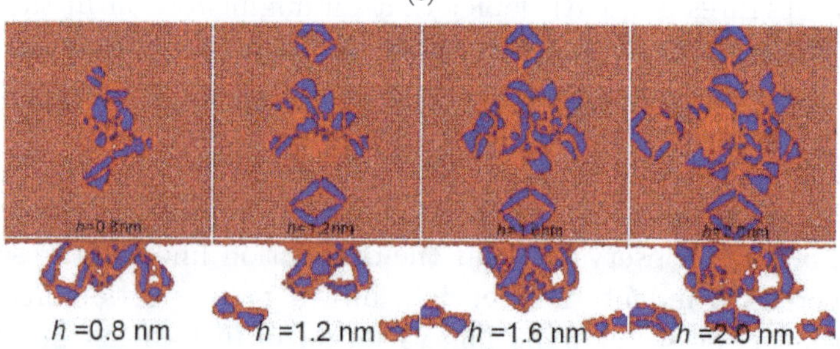

(d)

Figure 5.4. Snapshots of dislocation evolution under different water film thicknesses. (a) No water film, (b) water film thickness of 1.0 nm, (c) water film thickness of 2.0 nm, (d) water film thickness of 3.0 nm.

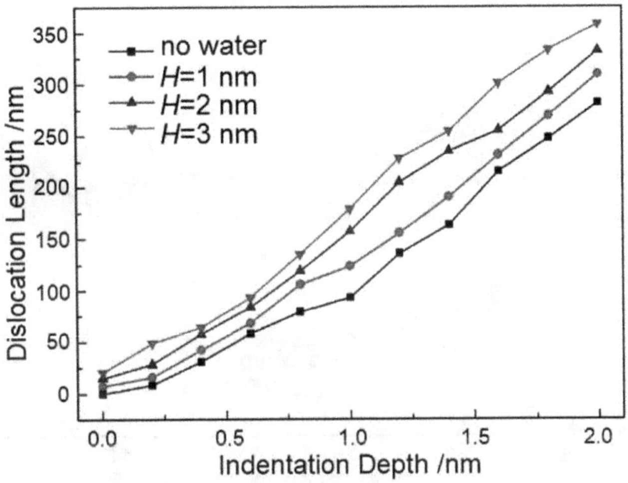

Figure 5.5. Relationship between the length of dislocation line and the indentation depth and water film thickness.

5.2.3.2 *Plastic deformation mechanism*

To thoroughly elucidate the mechanism underlying the water film's impact on the plastic deformation of single-crystal copper, initial calculations of the interaction between the indenter and the copper substrate were performed, as depicted in Figure 5.6. Observations from the figures indicate that the presence of a water film introduces a significant delay and diminishes the interaction between the indenter and the copper substrate compared to conditions without a water film. Furthermore, an increase in the water film's thickness correlates with an increased indentation depth, yet it attenuates the interaction between the copper and the indenter at equivalent depths.

The insets in Figure 5.6 display selected configurations of single-crystal copper surrounding the indenter at a water film thickness of 2.0 nm, while Figure 5.7 illustrate the quantity of water molecules within the indentation. The configuration diagrams reveal that during the initial phase of nanoindentation, a multitude of water molecules occupy the interfacial space between the copper substrate and the diamond indenter, partially obstructing direct contact between the indenter and the substrate. As the indentation depth escalates, some of the water molecules positioned between the indenter and the copper surface are compelled into the indentation pit.

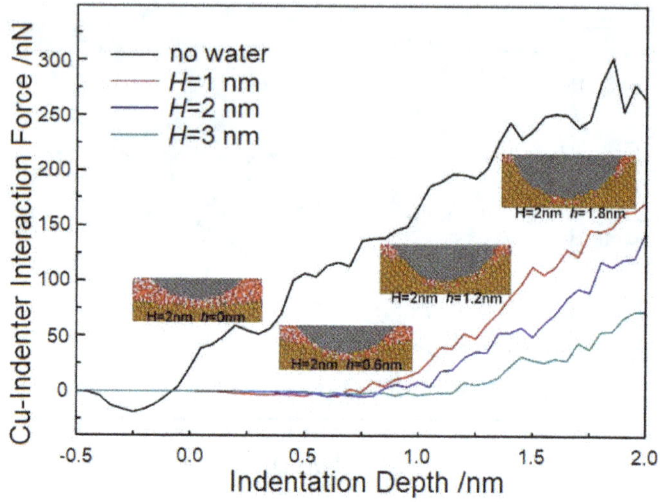

Figure 5.6. Interaction force between indenter and single-crystal copper for different water film thicknesses.

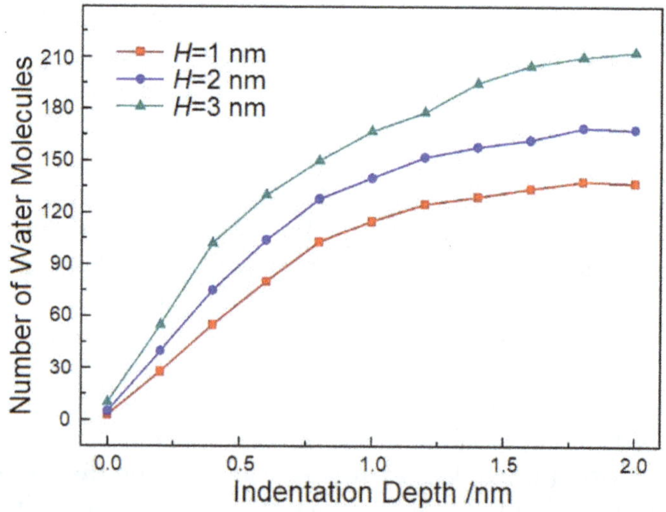

Figure 5.7. Number of water molecules in the indentation pit for different water film thicknesses.

A thicker water film results in a greater number of water molecules being forced into the pit, as shown in Figure 5.7, where the water molecule count within the pit escalates with both the indentation depth and the water film's thickness. Consequently, as Figure 5.7 demonstrates, the water molecule population within the pressure pit surges with the depth of indentation and the thickness of the water film. However, due to the interposition of these water molecules, the interaction between the indenter and the copper remains negligible until an indentation depth of 0.6 nm is reached, as observed in Figure 5.6. Subsequently, as the indentation depth and the load exerted by the indenter continue to increase significantly, the water molecules surrounding the indenter encounter greater resistance to further penetration into the pit. This results in a deceleration of the water molecule accumulation within the pit and a dispersion of those already present, as indicated in the insets of Figure 5.6, thereby initiating an increase in the interaction between the indenter and the copper substrate.

The findings indicate that water molecules exert a considerable obstructive effect on the contact between the indenter and the single-crystal copper film during nanoindentation. Intuitively, this obstructive effect would be expected to result in shallower indentations and a reduced number of atomic defects beneath the indenter in the presence of water compared to dry conditions. However, the outcomes presented in Figures 5.2–5.5 contradict this expectation. To reconcile this discrepancy, the authors computed the interaction forces between the indenter–water film and the copper–water film, as portrayed in Figure 5.8. It is evident that the trends of these two interaction forces are relatively analogous, gradually increasing from the onset of the indenter's downward pressure. Considering that the force between the indenter and the copper substrate in Figure 5.6 remains negligible for an extended period, it can be inferred that the water film initially bears the indenter's pressure, concurrently transmitting the pressure to the single-crystal copper surface. Once the indentation depth is substantial enough to breach the water film, and the water molecules within the pit are dispersed, the water film's integrity is compromised, and its influence is mitigated, leading to a reduction in the force between the water film and both the indenter and the copper substrate.

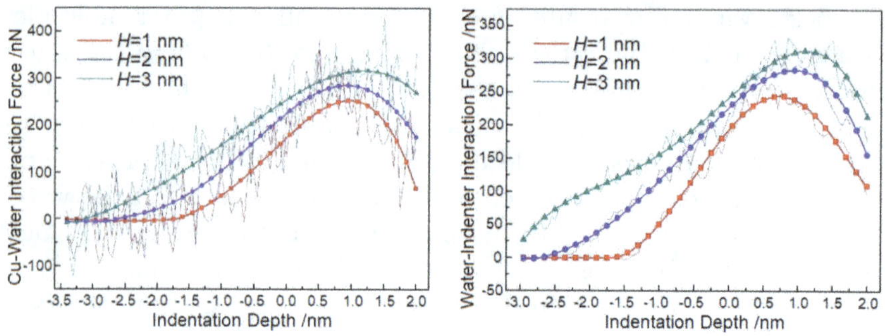

(a) (b)

Figure 5.8. Interaction forces between water film and copper substrate and indenter for different water film thicknesses. (a) Copper–water film interaction force, (b) indenter–water film interaction force.

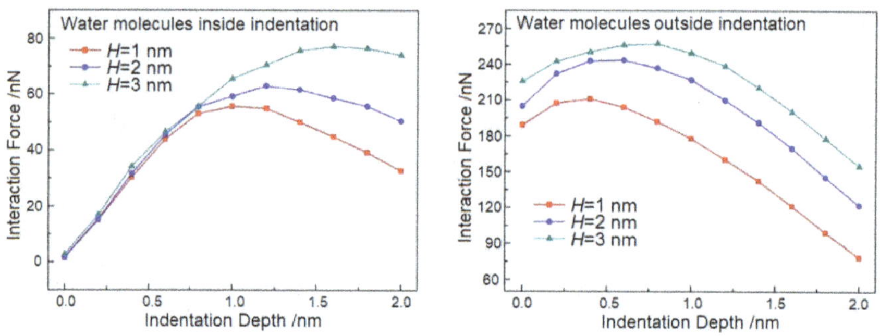

(a) (b)

Figure 5.9. Interaction force curves between the indenter and the water molecules inside and outside the pressure pit. (a) Water molecule-indenter force inside the pit, (b) water molecule-indenter force outside the pit.

Figure 5.9 presents an analysis of the interaction forces between the indenter and water molecules located both within and outside the indentation pit. The figure indicates that the interaction force between the water molecules within the pit and the indenter increases linearly with the indentation depth, showing no significant correlation with the thickness of the water film. Concurrently, the interaction force between the indenter and the water molecules external to the indentation pit initiates an increase from a substantial initial value. In essence, the escalation of these two forces is reflective of the nucleation, expansion, and reaction of a multitude of dislocations within the single-crystal copper, as depicted in Figure 5.3. Accordingly, during the incipient phase of plastic deformation in single-crystal copper, the water molecules positioned beneath the indenter serve as an intermediary for force transfer, conveying the pressure applied by the indenter to the copper surface, thereby instigating plastic deformation of the copper film in the absence of direct contact with the indenter. Conversely, water molecules external to the indenter are also subjected to lateral forces exerted by the indenter, channeling this force to the copper surface. This process impedes the propagation of subsurface dislocations to the surface while promoting the inward expansion of dislocations into the substrate's interior.

Upon the indenter's penetration through the water film, a diminution in the force between the indenter and the water molecules outside the pit is observed, as illustrated in Figure 5.9(b). This event enhances the direct interaction between the indenter and the copper film, culminating in substantial plastic deformation.

5.2.4 *Effect of indentation method on plastic deformation of single-crystal copper*

At the nanoscale, both displacement-controlled and load-controlled indentation methodologies are instrumental in dictating the maximum displacement and load during the indentation process, exerting a significant influence on the mechanical properties of materials [16–19]. In this study, displacement control implies that the indenter is lowered at a uniform velocity throughout the loading phase, with adopted velocities of 10 m/s and 50 m/s, culminating in a maximum indentation depth of 2.0 nm. Conversely, load control involves the application of an external force to the indenter, maintaining a constant increment of internal and external forces over time (the loading rate), with adopted force increments of 1 nN/ps and 3 nN/ps, capped at a maximum load of 288.5 nN.

5.2.4.1 *Load–depth curve*

Figures 5.10 and 5.11 demonstrate that under anhydrous film indentation conditions, all load–indentation curves exhibit a smooth linear progression when the indentation depth remains below 0.4 nm,

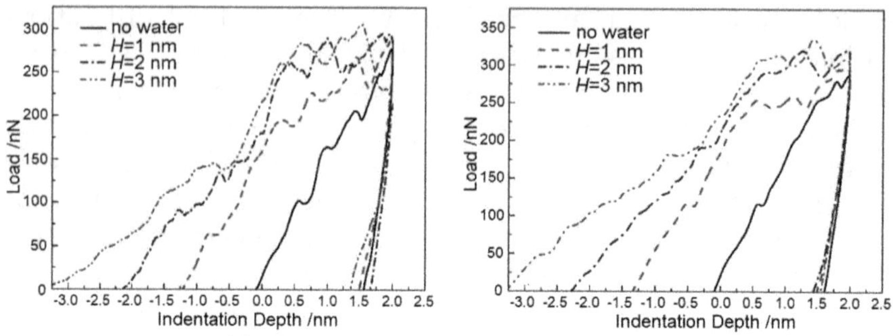

(a) (b)

Figure 5.10. Load–indentation depth curves for displacement-controlled loading. (a) Indentation speed $v = 10\,\mathrm{m}\cdot\mathrm{s}^{-1}$, (b) indentation speed $v = 50\,\mathrm{m}\cdot\mathrm{s}^{-1}$.

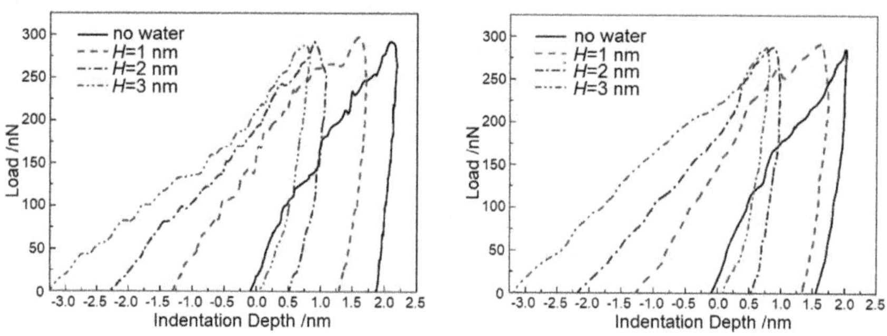

(a) (b)

Figure 5.11. Load–indentation depth curves for load-controlled loading. (a) Load rate $F = 1\,\mathrm{nN}\cdot ps^{-1}$, (b) load rate $F = 3\,\mathrm{nN}\cdot \mathrm{ps}^{-1}$.

aligning with the initial elastic deformation behavior of single-crystal copper. As the indenter applies increasing downward pressure, the load–displacement curves experience a sudden decline, indicative of plastic yielding. Subsequently, a surge in strain occurs due to escalating plastic deformation of the single-crystal copper. This quintessential initial plastic deformation behavior is corroborated by findings from related research [4, 13, 14, 20].

For the displacement-controlled nanoindentation process (Figure 5.10), the presence of a water film introduces notable fluctuations to the load–indentation curve before the indenter makes contact with the single-crystal copper film. These fluctuations intensify with increasing water film thickness, reflecting the dynamic interaction between the water film and the indenter. This interaction is predominantly attributed to the independent nature of water molecules at the nanoscale. Upon indenter contact with the copper film, the load–depth curve displays pronounced stress relaxation characteristics [20]. The presence of a water film also precipitates an earlier onset of plastic deformation in single-crystal copper; for instance, stress relaxation at a 2.0 nm water film thickness is observed at an indentation depth of 0.3 nm, compared to an indentation depth of 0.5 nm in the absence of water. Furthermore, the variability of the load–indentation curves at high velocities ($v = 50\,\text{m/s}$) is diminished compared to those at lower indentation velocities ($v = 10\,\text{m/s}$). The present author also calculates the maximum load applied to the indenter during the loading process, as depicted in Figure 5.12(a). This figure indicates that, for an equivalent maximum indentation depth, the maximum load escalates by 7.8% and 15.5% with the thickening of the water film and an increase in indentation speed, respectively.

Second, in the context of the load-controlled indentation process (Figure 5.11), the load–indentation depth curve exhibits no significant fluctuations. This consistency arises because the external force exerted on the indenter increments continuously over time. Figure 5.12(b) illustrates the maximum indentation depth derived from the load–indentation curve. Given that the maximum external force applied to the indenter at the culmination of the indentation

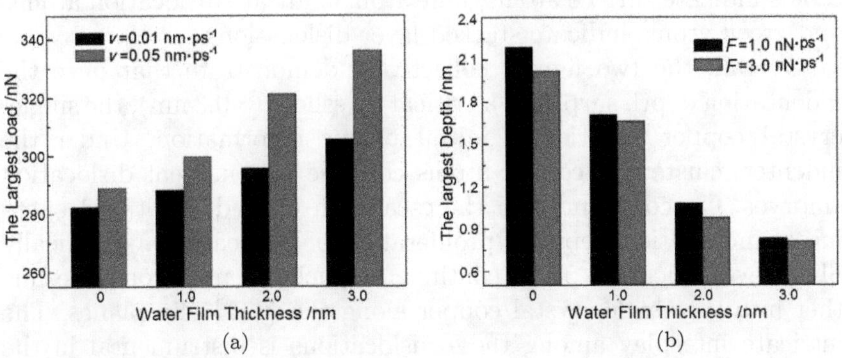

Figure 5.12. Maximum load during displacement-controlled press-in and maximum depth of press-in for load-controlled press-in. (a) Displacement-controlled press-in process, (b) load-controlled press-in process.

process is held constant, and considering the water film's substantial energy absorption during nanoindentation, the maximum indentation depth attenuates with the increment of the water film's thickness (with reductions of 61.2% and 59.3%, respectively). This outcome contrasts starkly with the behavior of the maximum load under displacement control conditions. Concurrently, the indentation depth reduction from 1.0 nm to 2.0 nm water film thickness is markedly more pronounced than the reduction observed when the film thickness increases from 2.0 nm to 3.0 nm. This disparity suggests that the influence of the water film on the nanoindentation behavior of single-crystalline Cu diminishes as the film's thickness augments. Furthermore, it is observable that an escalation in the loading rate precipitates a decrement in the maximum indentation depth. Similar conclusions were reached by Chang and Zhang [16] in their investigation of the loading rate's impact on the plastic behavior of silicon materials. Collectively, these findings underscore that both the loading rate/indentation speed and the presence of a water film exert a significant influence on the indentation behavior when a diamond indenter is applied to single-crystal copper.

5.2.4.2 *Dislocation evolution process*

Figures 5.13 and 5.14 show the dislocation nucleation and expansion processes within single-crystal copper at different water film thicknesses and indentation speeds/loading rates, respectively, where red atoms indicate surface atoms, blue atoms indicate dislocation atoms, and green atoms indicate stacked layer dislocations.

Overall, the two figures collectively demonstrate that once the indentation depth surpasses a critical threshold ($\sim$0.3 nm), the single-crystal copper experiences initial plastic deformation. Under the indenter, clusters of copper atoms coalesce into nascent dislocation embryos. Concomitant with the escalation of load and the deepening of indentation depth, a proliferation of dislocations, specifically Shockley dislocation pairs, continue to nucleate and propagate further into the single-crystal copper along the $\{111\}$ slip planes. The intricate interplay among these dislocations is instrumental in the genesis of defects, including dislocation rings and Lomer–Cottrell locks [5, 14, 21]. To provide a quantifiable assessment of the defects within the single-crystal copper at various nanoindentation junctures

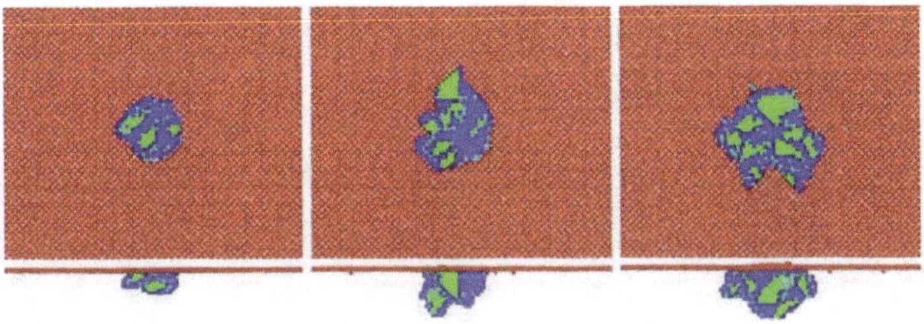

(a)

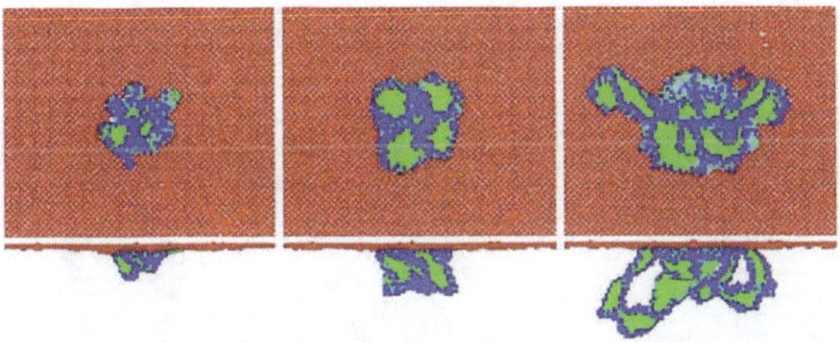

(b)

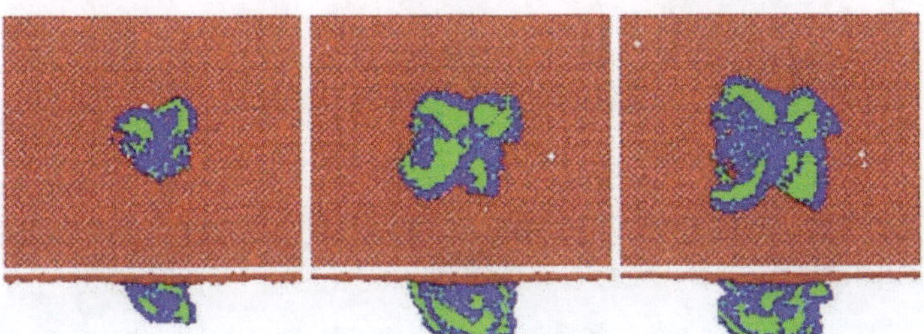

(c)

Figure 5.13. Evolution of dislocations within the single-crystal copper during displacement-controlled indentation. (a) Anhydrous, indentation speed $10\,\mathrm{m\cdot s^{-1}}$, depth of indentation 0.3 nm, 0.5 nm, and 0.7 nm in that order, (b) water film thickness of 2.0 nm, indentation speed of $10\,\mathrm{m\cdot s^{-1}}$, depth of indentation of 0.3 nm, 0.5 nm, and 0.7 nm in that order, (c) water film thickness of 2.0 nm, indentation speed of $50\,\mathrm{m\cdot s^{-1}}$, depth of indentation of 0.3 nm, 0.5 nm, and 0.7 nm, respectively.

or stages, this study undertakes a tally of dislocation line lengths throughout the indentation process, as depicted in Figure 5.15. The figure reveals that the dislocation line lengths, under diverse nanoindentation conditions, exhibit a continuous increment in tandem with

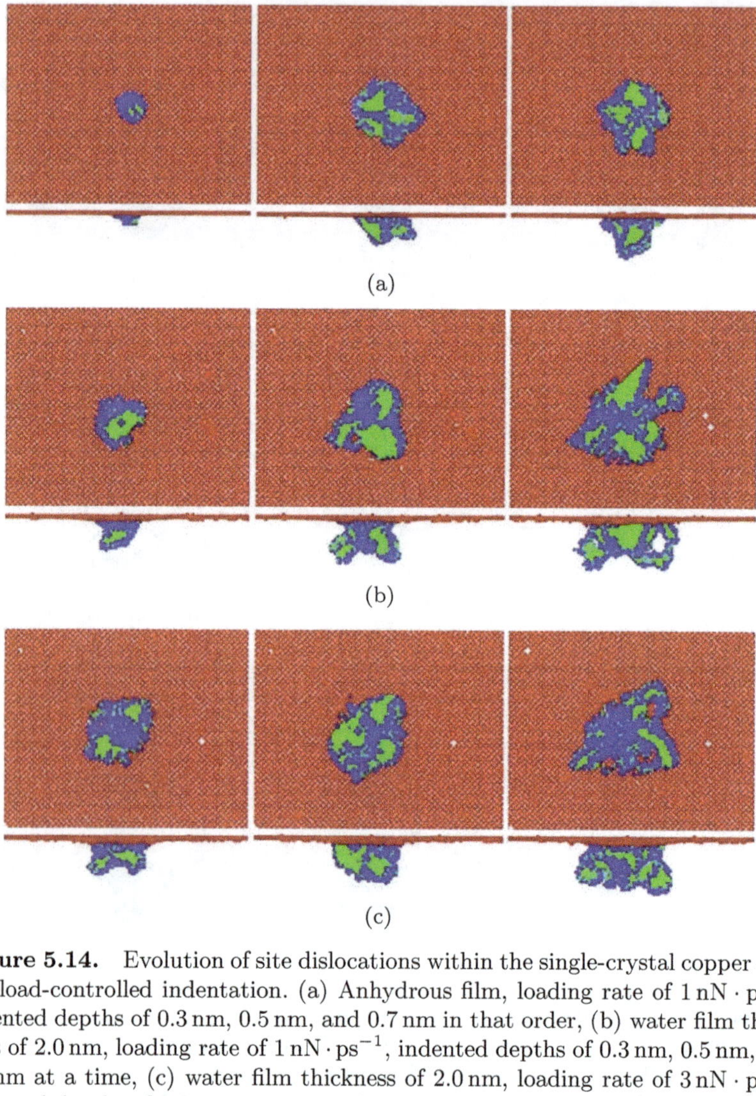

Figure 5.14. Evolution of site dislocations within the single-crystal copper during load-controlled indentation. (a) Anhydrous film, loading rate of $1\,\text{nN} \cdot \text{ps}^{-1}$, indented depths of 0.3 nm, 0.5 nm, and 0.7 nm in that order, (b) water film thickness of 2.0 nm, loading rate of $1\,\text{nN} \cdot \text{ps}^{-1}$, indented depths of 0.3 nm, 0.5 nm, and 0.7 nm at a time, (c) water film thickness of 2.0 nm, loading rate of $3\,\text{nN} \cdot \text{ps}^{-1}$, indented depths of 0.3 nm, 0.4 nm, and 0.5 nm at a time.

the growing indentation depth. This observation in Figures 5.13 and 5.14 substantiates the dislocation nucleation and expansion phenomena captured in the constitutive figures.

Figures 5.13(a) and 5.13(b) illustrate that, in comparison to the dislocation characteristics formed during the nanoindentation

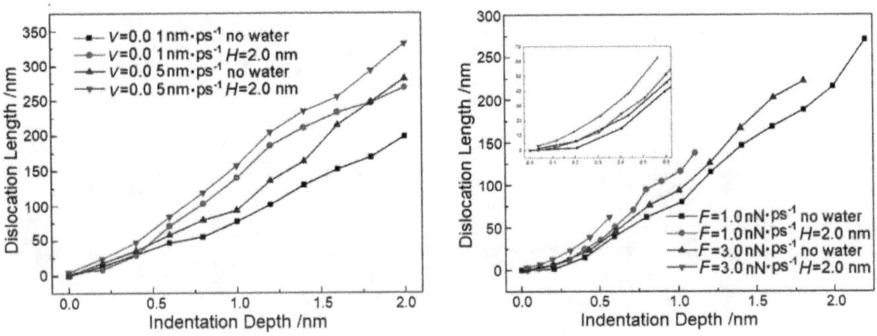

(a) (b)

Figure 5.15. Dislocation line lengths for different indentation conditions. (a) Displacement-controlled press-in process, (b) load-controlled press-in process.

process under anhydrous conditions, an increased number of dislocations are observed within the single-crystal copper at an indentation depth of 2.0 nm with an aqueous film of the same thickness. Furthermore, at an indentation depth of 0.7 nm, a greater formation of dislocation loops is evident, indicating that the plastic deformation of the single-crystal copper has entered a strain burst phase. Figure 5.15(a) reveals that the dislocation line length exhibits a more pronounced growth trend for water-containing films when the indentation speed is constant, suggesting that the presence of a water film enhances dislocation generation. This trend is corroborated in Figures 5.14(a), 5.14(b), and 5.15(b). Moreover, by integrating the dislocation line length variations depicted in Figure 5.15 with the dislocation structures shown in Figures 5.13(b), 5.13(c), 5.14(b), and 5.14(c), it is evident that for scenarios involving water-containing films, the growth trend of dislocation line length intensifies with higher indentation speeds or loading rates. This difference in the growth trend becomes increasingly pronounced as the indentation depth increases. In essence, irrespective of the presence of a water film, elevated indentation speeds or loading rates under both displacement-controlled and load-controlled nanoindentation methods accelerate a more extensive and pronounced dislocation distribution.

The aforementioned findings indicate that both the augmentation of water film thickness and the escalation of indentation speed or loading rate contribute to the nucleation and propagation of

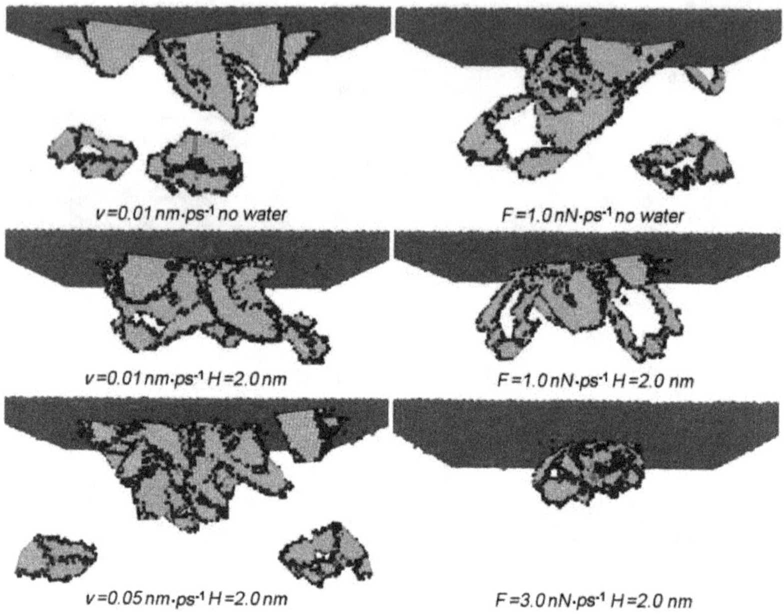

Figure 5.16. Dislocation configurations within single-crystal copper at the end of indentation.

dislocations during the initial stages of plastic deformation in nanoindentation. That is, they exert a significant influence on the incipient plastic deformation characteristics of single-crystal copper.

Figure 5.16 presents the dislocation configuration within single-crystal copper at the culmination of the nanoindentation process. In the scenario of displacement-controlled indentation, an extensive array of dislocations emanates from the indentation subsurface, penetrating into the copper substrate's interior in the absence of a water film. Notably, some dislocation rings encircle the stacking fault, delving deeper into the single-crystal copper and decoupling from the dislocation cores. As depicted in the left-side dislocation configuration of Figure 5.16, the presence of a water film and variations in indentation speed introduce a discernible divergence in the dislocation architecture. The dislocation density escalates with both the augmentation of water film thickness and the acceleration of indentation speed. Conversely, the right-side dislocation configuration in Figure 5.16 illustrates a precipitous reduction in dislocation density with the thickening of the water film and the elevation of the loading rate. These observations underscore that the introduction of a water film and the increment of indentation speed or loading rate exert a profound influence on the plastic deformation behavior of single-crystal copper.

5.2.4.3 *Plastic deformation mechanism*

In the displacement-controlled nanoindentation process without water film, the proliferation of dislocations with increasing indentation depth and speed is more pronounced at greater depths, as illustrated in Figure 5.15(a). An elevation in indentation speed makes it difficult to adjust rapidly, leading to localized deformation in the surface and subsurface of single-crystal copper surrounding the indenter. Consequently, a broader area of the copper surface and subsurface near the indenter undergoes deformation concurrently, with dislocations nucleating and expanding across this extensive region within the single-crystal copper, culminating in a large strain rate. This phenomenon also accounts for the diminished fluctuation observed in the load–indentation depth curve in Figure 5.10(b). Such a high strain rate induces a transient surge in dislocation energy within the single-crystal copper, which impedes the indenter's further penetration into the copper substrate. However, because the displacement-controlled nanoindentation process is indifferent to the magnitude of the external force applied to the indenter—given that the indenter is pressed down at a constant velocity—the external force on the indenter is persistently and transiently amplified to sustain the indentation. This adjustment provides the necessary external force to counteract the increased dislocation energy. Therefore, an increment in indentation velocity during nanoindentation fosters continuous dislocation nucleation and progression.

On the one hand, the presence of a water film impedes direct contact between the indenter and the single-crystal copper. On the other hand, interactions among water molecules consume a portion of the load. To further elucidate the role of the water film during indentation, Figure 5.17 presents an analysis of the surface damage configuration of single-crystal copper under varying indentation depths and water film thicknesses at an indentation speed of 50 m/s. It is observed that surface damage surrounding the indentation pits

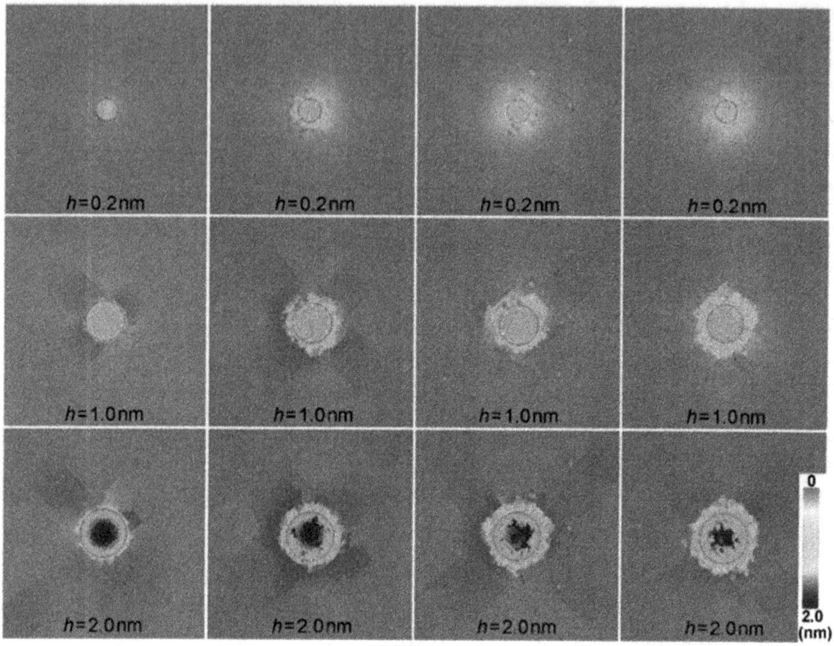

(a) (b) (c) (d)

Figure 5.17. Surface damage configurations of single-crystal copper under different conditions. (a) With no water film, (b) with 1.0 nm water film, (c) with 2.0 nm water film, (d) with 3.0 nm water film.

escalates with increasing indentation depths and water film thicknesses, with the emergence of the water film causing the pit interiors to become irregular. Water molecules surrounding the indenter act as a medium for force transfer, distributing the force exerted by the indenter over a larger area on the single-crystal copper surface. Simultaneously, some water molecules beneath the indenter are adsorbed onto the indenter's surface, effectively becoming part of the indenter. These molecules are pushed into the gaps between deformed copper atoms during the indenter's downward press, further contributing to dislocation generation. Despite the load dissipation due to water molecule interactions, the indenter is continuously adjusted to compensate for this load loss, ensuring a constant indentation speed and preventing a reduction in dislocation generation due to load consumption by the water film. An increase in indentation speed expedites the transfer of force from the water film to the copper surface, thereby hastening plastic deformation, a finding consistent with the results depicted in Figure 5.15(a). In conclusion, this study posits that alterations in the water film and indentation speed modify the mode of interaction between the indenter and the copper substrate, with the instantaneous adjustment of the load playing a pivotal role in the indentation process.

In the context of load-controlled nanoindentation, an increased loading rate in the absence of an aqueous film leads to a high strain rate within the single-crystal copper. This elevated strain rate accelerates dislocation generation and facilitates more rapid expansion into the regions within the single-crystal copper. Consequently, a higher loading rate results in a larger dislocation line length and load for the same depth of indentation, as previously discussed. Additionally, while dislocation density increases with indentation, high loading rates can also lead to excessive load depletion. The determined maximum load—where the maximum applied load is held constant in load-controlled indentation—is insufficient to supply adequate energy to further promote the indentation of the indenter in the final stages. This insufficiency results in a decrease in the final depth of indentation with an increase in the loading rate.

When a water film is present, the plastic deformation process of single-crystal copper mirrors that observed under displacement control conditions. That is, the water film enhances dislocation nucleation and expansion during the initial stages of indentation. Simultaneously, interactions among water molecules also consume a portion of the applied load. An increased loading rate elevates the strain rate, augmenting the plastic deformation of single-crystal copper at a given indentation depth. However, this rate increase can also lead to a rapid consumption of the load, thereby reducing the maximum indentation depth achieved at the end of the nanoindentation process.

From these findings, it can be inferred that during the load-controlled nanoindentation process, if the applied load is continuously supplied to ensure that the indenter reaches the maximum depth of indentation (2.0 nm) under load control, the total plastic deformation of single-crystal copper should be equivalent to that observed under displacement control. In essence, variations in nanoindentation methodologies merely reflect differences in the control of load and displacement (or velocity) during the indentation process, while the fundamental plastic deformation mechanism of single-crystal copper remains consistent.

5.2.5 *Stress relaxation and elastic recovery of single-crystal copper*

Under persistent stress—even when it falls short of the elastic limit—metallic materials can transition from elastic to plastic deformation through a gradual process that is intimately connected to the temperature at which the metal is subjected. At a given temperature, the phenomenon where a metal material, under continuous stress, undergoes a slow plastic deformation is termed metal creep. Creep represents the confluence of temperature and stress effects. Consequently, the instantaneous stress state within a material is typically associated with both the deformation occurring at that moment and the preceding deformation history. In materials that sustain a constant deformation, stress diminishes over time, a phenomenon known as stress relaxation, which can be considered a generalized form of creep. At the nanoscale, the phenomena of creep and stress relaxation in metallic crystals are pivotal factors influencing the mechanical, thermal, and electrical properties of materials. These phenomena are macroscopic expressions of the heat dispersion process within solid materials, being reflective not only of the material's bulk structure and microstructure but also significantly modulated by stress, temperature, and environmental media.

Our previous research has demonstrated that the creep behavior of single-crystal copper is contingent upon the loading modality:

(1) Under a displacement-controlled loading regimen, when the indenter is pressed into the substrate to a specific depth and then held stationary, the single-crystal copper substrate beneath the indenter experiences a decrement in load, indicating the onset of stress relaxation.
(2) In a load-controlled loading regimen, upon the indenter reaching a certain load upon penetration into the substrate, this load is maintained constant for a defined duration, yet the indenter's depth incrementally increases during this period, signifying the occurrence of creep. Despite disparities between the two loading methods, their fundamental nature is analogous, thus rendering

the essence of stress relaxation and creep phenomena under these conditions conceptually identical.

In this book, we further scrutinize the stress relaxation and elastic recovery characteristics of single-crystal copper under aqueous conditions through molecular dynamics simulations and displacement-controlled (constant loading rate) loading. The indenter was loaded and unloaded at a velocity and temporal frame of 10 m/s and 300 picoseconds, respectively, with its position held constant for an additional 50 picoseconds post-loading culmination, a phase uniformly denoted as the stress relaxation process within this book. Subsequently, the indenter was unloaded, as depicted in Figure 5.18. Figure 5.19 portrays the load–indentation curves under varying hydration conditions. While a detailed analysis of the indentation process has been previously presented, it is not reiterated here; instead, the focus is on examining the stress relaxation and unloading processes.

Figure 5.19 illustrates distinct variations in the loading and unloading curves of the single-crystal copper matrix during stress relaxation with different thicknesses of the water film. The negative load values observed during unloading suggest the presence of an adhesive force between the indenter particles and the single-crystal copper atoms as they separate from the substrate. It is evident that this adhesive force is most pronounced in the absence of a water film,

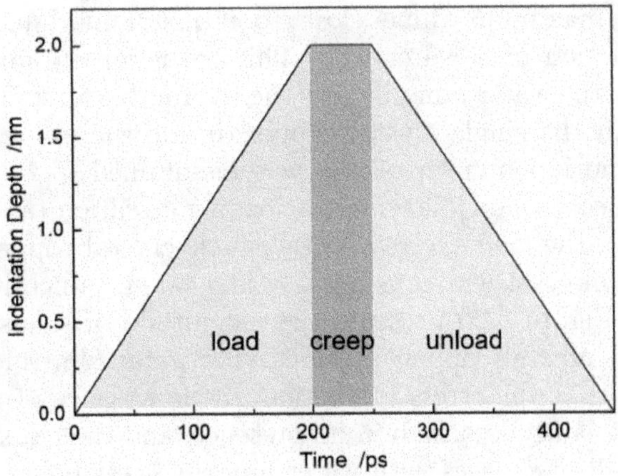

Figure 5.18. Schematic diagram of loading under displacement control.

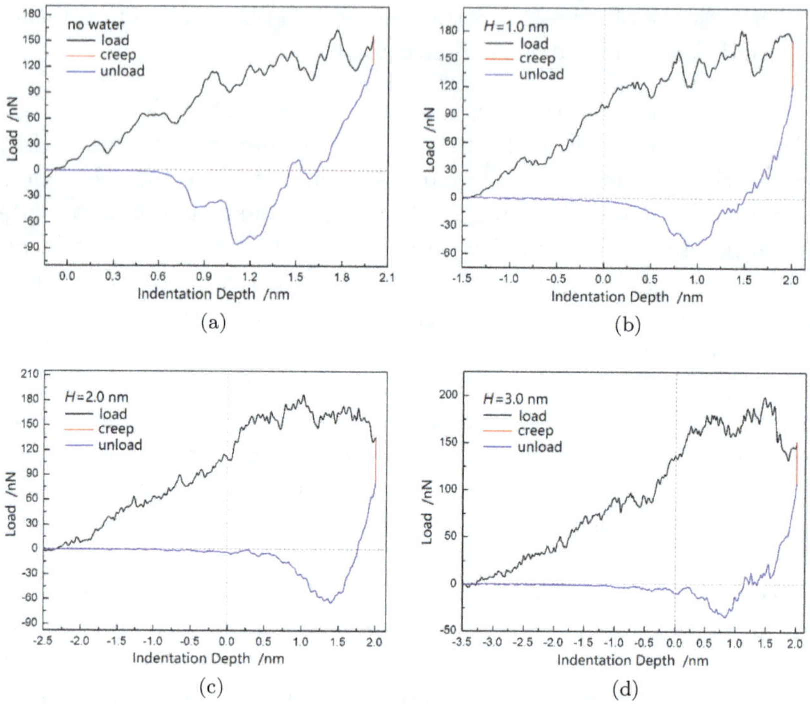

Figure 5.19. Load–indentation depth curves for different water film thicknesses. (a) No water film, (b) water film 1.0 nm, (c) water film 2.0 nm, (d) water film 3.0 nm.

reaching a maximum value close to 90 nN, and diminishes with the introduction of a water film. This reduction in adhesive force implies that the water film impedes the interactive forces between the indenter and the single-crystal copper during the unloading phase. Although the indenter's position was maintained for a brief duration of 50 picoseconds following the loading cessation (as depicted in Figure 5.18), the load exerted on the single-crystal copper substrate during this interval was considerably alleviated, indicative of stress relaxation. Figure 5.20 presents the magnitude of stress relaxation in the single-crystal copper under varying water film thicknesses. It is observed that the stress relaxation in the absence of an aqueous film is significantly less than in its presence, and the relaxation magnitude initially increases with water film thickness before reaching a decrement. This behavior indicates that the stress relaxation induced

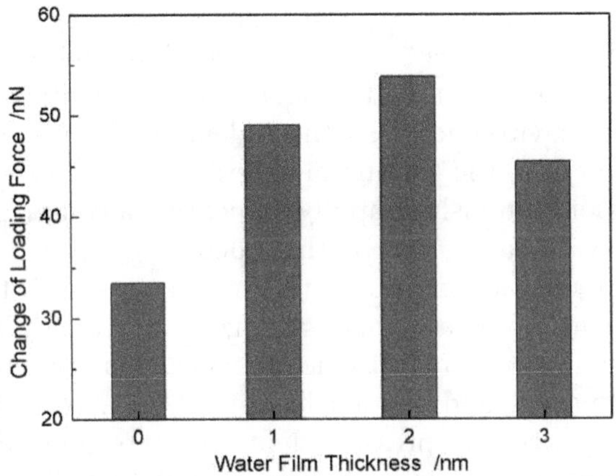

Figure 5.20. Stress relaxation of single-crystal copper at different water film thicknesses.

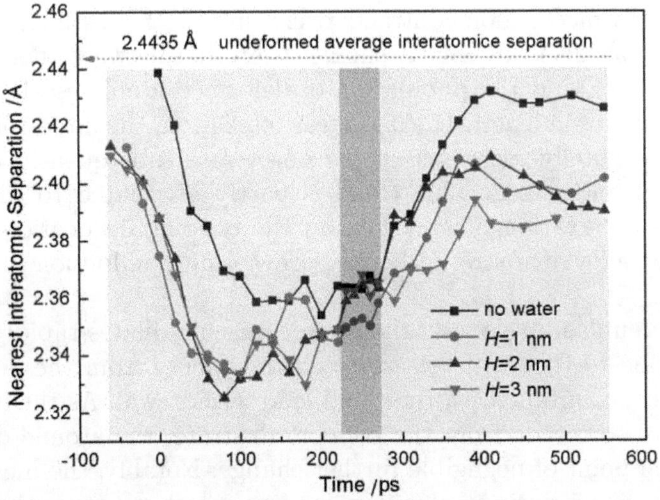

Figure 5.21. Comparison of the spacing between the nearest atoms in the indentation region.

by the water film does not adhere to a uniform trend, a phenomenon that is further elucidated in subsequent discussions.

Figure 5.21 delineates the calculation of the minimum distance between copper atoms within the truncation radius to further explore the elastic-plastic deformation of single-crystal copper throughout the loading, stress relaxation, and unloading processes under varying water film thicknesses. The average interatomic distance in the undeformed single-crystal copper is approximately 2.44 nm. In an anhydrous environment, as the indenter initiates contact with the copper matrix, the minimum interatomic distance within the deformed region diminishes rapidly, indicating a compression of copper atoms toward each other by the indenter.

Under anhydrous conditions, the initial pressing of the indenter into the single-crystal copper matrix is met with a swift reduction in the minimum interatomic distance, marking the onset of plastic deformation and the nucleation of numerous dislocations. Upon reaching a certain pressure depth, despite continued indentation, the minimum interatomic distance plateaus. This stagnation occurs as dislocations expand but without transformation of dislocations or defects, resulting in negligible load increments on the substrate. Conversely, the average interatomic distance in the plastically deformed region contracts over time, as the distance between deformed atoms is inherently smaller than that between their normal counterparts, and the population of deformed atoms escalates with the advancing indenter. During stress relaxation, the interatomic distance marginally increases at the onset and subsequently remains relatively unchanged. This trend is partly attributed to the modest stress relaxation magnitude and the ongoing dislocation expansion that alleviates stress, thereby preventing a reduction in atomic spacing.

In the unloading phase, the atomic spacing dilates rapidly at the outset due to the swift release of elastic energy from the deformed region, concomitant with the rapid load withdrawal. As the indenter continues to recede from the copper substrate, the atomic distance reaches a point of negligible further change. Notably, the increase in atomic spacing during unloading encompasses two facets: the reduction of the indenter's downward pressure, facilitating the release of dislocation energy and concluding the initial plastic deformation, and the removal of extrusion pressure, enabling the release of elastic energy and the gradual recovery of atomic elastic deformation to the perfect crystal structure.

In the presence of a water film, the temporal variation of the nearest neighbor distance of single-crystal copper atoms exhibits a trend analogous to that observed in anhydrous conditions. However, the atomic spacing begins to contract earlier during the pressing phase due to the water film's capacity to transmit the indenter's downward pressure to the copper surface. The atomic spacing at equivalent indented depths in subsequent stages of pressing is reduced compared to anhydrous scenarios. Observations from the later stages of indentation reveal that the water film induces a slight increase in atomic spacing, attributed to the compression of water molecules between copper atoms at larger indentation depths, dispersing some copper atoms under the rigidity of H–O bonds.

The impact of water film thickness on the nearest neighbor spacing of deformed region atoms during the entire indentation process does not follow a clear pattern. After 50 picoseconds of stress relaxation, the nearest neighbor spacing of copper atoms at a 1.0 nm water film thickness fluctuates within a range smaller than that of the anhydrous scenario. In contrast, at water film thicknesses of 2.0 and 3.0 nm, the atomic spacing increases and nearly matches the anhydrous case. During unloading, the spacing of the nearest copper atoms for 1.0 nm and 2.0 nm water films aligns, with the final stabilized stage exhibiting nearly identical spacing, whereas for a 3.0 nm water film, the atomic spacing is reduced, suggesting that thicker water films impede the elastic recovery of deformed atoms. Collectively, the interatomic spacing of single-crystal copper at the conclusion of unloading in the presence of a water film is constricted relative to anhydrous conditions, indicating that the water film inhibits the elastic recovery of single-crystal copper.

In Figure 5.22, the diagrams illustrate the changes in the indents' depth and width before and after removing the indenter under different water film thicknesses. It is evident that the size of the indents decreased to varying extents after removing the indenter, with the most notable changes occurring when there was no water film present. Moving on to Figure 5.23, the illustrations display the alterations in the dislocation configuration and elasticity of the indentation area of single-crystal copper from the end of loading (left), the beginning of unloading (middle), to the ending of unloading (right) under different water film conditions. It is apparent that the number of dislocations and stacking faults formed under the indentation region at the end of loading with an aqueous film is significantly higher than in the absence of a water film (as previously discussed in the nanoindentation section). The period between the end of loading and

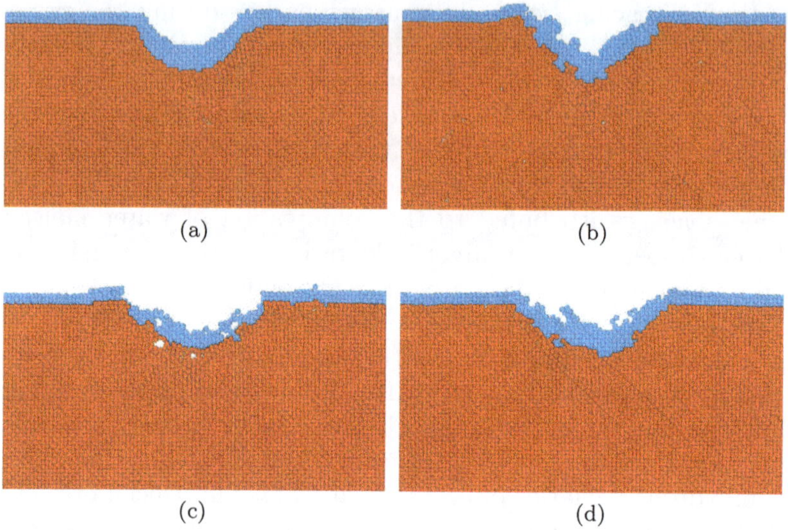

Figure 5.22. Configuration of single-crystal copper slices before (orange) and after (blue) abrasive unloading at different water film thicknesses (H). (a) No water film, (b) water film 1.0 nm, (c) water film 2.0 nm, (d) water film 3.0 nm.

the start of unloading represents a stress relaxation process. It is notable that the number of dislocations and stacking faults induced by indentation does not decrease due to stress relaxation. At the end of unloading, many dislocations and stacking faults disappear due to energy release, but defects remain when the water film thickness is large (e.g., $H = 3.0\,\text{nm}$). Additionally, the arrows in the diagram indicate the change in the indentation depth, with the depth being significantly smaller at the end of unloading than at the beginning of unloading, indicating the elastic recovery of the indentation region.

The graph in Figure 5.24 illustrates the maximum elastic recovery of single-crystal copper during the unloading process. The findings indicate that as the water film thickness increases from 0 (no film) to 2.0 nm, the maximum elastic recovery of single-crystal copper gradually decreases, reaching a decrease of up to 51.9%. Additionally, the maximum elastic recovery at a water film thickness of 3.0 nm shows a slight increase, although it is less than that at 1.0 nm. While the relationship between the maximum elastic recovery of single-crystal copper and the water film thickness is not identical to the nearest neighbor atomic spacing in Figure 5.21, the overall trend is similar.

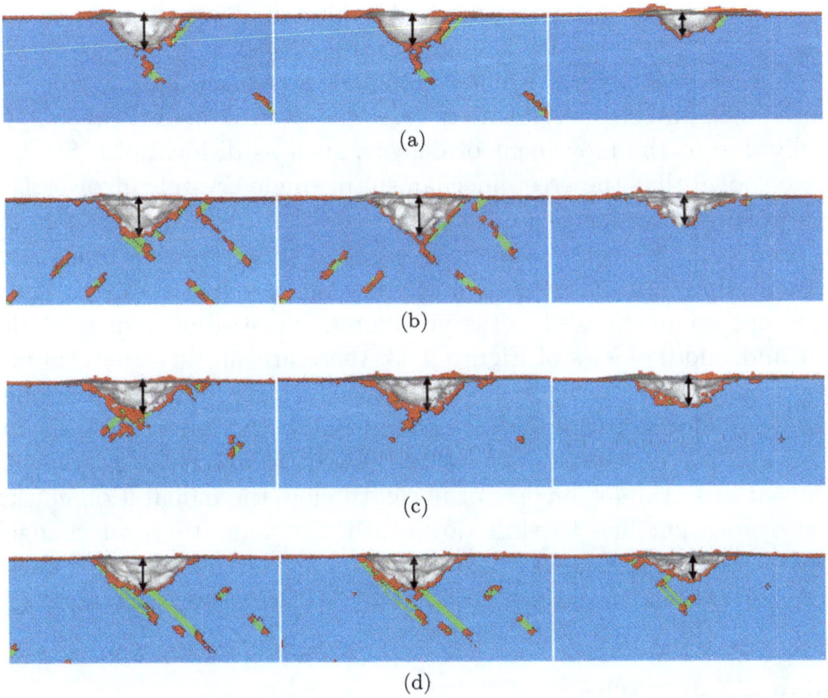

Figure 5.23. Defect configuration and elastic recovery in the indentation region. (a) No water film, (b) water film 1.0 nm, (c) water film 2.0 nm, (d) water film 3.0 nm.

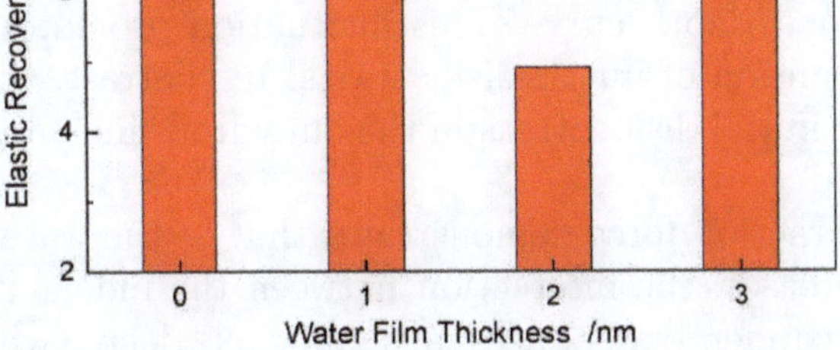

Figure 5.24. Maximum elastic recovery of the indentation area before and after unloading of the indenter.

In summary, the presence of a water film significantly reduces the elastic recovery capacity of single-crystal copper.

The behavior of solid materials under stress can be described as both creep and stress relaxation. In single-crystal metals, this is typically due to the movement of defects, such as dislocations. Studies have shown that the creep mechanism in single crystals involves lattice creep, where thermal movement leads to the removal of a column of atoms at a dislocation line, causing it to slip upward or downward by one slip plane. Even though the number of dislocation structures may appear unchanged between different illustrations, such as the left and middle views of Figure 5.23, there are subtle structural differences. For instance, in Figure 5.23(b), during stress relaxation, the original dislocation atoms (green) below the indentation disappear and the dislocation atoms (red) shift. The increase in stacking fault atoms at a 45° angle to the right below the indentation causes the dislocation line to shift downward. Similar structural changes have been observed under different water film conditions, and these transformations are consistent with the lattice creep mechanism.

5.2.6 *Summary*

The study investigated the impact of the water film and indentation method on the mechanical properties and elastic-plastic deformation characteristics of single-crystal copper during the nanoindentation process. It also analyzed the plastic deformation mechanism. Here's a summary of the findings:

(1) The water film significantly affects the initial plastic deformation of single-crystal copper. When compared to the anhydrous condition, the presence of a water film results in a notable fluctuation in the load–depth curve. This fluctuation promotes the growth and expansion of initial dislocations. In contrast, the expansion of many initial dislocations in the anhydrous indentation process is limited.

(2) The interaction force demonstrates that as the water film thickness increases, the interaction between the indenter and single-crystal copper experiences a significant delay, with its value being notably smaller than that of dry indentation. During the initial stages of plastic deformation, water molecules beneath

the indenter transmit the force from the indenter to the surface of the single-crystal copper, leading to increased plastic deformation. Meanwhile, water molecules outside the indenter prevent dislocations from extending to the surface and instead promote dislocation extension into the interior through lateral force transfer.

(3) Additionally, an increase in indentation speed (from 10 to 50 m/s) and loading rate (from 1 to 3 nN/ps) results in a higher transient strain rate, along with the nucleation and propagation of numerous dislocations in single-crystalline Cu. The presence of a water film consumes a portion of the loading energy due to intermolecular interactions, leading to a 61.2% and 59.3% reduction in the maximum compression depth in load controlled indentation, respectively. Conversely, displacement-controlled indentation processes exhibit the opposite result due to transient replenishment of the loading. The discrepancy between the two nanoindentation methods solely reflects the different control methods of load, displacement (or velocity) in the indentation process, while the plastic deformation mechanism of single-crystal copper remains fundamentally unchanged.

(4) Upon completion of the loading process, the indenter remains fixed, resulting in a significant decrease in the load on the single-crystal copper, known as stress relaxation. The extent of stress relaxation is considerably lower in the absence of a water film compared to when a water film is present. Moreover, the amount of stress relaxation initially increases and then decreases as the thickness of the water film is augmented. The existence of a water film causes premature reduction of the copper atom spacing and hinders the elastic recovery of single-crystal copper, leading to a maximum reduction of 51.9%.

5.3 Two-Body Abrasive Wear of Single-Crystal Copper with Aqueous Film

5.3.1 *Modeling and simulation methods*

In order to study the behavior of single-crystal copper in two-body abrasive wear under different conditions, we first created a model similar to the nanoindentation model, as shown in Figure 5.25.

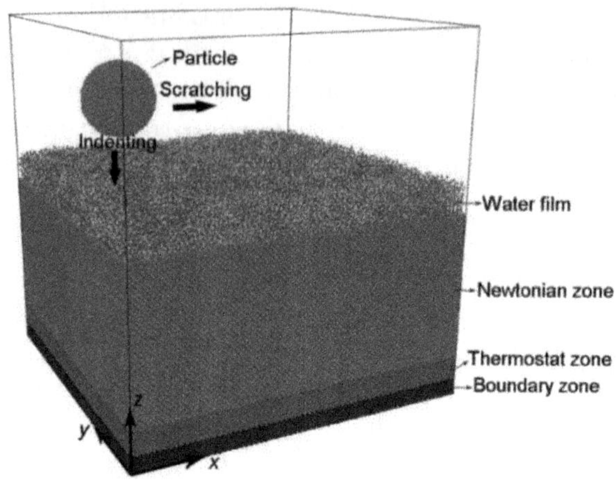

Figure 5.25. Initial model of two-body abrasive wear.

The model includes diamond nanospherical abrasives, a single-crystal copper matrix, and a water film on the surface (the water film is not included in the air wear simulation). The abrasive particles have a perfect diamond structure and are treated as rigid bodies during the simulation. The single-crystal copper substrate is constructed along the three crystal directions x-[100], y-[010], and z-[001], including the boundary region, heat bath region, and Newtonian region. The boundary region consists of five layers of atoms at the bottom of the substrate, the heat bath consists of eight layers of atoms to simulate heat dissipation, and the atoms in the Newtonian region follow Newton's law of motion. The abrasive particles were initially positioned 2.0 nm above the water film for the simulation. The reliability and accuracy of the simulation results depend on the interaction potential between atoms. The interaction potential between copper atoms uses the micro EAM potential method, which is a many-body potential method and effectively represents the formation of metal binding energy. The interaction between carbon atoms and copper atoms in diamond uses the Morse potential function, with specific parameters outlined in Table 5.2. The interaction between carbon atoms was not specified as the spherical abrasive was set up as a rigid body. The behavior of water molecules is modeled using the TIP4P model, and the interaction between oxygen atoms and copper and carbon atoms is modeled using the Lennard–Jones (12–6) potential,

Table 5.2. Lennard–Jones potential function and simulation process parameters.

Characterization	Parameterization
Abrasive grain	Diamond structure, spherical, radius 2.5 nm, 11976 atoms
Water film	Density $1.0\,\mathrm{g \cdot cm^{-3}}$, thickness ($H$): 0, 1.0, 2.0, 3.0 nm
Monocrystalline copper matrix	FCC structure, dimensions $68a \times 68a \times 35a$ (lattice constant $a = 0.3615\,\mathrm{nm}$), 656608 atoms
Morse potential function	$D_0 = 0.1\,\mathrm{eV}$, $\alpha = 1.7\,\mathrm{nm}^{-1}$, $r_0 = 0.22\,\mathrm{nm}$
Oxygen–oxygen TIP4P parameters	$\varepsilon = 0.1554\,\mathrm{kcal(mol)^{-1}}$, $\delta = 0.31655\,\mathrm{nm}$
Oxygen–copper LJ parameters	$\varepsilon = 0.2708\,\mathrm{kcal(mol)^{-1}}$, $\delta = 0.28877\,\mathrm{nm}$
Oxygen–carbon LJ parameters	$\varepsilon = 0.1\,\mathrm{kcal\ (mol)^{-1}}$, $\delta = 0.3275\,\mathrm{nm}$
Two-body wear process	Abrasive movement speed $10\,\mathrm{m \cdot s^{-1}}$, indented depth ($h$): -0.2, 0.1, 0.5, 1.0 nm

with parameters determined by the Lorentz–Berthelot mixing rule [9, 22] and outlined in Table 5.2.

The MD simulation utilized the Velocity–Verlet algorithm to integrate atom velocities in both the Newtonian and heat bath regions. The time step was set to 1 fs with periodic boundary conditions in the lateral x and y directions. The simulation comprised four phases: (1) a 50 ps system relaxation process allowing local energy optimization, (2) a nanoindentation process where abrasive nanoparticles moved along the $-z$ direction to penetrate the water film and press into the single-crystalline Cu substrate using a displacement-controlled method, (3) a 50 ps secondary relaxation phase following the nanoindentation, and (4) a two-body abrasion process where abrasive particles moved uniformly in the x direction. The motion speed of abrasive particles during both nanoindentation and wear processes was $10\,\mathrm{m \cdot s^{-1}}$. Simulation results were visualized using OVITO software [23].

Notably, previous studies highlighted significant differences in the nanoindentation behavior of single-crystal copper in the absence and presence of an aqueous film. The thickness of the aqueous film was found to greatly influence the plastic deformation process of the single-crystal copper matrix. Nanoindentation marks the beginning of the frictional wear process of a material, where abrasive particles or fragments become embedded in the material's surface. As a result, these particles can trigger various degrees of wear through different modes of motion, such as sliding or rolling. In our study, the authors simulated the nano-scratching process by using a displacement-controlled approach to press spherical abrasive particles into single-crystal copper to a depth of 1.0 nm. Subsequently, the abrasive particles underwent scratching along the horizontal direction, allowing the authors to uncover the deformation and wear mechanisms of the single-crystal copper. This investigation was carried out in the absence of an aqueous film and at aqueous film thicknesses of 1.0 nm, 2.0 nm, and 3.0 nm, respectively.

5.3.2 *Effect of water film on friction*

In Figure 5.26, we can see the friction and normal load on the abrasive particles for different water film thicknesses. The friction force is the combined force in the direction of the scratch where the single-crystal copper and the water film act with the abrasive particles, and the normal load is the combined force in the perpendicular direction. In this figure, we observe that the friction force and the normal load change with the distance of the scratch. The normal load decreases rapidly from a large value at the initial stage of the scratch, while the friction force increases rapidly from zero, indicating that the resistance of the abrasive particles significantly increases as they start to slide. When comparing, it is evident that the time taken for the friction to increase and the normal load to decrease increases with the thickness of the aqueous film, corresponding to the scratch length. Furthermore, the friction force and normal load remain dynamically stable as the scratch distance increases.

The average forces of the two figures in Figure 5.26 during the stabilization stage were calculated and are presented in Figure 5.27 and Table 5.3. These values were used to determine the coefficient of friction, as illustrated in Figure 5.28. In Figure 5.27(a), it is observed that the friction force on the abrasive particles notably increases with the thickness of the water film. The normal load shows minimal change in the presence of the water film but increases by at least

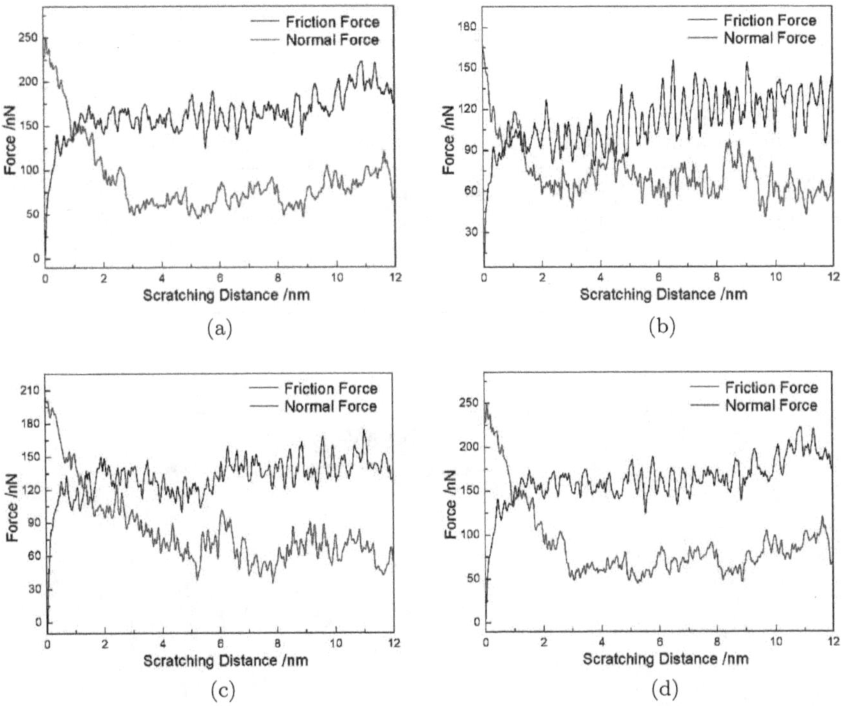

Figure 5.26. Friction and normal load acting on abrasive particles during scratching at different water film thicknesses. (a) No water film, (b) water film thickness of 1.0 nm, (c) water film thickness of 2.0 nm, (d) water film thickness of 3.0 nm.

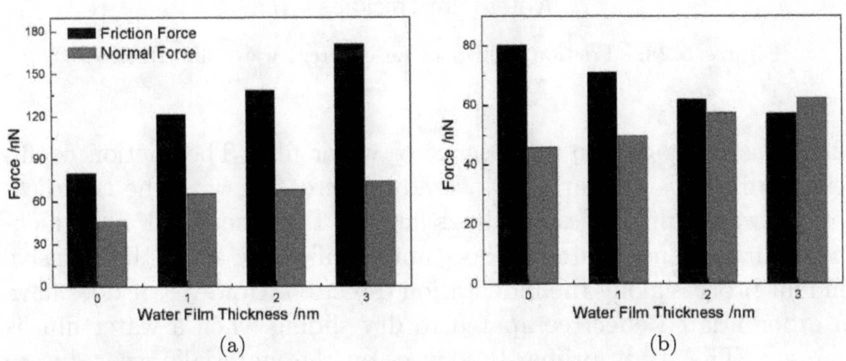

Figure 5.27. Average values of friction and normal load during the stabilization phase of a scratch. (a) Force on abrasive particles, (b) force on single-crystal copper.

Table 5.3. Average friction and normal load.

Water film thickness/nm	Force on abrasive particles/nN		Copper surface force/nN	
	Friction	Normal load	Friction	Normal load
0	80.3	46.1	80.3	46.1
1.0	121.8	65.8	71.2	49.9
2.0	138.3	68.5	62.2	57.7
3.0	171.0	74.3	57.5	62.5

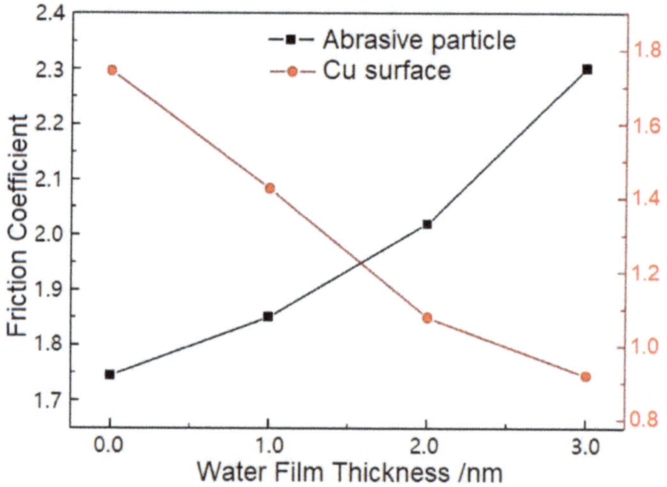

Figure 5.28. Friction coefficient for different water film thickness.

28 nN as compared to its absence of water film. The friction coefficients display a similar trend, steadily increasing with the thickness of the water film. These findings suggest that the water film thickness during nano-scratching does not significantly affect the abrasive indent process along the indentation direction. However, it does have a differentiated effect compared to dry sliding when a water film is present. The study primarily focuses on the material's wear due to friction. It discusses the impact of friction and positive pressures on the surface of single-crystal copper, taking into account abrasive effects and the presence of a water film. The research calculated the

average force during the stable motion phase of the abrasive and depicted the results in Figure 5.27(b). The findings reveal that the friction force on the copper surface decreases as the water film thickness increases, while the normal load exhibits an opposite trend. Additionally, the decrease in friction force outweighs the increase in normal load, ultimately leading to a decrease in the coefficient of friction, as illustrated in Figure 5.28 (indicated in red), when the water film thickness increases.

The thickening of the water film reduces the friction force on the copper surface by hindering the wear of the abrasive particles. From the law of action and reaction force stating that the effect of copper on the abrasive should be equal to the force of abrasive on the copper, the friction force on the copper surface is opposite to the friction force on the abrasive particles. This indicates that the water film significantly restricts the indentation of the abrasive particles, thus altering the trend of the friction force on the abrasive particles.

It's important to note that the friction coefficients obtained by the authors are greater than 1, surpassing values obtained from macroscopic experiments. At the nanoscale, the friction coefficient encompasses not only ploughing and adhesion effects [24, 25] but also the substantially enhanced hindering effect of the debris, given the larger volume ratio of abrasive to debris fragments [26]. The adhesion effect at the nanoscale surpasses the plough furrow cutting effect. Additionally, the contribution of the friction process due to the deformation of an accumulation of atoms forming debris during wear should not be underestimated in this research. The water film also plays a significant role. Therefore, the larger friction coefficient is reasonable when considering nano-wear [9, 26–28].

5.3.3 *Single-crystal copper surface morphology*

Figures 5.29 and 5.30 display the wear patterns on a single-crystal copper surface at the end of a nano-scratch, along with the distribution of ridges on both sides of the scratches. In the figures, yellow represents the surface and deformed atoms, green represents non-deformed copper atoms, blue represents oxygen atoms in water molecules, and red spheres indicate the abrasive particles.

In Figure 5.29, it is evident that a large number of surface atoms (yellow) have significantly changed their positions at the end of the

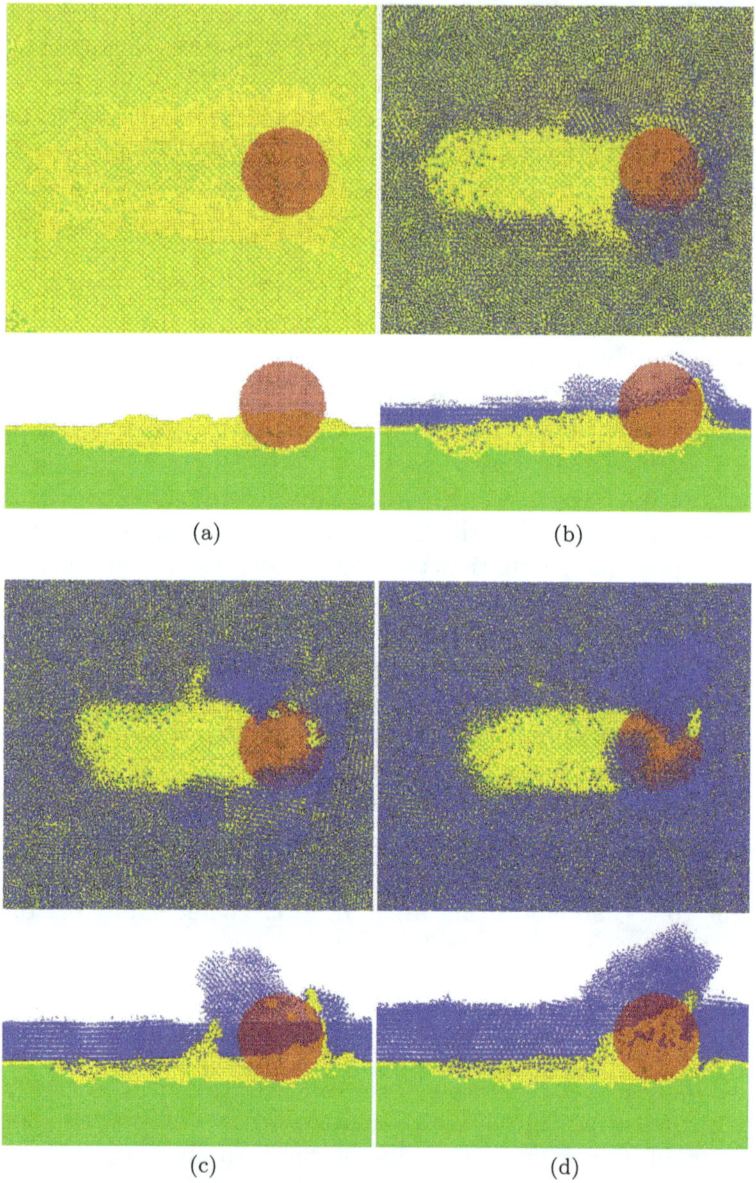

Figure 5.29. Top and bottom views in cross-section of nano-scratch morphology. (a) No water film, (b) water film thickness of 1.0 nm, (c) water film thickness of 2.0 nm, (d) water film thickness of 3.0 nm.

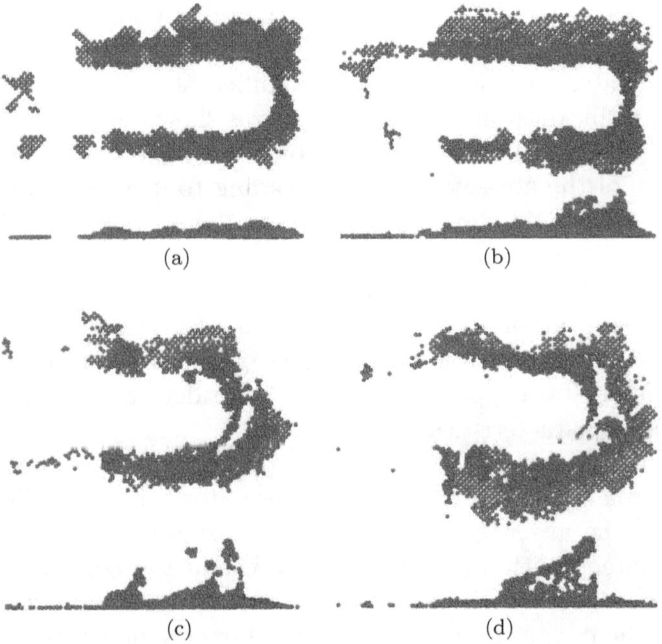

Figure 5.30. Top and bottom views in cross-section of the shape of the ridges on each side of the groove. (a) No water film, (b) water film thickness of 1.0 nm, (c) water film thickness of 2.0 nm, (d) water film thickness of 3.0 nm.

abrasion, indicating the deformation of the single-crystal copper. Viewed from the top, after the abrasive particles have scratched, a regularly shaped wear area appears on the copper surface, with a large number of worn-out copper atoms accumulating on both sides and at the front end, forming ridges and piles of abrasive debris. Through microscopic observation, it can be noted that the debris formed by the worn atoms increases with the scratching distance.

In air-lubricated sliding conditions, as shown in Figures 5.29(a) and 5.30(a), the grooves have a very regular shape throughout the scratched area, with uniform depth and almost uniform ridge height. However, in the presence of a water film, the groove distribution becomes irregular, with a notable atomic step at the bottom, varying ridge heights, and localized stacking. For instance, in Figures 5.30(c) and 5.30(d), there is only occasional debris at the tail end of the ridges, while at the front end, there is a large build-up of debris that almost reaches the height of the abrasive particles. Moreover, when abrasive particles scratch the single-crystal copper surface, a large number of water molecules accumulate at the front end. This accumulation of water molecules increases significantly with the increase in water film thickness and scratch length, as shown in Figures 5.29(b), 5.29(c), and 5.29(d). This build-up of water molecules increases the resistance of the abrasive particles, leading to increased friction. As a result, the frictions of the abrasive and the single-crystal copper exhibit opposite changes. Additionally, the accumulation of water molecules at the front end of the abrasive particles significantly hinders the build-up of abrasive debris during scratching, affecting the formation of ridges. Consequently, the grooves' internal structure, the single-crystal copper surface, and the ridges all display notable anisotropic characteristics.

Figure 5.31 depicts the top view of a single-crystal copper surface, showing wear patterns and depth where atoms locate indicated by colors. In the absence of water, the nano-scratched copper surface remains intact, with minor abrasions at the furrow edges and a regular atomic structure within. However, in the presence of a water film, wear on the copper surface outside the furrows increases, especially with thicker water films. Within the furrows, the distribution of atom colors becomes irregular, indicating an uneven and irregular structure. Consequently, both the surface and the furrow region of the single-crystal copper experience increased damage with the presence and thickening of the water film.

In order to understand the role of the water film in the nano-scratching process, we calculated the number of water molecules present in the center of the groove region, as shown in Figure 5.32. We found that as the water film thickness increases, the number of water molecules in the groove also increases. These water molecules remaining in the groove region act as a barrier, hindering direct interaction between the abrasive particles and the single-crystal copper, leading to reduced friction on the surface of the copper. Moreover, the presence of water molecules can transfer the scratch force of the abrasive particles to the surface of the copper, causing the formation of atomic steps in the groove similar to a mechanical corrosion effect. Additionally, more water molecules are found at the edge of the furrow than in the center, causing significant surface damage near the groove, as depicted in Figure 5.31. As a result, the water film has a significant influence on the scratching behavior of abrasive particles

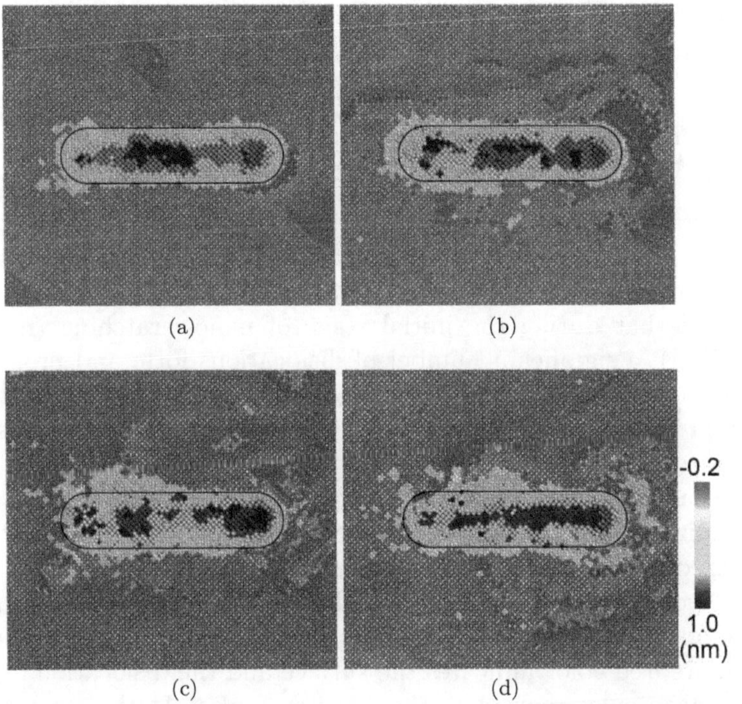

Figure 5.31. Top view of wear morphology on single-crystal copper surface. (a) No water film, (b) water film thickness of 1.0 nm, (c) water film thickness of 2.0 nm, (d) water film thickness of 3.0 nm.

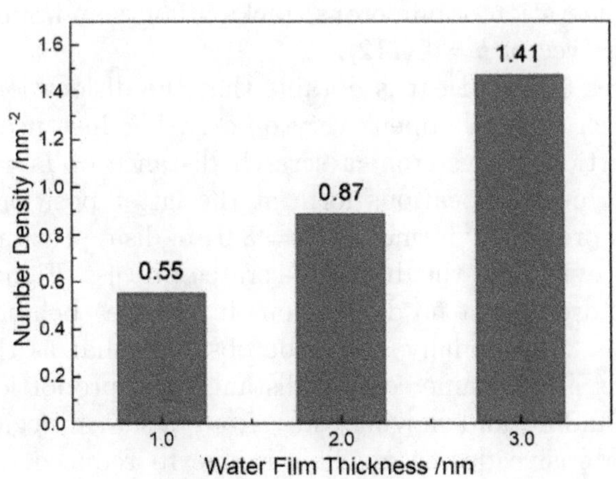

Figure 5.32. Surface density of water molecules located in the center region of the groove.

on single-crystal copper surfaces, resulting in anisotropy in surface damage, ridges, and grooves.

5.3.4 *Plastic deformation of single-crystal copper*

In order to understand how single-crystal copper deforms during nano-scratching, we examined a series of dislocation evolution configurations (Figure 5.33). In these figures, green lines represent Shockley dislocation pairs and blue arrows represent Burgers vectors. We observed that during the initial stage of nano-scratching (nanoindentation), a significant number of dislocations form and propagate along the $\{111\}$ slip surface. It was evident that there were more dislocation lines in the presence of a water film compared to the absence of a water film.

Throughout the nano-scratching process, we noted a distinct reorganization of the dislocation network. As the scratch distance increased, dislocations nucleated directly under the abrasive and expanded both along the scratch direction and at a 45° angle to the scratch direction, as depicted by the arrows in Figure 5.34. In the figure, red color indicates the surface and the dislocation atoms, while green color represents the stacking faults. As the scratch distance increased further, the concentrated dislocations were released at the midpoint of the groove. Meanwhile, the Burgers vector of the $\{111\}$ $\langle 110 \rangle$ slip system of the dislocations continued to react, resulting in the generation of numerous stacking faults and Shockley pairs with Burgers vector $b = \frac{a}{6}\langle 112 \rangle$.

In Figures 5.33(a–d), it is notable that the dislocation evolution in the scratch region happens very quickly. For instance, when the abrasive particles move from a scratch distance of $L = 3.0\,\text{nm}$ to $L = 6.0\,\text{nm}$, new dislocations form at the latter position while the dislocations previously formed at $L = 3.0\,\text{nm}$ disappear rapidly. This is especially evident in the dry nano-scratch process (Figure 5.33(a)), where there are almost no dislocations in the area behind the abrasive particles. Additionally, it can be observed that as the aqueous film thickens at the same scratch distance, the proportion of dislocations expanding and moving away from the wear region directly below the abrasive decreases. This is due to reduced friction acting on the single-crystal Cu surface, resulting in a decreased driving force for dislocation expansion. However, the proportion of

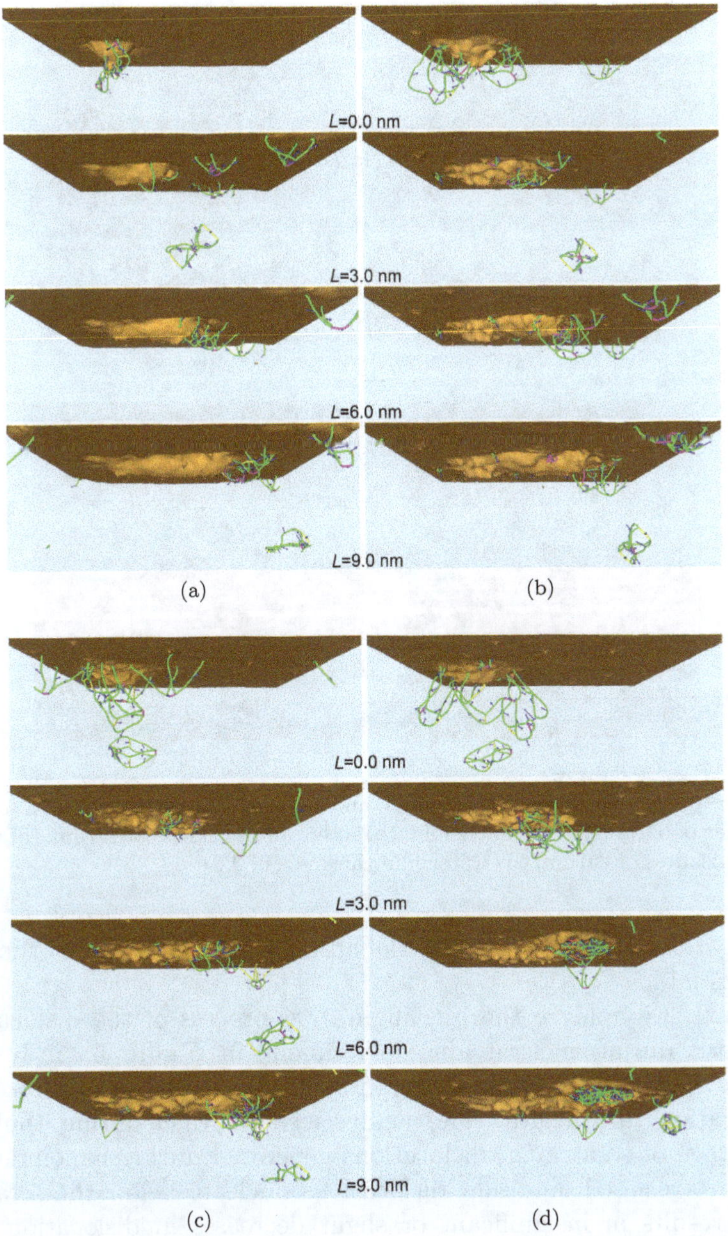

Figure 5.33. Evolution of dislocations within single-crystal copper with scratch length. (a) No water film, (b) water film thickness of 1.0 nm, (c) water film thickness of 2.0 nm, (d) water film thickness of 3.0 nm.

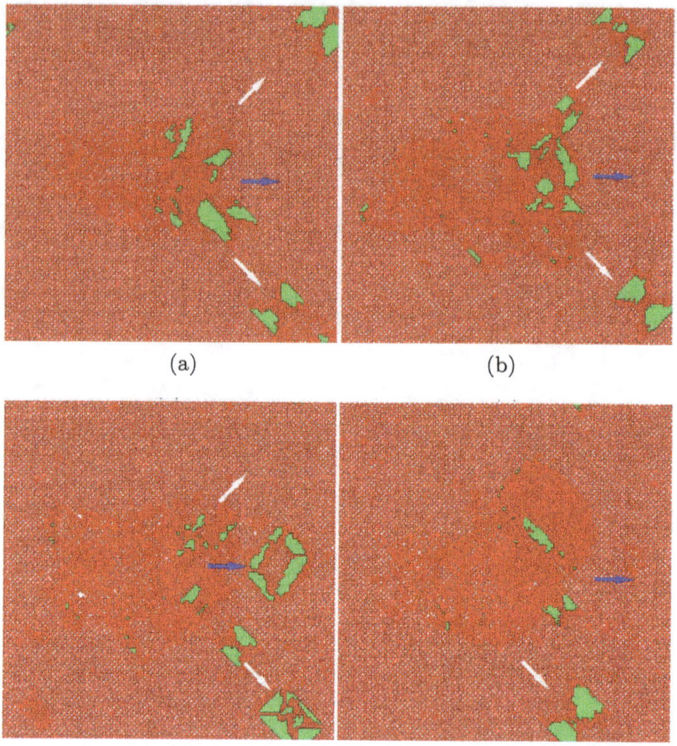

(a) (b) (c) (d)

Figure 5.34. Direction of dislocation and stacking fault expansion at a scratch distance of 6.0 nm. (a) No water film, (b) water film thickness of 1.0 nm, (c) water film thickness of 2.0 nm, (d) water film thickness of 3.0 nm.

dislocations nucleated around the abrasive increases due to the rise in normal load.

Next, we analyze the reconfiguration process of the dislocation network during a scratching experiment in Figure 5.35. It discusses how the presence of a water film affects the evolution of dislocation line length. The analysis shows that during the initial stage of scratching, dislocations generated during nanoindentation are released, and new dislocations nucleate below the scratch. This results in insignificant or slight decreases in dislocation line length for scratch distances less than 3.0 nm. As the scratch distance reaches 3.0 nm, the initially formed dislocations are mostly lost, and a reconfiguration of the dislocation network occurs between

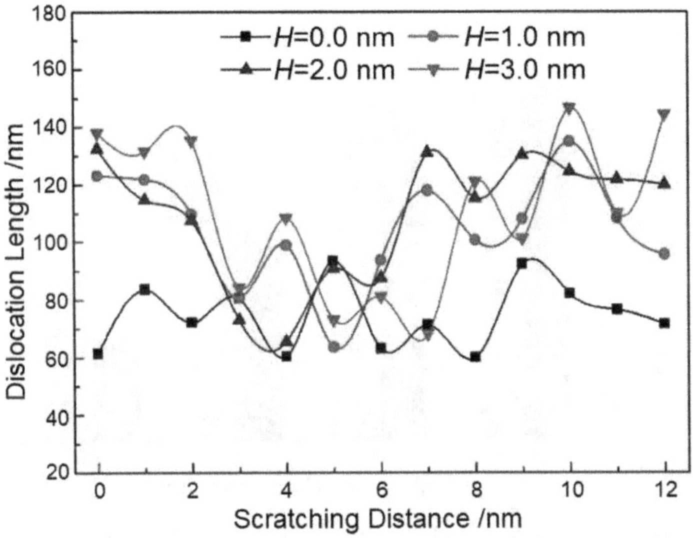

Figure 5.35. Variation of dislocation length with scratch distance.

3.0 nm and 7.0 nm. Moreover, when the scratch distance exceeds 7.0 nm, dislocation expansion and reaction lead to a rapid increase and subsequent fluctuation of the dislocation length. The presence of a water film causes an increase in the total dislocation length. It is noted that the change in dislocation line length constantly fluctuates in a range of 60 and 80 nm for a dry or wet scratching process. In summary, the increase of dislocation length reflects the continuous multiplication and reorganization of the dislocation network during the scratching process, and the decrease indicates the release and disappearance of dislocations.

5.3.5 *Evaluation of monocrystalline copper abrasion*

In the context of two-body wear processes such as scratches caused by abrasive particles on a material's surface, it is essential to assess the wear rate and the extent of material damage. Researchers have determined that a copper atom can be considered as removed or worn away if it shifts from its original position on a defect-free single-crystal copper surface by a distance of 0.18 nm (which is half the lattice constant of single-crystal copper) at the conclusion of the scratching process.

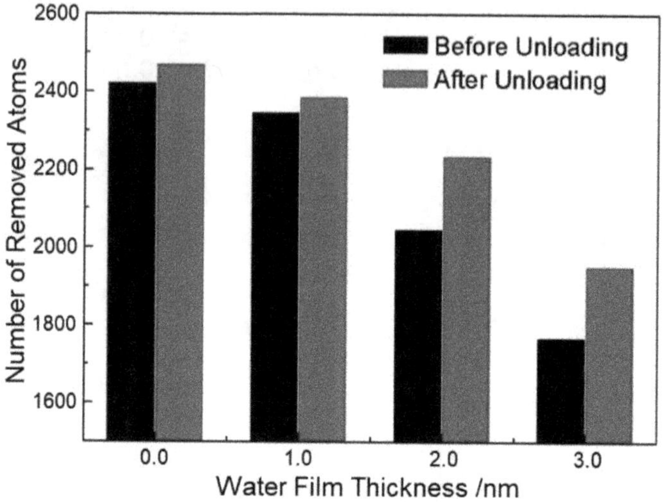

Figure 5.36. Number of copper atoms abraded before and after abrasive particle removal unloading.

The extent of wear on the nanomaterial is characterized by the number of copper atoms that have been worn away, and the change in the amount of single-crystal copper before and after the removal of the abrasive particles is illustrated in Figure 5.36. It is evident from the figure that the quantity of single-crystal copper removed prior to the removal of abrasive particles in the wear region decreases significantly with the presence and thickness of a water film, indicating that the presence and thickening of the water film diminishe the wear on the single-crystal copper. The presence of water molecules below and around the abrasive particles has two significant effects on the behavior of single-crystal copper surface. First, it hinders direct contact between the abrasive particles and the copper, thus reducing friction, as shown in Figure 5.27(b). Second, the accumulation of water molecules around the abrasive particles impedes the gathering of worn-out copper atoms on the surface. When the abrasive particles are removed, the change in the abraded amount of single-crystal copper remains consistent. This finding aligns with previous experiments where the water film acted as a lubricant during the scratching process [29]. Furthermore, it is worth noting that unloading the abrasive particles results in increased wear, particularly with a thicker water film. This occurs because some of the copper atoms underneath the abrasive particles are easily drawn out of the wear area, while others are present in the ploughing as an atomic step. Additionally, the elastic recovery of the single-crystal copper after the removal of the abrasive particles also leads to the extrusion of some atoms from the surface. Consequently, these results indicate that the presence of a water film diminishes the amount of wear on the single-crystal copper.

5.3.6 *Abrasive wear mechanism of single atomic layer of monocrystalline copper*

Researchers are increasingly interested in studying the removal and wear behavior of materials at the nanoscale. For instance, Luo and Dornfeld *et al.* [30, 31] conducted wear experiments using abrasive particles with a 50 nm diameter. They discovered that by adjusting experimental conditions, such as the depth of abrasion, it's possible to achieve precise removal of materials, reducing scratches to as little as 0.07 nm in depth. However, the frictional wear of materials under sub-nanoscale wear conditions differs from that of macroscale [32]. The present authors examined the effect of scratch depth (h) on the material wear, removal, or polishing process using scratch depths of −0.2, 0.1, 0.5, and 1.0 nm. Scratch depth refers to the depth of indentation of the abrasive particles on the material's surface. A negative value of −0.2 nm implies that the surface atoms of the abrasive particles do not indent into the substrate, resulting in no direct "contact" between abrasive particles and the material (non-contact condition). On the other hand, 0.1 nm indicates that the abrasive particles are in contact with the material at a single atomic layer.

Figure 5.37 illustrates the wear pattern of the single-crystal copper surface when the water film thickness is 1.0 nm. In the figure, red color represents abrasive particles, green depicts undeformed copper atoms, and yellow shows surface and deformed atoms. The top view displays only the monolayer of surface atoms, and water molecules are excluded. Figure 5.38 provides a cross-sectional slice view of the abrasion morphology of single-crystal copper.

The scratched region in Figure 5.37(a) reveals the emergence of new surface layers above the initial subsurface, with certain surface atoms positioned above the single-crystal copper surface. Furthermore, Figure 5.38(a) illustrates that numerous surface atoms

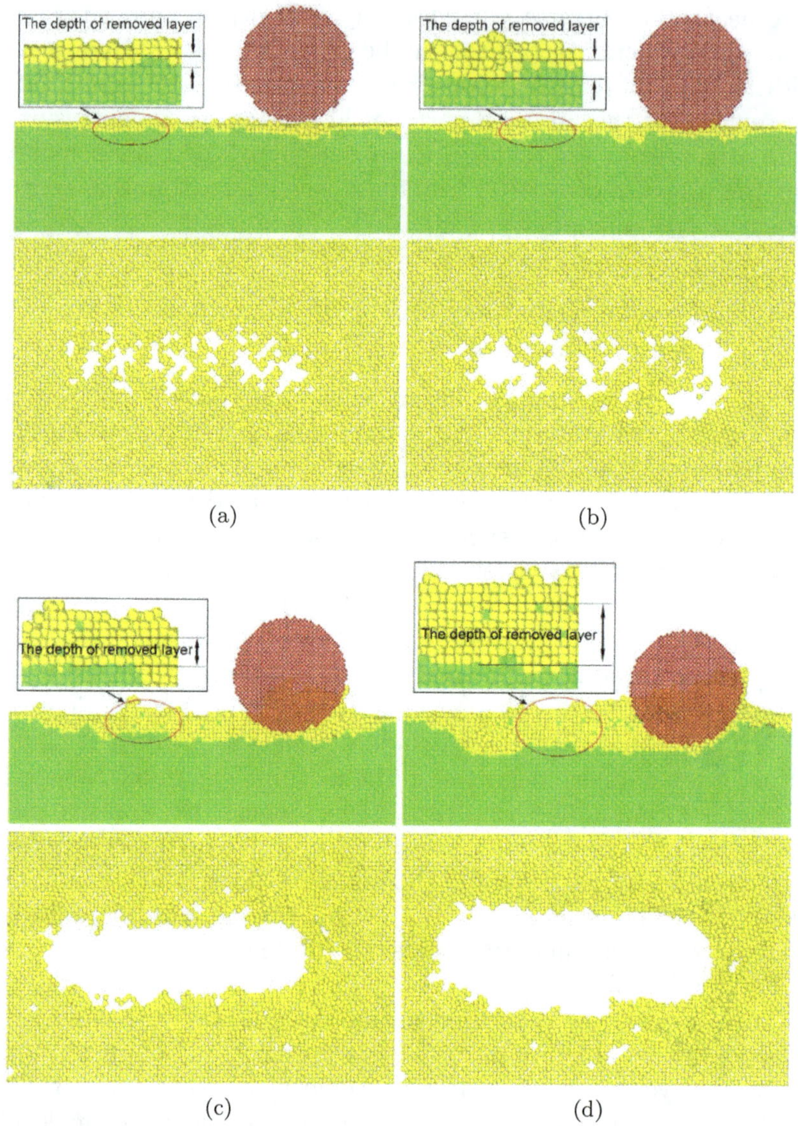

Figure 5.37. Cross-sectional view (top, xz plane) and top view (bottom, xy plane) of single-crystal copper wear at a water film thickness of 1.0 nm. (a) Scratch depth $h = -0.2$ nm, (b) scratch depth $h = 0.1$ nm, (c) scratch depth $h = 0.5$ nm, (d) scratch depth $h = 1.0$ nm.

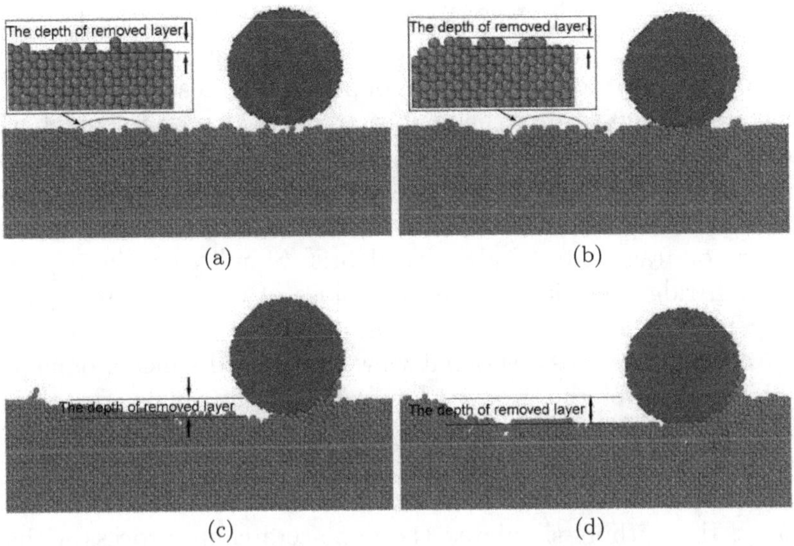

Figure 5.38. Cross-sectional slice view of single-crystal copper abrasion at a water film thickness of 1.0 nm. (a) Scratch depth $h = -0.2$ nm, (b) scratch depth $h = 0.1$ nm, (c) scratch depth $h = 0.5$ nm, (d) scratch depth $h = 1.0$ nm.

have shifted from their original positions, resulting in the creation of surface vacancies in the top view. Despite the abrasive particles not being in direct contact with the single-crystal copper surface at $h = -0.2$ nm, the single-atom layer on the single-crystal copper surface is removed. This particular type of single atomic layer removal also occurs at the scratch depth $h = 0.1$ nm, as depicted in Figures 5.37(b) and 5.38(b), when the roughness is more pronounced in localized areas of the newly formed surface. Nevertheless, as indicated by the arrows in Figures 5.37(a) and 5.37(b), the surface roughness post single-atomic-layer abrasion remains relatively consistent throughout the abraded region, typically within the thickness of one or two atomic layers. These findings suggest that the wear pattern of single-crystal copper surfaces under non-contact or single-atomic-layer contact conditions can reach near "0 nm flatness". The worn-out copper atoms appear either as single atoms or as clusters of several atoms, aligning with the outcomes observed for single-crystal silicon at an indentation depth of 0.1 nm. This result is consistent with the result for single-crystal silicon at an indentation depth of 0.1 nm [32].

In Figures 5.37(c), 5.37(d), 5.38(c), and 5.38(d), the increased depth of the scratches results in abrasive particles coming into contact with more monocrystalline copper atoms, leading to a larger wear area. Almost all of the copper atoms that are in contact with the abrasive particles are removed from the surface due to wear. As the abrasive is scratched forward, a significant number of deformed atoms gather at the leading end of the abrasive particles, forming clusters or fragments of abrasive debris. Simultaneously, a groove and high ridges are formed on both sides of the wear area, particularly notable at a depth of 1.0 nm of the scratch, as depicted in Figure 5.30. The cross-sectional view reveals a distinct atomic step at the bottom of the groove. The wear on the single-crystal copper surface significantly increases with greater scratch depth, unlike non-contact and single-atomic-layer contact.

To assess the impact of aqueous film thickness on wear and surface quality, the authors simulated the nano-scratching process without an aqueous film and with aqueous film thicknesses of 0.3 nm (single-layer) and 0.5 nm (double-layer). The wear morphology is illustrated in Figures 5.39, 5.40, and 5.41, respectively. The wear morphology of single-crystalline Cu with abrasive particles at different scratch depths resembles the wear of single-crystalline Cu under smaller scratches with a water film thickness of 1.0 nm. That is, the wear of a single atomic layer was observed after scratching the surface, similar to the wear of single-crystal copper under smaller scratches with a water film thickness of 1.0 nm.

In the case of the nano-scratch without an aqueous film (see Figure 5.39), wear at a depth of 0.1 nm leads to single-atom layer wear. Under the non-contact condition ($h = -0.2$ nm), the single-crystal copper surface still retains its original intact crystal structure. As the scratch depth increases to 1.0 nm (Figure 5.39(c)), the abrasive particles come into direct contact with the single-crystal copper substrate, resulting in the formation of a large wear region. Nearly all the copper atoms that come into contact with the abrasive particles are abraded away. This process leads to the formation of deeper grooves on the single-crystal copper surface, with a large number of copper atoms accumulating in front of the abrasive to form chips. By comparing the top views of Figures 5.37(a), 5.37(b), 5.39(b), and 5.41, it becomes evident that surface vacancies caused by atomic wear increase as the water film thickness decreases (from 1.0 nm to

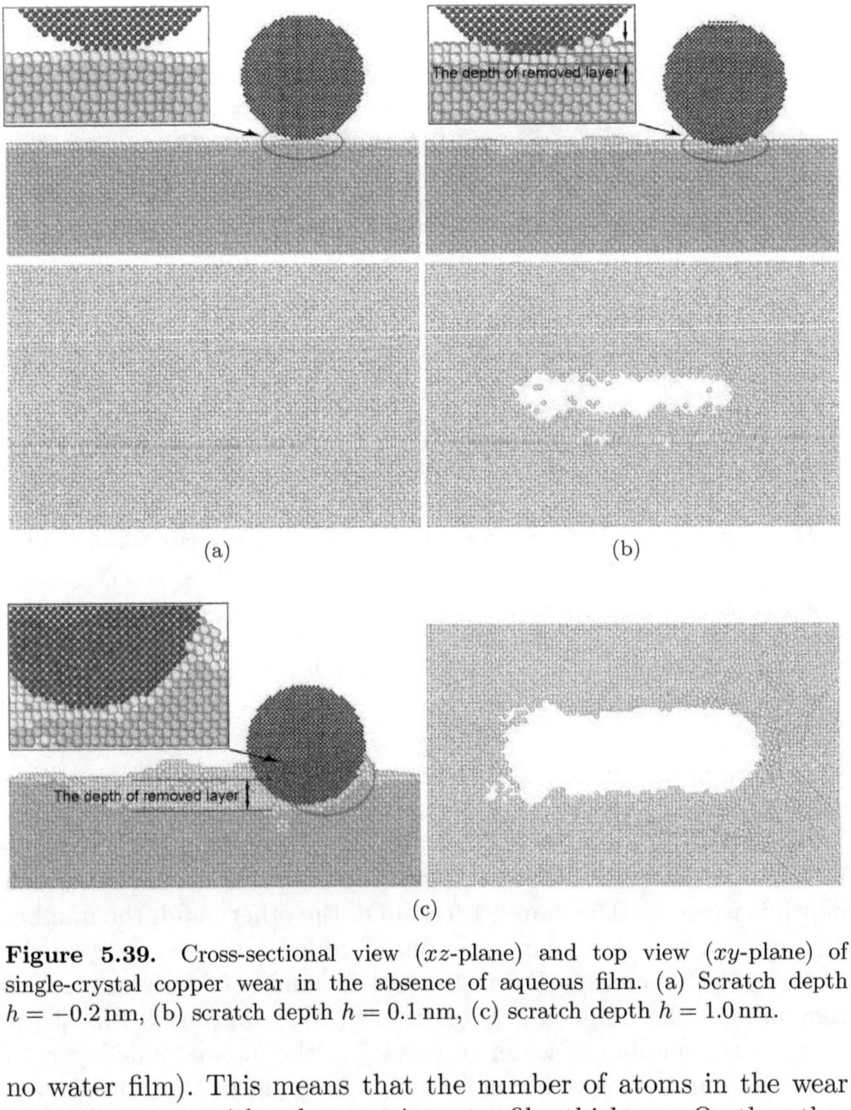

Figure 5.39. Cross-sectional view (xz-plane) and top view (xy-plane) of single-crystal copper wear in the absence of aqueous film. (a) Scratch depth $h = -0.2\,\text{nm}$, (b) scratch depth $h = 0.1\,\text{nm}$, (c) scratch depth $h = 1.0\,\text{nm}$.

no water film). This means that the number of atoms in the wear region increases with a decrease in water film thickness. On the other hand, the area of the wear region of the single-crystal copper at the same scratch depth slightly decreases with a decrease in water film thickness.

The investigation into the wear mechanism of single-crystal copper involved extracting the wear morphology with a water film (Figure 5.42) and calculating the number of water molecules

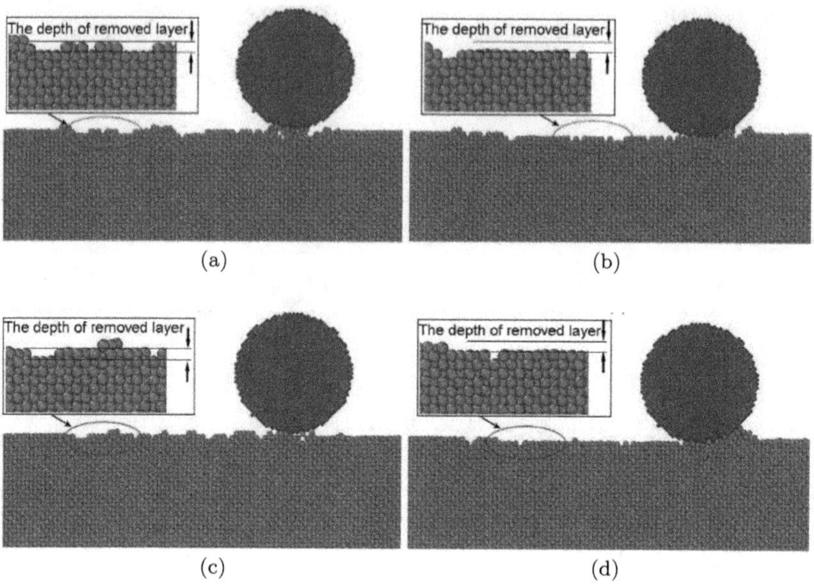

Figure 5.40. Cross-sectional (xz-plane) slice view of single-crystal copper abrasion at a water film thickness of 0.3 nm. (a) Scratch depth $h = -0.2$ nm, (b) scratch depth $h = 0.1$ nm, (c) scratch depth $h = 0.5$ nm, (d) scratch depth $h = 1.0$ nm.

remaining in the wear area (Figure 5.43). Figure 5.42 shows that a small number of water molecules accumulate at the front end of the abrasive particles at scratch depths of −0.2 nm and 0.1 nm. The accumulation of water molecules significantly increases when the scratch depth is raised to 0.5 nm and 1.0 nm. On the other hand, the number of water molecules remaining in the abraded area decreases as the scratch depth increases. For instance, the number of water molecules remaining under non-contact conditions ($h = -0.2$ nm) is almost the same as the number of water molecules in the monoatomic layer. In fact, the number of water molecules remaining in the non-contact condition ($h = -0.2$ nm) is almost four times higher than that in single-atomic-layer contact (Figure 5.43). These findings suggest that the water film plays a crucial role in monolayer wear at small scratch depths.

In the course of nano-scratching under non-contact conditions, the wear cross-section depicted in Figure 5.42 reveals that the area between the abrasive particles and single-crystal copper is filled with

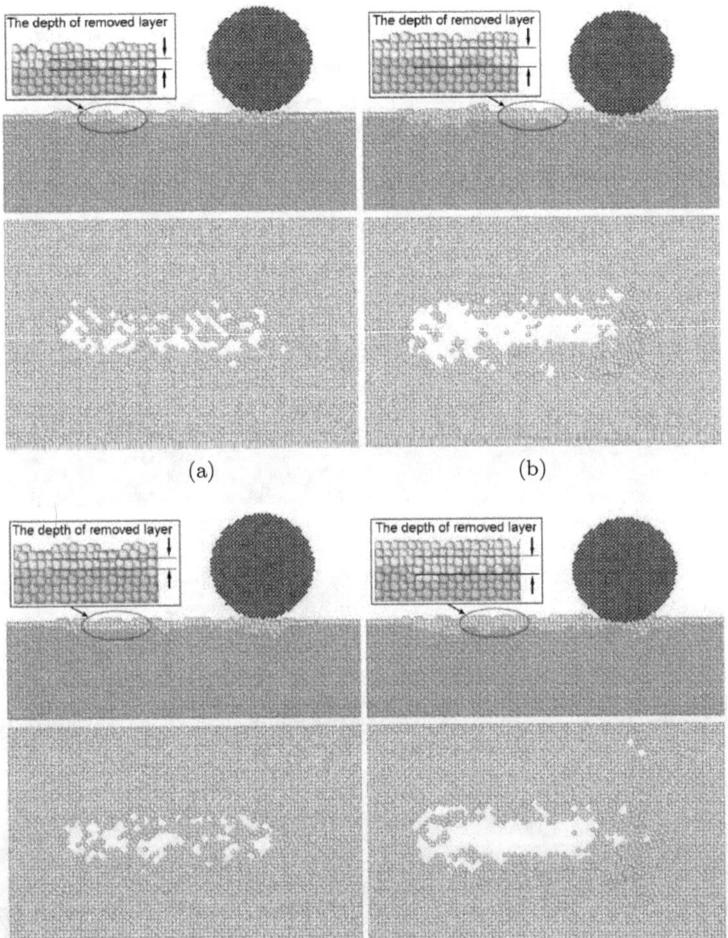

Figure 5.41. Cross-sectional view (top, xz-plane) and top view (bottom, xy-plane) of single-crystal copper abrasion at a water film thickness of 0.5 nm. (a) Scratch depth $h = -0.2$ nm, (b) scratch depth $h = 0.1$ nm, (c) scratch depth $h = 0.5$ nm, (d) scratch depth $h = 1.0$ nm.

numerous residual water molecules. Additionally, the average interaction force from Table 5.4 and Figure 5.44 suggests that the force between the abrasive and single-crystal copper (0.18 nN) is significantly less than that between the water film and single-crystal copper and between the water film and abrasive (41.52 nN and

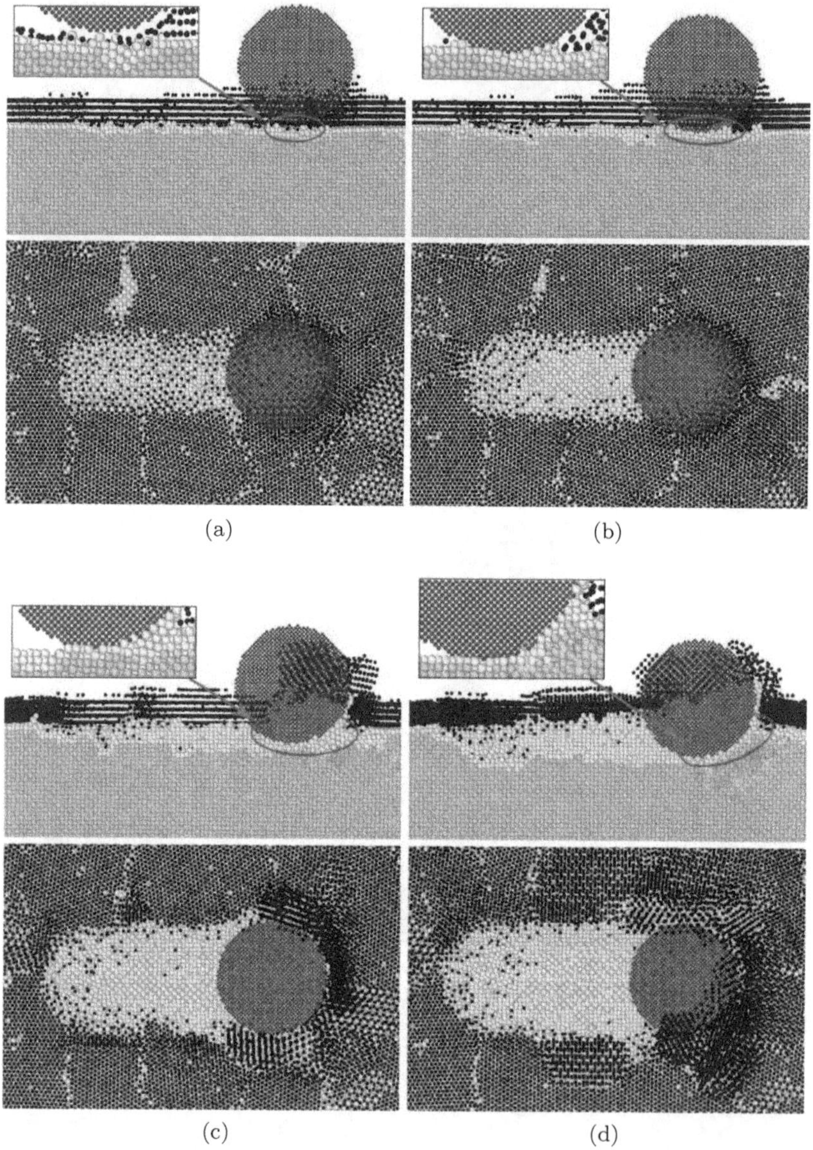

Figure 5.42. Cross-sectional view (top, xz-plane) and top view (bottom, xy-plane) of single-crystal copper abrasion at a water film thickness of 1.0 nm. (a) Scratch depth $h = -0.2$ nm, (b) scratch depth $h = 0.1$ nm, (c) scratch depth $h = 0.5$ nm, (d) scratch depth $h = 1.0$ nm.

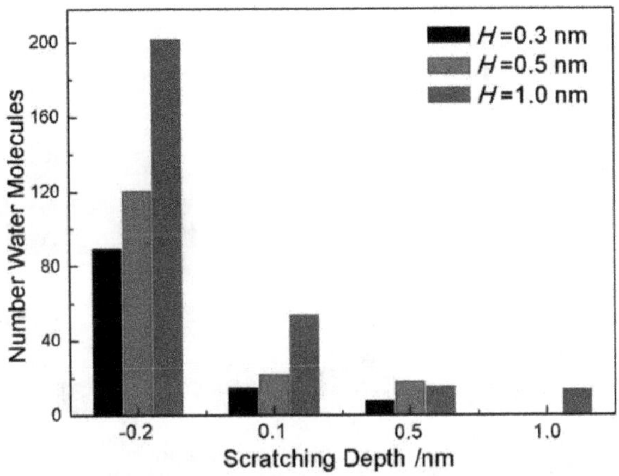

Figure 5.43. Number of water molecules remaining in the wear region of single-crystal copper.

Table 5.4. Average forces between different components in the direction of the scratch.

Water film thickness/nm	Scratch depth/nm	Water–copper force/nN	Water–abrasive force/nN	Abrasive–copper force/nN
0.3	−0.2	7.70	9.24	1.23
	0.1	10.41	11.26	21.96
	0.5	4.48	5.40	86.94
0.5	−0.2	17.29	16.17	0.73
	0.1	16.28	17.80	20.12
	0.5	7.02	6.52	90.08
1.0	−0.2	41.52	46.08	0.18
	0.1	39.44	51.94	18.44
	0.5	19.98	30.69	95.59
	1.0	4.24	13.12	147.98

46.08 nN, respectively). This implies that the wear of the monoatomic layer is primarily due to the interaction between the water film and the single-crystal copper. Figure 5.45 depicts the wear process of a single atomic layer for a water film thickness of 1.0 nm and a scratch depth of −0.2 nm. In the diagram, the red color represents the oxygen atoms in the water molecules, the gray color denotes the abrasive particles, the earthy yellow color represents the copper substrate,

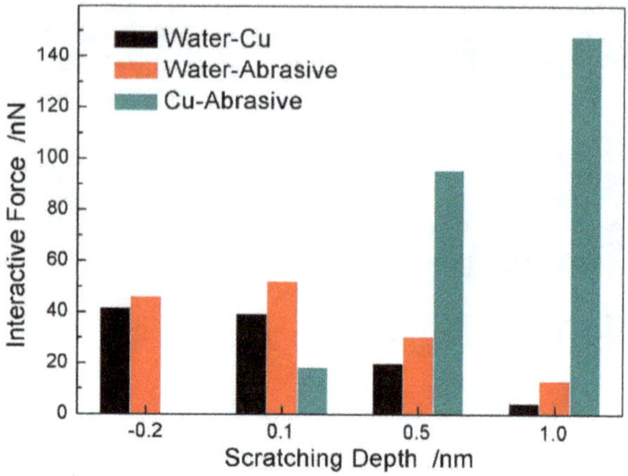

Figure 5.44. Interaction forces between components at different scratch depths for a water film thickness of 1.0 nm.

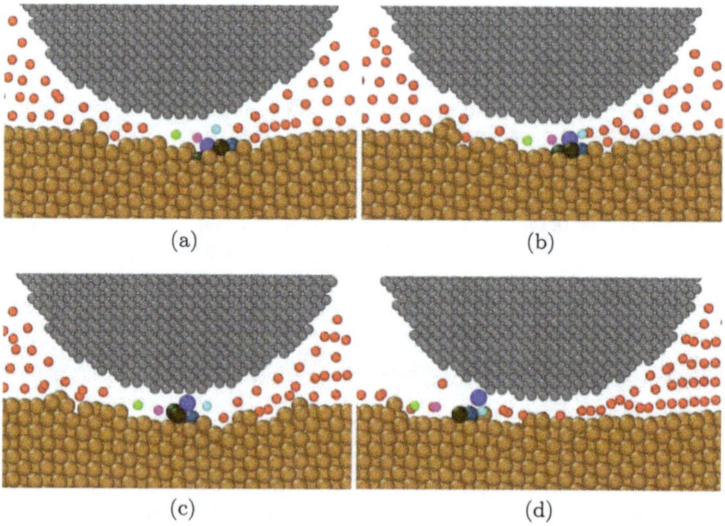

Figure 5.45. Transient configurations of single-crystal copper wear at different moments ($h = -0.2$ nm). (a) $t = 5$ ps, (b) $t = 15$ ps, (c) $t = 35$ ps, (d) $t = 100$ ps.

and a few other atoms labeled with various colors are employed to elucidate the detailed process of atom removal. When abrasive particles begin to move horizontally, the copper atoms beneath them become slightly deformed. Over time, from 5 ps to 15 ps to 35 ps, the surrounding oxygen atoms (depicted as water molecules colored by light blue) gradually attract or push the copper atoms. When the force between the water and copper is strong enough to break the metallic bonds between the copper atoms, the atoms are displaced from their original position and may stick to the abrasive particle. By 100 ps, the blue copper atoms have worn away and are located above the original copper surface (Figure 5.45(d)). This wearing process continues during the nano-scratching, and the removal and adhesion of copper atoms on the surface only affect the surface microstructure. Any changes in surface roughness are limited to the single-atom scale [30–32]. Similar surface atomic wear processes also occurred during nano scratches with a depth (h) of 0.1 nm. At the same time, the force between the abrasive particles and the Cu substrate increased to 18.44 nN (Table 5.4 and Figure 5.37), indicating a significant increase in the adhesion of the abrasive particles to the Cu atoms. This led to more wear atoms and a larger wear region. Additionally, Figure 5.46 shows the migration of wear atoms, where copper atoms or small clusters of atoms can be seen to have moved to zones 3 and 6 from their original locations, i.e., zones 2 and 5. However, no abrasive chips were formed. Less atoms migrated at a scratch depth of −0.2 nm, as shown in Figure 5.46(a). These results strongly suggest that adhesion is the main cause of surface atom wear during nanowear in non-contact and single-atom-layer contact.

As the scratch depth increases to 0.5 nm and 1.0 nm, most of the water molecules beneath the abrasive particles are pushed out of the wear area, as depicted in Figures 5.42 and 5.43. This direct contact between the abrasive and the copper substrate causes a rapid decrease in the interaction forces between water and copper, as well as between water and abrasive, while the force between the abrasive particles and the copper substrate dramatically increases. Table 5.4 and Figure 5.44 illustrate this change. Notably, at a scratch depth of 1.0 nm, the interaction force between abrasive particles and copper reaches 147.98 nN, which is significantly larger than the other two forces (4.24 nN and 13.12 nN) and very close to the abrasive-to-copper force of 158 nN in the case of an aqueous-free film. This suggests that the direct interaction between the abrasive particles and the copper substrate predominantly influences the wear of single-crystal copper at this scratch depth. As the abrasive particles scratch the surface, an amorphous layer forms on the copper surface, the

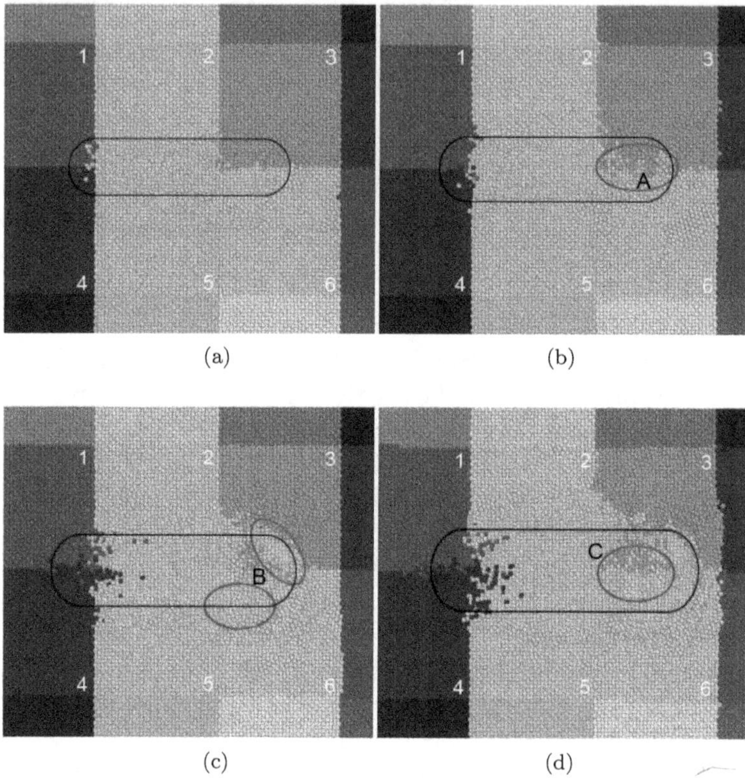

Figure 5.46. Migration configuration of surface copper atoms at a water film thickness of 1.0 nm. (a) Scratch depth $h = -0.2$ nm, (b) scratch depth $h = 0.1$ nm, (c) scratch depth $h = 0.5$ nm, (d) scratch depth $h = 1.0$ nm.

deformation region expands, and larger abrasive chips appear on the abrasive front, as shown in Figures 5.46(c) and 5.46(d). Consequently, the authors concluded that at greater scratch depths, material wear or removal is primarily controlled by ploughing action, with a small amount of adhesion. The ploughing action results in plastic deformation of single-crystal copper not only on the surface but also within the substrate matrix. This process causes an increase in surface roughness and a deterioration in surface quality, as illustrated in Figures 5.37(c), 5.37(d), 5.42(c), and 5.42(d).

In Figure 5.47, the wear morphology of single-crystal copper is displayed at single and double water film thicknesses. It is evident that the distribution of water molecules around the abrasive particles

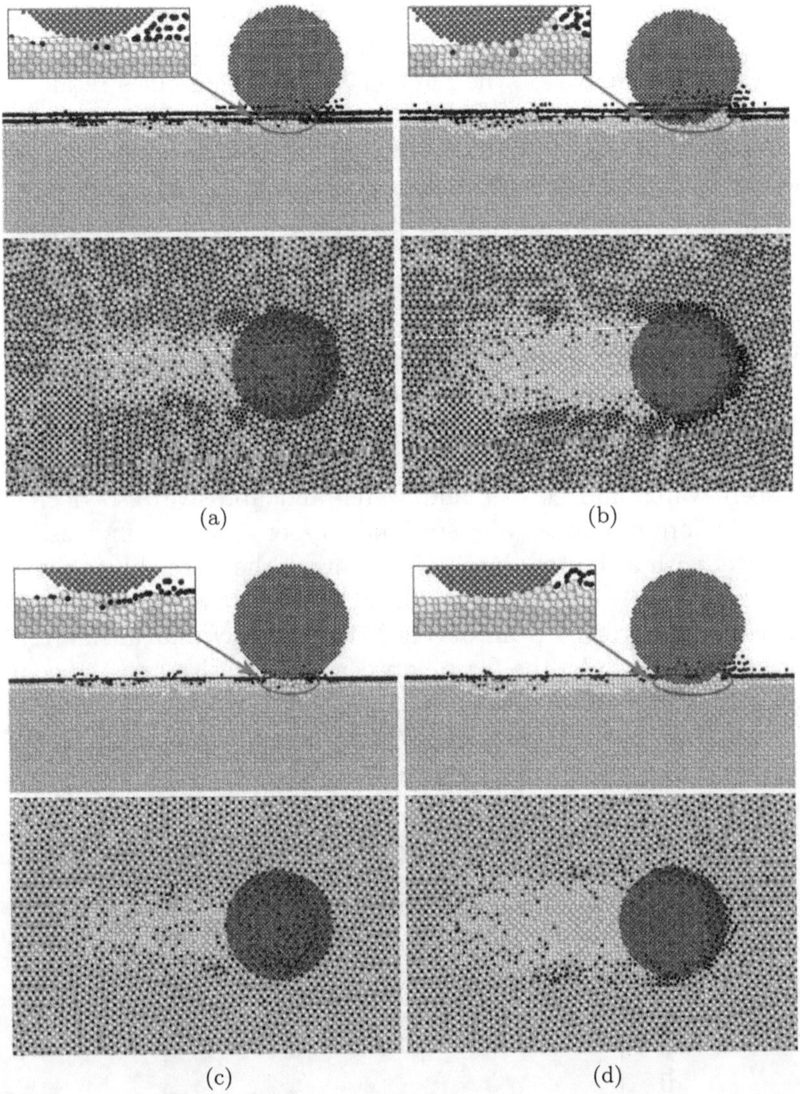

Figure 5.47. Cross-sectional view (top, xz-plane) and top view (bottom, xy-plane) of single-crystal copper abrasion when the water film is single and double-layered. (a) Water film thickness of 0.5 nm, $h = -0.2$ nm, (b) water film thickness of 0.5 nm, $h = 0.1$ nm, (c) water film thickness of 0.3 nm, $h = -0.2$ nm, (d) water film thickness of 0.3 nm, $h = 0.1$ nm.

at shallow indentation depths resembles that at a water film thickness of 1.0 nm (Figures 5.42(a) and 5.42(b)). However, as the thickness of the water film increases, the number of water molecules retained in the wear region decreases (Figure 5.43). The insets of Figures 5.47(a) and 5.47(c) indicate the retention of some water molecules below the abrasive particles, while a greater quantity of water molecules is present in the space in front of the abrasive. These water molecules collectively play a crucial role in facilitating the wear process of the monatomic layer. In contrast, when the scratch depth reaches 0.1 nm, there are scarcely any water molecules left underneath the abrasive, and the abrasive particles begin to make contact with the copper substrate, similar to dry nano-scratches (Figure 5.41(b)). The data presented in Table 5.4 and Figures 5.48 and 5.49 indicates that as the water film thickness decreases to 0.5 nm and 0.3 nm, the forces between water and copper and water and abrasive also decrease, while the forces between abrasive and copper substrate increase. For instance, with a scratch depth of 0.1 nm, as the water film thickness reduces from 1.0 nm to 0.5 nm to 0.3 nm, the force between water film and copper decreases from 39.44 nN to 16.28 nN and ultimately to 10.41 nN. Meanwhile, the force between the abrasive and the copper matrix increases from 18.44 nN to 20.12 nN and then to 21.96 nN. It's apparent that the force between the abrasive particles and the single-crystal copper almost reaches two times the force between the

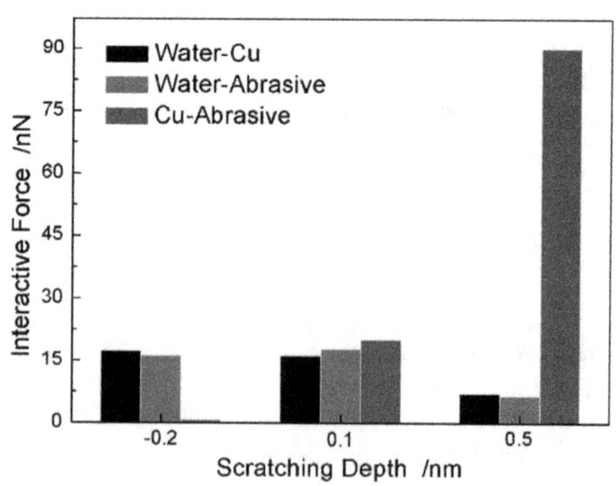

Figure 5.48. Interaction forces between components at different scratch depths for a water film thickness of 0.5 nm.

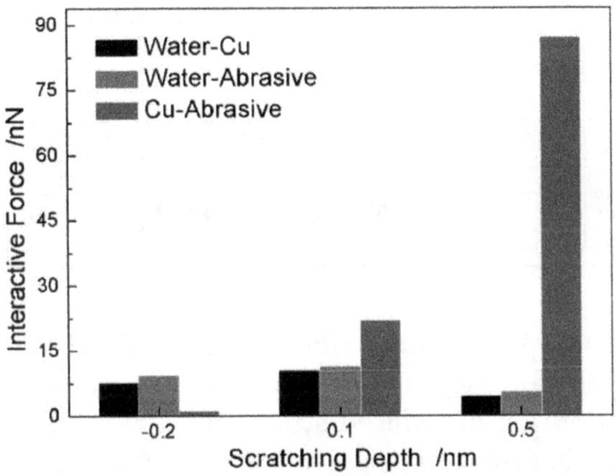

Figure 5.49. Interaction forces between components at different scratch depths for water film thickness of 0.3 nm.

water film and the copper. These outcomes suggest that a decrease in the water film thickness reduces its impact on the wear of the single atomic layer of single-crystal copper but increases the adhesion of the abrasive particles to the copper atoms.

It is widely recognized that while material wear is nearly unavoidable in most conditions, minimizing surface damage from wear and reducing material defects are always desirable. The ultimate goal of wear reduction is to achieve "0 nm roughness and 0 defects under 0 external force." Research has shown that non-contact and single-layer contact conditions can be achieved during single-atomic layer wear or material removal, resulting in high surface quality, particularly when the water film is thicker. However, when the scratch depth increases to 0.5 nm and 1.0 nm, the formation of plough grooves, ridges, abrasive front end abrasive debris build-up, and increased surface roughness lead to degradation of surface quality. Analysis of the dislocation line length at the end of the scratch in single-crystal copper illustrates plastic deformation. At scratch depths of −0.2 nm and 0.1 nm, no dislocations are observed, indicating no plastic deformation in the single-crystal copper, consistent with relevant experimental findings. The findings align with the results of similar experiments [32, 33]. As the scratch depth continued to increase, dislocations nucleated and expanded. For instance, the length of the dislocation line sharply increased from 28.57 nm to 108.01 nm as the scratch depth went from

Table 5.5. Dislocation line lengths within single-crystal copper.

Water film thickness/nm	0.3			0.5			1.0			
Scratch depth/nm	−0.2	0.1	0.5	−0.2	0.1	0.5	−0.2	0.1	0.5	1.0
Dislocation line length/nm	0	0	65.55	0	0	42.33	0	0	28.57	108.01

0.5 nm to 1.0 nm. Additionally, the thickening of the water film led to a decrease in the force exerted by the abrasive particles on the single-crystal Cu, resulting in reduced stress on the material and a subsequent decrease in dislocations as the water film thickened (Table 5.5).

5.3.7 *Summary*

Investigation was conducted on the nano-scratch behavior of spherical abrasive particles on single-crystal copper surfaces with varying water film thickness, as well as the impact of different scratch depths on the wear of single-crystal copper under an ultrathin water film to remove single-layer material. The key findings are as follows:

(1) Dry sliding of abrasive particles resulted in aggravated wear on the copper surface as the sliding speed increased from 50 m/s to 200 m/s, causing a 9.6% rise in friction coefficient. However, this also led to rapid release of local stress concentration, thereby reducing defects in the copper. Higher normal load (64–160 nN) increased damage to the copper surface by abrasive particles and internal defects within the copper. Friction coefficient is also increased by 13.4%.
(2) The presence and thickening of the water film significantly increased friction on the abrasive particles. However, the water film acted as a barrier, preventing direct interaction between the abrasive particles and the single-crystal copper, thereby reducing friction on the copper. In the absence of a water film, the wear on the copper surface was regular and less damaged. With the presence of a water film, the accumulation of water molecules around the abrasive particles led to increased ridge anisotropy and surface damage on both sides of the plough groove, resulting in the formation of irregular atomic steps inside the groove.

(3) In the process of nano-scratching, the dislocation network undergoes rapid continuous reorganization, with the creation and disappearance of dislocations happening simultaneously. The presence of a water film increases the number of dislocations within single-crystal copper. Interestingly, the wear amount of single-crystal copper decreases as the water film appears and thickens, suggesting that the water film provides some degree of lubrication during the two-body wear of single-crystal copper.

(4) Moreover, under non-contact ($h = -0.2$ nm) and monolayer contact ($h = 0.1$ nm) conditions, nanomaterial abrasion can lead to the removal of a single layer, resulting in a relatively flat new surface free of deformation layers and plastic defects. However, at greater scratch depths (0.5 nm and 1.0 nm), the abrasive particles displace water molecules between them and the abrasive particle directly interacts with the single-crystal copper, leading to surface damage, accumulation of abrasive fragments, and severe plastic deformation of the subsurface and interior.

(5) Furthermore, the wear or removal of the single-atom layer during nano two-body wear is governed by interatomic adhesion. The water film can transfer the force of abrasive particles to the single-crystal copper's surface, causing the migration and removal of atoms from the surface. At larger scratch depths, material wear is primarily attributed to ploughing action.

5.4 Three-Body Abrasive Wear of Single-Crystal Copper Water-Bearing Films

5.4.1 *Introduction*

In the process of three-body wear, chips or abrasive particles move in various ways in relation to the upper and lower solid surfaces, such as sliding, rolling, or a combination of both. These different movements result in varying degrees of wear and damage to the material surface [30, 31, 34]. Although there have been numerous experimental studies on three-body abrasive wear, the majority have not taken into account the external environment in which the material is situated, such as humidity or water-containing working conditions. The presence of water significantly impacts the mechanical properties, as well as the friction and wear properties of materials. Research has revealed that several factors affect the motion mode of abrasive particles or chips, including the nature of the abrasive particle itself (geometry, hardness, etc.), material properties (varying surface roughness, hardness, defects, etc.), and external conditions (pressure, velocity, lubricating fluid, etc.). Nevertheless, it remains uncertain whether the influence of these factors on abrasive motion mode changes in the case of higher humidity or water-containing working conditions. As a result, studying the friction and wear behavior of materials under high humidity or water-containing conditions is essential to provide more informative results for practical applications.

In this section, we discuss the nano three-body abrasive wear of single-crystal copper under water-containing conditions. We examine the effects of abrasive shape, applied load, and driving speed of the upper and lower plates of single-crystal copper on the ellipsoidal abrasive mode of motion. This analysis aims to understand the abrasive wear behavior of single-crystal copper under water-containing conditions and the inter-component interaction mechanism.

5.4.2 *Modeling and simulation methods*

The illustration in Figure 5.50 depicts the initial model of single-crystal copper three-body abrasive wear. The model comprises two smooth, defect-free single-crystal copper flat plates positioned at the top and bottom, ellipsoidal diamond abrasive particles sandwiched between plates, and an ultrathin aqueous film adsorbed on the single-crystal copper surface. The dimensions of the single-crystal copper plate along the x-[100], y-[010], and z-[001] crystal directions are $8.0 \times 2.5 \times 4.0$ nm, respectively. The two plates can build a contact by the sandwiched abrasive particle under macroscopic conditions and consist of a boundary layer, a hot bath layer, and a Newtonian layer. The boundary layer serves as a rigid body to ensure system stability, the hot bath layer controls the system's temperature, and the Newtonian layer is where substrate abrasion and deformation occur. In the Newtonian layer, the atoms follow the Newtonian law of motion. When studying the effect of abrasive particle shape on three-body wear, the present researchers varied the short and long axis ratio (α) of the diamond abrasive particles from 0.70 to 0.90 (Table 5.6).

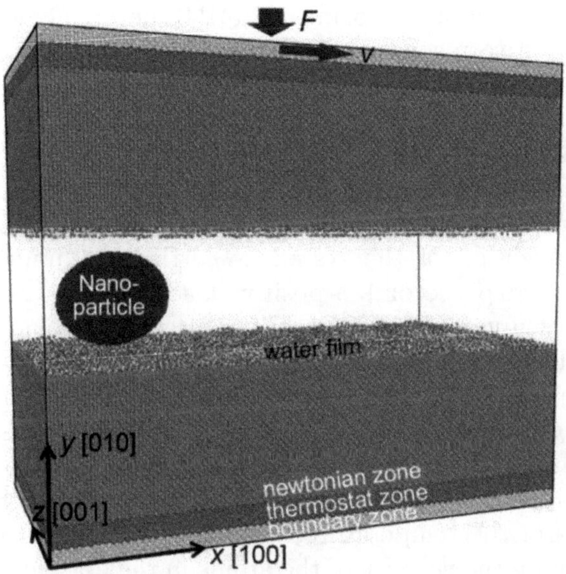

Figure 5.50. Initial model of single-crystal copper three-body abrasive wear.

Table 5.6. Semi-axis lengths of ellipsoidal abrasive particles.

$\gamma = b/a$	0.70	0.80	0.85	0.90
a (x-direction)/Å	60	60	60	60
b (y-direction)/Å	42	48	51	54
c (z-direction)/Å	60	60	60	60

It was observed that as the axis ratio decreases, the abrasives become sharper, and conversely, as the ratio increases, the abrasives resemble a more spherical shape. Manipulating the size of the axis length in the y direction is the sole method used to alter the shape of the abrasive particles in this study, leading to variations in the so-called "sharpness" and the abrasive volume.

The EAM potential function is used to model the interaction force between copper atoms, while the interaction between copper and carbon atoms is represented using the Morse potential function. The abrasive particles are treated as a rigid body, and the force between carbon atoms is not computed. The behavior inside the water molecule is described using the TIP4P potential model, and the interactions between copper atoms, carbon atoms, and oxygen atoms in the water molecule are modeled using the Lennard–Jones potential function. Detailed potential function models and action parameters can be found in Table 5.6.

The abrasion process of diamond abrasive particles between single-crystal copper surfaces involves three main stages: First, the energy optimization of the system is carried out under the NVT ensemble for 100 ps; second, a positive load is applied to the single-crystal copper upper plate along the $-y$ direction until equilibrium is reached in the vertical indentation process; third, the upper and lower plates are moved horizontally along the x and $-x$ directions at a constant driving speed (v). The second and third stages of the wear process are simulated using the NVE ensemble, and the atomic motion is integrated using the Velocity–Verlet algorithm with a time step of 1 fs. To control the temperature, the Langevin hot bath method is used to regulate the velocity of the atoms in the hot bath layer, with a target temperature of 300 K. The simulation system is laterally oriented in the xz-plane using cyclic boundary conditions. The results of the simulation were visualized using OVITO software.

5.4.3 *Influence of abrasive shape on nano three-body abrasive wear of single-crystal copper*

In the initial model presented for non-equilibrium MD simulation of three-body abrasive wear under ultra-thin water film conditions (Figure 5.50), a water film thickness of $H = 0.35$ nm, a single-crystal copper upper and bottom plate with a driving speed of $50\,\mathrm{m \cdot s^{-1}}$, and an applied positive load of 80 nN were utilized. Given the significant impact of the mode of motion of the abrasive particles on the wear of the single-crystal copper substrate, the initial part of this section focuses on examining the relationship between the abrasive particles' shape and their mode of motion.

Figure 5.51 depicts the transient configurations of abrasive particle motion and single-crystal copper wear at various time points. It is observed that as the axis ratio (α) increases from 0.70 to 0.90, the shape of the nanoparticles gradually shifts from sharp ellipsoid to spherical. A comparison of the configurations at the initial

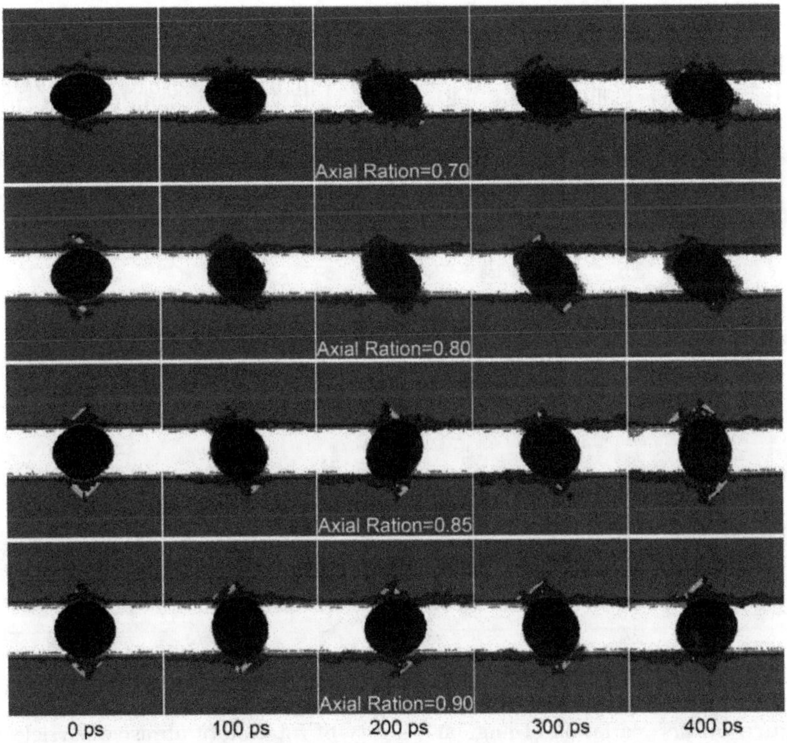

Figure 5.51. Transient configurations of abrasive particle motion and single-crystal copper wear for different axis ratios.

two time points (0 ps and 100 ps) reveals that ellipsoidal abrasive particles with varying axis ratios undergo a rolling process. Furthermore, it indicates that with an increase in the axis ratio of the abrasive particles, the tip comes into contact with the substrate surface earlier, suggesting a significant increase in the angle of abrasive rolling with an increasing axial ratio. After 100 picoseconds, the transient states of the two more elongated abrasive particles with axial ratios of 0.70 and 0.80 stabilize, indicating a transition from initial rolling to a consistent sliding motion which persists after 200 picoseconds. Conversely, the two ellipsoidal abrasive particles with larger axial ratios ($\alpha = 0.85$ and $\alpha = 0.90$) continue to undergo instantaneous changes in their state throughout the wear process, suggesting that they remain in a rolling mode.

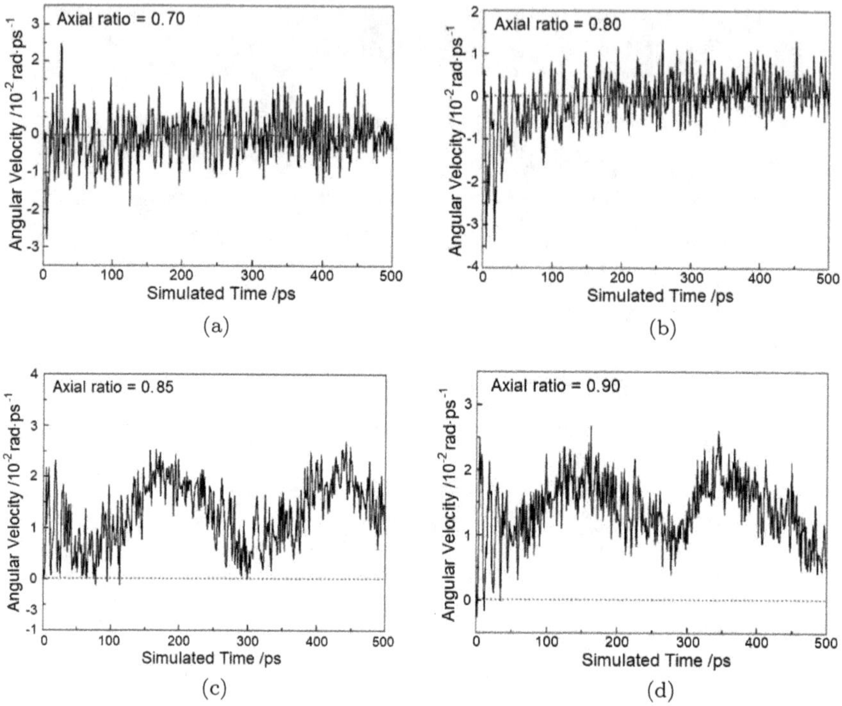

Figure 5.52. Variation of angular velocity of rotation of abrasive particles as a function of time. (a) $\alpha = 0.70$, (b) $\alpha = 0.80$, (c) $\alpha = 0.85$, (d) $\alpha = 0.90$.

In Figure 5.52, the evolution of the angular velocity of rotation of the nano-abrasive particles is depicted with simulation time. In Figures 5.52(a) and 5.52(b), it is observed that the rotation angular velocity of the sharp abrasive particles exhibits considerable fluctuation only at the beginning of the motion, attributed to the transient action at the start of wear. Subsequently, it maintains a dynamic equilibrium near the value of 0, validating the characteristics of the abrasive's motion in Figure 5.51. Concurrently, for both ellipsoidal abrasives moving in a sliding mode, the absolute value of the initial non-zero angular velocity increases with the increase in the abrasive's axial ratio, indicating that the relatively blunt abrasive necessitates a larger initial angular velocity of rotation to balance normal load and friction. As for Figures 5.52(c) and 5.52(d), the rotational angular velocities of the two ellipsoidal abrasives with larger axial ratios exhibit periodic fluctuations resembling sinusoidal curves during the three-body wear process, with values greater than 0, a characteristic expected in the process of rolling motion of the abrasive particles.

When the abrasive axial ratio increases from 0.85 to 0.90, the minimal value of the sine curve increases while the amplitude decreases. This indicates that the rotational variability of the abrasive at different positions decreases and the angular velocities gradually approximate for two particles. These results suggest that spherical abrasive particles ($\alpha = 1.0$) also exhibit a rolling motion under the same wear conditions, accompanied by a constant angular velocity of rotation. Comparison with relevant literature reveals that the motion mode of abrasive particles in the presence of ultra-thin water film is the same as that in three-body abrasive wear under water-free conditions, with the axial ratio of abrasive particles sliding and rolling transitions in the water-free case being 0.83 [35].

In Figure 5.53, we observe the transient configuration of the upper and bottom surfaces of single-crystal copper at the conclusion of three-body wear. The dark blue balls in the figure represent the abrasive particles, with their shapes varying depending on the viewing angle. The red and yellow colors represent oxygen and hydrogen atoms in the water molecules, while the cyan and light blue colors represent the surface copper atoms.

In Figures 5.53(a) and 5.53(b), we note that only a few water molecules remain in the wear area, while more water molecules are gathered at the front end of the abrasive. This suggests that most of the water molecules originally adsorbed on the surface of the single-crystal Cu were extruded from the wear area by the abrasive particles and pushed to the front end during the three-body wear process. In contrast, Figures 5.53(c) and 5.53(d) show that many water molecules were not extruded from the worn zone but remained there, with no water molecules aggregated at the front end of the abrasive. This observation reflects the difference in abrasive movement.

The statistical results regarding the number of water molecules remaining in the worn zone are depicted in Figure 5.54. The figure illustrates that the number of water molecules remaining in the worn zone is significantly higher under rolling motion compared to sliding motion. Changes in the shape of the abrasive particles result in variations in the contact area between the particles and the water film or the single-crystal copper surface. Additionally, under sliding motion, there is a marginal decrease in the number of water molecules

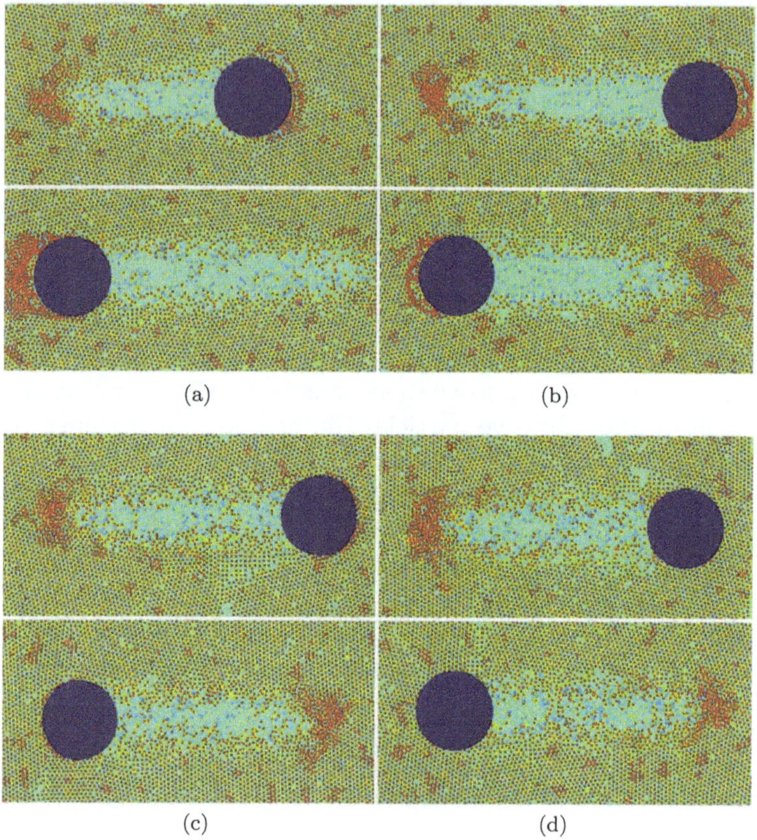

Figure 5.53. Transient configurations of the upper (top) and lower (bottom) surfaces of single-crystal copper at the end of wear. (a) $\alpha = 0.70$, (b) $\alpha = 0.80$, (c) $\alpha = 0.85$, (d) $\alpha = 0.90$.

remaining in the worn zone with an increase in the ratio of the abrasive axes, while under rolling motion, there is a slight increase in the number of water molecules remaining with the increase in the ratio of the abrasive axes.

It is important to note that at an abrasive axial ratio of 0.70 (Figure 5.53(a)), the moving distance between the upper and lower surfaces of the single-crystal copper is not uniform. Specifically, the front end of the sharp abrasive particles preferentially embeds into the substrate, resulting in a greater embedding depth. Additionally, random transient fluctuations of the acting force cause a change in

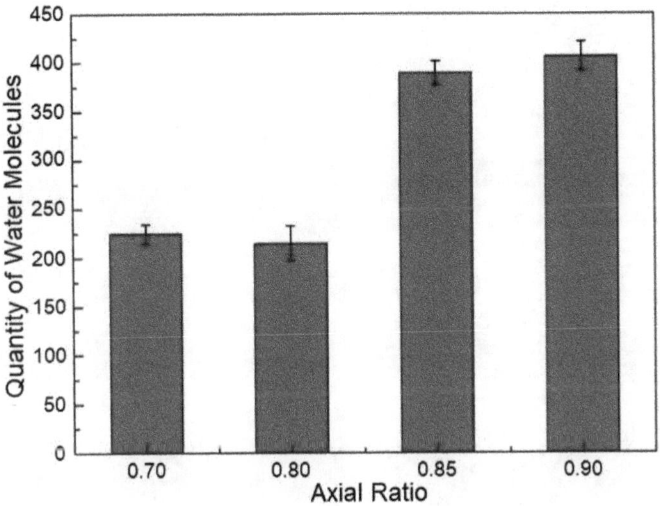

Figure 5.54. Number of water molecules remaining in the worn zone.

the transient configuration of the abrasive particles, leading to random embedding of abrasive in the upper plate and hindering its sliding related to the upper plate.

The arrangement of water molecules impacts not only the movement of abrasive particles but also the friction and wear of single-crystal copper. Figure 5.55 illustrates the wear patterns on the upper and bottom plates of single-crystal copper after wear. From Figures 5.55(a) and 5.55(b), it is evident that the sliding abrasive particles create regular wear marks on the upper and bottom surfaces of single-crystal copper. Additionally, a large number of copper atoms in the worn zone are detached from their original position, forming ridges and accumulations of debris, similar to the outcomes of two-body wear (references [9] and [36]). Similarly to two-body wear, the accumulation of abrasive debris at the front end is more notable when a water film is present, but the height of the ridges on each side of the wear mark is reduced. This implies that the presence of a water film impedes the build-up of debris on both sides of the ridge [35]. Moreover, the rolling motion of the abrasive causes irregular grooves, ridges, and uneven distribution of worn atoms in the worn zone, in contrast to the results of dry three-body abrasive wear [35]. This indicates that the water film, particularly the water

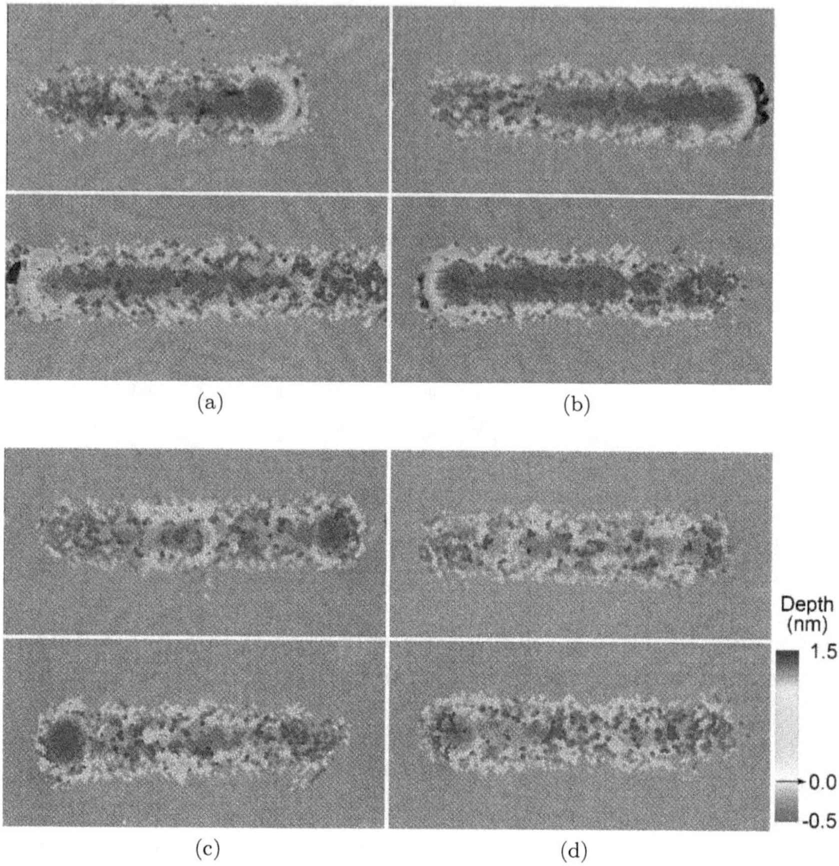

Figure 5.55. Wear morphology of single-crystal copper upper and bottom plates at the end of wear. (a) $\alpha = 0.70$, (b) $\alpha = 0.80$, (c) $\alpha = 0.85$, (d) $\alpha = 0.90$.

molecules remaining in the worn zone, significantly influences the abrasive wear process.

The wear damage of single-crystal copper is determined by calculating the number of worn or removed copper atoms and the length of the dislocation line of the shaped nucleus within the single-crystal copper matrix. The assessment of the number of abraded copper atoms in this section aligns with Figure 5.36. From Figure 5.56, it is evident that the number of abraded copper atoms resulting from the rolling of abrasive particles is fewer than those caused by the sliding motion of abrasives. However, the length of dislocation lines within

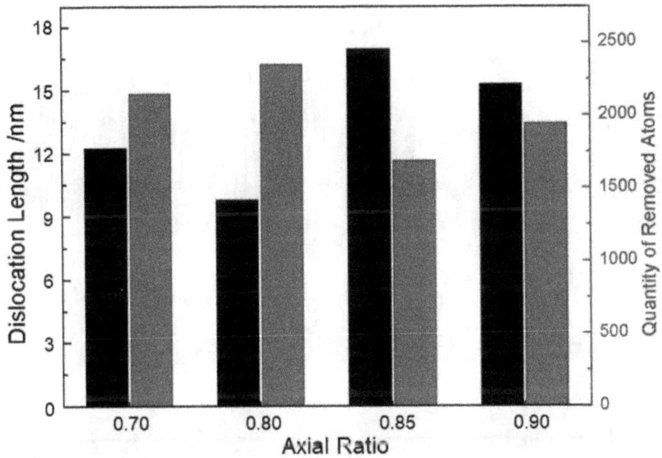

Figure 5.56. Number of worn copper atoms versus length of dislocation lines within single-crystal copper.

the single-crystal copper demonstrates the opposite behavior. This suggests that the rolling motion of ellipsoidal abrasives diminishes wear on the surface of the single-crystal copper but intensifies damage within it (e.g., as illustrated in Figure 5.51). In the sliding process, the initial rotation angle increases with the axis ratio, thereby enhancing the cutting ability of the abrasive particles to progress while reducing the depth of the worn surface. Conversely, during the rolling process, the primary wear occurs in the abrasive particles in a stationary state (as depicted in Figure 5.51), and the decrease in the sharpness of the abrasive particles (increasing the axis ratio) enlarges the contact area of the abrasive particles with the surface of the single-crystal copper. This reduces the local stress concentration and formation of defects but escalates the extent of surface damage. Figure 5.56 illustrates that as the abrasive axial ratio increases, the number of worn copper atoms also increases while the dislocation line length decreases for the same abrasive motion. These results lead to the conclusion that the main cause of wear on the single-crystal copper substrate during abrasive sliding is the polishing or groove effect, whereas the main mechanism of wear during abrasive rolling is the rolling or ball effect, i.e., extrusion or gouging.

The friction coefficient (μ) was calculated during the wear process of three bodies, as illustrated in Figure 5.57. The graph indicates

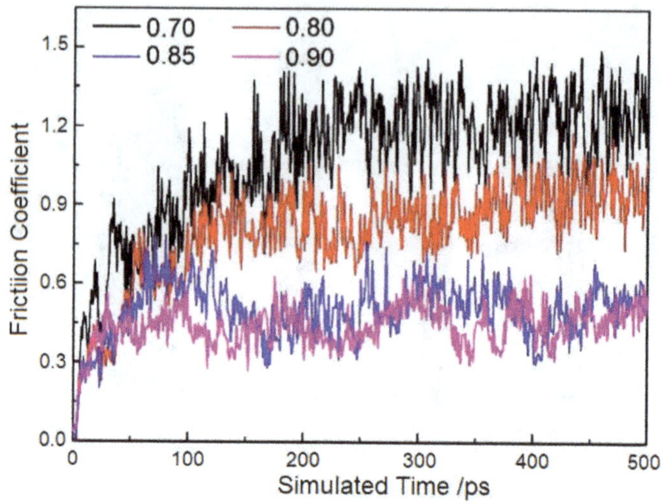

Figure 5.57. Curve of variation of coefficient of friction for different abrasive axial ratios.

that the friction coefficient increases rapidly from zero and eventually stabilizes as the simulation progresses. The sharper the abrasive particles, the longer it takes for the friction coefficient to increase, suggesting that it requires more time to adjust the kinematic state of sharp abrasives during the early stages of wear. In the stable stage of the friction coefficient, the average value decreases from 1.20 to 0.90 when the axial ratio of the abrasive material increases from 0.70 to 0.80. Moreover, the coefficient of friction of the sliding condition is much larger than that of the rolling condition. Specifically, the coefficient of friction of abrasive with an axial ratio of 0.70 decreases from 1.20 to 0.45 relative to the coefficient of friction of abrasive with an axial ratio of 0.90, which represents a reduction of 62.5%. These results indicate that the transition of abrasive particles from sliding to rolling can lead to a reduction in material friction and wear, thereby confirming the rolling mechanism of wear reduction [37–40].

The mode of motion of abrasive particles sandwiched between solid surfaces is influenced by the interaction between different components. In order to understand this interaction, the variation of the interaction force over time between the abrasive particles (S), the single-crystal copper upper and bottom plates (B), and the water film (W) along the x (horizontal) and y (vertical load) directions were examined in Figures 5.58–5.61. The black and red curves represent the interaction force acting on the upper and bottom plates and the water film adsorbed on them, respectively. Positive and negative values indicate the direction of the force. A constant load of 80 nN is applied to the upper plate of single-crystal copper, resulting in the bottom plate being subject to an 80 nN force as well. This makes the force acting on the upper and bottom plates or the upper and bottom water film symmetrically and basically equal in numerical value.

In a system where abrasive particles move in a sliding manner (Figures 5.58 and 5.59), the following is notable along the direction of positive load loading:

(1) The force F_y_S-B between the ellipsoidal abrasive particles and the single-crystal copper matrix initially exhibits small fluctuations, followed by a rapid, nearly linear increase from 0, eventually stabilizing.
(2) The water film-abrasive force F_y_W-S decreases from an initial extreme value (approximately equal to the applied load of 80 nN) to zero, then experiences a slight increase in the opposite direction of the initial interaction force, and eventually stabilizes at around 15 nN.
(3) Both the force F_y_W-B between the water film and the single-crystal copper plate and the force F_y_W-S between the water film and the abrasive particles exhibit similar variation characteristics.

In conjunction with the three-body wear configuration, an analysis of the water film distribution characteristics and the variations in the positive interaction force reveals the following: During the initial phase of three-body abrasive wear, the water film has not yet been fully disrupted. Consequently, the external force applied to the single-crystal copper plates is transmitted to the abrasive particles through the water film medium rather than through direct contact between the abrasive particles and the single-crystal copper plates. As the relative movement progresses, the water film is progressively expelled from the wear region (Figures 5.53(a) and 5.53(b)). Subsequently, the abrasive particles engage directly with the single-crystal

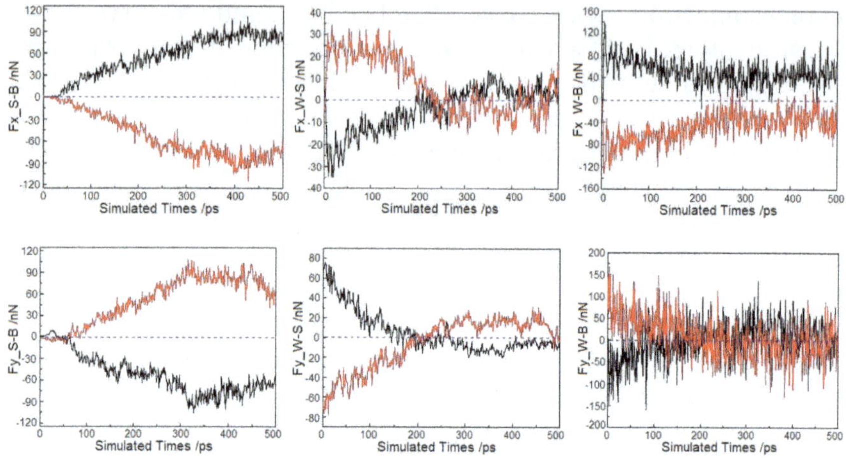

Figure 5.58. Interaction forces in the abrasive system with an axial ratio of 0.70.

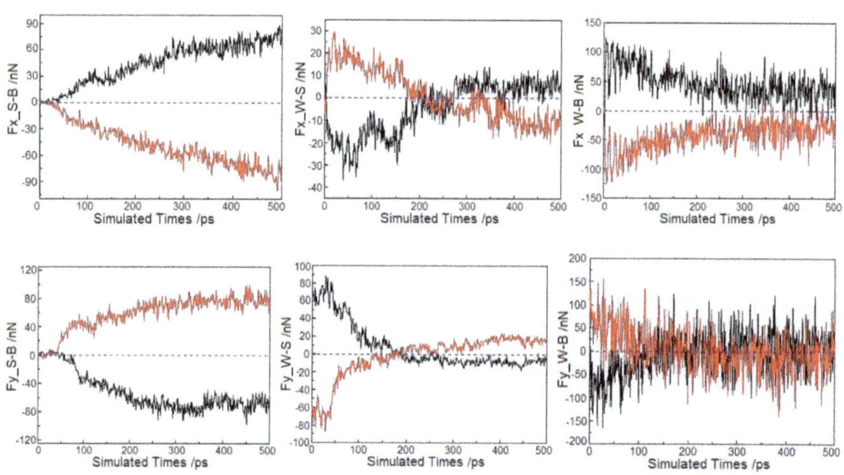

Figure 5.59. Interaction forces in the abrasive system with an axial ratio of 0.80.

copper plates, leading to a reduction in the force transmitted through the water film.

Along the horizontal motion direction of the upper plate, the force F_x S-B between the abrasive particles and the single-crystal copper plate exhibits a trend similar to that of F_y S-B. The force F_x W-S between the water film and the abrasive particles increases rapidly

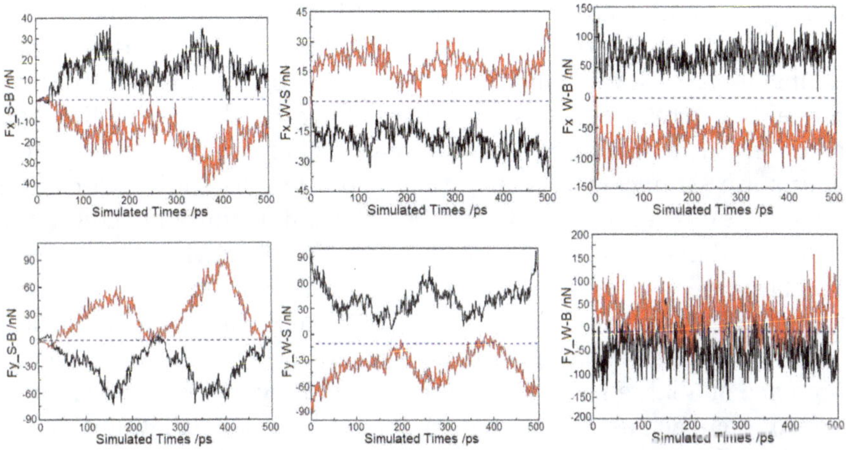

Figure 5.60. Interaction forces in the abrasive regime with an axial ratio of 0.85.

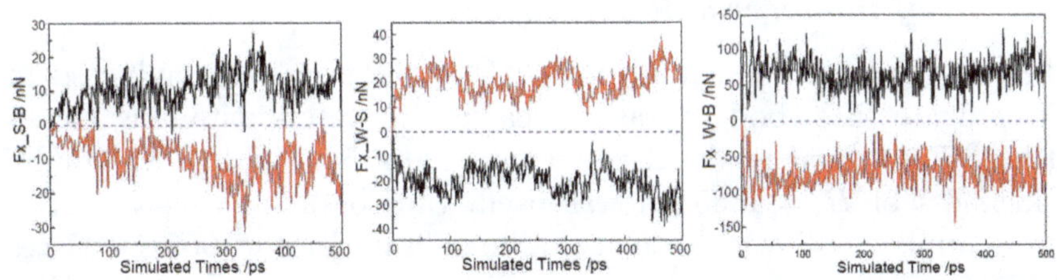

Figure 5.61. Interaction forces in the abrasive regime with an axial ratio of 0.90.

from 0 to its maximum value (approximately 25 nN) at the onset of motion, following which its behavior parallels that of F_y W-S. Conversely, the force F_x W-B between the water film and the single-crystal copper plate gradually decreases, stabilizing after 250 ps in the simulation. Throughout this period, the direction of F_x W-B remains constant, and its magnitude exceeds that of F_y W-B when stable. The observed change in the direction of the force between the water film and the abrasive particles indicates a transition from a repulsive to an attractive force.

In the case of abrasive particles moving in a rolling mode (Figures 5.60 and 5.61), the contact area of the abrasive with the copper plate and the water film with the copper plate varies continuously during motion, resulting in varying localized forces F_S-B and F_W-S, especially in the direction of positive loading. The curves of the acting forces F_x_W-S and F_y_W-S in the rolling motion state do not exhibit characteristics of reduced force or change in direction as observed in the sliding motion state. Their average values after 250 ps are significantly greater than those for the sliding case. The forces F_x_W-B and F_y_W-B between the water film and the copper plate are larger than the corresponding forces in the sliding case. Furthermore, it is noted that the forces between the water film and the abrasive particles, and the water film and the copper plate are consistently larger throughout the rolling process, mainly due to the presence of a large number of water molecules in the abrasive rolling system that act as a lubricant in the wear region (Figures 5.53(c) and 5.53(d)).

5.4.4 *Influence of load on three-body abrasive wear of monocrystalline copper*

In this study, a non-equilibrium MD simulation was conducted to investigate three-body abrasive wear under water-containing conditions. The initial model depicted in Figure 5.50 with a water film thickness of $H = 0.35$ nm was used. Horizontal driving velocities were applied to both single-crystalline Cu upper and bottom plates at 50 m/s, along with an applied positive load ranging from 50 nN to 100 nN, and an axial ratio of the abrasive particles of 0.80.

Figure 5.62 presents the transient configurations of abrasive particle motion and single-crystal copper wear at different wear stages. It was observed that the abrasive particles exhibited similar rolling motion patterns within the first 100 ps under different applied loads, suggesting that the initial positive loading had minimal effect on the abrasive motion mode at the early stage of three-body wear. However, as the wear progressed, it was noted that the abrasive particles displayed varied transient configurations at different motion moments under applied loads of 50 nN and 60 nN while maintaining the initial rolling motion. Subsequently, as the load increased to 80 nN and 100 nN, the abrasive particles transitioned from the initial rolling motion to a sliding motion mode after 200 ps.

The evolution of the angular velocity of rotation of the abrasive particles for different loads is illustrated in Figure 5.63. As depicted

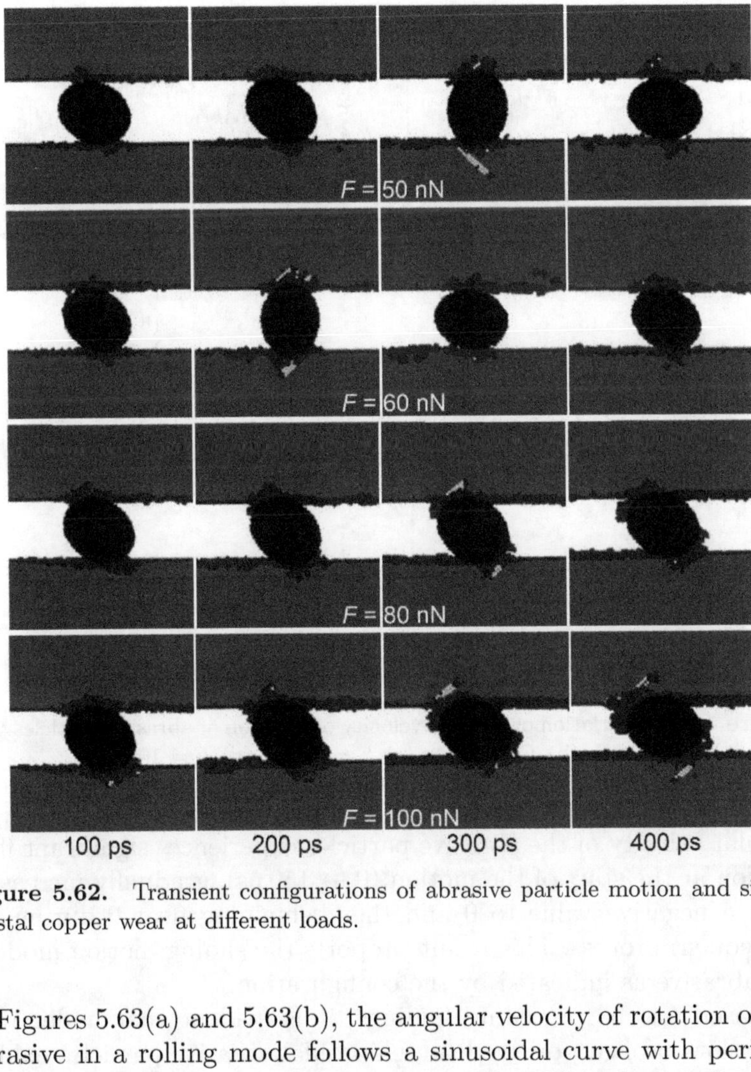

Figure 5.62. Transient configurations of abrasive particle motion and single-crystal copper wear at different loads.

in Figures 5.63(a) and 5.63(b), the angular velocity of rotation of the abrasive in a rolling mode follows a sinusoidal curve with periodic variations during the three-body wear process. This observation confirms the mode of rotation of the abrasive particles in the wear configuration. Additionally, at smaller loads (50 nN), the angular velocity fluctuates more at around 100 ps and is closer to 0. Furthermore, the configuration of the abrasive particles at 100 ps and 200 ps closely resembles each other compared to Figure 5.62, suggesting a slower rolling process or intermittent rolling–sliding mixed motion during this period. Figures 5.63(c) and 5.63(d) reveal that the rotational

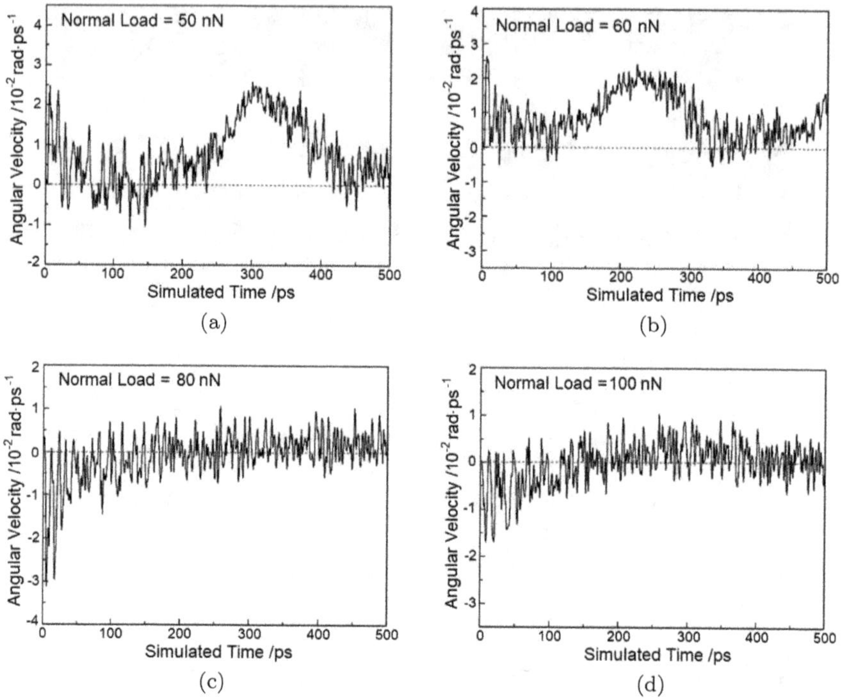

Figure 5.63. Variation of angular velocity of rotation of abrasive particles with time. (a) $F = 50\,\text{nN}$, (b) $F = 60\,\text{nN}$, (c) $F = 80\,\text{nN}$, (d) $F = 100\,\text{nN}$.

angular velocity of the abrasive particles experiences significant fluctuation at the start of the motion ($0 \rightarrow 150\,\text{ps}$), gradually increasing from a negative value to 0 and then remaining near 0 during the subsequent process. This result supports the sliding motion mode of the abrasive as indicated by the configuration.

Based on the preceding findings, it is evident that under relatively low applied positive load, ellipsoidal abrasive particles exhibit rolling motion at a specific driving speed. In contrast, under a higher applied positive load, these particles transition to sliding motion. It can be inferred that as the applied load continues to increase, the abrasive particles will persist in sliding motion due to their tendency to become embedded in the substrate under high pressure, thereby significantly increasing resistance to rolling.

If the applied pressure decreases, will the abrasive particles still exhibit rolling motion under low pressure? To explore this question,

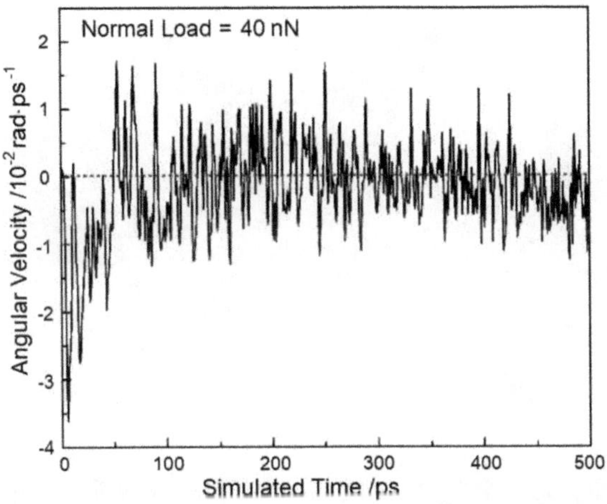

Figure 5.64. Angular velocity of rotation of abrasive particles for a load of 40 nN.

the wear process of three-body abrasive under a normal load of 40 nN was simulated, with all conditions held constant except for the pressure. The resulting rotation angular velocity curve of the abrasive, as depicted in Figure 5.64, mirrors the pattern seen in the rotation angle curve of sliding abrasive obtained from Figure 5.63. This alignment suggests that the ellipsoidal abrasive particles continue to move in a sliding manner under this applied load. The findings indicate a shift in the influence of pressure on the movement of abrasive particles, as demonstrated in Figures 5.62 and 5.63. In the three-body wear process, the interaction between the substrate material and the abrasive drives the rolling motion of abrasives. This frictional force is directly correlated to the applied normal load. When the normal load is low, the contact area between the substrate and abrasive is minimal, resulting in insufficient frictional force to propel the abrasive material into a rolling motion.

In Figure 5.65, we can see the transient configuration of a single-crystal copper surface after three-body wear. Dark blue represents abrasive particles, red and yellow represent oxygen and hydrogen atoms in water molecules, and cyan and light blue represent surface copper atoms. In Figures 5.65(a) and 5.65(b), we observe abrasive particles rolling from the upper and bottom surfaces of the single-crystal copper under a small applied load, forming a worn region.

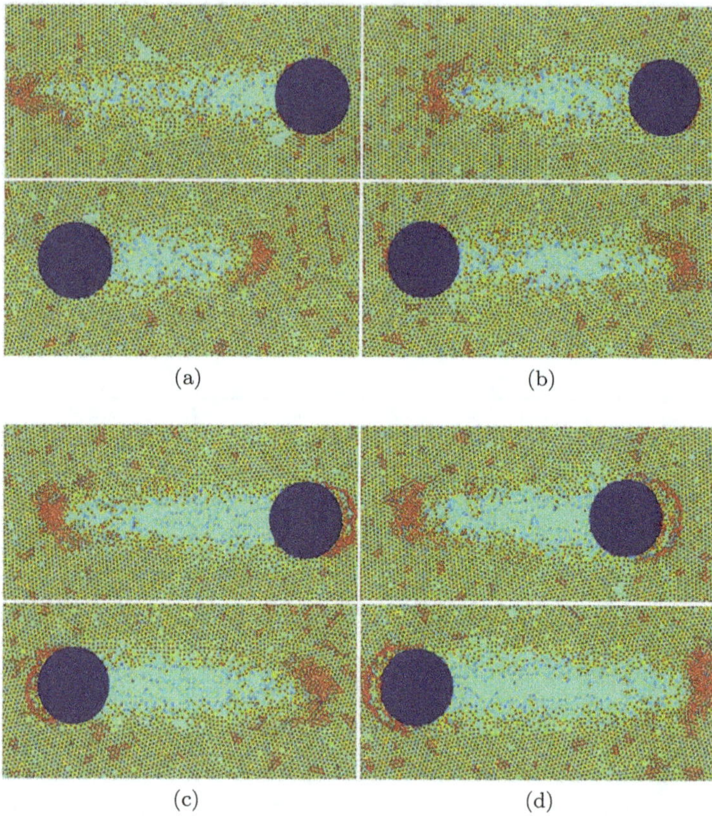

Figure 5.65. Transient configurations of the upper and bottom surfaces of single-crystal copper at the end of wear. (a) $F = 50\,\text{nN}$, (b) $F = 60\,\text{nN}$, (c) $F = 80\,\text{nN}$, (d) $F = 100\,\text{nN}$.

Some water molecules originally adsorbed in the worn region were extruded, while others remained. In Figures 5.65(c) and 5.65(d), in a sliding wear system under a large load, the abrasive slides over the surface to form a clear and regular worn zone, with almost all water molecules in the worn zone being extruded and collecting in large quantities at the front of the abrasive.

A comparison between sliding and rolling wear surface configurations reveals that there are more water molecules remaining in the wear area during rolling motion than in sliding motion. This difference is due to the change in load, leading to an increase in the contact area between the abrasive particles and the water film or the surface of the single-crystal copper. Furthermore, when the applied load is less than or equal to 40 nN, the sliding of abrasive particles causes more water molecules on the surface to be extruded from the wear area, resulting in a significantly reduced wear area due to the reduced contact area between the abrasive and the single-crystal copper surface.

The images in Figure 5.66 display the wear patterns on the upper and bottom plates of the single-crystal copper after the wearing process. Figures 5.66(a) and 5.66(b) demonstrate that the rolling of abrasive particles led to the formation of uneven wear patterns on the single-crystal copper surface. Additionally, the wear atoms in the wear region exhibited an anisotropic distribution. At a load

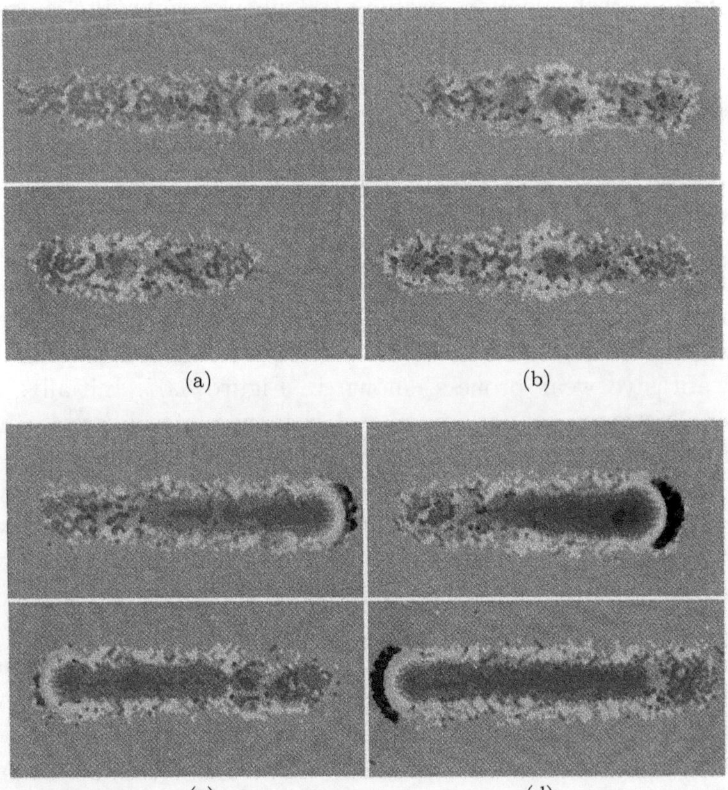

Figure 5.66. Wear morphology of single-crystal copper upper and lower plates at the end of wear. (a) $F = 50\,\text{nN}$, (b) $F = 60\,\text{nN}$, (c) $F = 80\,\text{nN}$, (d) $F = 100\,\text{nN}$.

of 60 nN, the local depth of the worn region and the accumulation of wear atoms were slightly greater than those at a load of 50 nN. In Figures 5.66(c) and 5.66(d), the sliding of abrasive particles produced regular groove patterns on both surfaces of the single-crystal copper. A large number of copper atoms were worn away from their original positions, and some of them accumulated on the sides of the groove to form ridges, while others formed chips at the front of the abrasive material. These wear patterns were similar to the results obtained from nano-scratch wear tests [9, 36]. Moreover, the depth and width of the grooves, as well as the accumulation of abrasive chips on both sides and the front, increased with higher applied loads. Upon comparison of the wear morphology with and without the presence of water film, it is observed that the accumulation of chips at the front end of the abrasive is more pronounced when a water film is present. Additionally, there is a decrease in the height of the ridges on both sides of the grooves, indicating that the aggregation of the water film impedes the accumulation of chips on both sides of the grooves. These findings suggest that the sliding abrasive action is more prominent in the presence of a water film. Furthermore, it is noted that sliding ellipsoidal abrasive particles lead to greater wear of the substrate material but create a more regular worn zone, whereas the opposite is true for rolling.

In order to further understand the frictional and wear behavior, we also calculated the dynamic friction coefficient of the three-body abrasive wear process, shown in Figure 5.67. Initially, when the single-crystal copper upper and bottom plates began to move, the friction force acting on the abrasive particles gradually increased from zero, leading to an increase in the friction coefficient, which eventually stabilized. During the three-body wear process, when abrasive particles are in a rolling motion, the normal forces between the abrasive particles and the single-crystal copper surface, as well as the water film adsorbed on the surface, are continuously changing (Figures 5.60 and 5.61). Specifically, as the force between the abrasive and the single-crystal copper surface increases, the force between the abrasive and the water film decreases. This occurs because the increased force between the abrasive and the single-crystal copper results in the expulsion of the water film from the interface between them. Additionally, as the abrasive rolls, the contact area with both the single-crystal copper and the water film changes continuously,

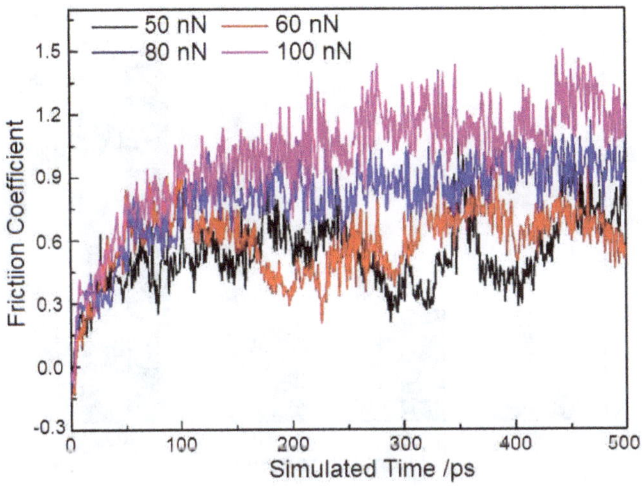

Figure 5.67. Curve of friction coefficient at varied loads.

leading to variations in the frictional force experienced (Figures 5.60 and 5.61). The fluctuations in these forces result in significant variability in the coefficient of friction, as shown in Figure 5.67. In contrast, for abrasive particles that move by sliding, the frictional force and normal pressure are relatively stable, leading to a more stable coefficient of friction. Figure 5.67 also indicates that the average value of the friction coefficient during the stable phase increases with the external load. This increase is attributed to the deeper embedding of abrasive particles into the substrate and the larger contact area under higher external loads, which results in more severe wear of the substrate (Figure 5.66).

In investigating the impact of ellipsoidal abrasive particles on single-crystal copper under varying positive loads, we calculated the number of copper atoms removed during the wear process and the length of dislocation lines within the single-crystal copper. Figure 5.68 illustrates that the number of copper atoms worn due to the rolling process is substantially lower than in the case of sliding. This suggests that rolling abrasive particles cause less wear on single-crystal copper compared to sliding. Additionally, the number of single-crystal copper atoms worn or removed, whether due to sliding or rolling, increases with the applied load. This is attributable to the larger contact area between the abrasive and the single-crystal

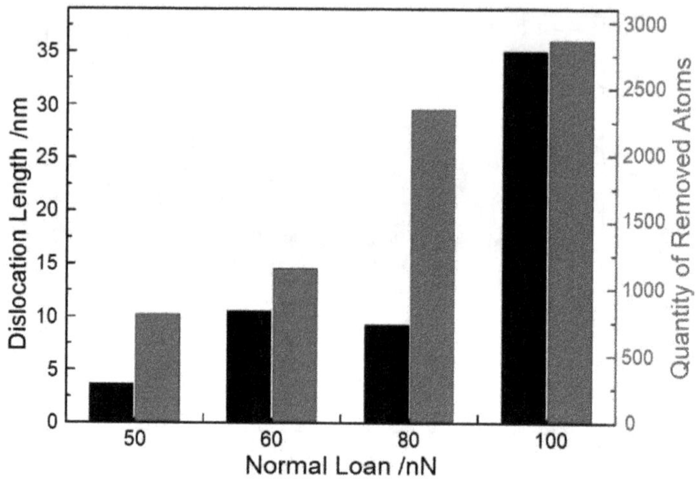

Figure 5.68. Number of worn copper atoms versus length of dislocation lines within single-crystal copper.

copper for greater applied loads. This finding aligns with the friction coefficient trend shown in Figure 5.67. Furthermore, the length of dislocation lines within the single-crystal copper under different applied loads indicates that under rolling conditions, the dislocation lines increase with heavier loads. Our analysis in Section 5.5.3 demonstrates that the primary friction and wear during the rolling process of the abrasive occur when the abrasive particles are in a standing state. Moreover, an increase in the applied load amplifies the contact between the abrasive and the surface of the single-crystal copper, leading to greater local stress concentration and defect formation. Consequently, this results in the creation of more internal defects. When the applied load is increased from 60 nN to 80 nN, the motion of the ellipsoidal abrasive particles shifts from rolling to sliding. Concurrently, the dislocation lines within the single-crystal copper decrease, indicating that the sliding of the abrasive particles under the applied pressure primarily results in the removal of a substantial number of surface atoms without leading to significant local stress concentration. However, as the applied load further increases to 100 nN, the dislocation lines within the single-crystal copper matrix undergo a dramatic increase in both quantity and length, more than doubling in size. At this stage, the sliding of the abrasive particles not only inflicts significant damage to the surface of the single-crystal copper but also generates numerous defects within the material.

The impact of the applied load on the wear of single-crystal copper is different from the effect of the shape of the abrasive material, which follows a consistent trend. As the applied load gradually increases, the ellipsoidal abrasive particles experience a transition from sliding to rolling and then back to sliding. This transition leads to sustained wear on the surface of the single-crystal copper and the formation of internal defects. Both sliding and rolling states of the abrasive material are affected by the increase in the applied load, but the pressure is reduced in the small range of rolling changed to sliding.

5.4.5 *Influence of driving speed on single-crystal copper three-body abrasive wear*

In the study of three-body abrasive wear, the movement of the single-crystal copper upper and bottom plates horizontally is the main driving factor of the entire wear process. Therefore, the speed at which the upper and bottom plates move is a crucial factor that influences the pattern of abrasive particle motion and the wear of single-crystal copper. A non-equilibrium MD simulation study of three-body abrasive wear under ultra-thin water film conditions was conducted using the initial model depicted in Figure 5.50. The water film had a thickness of $H = 0.35\,\text{nm}$, an applied positive load of 80 nN was exerted on the single-crystal Cu upper plate, and the axial ratio of the abrasive particles was 0.80.

Figure 5.69 displays the instantaneous configurations of three-body abrasive wear at different time points. It can be observed from the figure that when the driving speed is 10 m/s and 50 m/s, the abrasive particles initially undergo a short time rolling process ($0 \rightarrow 100\,\text{ps}$) at the onset of the wear process and subsequently continue to slide at 100 ps during the ongoing wear process. Additionally, the inclination angle of the abrasive sliding significantly increases with the rise in driving speed. However, as the driving speed reaches 70 m/s and 90 m/s, the abrasive particles maintain a rolling motion throughout the wear process.

In Figure 5.70, the graph illustrates the changes in rotational angular velocity of the abrasive particles over time at different driving speeds. At a driving speed of 10 m/s, the angular velocity fluctuates

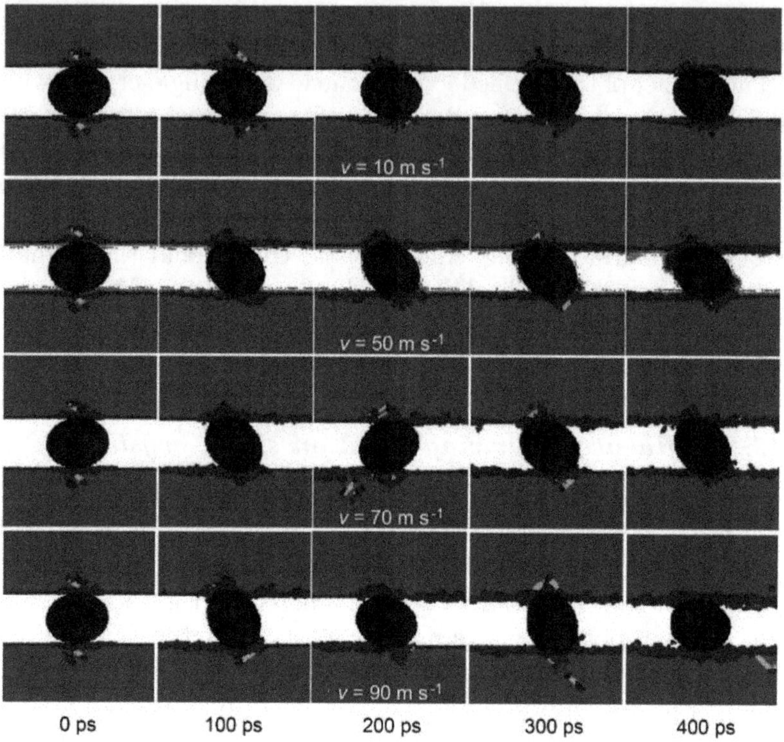

Figure 5.69. Transient configurations of abrasive particle motion and single-crystal copper wear at varied driving speeds.

around 0, indicating minimal rotational motion. However, at a driving speed of 50 m/s, the angular velocity initially fluctuates greatly, gradually increasing from a negative value to 0, and then staying near 0. This suggests that there is no significant rotational motion of the abrasive particles, confirming the sliding motion mode obtained from the configuration. Furthermore, when the abrasive moves in a rolling mode, the angular velocity demonstrates a periodic sinusoidal curve during the three-body wear process, with the frequency of curve fluctuation increasing as the driving speed increases. These results indicate a transition in the motion mode of abrasive particles from sliding to rolling as the driving speed increases.

Figure 5.71 depicts the transient configuration of the single-crystal copper surface after three-body abrasive wear. In Figure 5.71(a), at a driving speed of 10 m/s, the scratch region on the upper and bottom

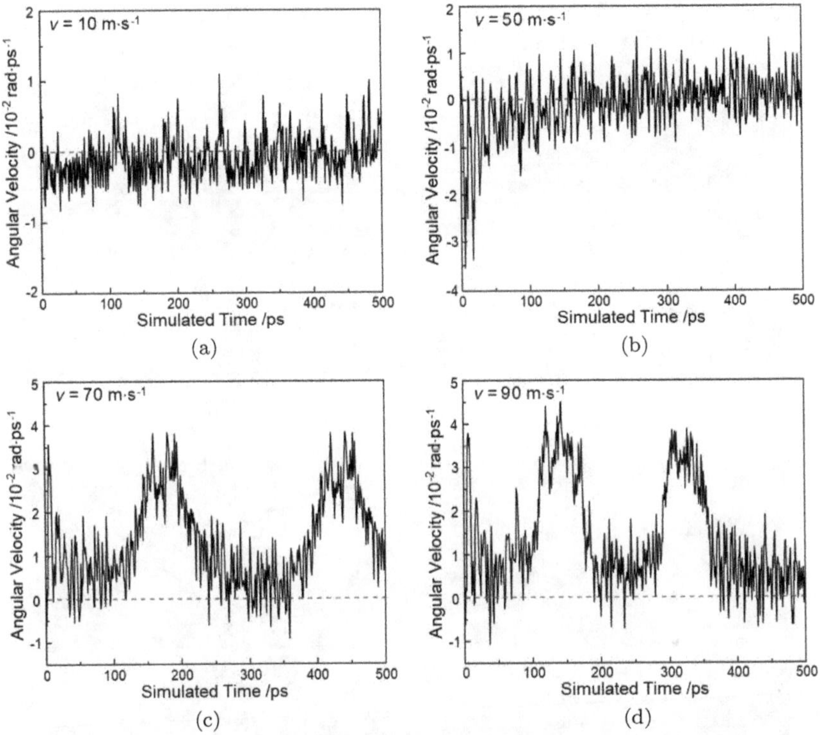

Figure 5.70. Variation of angular velocity of rotation of abrasive particles with time. (a) $v = 10\,\mathrm{m}\cdot\mathrm{ps}^{-1}$, (b) $v = 50\,\mathrm{m}\cdot\mathrm{ps}^{-1}$, (c) $v = 70\,\mathrm{m}\cdot\mathrm{ps}^{-1}$, (d) $v = 90\,\mathrm{m}\cdot\mathrm{ps}^{-1}$.

surfaces of the single-crystal copper is very short due to the slower movement speed. Additionally, more water molecules remain in the scratch region and aggregate in the front end of the abrasive particles. In Figure 5.71(b), at a driving speed of 50 m/s, the abrasive forms an obvious, regularly distributed wear area with debris build-up in the front. Most water molecules are displaced from their initial adsorption position and aggregate in the front of the abrasive material. Conversely, in Figures 5.71(c) and 5.71(d), when abrasive particles roll between the surfaces of the single-crystal copper, a longer worn area with irregular distribution of worn depth and width is formed. More water molecules remain in the wear area, even though some are extruded by the abrasive and end up in the front of the abrasive. A comparison shows that rolling wear retains more water molecules in the worn area compared to sliding wear.

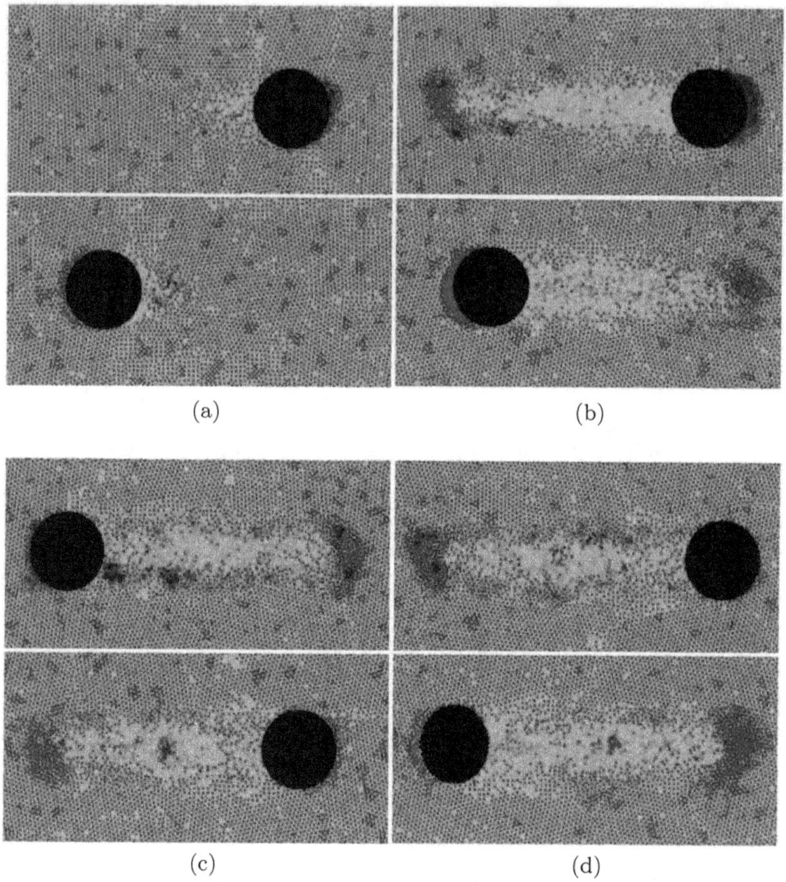

(a) (b) (c) (d)

Figure 5.71. Transient configurations of the upper (top) and lower (bottom) surfaces of single-crystal copper at the end of wear. (a) $v = 10\,\mathrm{m \cdot s^{-1}}$, (b) $v = 50\,\mathrm{m \cdot s^{-1}}$, (c) $v = 70\,\mathrm{m \cdot s^{-1}}$, (d) $v = 90\,\mathrm{m \cdot s^{-1}}$.

The wear morphology of the upper and bottom surfaces of single-crystal copper at the end of wear is depicted in Figure 5.72. In Figure 5.72(a), at low speeds, the single-crystal copper surface formed a very short and shallow scratch. However, at a driving speed of 50 m/s, the sliding of abrasive particles caused a large number of copper atoms to wear away, resulting in the formation of a regular wear groove. The abraded copper atoms accumulated at regular ridges and debris, as shown in Figure 5.72(b). The wear pattern formed by abrasive sliding is distinct from that formed by abrasive rolling, as shown

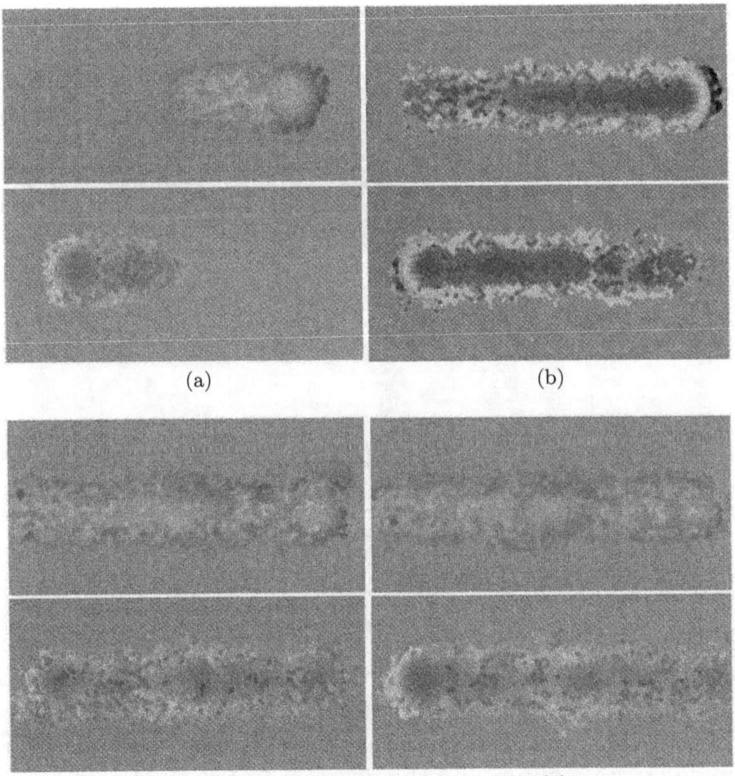

Figure 5.72. Wear morphology of single-crystal copper upper and bottom plates at the end of wear. (a) $v = 10\,\text{m}\cdot\text{s}^{-1}$, (b) $v = 50\,\text{m}\cdot\text{s}^{-1}$, (c) $v = 70\,\text{m}\cdot\text{s}^{-1}$, (d) $v = 90\,\text{m}\cdot\text{s}^{-1}$.

in Figures 5.72(c), 5.72(d). Abrasive rolling leads to the formation of irregular grooves with varying depths and an anisotropic distribution of the wear atoms. It is apparent that the driving speed has no significant effect on the wear pattern.

In Figure 5.73, the relationship between the coefficient of friction and simulation time at different driving speeds is depicted. At low speeds, abrasive particles move slowly, causing a rapid increase in the coefficient of friction at a driving speed of 10 m/s. This coefficient then increases slowly during the following simulation process, indicating that the abrasive particles have not reached an equilibrium stage within 500 ps. For a driving speed of 50 m/s, the coefficient of friction

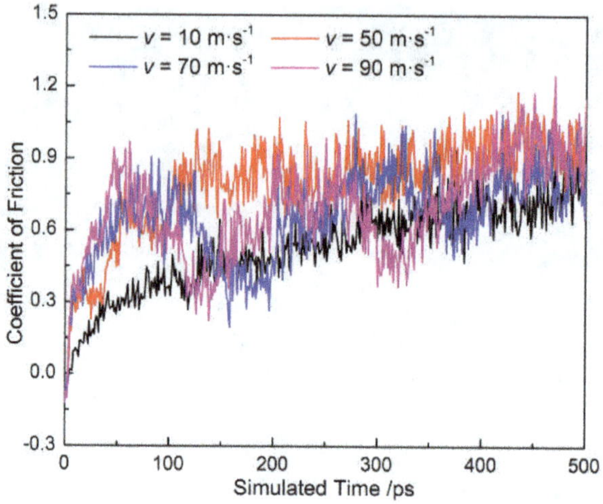

Figure 5.73. Friction coefficient variation curves at varied driving speeds.

initially increases at a faster rate and then stabilizes after 100 ps. At driving speeds of 70 m/s and 90 m/s, the abrasive particles move in a rolling manner, leading to dynamic changes in the force between the abrasive and the single-crystal copper surface, as well as the water film. This is due to the increase in force between the abrasive and the single-crystal copper and the decrease in force between the abrasive and the water film, pushing out the water molecules, while the area of interaction with the single-crystal copper and the water film is constantly changing as the abrasive rolls. Consequently, the coefficient of friction exhibits a notable jagged or sinusoidal curve, with the fluctuation period decreasing as the driving speed increases.

When comparing the friction coefficient curves, it is evident that the friction coefficient is highest at a driving speed of 50 m/s. This indicates that the frictional wear caused by abrasive particles on the single-crystal copper surfaces and subsurfaces is most severe at this speed. Conversely, the wear on the single-crystal copper surfaces is minimal at a driving speed of 10 m/s, as shown in Figure 5.72. It is apparent that as the driving speed increases, the motion of ellipsoidal abrasive particles transitions from sliding to rolling. The wear caused by sliding abrasive particles on single-crystal copper surfaces significantly increases with higher driving speeds, while the wear caused by rolling abrasive particles is relatively reduced.

5.4.6 *Summary*

The three-body abrasive wear of single-crystal copper under conditions containing water film has been investigated. This study examines the influence of abrasive shape, applied load, and the driving speed of the upper and bottom plates on the ellipsoidal abrasive motion mode. The research focuses on the friction and wear behaviors, as well as the microscopic mechanisms of single-crystal copper under these conditions. The principal findings are as follows:

(1) As the axial ratio of the abrasive particles increases, the shape of the abrasive transitions from flattened ellipsoids to spherical. Under consistent conditions, when the axial ratio exceeds 0.80, the motion of the abrasive shifts from sliding to rolling. An increase in the applied load causes the movement of ellipsoidal abrasive particles to change from sliding at low loads ($F < 50\,\mathrm{nN}$) to rolling at medium loads ($50\,\mathrm{nN} \leq F < 80\,\mathrm{nN}$) and then back to sliding at high loads ($F \geq 80\,\mathrm{nN}$). Additionally, an increase in the driving speed of the single-crystal copper plates results in a transition of the ellipsoidal abrasive particles' motion from sliding at low speeds ($v \leq 50\,\mathrm{m/s}$) to rolling at high speeds ($v > 50\,\mathrm{m/s}$).

(2) During the rolling mode of abrasive particle movement, the contact area between the abrasive, the single-crystal copper matrix, the water film, and the copper surface continuously changes. Consequently, the force also undergoes cyclical variations. A significant amount of water molecules remains in the wear area, preventing direct contact between the abrasive particles and the single-crystal copper surface. This results in a substantial lubrication effect, greatly reducing material friction and wear; the coefficient of friction can be reduced by up to 62.5%. However, transient local stress concentrations can increase the formation of defects within the single-crystal copper.

(3) In the sliding mode of abrasive particle movement, initially, the external force applied to the upper and bottom plates is transmitted through the water film medium to the abrasive particles. As the distance of movement increases, the water film is gradually expelled from the worn area, leading to direct interaction between the abrasive particles and the single-crystal copper

surface. This direct contact increases friction and wear on the copper surface but reduces internal plastic deformation of the single-crystal copper.

(4) The sliding motion of ellipsoidal abrasive particles under low applied loads does not cause significant surface damage or internal defects in single-crystal copper. Conversely, sliding under high applied loads not only results in severe surface abrasion of single-crystal copper but also induces a substantial number of internal defects within the copper matrix.

5.5 Theoretical Modeling of Three-Body Abrasive Wear Containing Water Films

5.5.1 *Introduction*

Various forms of material wear include adhesive wear, fatigue wear, corrosive wear, and abrasive wear. When materials are reduced to micron and nanometer scales, the surface force, represented by adhesive force and electrostatic force, significantly increases due to the heightened surface-to-volume ratio. This increase becomes the primary controlling factor for material deformation and wear [41, 42]. In the friction and wear process of micro and nanomaterials, initial adhesive wear transforms into three-body abrasive wear as chips form and accumulate, becoming the most important form of material wear. This transformation accelerates the generation of chips, which can lead to material deformation, damage, and even failure [43, 44]. Three-body wear involves chips or particles moving in different ways relative to the upper and bottom solid surfaces, such as sliding and rolling, causing varying degrees of wear and damage to the material surface [30, 31, 34].

The various modes of movement lead to differing levels of wear and damage to the material's surface. Several factors influence the movement of abrasive particles or chips, such as the characteristics of the particles themselves (including geometry and hardness), material properties (e.g., varying surface roughness, hardness, and defects), and external conditions (such as pressure, velocity, and the presence of lubricating fluids). In recent years, experimental studies of abrasive wear have utilized methods such as *in situ* observation, x-ray scattering, and Raman spectroscopy. However, studying abrasive wear at the nanoscale presents significant challenges due to limitations in existing experimental techniques. This has resulted in an unclear understanding of the relationship between abrasive particle motion and influencing factors. Additionally, theoretical models of three-body abrasive wear have been underexplored. While Fang *et al.* [51–53] proposed a mechanical theoretical model to assess the transition between roll and slip motion of abrasive particles, understanding the motion criteria for more complex abrasive shapes remains challenging.

According to the theoretical model of abrasive motion under air wear conditions, we will establish the arm ratio of a pair of positive forces and friction forces (e/h) acted on the upper and bottom plate in the motion of ellipsoidal abrasive under water-containing conditions, respectively. By comparing the value of e/h with the coefficient of friction, we will theoretically predict the motion mode of ellipsoidal abrasive in the three-body wear process. Additionally, we analyze the geometric coefficients, thickness of the water film, and the coefficient of elastic recovery to examine the main influencing factors.

5.5.2 *Theoretical modeling of abrasive motion mode under water-bearing conditions*

5.5.2.1 *Spherical abrasive roll–slip judgment*

Fang *et al.* [51–53] developed a theoretical prediction model for the roll–slip criterion of spherical abrasives in air friction wear conditions. The model was created through experiments, theoretical derivations, and molecular dynamics simulation studies, as shown in Figures 4.23 and 4.24 and Eq. (4.11). Building on this work, it is essential to extend the model to encompass abrasive motion modes in three-body wear under water-bearing conditions.

Figure 5.74 illustrates the forces and geometrical parameters of spherical abrasives in three-body abrasive wear under water-bearing conditions. The blue areas between the abrasive particles and the substrate surface represent the presence of a water film. The abrasive experiences normal load (N_1 and N_2) and horizontal friction (F_1 and F_2) when in contact with the upper and bottom surfaces. The friction and normal load moments, denoted as $F \cdot h$ and $N \cdot e$, respectively, can be obtained when the system is in dynamic equilibrium. In this state, $F \cdot h$ serves as the driving moment for the abrasive to roll, while $N \cdot e$ acts as the resisting moment. Therefore, during rolling

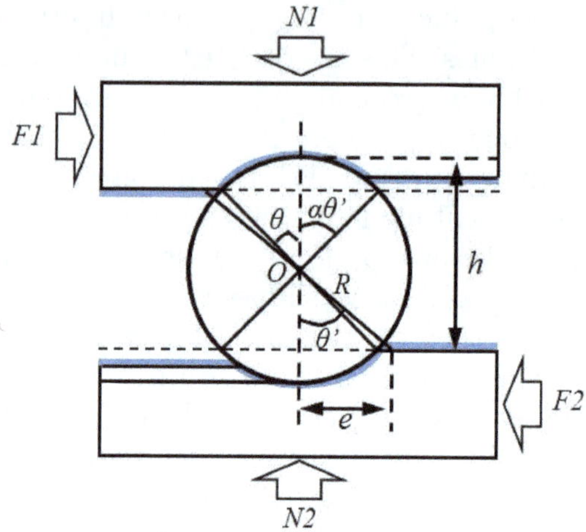

Figure 5.74. Geometrical parameters of spherical abrasives under water-bearing conditions.

motion, the frictional moment exceeds the normal load moment, i.e.,

$$F \cdot h > N \cdot e \tag{5.1}$$

Conversely, when the abrasive is in sliding motion, the frictional torque is less than the normal load torque, i.e.,

$$F \cdot h < N \cdot e \tag{5.2}$$

The roll–slip criterion of abrasive particles in three-body wear can be obtained from the definition of the coefficient of friction $\mu = F/N$. We can get equations (5.3) and (5.4) [51–53]:

$$\frac{e}{h} < \mu_r \quad \text{for rolling} \tag{5.3}$$

$$\frac{e}{h} > \mu_s \quad \text{for sliding} \tag{5.4}$$

where μ_r is the rolling friction coefficient and μ_s is the sliding friction coefficient.

In the process of nano three-body abrasive wear, as shown in Figure 5.74, the water film will obstruct direct contact between the abrasive particles and the substrate surface. This causes a slight change in the force point of the abrasive particles. Additionally, the water film can transfer the force of the abrasive to the substrate surface, leading to friction and wear of the substrate. Under these conditions, the relationship between the force arm and the indentation angle θ changes to θ', and the elastic recovery angle $\alpha\theta$ changes to $\alpha\theta'$. This results in a geometric expression of the force arm relationship under water-bearing conditions as follows:

$$e_c = \frac{8R}{3\pi(\sin^2(\theta') + \sin^2(\alpha\theta'))}(\sin^3(\theta') - \sin^3(\alpha\theta')) \tag{5.5}$$

$$h_c = \frac{4}{3}\left(\frac{R\sin^3(\theta')}{\theta' - \sin(\theta')\cos(\theta')}\right) \tag{5.6}$$

5.5.2.2 *Ellipsoidal abrasive roll–slip judgment*

In order to investigate the impact of abrasive particle shape on three-body wear in water-bearing conditions, the author explores the alteration of geometrical parameters for spherical abrasives to obtain ellipsoidal abrasives with three different axial ratios (a, b, and c, where $a = c$) as detailed in Table 5.6. For the sake of the study, the contact point between the abrasive particles and the substrate surface is considered to be equivalent to a spherical surface with the same radius of curvature. After calculations and verification, it is evident that the error between the equivalent surface and the actual surface is minimal, demonstrating the viability of the method. Several assumptions are also made in the treatment process, including treating the abrasive particles as ellipsoidal rigid bodies, disregarding the weight of the abrasive particles and the water film, neglecting gravitational forces, assuming the movement of abrasive particles in a plane perpendicular to the material surface at the parallel direction of driving force (or frictional force), discounting rotation along the vertical axis, and assuming uniform distribution of friction and normal load on the solid contact surface.

In water-bearing conditions, the relationship between the ellipsoidal abrasive roll–slip criterion is established using the equivalent circle method, consistent with dry wear conditions. In this context, the geometric relationship between the ellipsoidal abrasive and the equivalent circle, as well as the relevant parameters, is depicted in Figure 5.75. The blue area in the figure represents the water film.

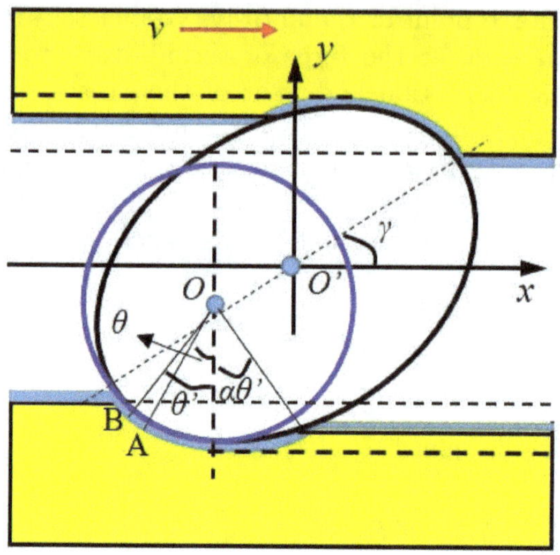

Figure 5.75. Geometric parameters of ellipsoidal abrasive under water-bearing conditions.

As the ellipsoidal abrasive moves, the water film serves a dual role. It hinders direct action between the abrasive and the substrate surface while also accumulating in the front end of the abrasive, thereby obstructing its movement. Consequently, the action point shifts from point A to point B in the absence of water, altering the abrasive's indentation angle from θ to θ' and changing the angle of elastic recovery from $\alpha\theta$ to $\alpha\theta'$.

Next, we outline the detailed derivation of the criteria for ellipsoidal abrasive motion in conditions involving water. This is analogous to establishing the relationship for ellipsoidal motion in conditions involving air abrasion. The detailed derivation process can be found in Chapter 4.4.3 of this book, and a brief summary is provided in the following. The parametric equations for the ellipse are given in equation (5.7), and the radius of curvature at any point is given in equation (5.8):

$$\begin{cases} x = a\cos(\beta) \\ y = b\sin(\beta) \end{cases} \tag{5.7}$$

$$R = \frac{[(a^2 - b^2)\sin^2(\beta) + b^2]^{\frac{3}{2}}}{ab} \tag{5.8}$$

Then, the coordinates of the center of the curvature circle (x_0, y_0) can be expressed as follows:

$$\begin{cases} x_0 = x - \dfrac{y'(1+y'^2)}{y''} = \dfrac{\cos^3(\beta)(a^2-b^2)}{a} \\ y_0 = y + \dfrac{1+y'^2}{y''} = \dfrac{\sin^3(\beta)(b^2-a^2)}{b} \end{cases} \tag{5.9}$$

where y' is the first-order derivative of the elliptic equation and y'' is the second-order derivative of the elliptic equation.

When the ellipsoid motion turns over an angle of γ, the elliptic equation becomes

$$\begin{cases} x_1 = a\cos(\beta)\cos(\gamma) - b\sin(\beta)\sin(\gamma) \\ y_1 = a\cos(\beta)\sin(\gamma) + b\sin(\beta)\cos(\gamma) \end{cases} \tag{5.10}$$

The curvature circle center (x_{01}, y_{01}) of the rotated ellipse is thus obtained:

$$\begin{cases} x_{01} = \dfrac{\cos^3(\beta)(a^2-b^2)}{a}\cos(\gamma) - \dfrac{\sin^3(\beta)(b^2-a^2)}{b}\sin(\gamma) \\ y_{01} = \dfrac{\cos^3(\beta)(a^2-b^2)}{a}\sin(\gamma) + \dfrac{\sin^3(\beta)(b^2-a^2)}{b}\cos(\gamma) \end{cases} \tag{5.11}$$

Combined with the equivalent circle method shown in Figure 5.75 and the curvature trajectory of the rotating ellipse, the expressions for the resistance arm e_e and the driving force arm h_e of the motion of the ellipsoidal abrasive particles can be obtained, and the moment relationship in the ellipsoidal abrasive motion mode criterion can be expressed as follows:

$$\frac{e_e}{h_e} = \frac{e_c + x_{01}\text{sign}(x_1)}{h_c + y_{01}\text{sign}(y_1)} \tag{5.12}$$

where the sign functions sign(x) and sign(y) are introduced to represent the direction of rotation of the ellipsoidal abrasive. Equations (5.5), (5.6), and (5.11) are brought into Eq. (5.12) to obtain the moment relations in the geometrically parameterized ellipsoidal abrasive motion mode criterion:

$$\frac{e_e}{\hbar_e} = \frac{\frac{8[(a^2-b^2)\sin^2(\beta)+b^2]^{\frac{3}{2}}(\sin^3(\theta')-\sin^3(\alpha\theta'))}{3\pi ab(\sin^2(\theta')+\sin^2(\alpha\theta'))} + A\cdot \mathrm{sign}(x_1)}{\frac{4[(a^2-b^2)\sin^2(\beta)+b^2]^{\frac{3}{2}}\sin^3(\theta')}{3ab[\theta'-\sin(\theta')\cos(\theta')]} + B\cdot \mathrm{sign}(y_1)} \tag{5.13}$$

where

$$A = \frac{\cos^3(\beta)(a^2-b^2)}{a}\cos(\gamma) - \frac{\sin^3(\beta)(b^2-a^2)}{b}\sin(\gamma) \tag{5.14}$$

$$B = \frac{\cos^3(\beta)(a^2-b^2)}{a}\sin(\gamma) + \frac{\sin^3(\beta)(b^2-a^2)}{b}\cos(\gamma) \tag{5.15}$$

where a is the length/nm of the long axis of the ellipse (x direction), b is the length/nm of the short axis of the ellipse (y direction), γ is the angle of rotation of the ellipse/°, β is the parameter of the elliptic equation, and α is the elastic coefficient of return of the material.

In Equation (5.13), when the long and short axes a and b are equal, the ellipsoidal abrasive becomes a spherical abrasive. At this point, the geometric relationship of the roll–slip criterion agrees with the ratio of Equations (5.5) and (5.6). This result aligns with the findings of Sun *et al.* [53] and verifies the correctness of the ellipsoidal abrasive roll–slip criterion geometrical analysis model under specific conditions.

5.5.3 *Discussion and analysis of abrasive roll–slip criteria under water-bearing conditions*

5.5.3.1 *Influence of angular coefficients on the force–arm relationship*

From equation (5.13), it can be seen that the abrasive roll–slip relationship is related to the rotation angle γ, the indentation angle θ', the elastic recovery angle $\alpha\theta'$, and the elliptic parameter equation angle β. Figures 5.76, 5.77, and 5.78 show the relationship curves between e/h and different angular parameters, which have an important influence on the e/h, which in turn influences the relationship between e/h and the size of the coefficient of friction μ.

When the rotation angle is less than 45°, the value of e/h increases as the rotation angle increases, and vice versa. The value of e/h

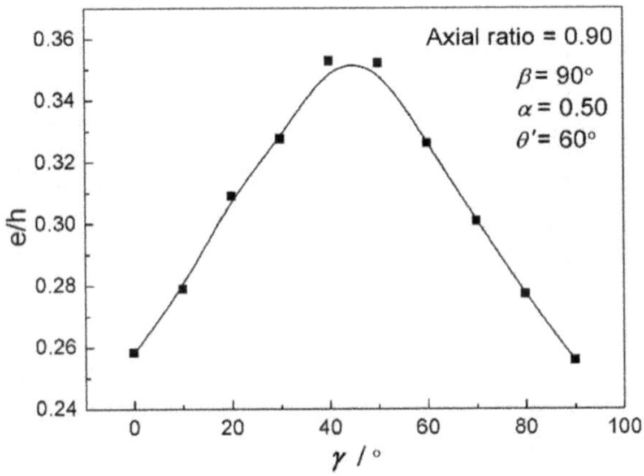

Figure 5.76. e/h vs. angle of abrasive rotation.

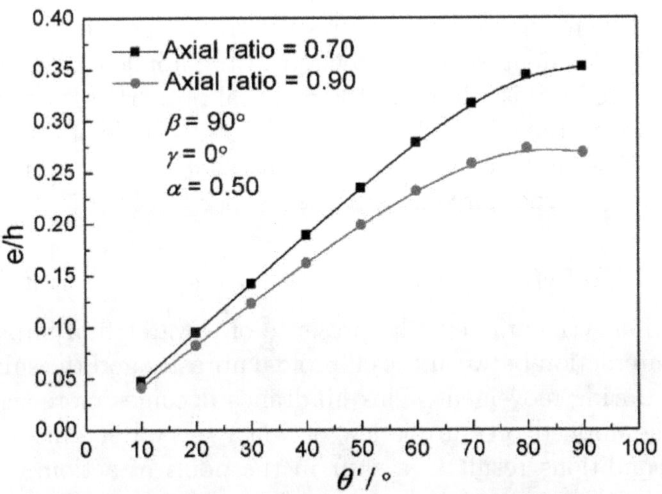

Figure 5.77. e/h vs. angle of abrasive indentation.

increases and decreases monotonically with the pressing angle and the elastic recovery coefficient, indicating that the deeper the abrasive material is pressed into the substrate, the weaker the material's ability to recover elastically, making the abrasive more likely to be in a sliding state. Furthermore, the e/h value is smaller when the abrasive axial ratio is larger under the same conditions, showing that sharper abrasives are less likely to roll, which is consistent with the

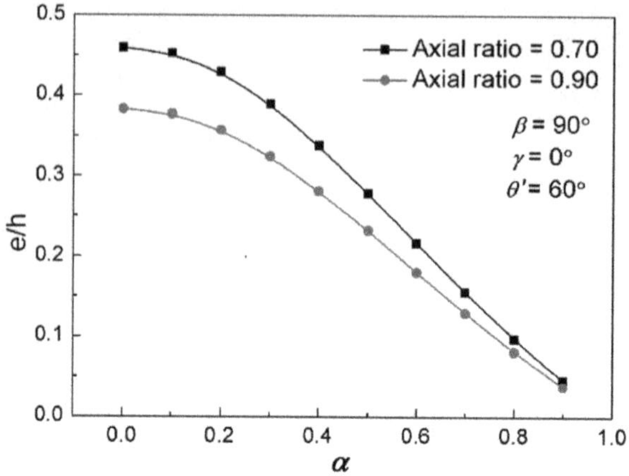

Figure 5.78. e/h versus substrate elastic recovery coefficient.

molecular dynamics simulation results from the previous chapter. The angle of indentation θ' is related to the applied load, the angle of elastic recovery is related to the material properties, and the angle β is related to the ellipsoid shape. The motion mode of the ellipsoid-shaped abrasive depends not only on its shape but also on material properties, external force applied, and other factors.

5.5.3.2 *Effect of water film thickness on indentation angle*

As previously mentioned, the presence of a water film impedes the direct interaction between the ellipsoidal abrasive and the substrate's surface during movement. This hindrance becomes more significant when the applied pressure is low or when the water film is thicker. These conditions result in a shift in the point of action, causing a change in the angle of indentation from θ to θ'. In a straightforward scenario where the water film is consistently distributed during abrasive movement and maintains a constant thickness H_N, the force–arm relationship e/h with varying angular parameters (refer to Section 5.5.2) remains the same as when there is no water film present. In reality, the contact between the ellipsoidal abrasive particles and the substrate surface continuously changes during the movement process, particularly during rolling. This results in different contact areas under the same normal load, subjecting the substrate surface to varying forces and causing different friction forces on the water film. Consequently, the water film between the abrasive and the surface is unevenly distributed and its thickness fluctuates over time.

The fluidic properties of water films are weakened and the solid-like properties are enhanced at the nanoscale or atomic scale. Even individual water molecules or clusters of molecules exhibit very strong solid-like properties during nanoindentation and abrasion processes. For instance, in the study of single-crystal copper nanoindentation process, it was observed that water molecules under the indenter can transmit pressure to the surface of single-crystal copper, leading to elastic and plastic deformation of the substrate. In the two-body wear process of single-crystal copper, water molecules that are not displaced from the worn region by the abrasive particles cause some damage to the surface and interior of the single-crystal copper.

The presence of water films can significantly complicate the three-body wear process due to its solid nature. These water films exist in two primary forms:

(1) When abrasive particles are pressed into the single-crystal copper substrate and driven by horizontal force, some water molecules remain in the indentation area and worn area, creating an ultra-thin water film between the abrasive particles and the substrate surface. This water film can transfer forces to the single-crystal copper surface, leading to elastic and plastic deformation even when direct contact is not established between the copper and abrasive particles. The mismatch in physical and mechanical properties between the water film and the single-crystal copper substrate causes changes in the angle of indentation and elastic recovery of the abrasive particles.
(2) When abrasive particles are subjected to horizontal driving force, water molecules are extruded from the indentation area, creating an accumulation at the front end of the ellipsoidal abrasive as indicated in the blue area of Figure 5.79. This accumulation of water molecules or clusters impedes the forward motion of the abrasive particles, leading to increased friction in the horizontal direction.

In the actual wear process, the thickness of the water film is influenced by the normal load N. Greater pressure results in a denser

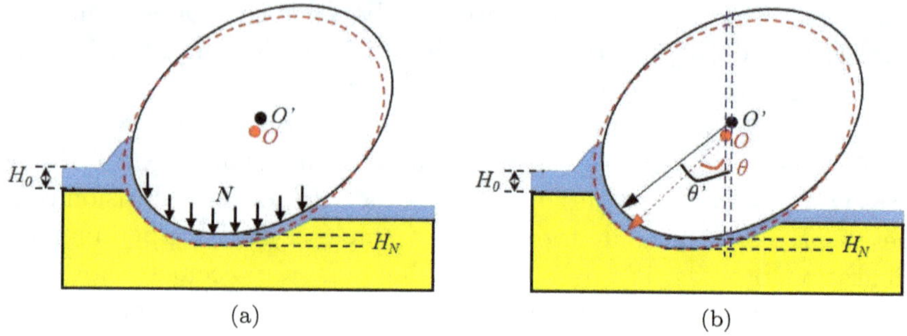

Figure 5.79. Analysis of water film forces, thickness, and angle variations. (a) Change in water film thickness due to force, (b) change in indentation angle.

water film and a smaller thickness H_N. Additionally, the relationship between water film thickness and the initial water film thickness H_0 (Figure 5.79) is such that the greater the H_0, the smaller the average force each water molecule can withstand, leading to a greater thickness H_N, expressed by the following formula:

$$\begin{cases} H_N \propto \dfrac{1}{N} \\ H_N \propto H_0 \end{cases} \tag{5.16}$$

The thickness of the water film during the wear process is influenced by the initial water film thickness (H_0) and the normal load (N). This, in turn, impacts the depth and angle of indentation of the abrasive particles, as shown in Figure 5.79(b). The presence of the water film causes the center of gravity and equivalent point of action of the abrasive particles to shift upward, leading to a change in the angle of indentation. This relationship can be expressed by the following equation:

$$\begin{cases} \varphi = (\theta' - \theta) \propto \dfrac{1}{N} \\ \varphi = (\theta' - \theta) \propto H_0 \end{cases} \tag{5.17}$$

(3) The effect of water film aggregation on the abrasive's mode of movement

In Figure 5.80, the horizontal motion of abrasive particles is influenced by several forces, including the horizontal friction force

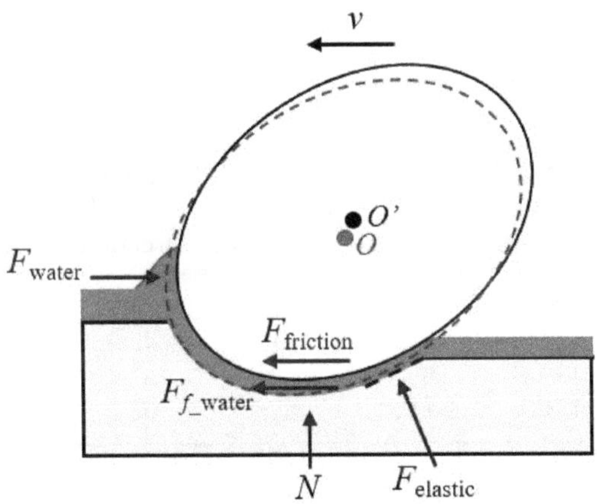

Figure 5.80. Abrasive particle force analysis.

($F_{friction}$) exerted by the surface of the single-crystal copper, the force of the water film below the abrasive (F_{f_water}), the resistance of the water molecules gathered at the front end (F_{water}), and the elastic recovery force ($F_{elastic}$) generated by the single-crystal copper at the rear of the abrasive. The friction force initiates the rolling of abrasive particles, while the resistance force (F_{water}) exerted by the water film acts against this motion. On the other hand, the elastic recovery force ($F_{elastic}$) aids in promoting the rolling of the abrasive.

When the water film is thicker, the motion of abrasive particles leads to an accumulation of water molecules at the front end, inhibiting the rolling or sliding of the abrasive particles. Table 5.7 illustrates that the two-body wear process of abrasive particles experiences a significant increase in force with the thickening of the water film, for instance, when the water film thickness (H) is 3.0 nm, the horizontal force on the abrasive is more than two times that of the water film. In contrast, the force on the single-crystal copper surface reduces with the increase in water film thickness. This reduction occurs because the force on the copper surface primarily stems from the action of abrasive particles. The presence of the water film prevents direct interaction between the abrasive particles and the copper surface, resulting in a reduced force between the two. The horizontal force acting on the abrasive is a result of the combined interaction between the copper surface and the water film. The resistance to the

Table 5.7. Interaction forces between different components.

Water film thickness/nm	Force on abrasive particles/nN		Copper surface force/nN		Water film-abrasive force/nN	
	Horizontal direction	Normal load direction	Horizontal direction	Normal load direction	Horizontal direction	Normal load direction
0	80.3	46.1	80.3	46.1	0	0
1.0	121.8	65.8	71.2	49.9	50.6	15.9
2.0	138.3	68.5	62.2	57.7	76.1	10.8
3.0	171.0	74.3	57.5	62.5	113.5	11.8

advancement of the abrasive is primarily caused by the water film at the front end and the force on the abrasive from the remaining water molecules in the worn area. The force on the copper surface decreases on abrasive particles, while the total interaction force on the abrasive particles increases significantly, partly due to the contribution of the water film, as illustrated in Table 5.7 (water film-abrasive forces). On the one hand, Figure 5.29 shows that the water molecules gathered in the front of the abrasive increased significantly with the initial thickness of the water film, and even the accumulation of the water molecules flooded the abrasive particles, so this part of the hindering force is not negligible, and it produces a large hindering effect on the rolling of the ellipsoidal abrasive particles. On the other hand, even though the action force $F_{\text{f\_water}}$ is in the same direction as the main driving force of rolling F_{friction}, the presence of $F_{\text{f\_water}}$ leads to the decrease of F_{friction}. In contrast, the hindering effect of water film build-up is stronger than the contribution from the friction force.

5.5.4 *Influence of elastic recovery on the movement mode of abrasive particle*

In abrasive wear at the nanoscale, the scale and surface effects play a significant role in the elastic recovery behavior, which can greatly impact the motion pattern of the abrasive. Nanoindentation results indicate that during the indentation process, internal defects in single-crystal copper move toward the substrate's surface as the force of the indenter decreases, leading to stress relaxation within the material. Additionally, atoms in the indentation pits exhibit varying degrees of elastic recovery, as illustrated in Figures 5.22 and 5.23. Similarly, when abrasive particles slide or roll on the surface of single-crystal copper, they cause substantial plastic deformation accompanied by some degree of elastic recovery. As depicted in Figure 5.80, the copper surface atoms behind the abrasive particles exert a transient push on the abrasive due to elastic recovery, known as the elastic recovery force (F_{elastic}). To quantify the strength of the elastic recovery effect, the analysis in this section is based on the nanoindentation model. Wang *et al.* [55] proposed two methods for calculating the elastic recovery coefficient in their study of NiAl alloys, detailed in Eqs. (5.18) and (5.19), where D and W are illustrated in Figure 5.81:

$$\alpha_L = \frac{D_0 - D_1}{D_0}\% \tag{5.18}$$

$$\alpha_W = \frac{W_0 - W_1}{W_0}\% \tag{5.19}$$

In Figure 5.82, the elastic recovery of a single-crystal copper matrix is shown before and after abrasive unloading with different water film thicknesses (α_L and α_W). It is observed that the presence of a water film results in smaller elastic recovery compared to the absence of a water film, which is consistent with the findings in Figure 5.24. Additionally, the elastic recovery in the lateral direction is almost three times that in the longitudinal direction, indicating a greater aggregation of elastic energy along the direction of abrasive indentation and thus a greater release of elastic recovery.

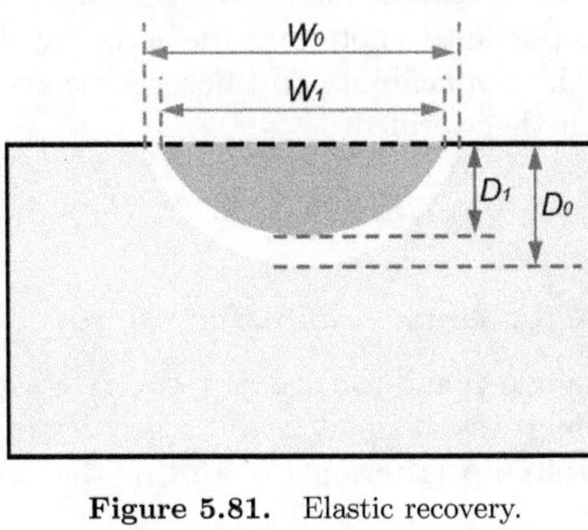

Figure 5.81. Elastic recovery.

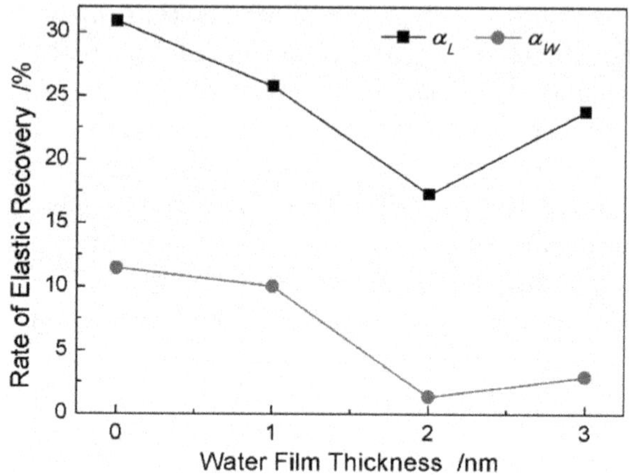

Figure 5.82. Elastic recovery of single-crystal copper at different water film thicknesses.

The elastic recovery of single-crystal copper exceeds 30% under anhydrous conditions, while even the smallest elastic recovery under aqueous films reaches 18%, which is significantly lower than what is achievable under macroscopic conditions. This suggests that the elastic recovery behavior of the material has a strong scale effect and emphasizes the importance of considering elastic recovery at the nanoscale. The force F_{elastic} induced by the elastic recovery of the matrix material during abrasive wear may also influence the motion pattern of the abrasive particles. A functional relationship is established between the elastic recovery rate of single-crystal copper (α_L) and the initial thickness of the aqueous film ($H_0 \leq 2.0\,\text{nm}$) in Eq. (5.20). It's important to note that the accuracy of this functional relationship and its applicability to thicker aqueous film conditions still need to be further verified:

$$\alpha_L = -0.06775H_0 + 0.31398 \quad H_0 \leq 2.0 \tag{5.20}$$

5.5.5 *Multifactor coupled abrasive rol–slip criterion*

Introducing the angle φ and the elastic recovery coefficient α_L when the water film thickness is small into the geometric model of ellipsoidal abrasive roll–slip criterion Eq. (5.13), the roll–slip criterion model after coupling the factors of elastic recovery and water film thickness is obtained as follows:

$$\frac{e_e}{h_e} = \frac{\frac{8[(a^2-b^2)\sin^2(\beta)+b^2]^{\frac{3}{2}}(\sin^3(\theta-\varphi)-\sin^3(\alpha_L(\theta-\varphi)))}{3\pi ab(\sin^2(\theta-\varphi)+\sin^2(\alpha_L(\theta-\varphi)))} + A\cdot\mathrm{sign}(x_1)}{\frac{4[(a^2-b^2)\sin^2(\beta)+b^2]^{\frac{3}{2}}\sin^3(\theta-\varphi)}{3ab[(\theta-\varphi)-\sin(\theta-\varphi)\cos(\theta-\varphi)]} + B\cdot\mathrm{sign}(y_1)} \tag{5.21}$$

where the functional form of the angle φ is

$$\varphi = f\left(H_0, \frac{1}{N}\right) \tag{5.22}$$

The indentation angle θ is [54, 56]

$$\theta = ar\cos\left(\sqrt{1 - \frac{Na^2b^2}{\pi H[(a^2-b^2)\sin^2(\beta)+b^2]}}\right) \tag{5.23}$$

where a is the length of the long axis of the ellipse (x-direction)/nm, b is the length of the short axis of the ellipse (y-direction)/nm, γ is the angle of rotation of the ellipse/°, β is the angle of the elliptic parameter equations, α_L is the elasticity coefficient of recovery, N is the applied normal load/nN, H is the hardness of the material/GPa, β is the angle of elliptic parameter equations, and H_0 is the initial thickness of the water film/nm.

5.5.6 *Summary*

The geometric model of force–arm relationship e/h during the motion of ellipsoidal-shaped abrasive was established based on the force–arm relationship model and roll–slip criterion of spherical abrasive motion mode. This model enabled the theoretical prediction of abrasive motion mode. The key findings of the research are as follows:

(1) A geometrical analytical model was developed to predict the abrasive movement mode by comparing the force–arm relationship e/h with the magnitude of the friction coefficient for spherical and ellipsoidal abrasive particles moving between substrate materials under water film conditions.

(2) The influence of angle parameters, water film thickness and accumulation, and elastic recovery on the abrasive rolling–slip model was analyzed. The results indicate that when the rotation angle is less than 45°, the e/h value increases with the rotation angle, and vice versa. The e/h value increases monotonically with the indented angle and decreases with the elastic recovery coefficient, suggesting that a greater depth of abrasive penetration into the substrate and weaker material elastic recovery results in a greater tendency of the abrasive to transition into a sliding state. Under the same conditions, a larger abrasive axis ratio corresponds to a smaller e/h value, demonstrating that sharper abrasives are less conducive to rolling. As the normal pressure increases, the water film thickness decreases. The water film thickness is positively correlated with the initial water film thickness. Water molecules accumulate at the front end of the abrasive significantly impeding the rolling of ellipsoidal abrasives. The force $F_{\text{f\_water}}$ leads to a reduction in F_{friction}; comparatively, the hindering effect of water film accumulation is stronger than the contribution from the friction.

(3) In the presence of a water film, the elastic recovery rate of single-crystal copper is lower than in the absence of a water film, with the lateral elastic recovery rate being nearly three times that of the longitudinal elastic recovery rate. This suggests that greater elastic energy accumulates along the direction of abrasive penetration, resulting in a larger release of elastic recovery.

(4) Under dry conditions, the elastic recovery rate of single-crystal copper exceeds 30%. In the presence of a water film, even the minimum elastic recovery rate reaches 18%, which is significantly higher than under macro-scale conditions. Based on the elastic recovery rate α_L of single-crystal copper, a functional relationship between the elastic recovery rate and the initial water film thickness ($H_0 \leq 2.0\,\text{nm}$) has been established.

(5) In the theoretical geometric model for the rolling–slip criterion of ellipsoidal abrasives, the angle φ and the elastic recovery coefficient α_L for small water film thicknesses were incorporated, resulting in a theoretical rolling–slip criterion model that accounts for the coupling effects of elastic recovery and water film thickness.

References

[1] Voyiadjis GZ, Yaghoobi M. Large scale atomistic simulation of size effects during nanoindentation: dislocation length and hardness. *Mater Sci Eng A*, 2015, 634: 20–31.

[2] Zhang L, Zhao H, Dai L, *et al.* Molecular dynamics simulation of deformation accumulation in repeated nanometric cutting on single-crystal copper. *RSC Adv*, 2015, 5: 12678–12685.

[3] Li L, Song W, Xu M, *et al.* Atomistic insights into the loading–Unloading of an adhesive contact: A rigid sphere indenting a copper substrate. *Comput Mater Sci*, 2015, 98: 105–111.

[4] Liang H, Woo C, Huang H, *et al.* Dislocation nucleation in the initial stage during nanoindentation. *Philos Mag*, 2003, 83: 3609–3622.

[5] Gao Y, Ruestes CJ, Urbassek HM. Nanoindentation and nanoscratching of iron: Atomistic simulation of dislocation generation and reactions. *Comput Mater Sci*, 2014, 90: 232–240.

[6] Daw MS, Baskes MI. Embedded-atom method: Derivation and application to impurities, surfaces, and other defects in metals. *Phys Rev B*, 1984, 29: 6443.

[7] Daw MS, Foiles SM, Baskes MI. The embedded-atom method: a review of theory and applications. *Mater Sci Rep*, 1993, 9: 251–310.

[8] Girifalco LA, Weizer VG. Application of the Morse potential function to cubic metals. *Phys Rev*, 1959, 114: 687.

[9] Ren J, Zhao J, Dong Z, *et al.* Molecular dynamics study on the mechanism of AFM-based nanoscratching process with water-layer lubrication. *Appl Surf Sci*, 2015, 346: 84–98.

[10] Boda D, Henderson D. The effects of deviations from Lorentz–Berthelot rules on the properties of a simple mixture. *Mol Phys*, 2008, 106: 2367–2370.

[11] Al-Matar AK, Rockstraw DA. A generating equation for mixing rules and two new mixing rules for interatomic potential energy parameters. *J Comput Chem*, 2004, 25: 660–668.

[12] Werder T, Walther J, Jaffe R, *et al.* On the water–carbon interaction for use in molecular dynamics simulations of graphite and carbon nanotubes. *J Phys Chem B*, 2003, 107: 1345–1352.

[13] Sun K, Shen W, Ma L. The influence of residual stress on incipient plasticity in single-crystal copper thin film under nanoindentation. *Comput Mater Sci*, 2014, 81: 226–232.

[14] Ziegenhain G, Urbassek HM, Hartmaier A. Influence of crystal anisotropy on elastic deformation and onset of plasticity in nanoindentation: a simulational study. *J Appl Phys*, 2010, 107: 061807.

[15] Kracke B, Damaschke B. Measurement of nanohardness and nanoelasticity of thin gold films with scanning force microscope. *Appl Phys Lett*, 2000, 77: 361–363.
[16] Chang L, Zhang L. Mechanical behaviour characterisation of silicon and effect of loading rate on pop-in: a nanoindentation study under ultra-low loads. *Mater Sci Eng A*, 2009, 506: 125–129.
[17] Schuh C, Nieh T. A nanoindentation study of serrated flow in bulk metallic glasses. *Acta Mater*, 2003, 51: 87–99.
[18] Ma Z, Long S, Pan Y, *et al.* Loading rate sensitivity of nanoindentation creep in polycrystalline Ni films. *J Mater Sci*, 2008, 43: 5952–5955.
[19] Schuh CA, Lund AC. Application of nucleation theory to the rate dependence of incipient plasticity during nanoindentation. *J Mater Res*, 2004, 19: 2152–2158.
[20] Schuh CA. Nanoindentation studies of materials. *Mater Today*, 2006, 9: 32–40.
[21] Ziegenhain G, Hartmaier A, Urbassek HM. Pair vs many-body potentials: Influence on elastic and plastic behavior in nanoindentation of fcc metals. *J Mech Phys Solids*, 2009, 57: 1514–1526.
[22] Shi J, Zhang Y, Sun K, *et al.* Effect of water film on the plastic deformation of monocrystalline copper. *RSC Adv*, 2016, 6: 96824–96831.
[23] Stukowski A. Visualization and analysis of atomistic simulation data with OVITO–the Open Visualization Tool. *Model Simul Mater Sci Eng*, 2009, 18: 015012.
[24] Bowden F, Tabor D. *The Friction and Lubrication of Solids*. Oxford: Clarendon Press, 1964.
[25] Halling J. *Principles of Tribology*. London: Macmillan Press, 1975.
[26] Zhu P-z, Hu Y-z, Ma T-b, *et al.* Molecular dynamics study on friction due to ploughing and adhesion in nanometric scratching process. *Tribol Lett*, 2011, 41: 41–46.
[27] Zhu P-z, Hu Y-z, Ma T-b, *et al.* Study of AFM-based nanometric cutting process using molecular dynamics. *Appl Surf Sci*, 2010, 256: 7160–7165.
[28] Bhushan B. *Springer Handbook of Nanotechnology*. Berlin: Springer-Verlag GMBH Germany, 2017.
[29] Chen X, Zhao Y, Wang Y, *et al.* Nanoscale friction and wear properties of silicon wafer under different lubrication conditions. *Appl Surf Sci*, 2013, 282: 25–31.
[30] Luo J, Dornfeld DA. Material removal mechanism in chemical mechanical polishing: theory and modeling. *IEEE Trans Semicond Manuf*, 2001, 14: 112–133.

[31] Luo J, Dornfeld DA. Material removal regions in chemical mechanical planarization for submicron integrated circuit fabrication: coupling effects of slurry chemicals, abrasive size distribution, and wafer-pad contact area. *IEEE Trans Semicond Manuf*, 2003, 16: 45–56.
[32] Si L, Guo D, Luo J, *et al.* Monoatomic layer removal mechanism in chemical mechanical polishing process: A molecular dynamics study. *J Appl Phys*, 2010, 107: 064310.
[33] Tsujimura M. The way to zeros: The future of semiconductor device and chemical mechanical polishing technologies. *Jpn J Appl Phys*, 2016, 55: 06JA01.
[34] Kasai T, Bhushan B. Physics and tribology of chemical mechanical planarization. *J Phys Condens Matter*, 2008, 20: 225011.
[35] Fang L, Sun K, Shi J, *et al.* Movement patterns of ellipsoidal particles with different axial ratios in three-body abrasion of monocrystalline copper: a large scale molecular dynamics study. *RSC Adv*, 2017, 7: 26790–26800.
[36] Shi J, Chen J, Fang L, *et al.* Atomistic scale nanoscratching behavior of monocrystalline Cu influenced by water film in CMP process. *Appl Surf Sci*, 2018, 435: 983–992.
[37] Tang Z, Li S. A review of recent developments of friction modifiers for liquid lubricants (2007–present). *Curr Opin Solid State Mater Sci*, 2014, 18: 119–139.
[38] Dai W, Kheireddin B, Gao H, *et al.* Roles of nanoparticles in oil lubrication. *Tribol Int*, 2016, 102: 88–98.
[39] Wu Y, Tsui W, Liu T. Experimental analysis of tribological properties of lubricating oils with nanoparticle additives. *Wear*, 2007, 262: 819–825.
[40] Ewen JP, Gattinoni C, Thakkar FM, *et al.* Nonequilibrium molecular dynamics investigation of the reduction in friction and wear by carbon nanoparticles between iron surfaces. *Tribol Lett*, 2016, 63: 38.
[41] Zhang WM, Meng G. Advances in the wear characteristics of microelectromechanical systems. *J Tribol*, 2005, 25: 489–494 (in Chinese).
[42] Tanner DM, Dugger MT. *Reliability, testing, and characterization of MEMS/MOEMS II.* International Society for Optics and Photonics, 2003, 22–41.
[43] Alsem DH, Dugger MT, Stach EA, *et al.* Micron-scale friction and sliding wear of polycrystalline silicon thin structural films in ambient air. *J Microelectromech Syst*, 2008, 17: 1144–1154.
[44] Shen S, Meng Y, Zhang W. Characteristics of the wear process of side-wall surfaces in bulk-fabricated Si-MEMS devices in nitrogen gas environment. *Tribol Lett*, 2012, 47: 455–466.

[45] Kim D, Oh S. Atomistic simulation of structural phase transformations in monocrystalline silicon induced by nanoindentation. *Nanotechnology*, 2006, 17: 2259.
[46] Anantheshwara K, Lockwood A, Mishra RK, *et al.* Dynamical evolution of wear particles in nanocontacts. *Tribol Lett*, 2012, 45: 229–235.
[47] Gupta MC, Ruoff AL. Static compression of silicon in the [100] and in the [111] directions. *J Appl Phys*, 1980, 51: 1072–1075.
[48] Clarke D, Kroll M, Kirchner P, *et al.* Amorphization and conductivity of silicon and germanium induced by indentation. *Phys Rev Lett*, 1988, 60: 2156–2159.
[49] Mann A, Heerden DV, Pethica J, *et al.* Contact resistance and phase transformations during nanoindentation of silicon. *Philos Mag A*, 2002, 82: 1921–1929.
[50] Oliver MR. *Chemical-Mechanical Planarization of Semiconductor Materials*. New York: Springer Science & Business Media, 2013.
[51] Fang L, Zhao J, Li B, *et al.* Movement patterns of ellipsoidal particle in abrasive flow machining. *J Mater Process Technol*, 2009, 209: 6048–6056.
[52] Fang L, Kong X, Su J, *et al.* Movement patterns of abrasive particles in three-body abrasion. *Wear*, 1993, 162: 782–789.
[53] Sun K, Fang L, Yan Z, *et al.* Atomistic scale tribological behaviors in nano-grained and single crystal copper systems. *Wear*, 2013, 303: 191–201.
[54] Zhu XZ. *Three-Dimensional Large-Scale Molecular Dynamics Simulation of the Motion Characteristics of Nanoscale Ellipsoidal Abrasives in the Three-Body Abrasive Wear of Single-Crystal Copper*. Xi'an: Xi'an Jiaotong University, 2016.
[55] Wang CH, Fang TH, Cheng PC, *et al.* Simulation and experimental analysis of nanoindentation and mechanical properties of amorphous NiAl alloys. *J Mol Model*, 2015, 21: 1–10.
[56] Fang L, Zhao J, Li B, *et al.* Movement patterns of ellipsoidal particle in abrasive flow machining. *J Mater Process Technol*, 2009, 209: 6048–6056.

Chapter 6

Molecular Dynamics of Chemical Mechanical Polishing Processes

6.1 General

This chapter delves into the nanoscale fundamental theoretical aspects of the chemical mechanical polishing (CMP) process in chip fabrication. Given that the polishing process entails a combination of mechanical and chemical interactive effects, a comprehensive study of the entire process is complex and involves numerous factors. To simplify the research process, the chemical action is modeled by synthesizing an amorphous SiO_2 film on the surface of single-crystal silicon. Subsequently, the mechanical and tribological behaviors are examined to elucidate the influence of mechanical action on the CMP process. While this research method deviates somewhat from actual CMP, it represents an advancement over approaches that neglect chemical action and consider only mechanical effects. Moreover, it lays the groundwork for future studies that will comprehensively account for both chemical and mechanical actions.

The content of this chapter specifically involves the following: (1) through the spherical nanoindentation simulation, to explore the effect of different thicknesses of amorphous SiO_2 film on the mechanical behavior of SiO_2 film/monocrystalline silicon bilayer composites in low-dimensional scale, to analyze the plastic deformation behavior and evolution process of amorphous SiO_2 film/monocrystalline silicon, as well as the effect of the process parameter on the plastic deformation behavior, and the study can provide better understanding of the mechanical behavior and corresponding plastic deformation mechanism and the connection between the two in nanoscale amorphous/crystalline composites; (2) through the nanoindentation creep behavior and the corresponding plastic deformation mechanism of amorphous SiO_2 crystalline composites at the nanoscopic scale, the mechanical behavior, the corresponding plastic deformation mechanism, and the connection between the two; (3) through the simulation of the nanoindentation creep behavior and stress relaxation behavior of amorphous SiO_2 membrane/monocrystalline silicon bilayer composites, to study the effects of different creep loads on creep and different indentation depths on the stress relaxation of composite specimens, and to analyze the time-dependent amorphous SiO_2 membrane and monocrystalline silicon matrix plastic deformation characteristics. (4) Simulate the two-body abrasive wear of amorphous SiO_2 abrasive on amorphous SiO_2 film/monocrystalline silicon specimens, analyze the tribological properties of monocrystalline silicon with different SiO_2 film thicknesses, investigate the effect of abrasive size on the wear, and evaluate the material removal efficiency and the surface quality, to reveal the mechanism of the material removal on the atomic scale as well as the intrinsic connection between the three.

6.2 Indentation Behavior of SiO_2/Si Bilayer Nanomaterials

6.2.1 *Introduction*

The advancement of information technology and ultra-precision technology has led to higher integration levels of large-scale integrated circuits. This has been accompanied by an increase in the diameter of silicon wafers and a gradual decrease in the characteristic line width of the chip [1]. Simultaneously, there is a growing popularity of micro/nano-electromechanical devices (MEMS/NEMS) in various fields, such as military, medical, biological, environmental control, and sensors [2, 3]. Due to these developments, the processing and manufacturing of large-scale integrated circuits and MEMS devices have become subjected to more stringent requirements. For instance, the silicon wafers used in chip processing now need to exhibit atomic-level flatness with zero sub-surface defects. Chemical mechanical polishing is currently the primary processing technology capable of meeting this requirement. It involves the oxidation of the surface of single-crystal silicon by chemical reagents in the polishing solution, resulting in the formation of a thin amorphous SiOx film [4, 5]. However, recent studies on chemical mechanical polishing have focused more on the influence of macroscopic process parameters on wafer surface quality and polishing efficiency, neglecting the examination of the surface-oxidized single-crystal silicon at the nanoscale. Therefore, there is a significant need to comprehend the mechanical properties and deformation behavior of this composite material.

Nanoindentation experiments and theoretical studies [6, 7] have shown that monocrystalline silicon undergoes a transition from the Si-I phase to the denser Si-II and bct-5 phases during loading. Additionally, the Si-II phase transforms into a mixture of Si-III and Si-XII during slow unloading, and into an amorphous phase during fast unloading. Jiang *et al.* [6, 8] systematically investigated the effects of indenter angle and maximum load on the load–displacement curves of nanoindentations. They found that the Si-XII/Si-III phases appeared during unloading only when enough Si-II phases were generated during loading.

Amorphous SiO_2 is a random mesh structure consisting of SiO_4 tetrahedra. Jin *et al.* [9] found that the plastic deformation of SiO_2 is related to the change in the coordination number of silicon atoms, and the number of silicon atoms with coordination numbers 5 and 6 increases, while the number of silicon atoms with coordination number 4 decreases when the material is compressed. Mott [10] proposed that changes in the Si–O bonds in the SiO_4 tetrahedra lead to plastic flow of SiO_2, achieved by changing the size of the –Si–O–Si–O-ring (number of silicon atoms in the ring). Since the depth of indentation of individual abrasives into the substrate in chemical mechanical polishing is about 1 nm, which is on the order of magnitude of the thickness of the amorphous SiO_2 film generated on the surface of monocrystalline silicon, any testing of the properties of SiO_2/Si bilayers composites needs to take into account the mechanical behavior and plastic deformation of both the SiO_2 film and the monocrystalline silicon. However, there are fewer studies and reports on this subject.

Molecular dynamics simulation and nanoindentation are still used to investigate the mechanical behavior of nano-scale amorphous SiO_2/monocrystalline silicon bilayer composites, to analyze the effects of amorphous SiO_2 membrane thickness and different indentation depths on the mechanical properties of the composites, and to reveal the relationship between the mechanical behavior of the materials and the plastic deformation of the amorphous SiO_2 membranes and the monocrystalline silicon matrix, which can help better understand the mechanical behaviors and corresponding plastic deformation evolution of the amorphous membrane/crystal matrix materials mechanical behavior and the corresponding plastic deformation evolution.

6.2.2 *Behavior of nanoindentation under small indentation*

6.2.2.1 *Modeling*

The simulation system comprises a spherical diamond indenter, an amorphous SiO_2 film, and a single-crystal silicon substrate, as depicted in Figure 6.1(a). The spherical indenter, with a radius of 6 nm and consisting of 158,921 atoms, is modeled as a rigid body. The dimensions of the single-crystal silicon matrix are $30 \times 30 \times 16\,\text{nm}^3$, containing 671,982 atoms. The crystallographic orientations of the single-crystal silicon along the X, Y, and Z axes are [100], [010], and [001], respectively. The amorphous SiO_2 film was prepared by quenching molten crystalline SiO_2 [11]. Specifically, α-quartz crystals (crystalline SiO_2) were heated to 7000 K and maintained at that temperature for 200 ps to ensure complete melting. Subsequently, the system was cooled to 3000 K, 1500 K, and finally 300 K at a cooling rate of 5 K/ps, with each temperature held for 200 ps to equilibrate the system, resulting in the formation of amorphous SiO_2.

The nearest neighbor radial distribution function (RDF) of the prepared SiO_2 film, with a thickness of 1.0 nm, is shown in Figure 6.1(b).

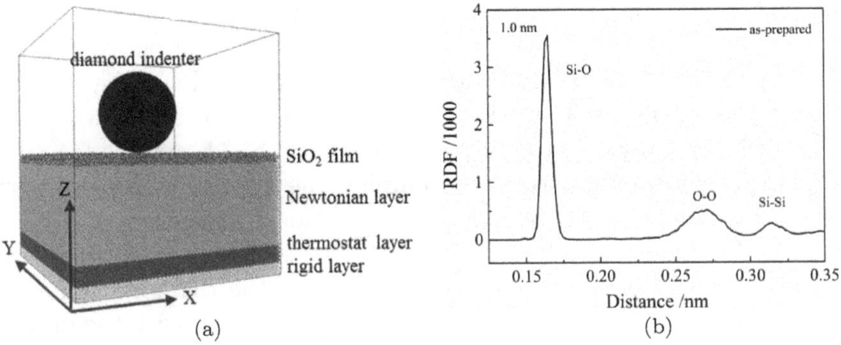

Figure 6.1. Nanoindentation modeling. (a) Schematic of the model, (b) RDF plot of SiO_2 film.

The average distances between Si–O, O–O, and Si–Si atoms were found to be 0.163 nm, 0.269 nm, and 0.315 nm, respectively, aligning with experimental results [12]. This consistency verifies the effectiveness of our SiO_2 membrane preparation method and confirms the amorphous nature of the film. The SiO_2 membranes, with thicknesses of 0.4, 0.6, 0.8, 1.0, 1.4, and 2.0 nm, contain 22,932, 35,084, 47,040, 58,996, 83,104, and 119,168 atoms, respectively.

The prepared amorphous SiO_2 film was imported into the input file (IN file) of the nanoindentation simulation system and positioned on the surface of a single-crystal silicon substrate. The entire simulation system was then relaxed. Besides the diamond indenter, the simulation system was divided into three regions: a fixed layer, a hot bath layer, and a Newtonian layer. The fixed layer, 1.0 nm thick, was located at the bottom of the model to ensure stability. The hot bath layer, 1.5 nm thick, simulated the heat generated during the nanoindentation process, maintaining the system at an operational temperature of 300 K. The remaining atoms in the bilayered substrate comprised the Newtonian layer, where all atoms were free to move according to Newton's laws of motion. Periodic boundary conditions were applied in the X and Y directions, while free boundary conditions were applied in the Z direction.

The selection of an appropriate potential function is critical for ensuring both the accuracy and efficiency of MD simulations. The three-body Tersoff potential function was initially developed to describe the interactions within silicon, facilitating the study of various silicon phase structures and phase transitions. Munetoh *et al.* [13] extended the original Tersoff potential function using first-principles calculations to develop a Tersoff potential suitable for the Si–O system. This extended potential has been successfully applied to simulations involving amorphous SiO_2 and Si–O systems [14, 15]. Consequently, in this study, we employ the extended Si–O Tersoff potential function to simulate the interactions between Si–Si, O–O, and Si–O atoms in a bilayer matrix. The interactions between Si atoms in the bilayer matrix and C atoms in the diamond indenter are described using the Morse potential function [16], with the potential energy expressed by Eq. (2.26). In this equation, $D_0 = 0.435\,\text{eV}$, $\alpha = 4.6487\,\text{Å}^{-1}$, $r_0 = 1.9475\,\text{Å}$, and r represent the adhesion energy, elastic modulus, equilibrium distance, and instantaneous distance between atoms, respectively. The C–O atomic interactions are described by the standard Lennard–Jones potential function with parameters $\varepsilon = 0.1\,\text{eV}$ and $\sigma = 3.275\,\text{Å}$.

Before applying nanoindentation loading, the system underwent local energy minimization using the steepest descent algorithm, followed by structural relaxation. This was achieved using the NVT ensemble with a Nosé–Hoover thermostat at 300 K for up to 100 ps, ensuring robust bonding of the amorphous SiO_2 film to the single-crystal silicon substrate and minimizing the system's energy. Once equilibrated, the nanoindentation process was simulated using the NVE ensemble under a Langevin thermostat. The Newtonian equations of motion were integrated using the Velocity–Verlet algorithm with a time step of 0.5 fs. The indentation process was displacement-controlled, with the indenter loaded and unloaded along the Z-axis at a velocity of 25 m/s, reaching a maximum indentation depth of 3.2 nm. The MD simulations were performed using the open-source software LAMMPS [17], and the results were visualized with OVITO post-processing software [18].

6.2.2.2 *Mechanical property test*

To investigate the mechanical properties and deformation behavior of monocrystalline silicon surfaces coated with amorphous SiO_2 films of varying thicknesses ($L = 0.4, 0.6, 0.8, 1.0, 1.4,$ and $2.0\,\text{nm}$), a series of nanoindentation simulations were conducted. The load-displacement indentation curves at different maximum

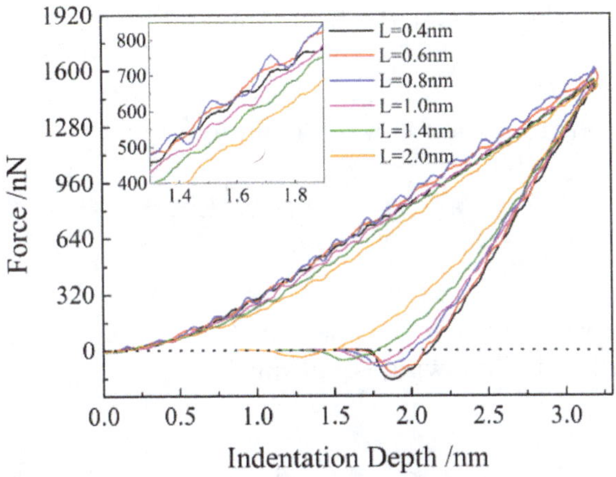

Figure 6.2. Nanoindentation curves of single-crystal silicon surface coated with different thicknesses of SiO_2 (L).

indentation depths were obtained, with Figure 6.2 presenting the curve at a maximum depth of 3.2 nm. Compared to the loading curves, the unloading curves exhibit a distinct hysteresis, indicating significant plastic deformation during the indentation process. Additionally, all loading curves display a characteristic serrated pattern, which is indicative of localized inhomogeneous plastic deformation caused by sudden bursts of plasticity during displacement-controlled loading [19]. This serrated phenomenon has also been observed in the nanoindentation of metallic thin films [20, 21] and bulk amorphous metallic glass [22]. The inset in Figure 6.2 provides a magnified view of the localized loading curves, revealing that the serrated feature is more pronounced for SiO_2 thicknesses of 0.6 nm and 0.8 nm.

Based on Oliver and Pharr's method [23], the nanoindentation mechanical properties (indentation modulus and hardness) of bilayer composites can be derived from a complete loading, unloading cycle. The mechanical properties can be obtained by nonlinearly fitting the nanoindentation unloading curves with the following relation:

$$P = B(h - h_f)^m \tag{6.1}$$

where P is the indentation load/nN, h is the indentation depth/nm, h_f is the depth of plastic deformation remaining after unloading of the curve/nm, and B, m are the fitting parameters. The contact stiffness (S) is the slope at the beginning of the unloading curve and is the dP/dh value at the maximum indentation depth (h_{max}). The contact depth (h_c) between the spherical indenter and the specimen can be derived from the following equation:

$$h_c = h_{\text{max}} - 0.75\frac{P_{\text{max}}}{S} \tag{6.2}$$

where P_{max} is the maximum indentation load/nN. Based on the above measurements, the hardness of the sample to be tested can be obtained by the following equation:

$$H = \frac{P_{\text{max}}}{A} \tag{6.3}$$

where A is the projected contact area/nm^2 as a function of contact depth. The effective indentation modulus (E_{eff}) can be calculated by the following equation:

$$E_{\text{eff}} = \frac{S}{2}\sqrt{\frac{\pi}{A}} \tag{6.4}$$

We know that E_{eff} is defined by the following equation:

$$\frac{1}{E_{\text{eff}}} = \frac{1-v^2}{E} + \frac{1-v_i^2}{E_i} \tag{6.5}$$

The effective indentation modulus accounts for both the elastic displacement of the indenter and the specimen. E, E_i, v, and v_i are the modulus and Poisson's ratio of the specimen and indenter, respectively. For the bilayer composites, varying thicknesses of the SiO_2 film and a constant thickness of monocrystalline silicon theoretically influence Poisson's ratio of the composites. However, because the thickness of the SiO_2 film is much smaller than that of the monocrystalline silicon matrix, the film's impact on Poisson's ratio of the composites is relatively minor. Consequently, there is no significant difference in Poisson's ratios of the composite specimens. Therefore, we use a constant Poisson's ratio of 0.21 for the entire nanoindentation simulation (with Poisson's ratios of 0.22–0.23 for monocrystalline silicon and 0.17 for amorphous SiO_2) [24, 25]. In the simulations, the indenter is defined as a rigid body, resulting in an infinite value for its modulus. Thus, the value of the last term on the right-hand side of Eq. (6.6) is zero.

The relationship between indentation modulus and hardness with indentation depth can be derived from the aforementioned equation, as illustrated in Figures 6.3 and 6.4, respectively. It is evident that, as the indentation depth increases, the influence of the matrix progressively becomes more significant, resulting in a gradual increase in the measured indentation modulus and hardness of the composites. For

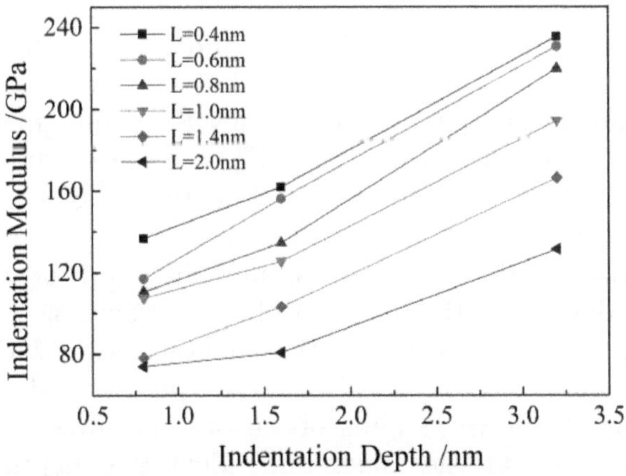

Figure 6.3. Indentation modulus variation curve.

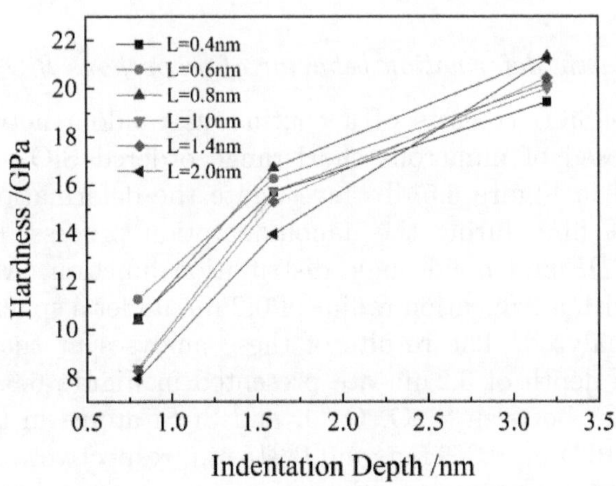

Figure 6.4. Indentation hardness variation curve.

instance, the modulus and hardness of a specimen with a film thickness of 2.0 nm increased from 79.9 GPa and 8.0 GPa at an indentation depth of 0.8 nm to 131.3 GPa and 21.2 GPa at 3.2 nm, respectively. However, at the same indentation depth, the indentation modulus decreased significantly with increasing SiO_2 film thickness. For example, at an indentation depth of 0.8 nm, the moduli of specimens with film thicknesses of 0.4 nm and 2.0 nm were 136.4 GPa and 79.9 GPa, respectively. This trend in the indentation modulus is consistent with the experimental results of the composites by Xu *et al.* [25], while the variation in hardness at the same indentation depth is more complex. At shallow indentation depths (0.8 nm and 1.6 nm), the hardness of the specimens (except for those with film thicknesses of 0.6 nm and 0.8 nm) decreases slightly with increasing SiO_2 film thickness. Conversely, at an indentation depth of 3.2 nm, the hardness increases gradually with the increase of SiO_2 film thickness on the surface. This phenomenon occurs because, at larger indentation depths, the Si–O chemical bonds within the thinner SiO_2 film begin to break, causing a reduction in the hardness of the specimens.

Based on the atomic configuration diagram of the SiO_2 film, it is concluded that throughout the indentation process, only compaction and thinning of the SiO_2 films of various thicknesses occur without rupture, even at the maximum indentation depth of 3.2 nm. Detailed analyses of SiO_2 deformation are presented in the subsequent paragraph.

6.2.2.3 *Plastic deformation behavior of amorphous SiO_2 films*

Amorphous SiO_2 consists of a continuous random network structure composed of numerous short-range ordered SiO_4 tetrahedra, as depicted in Figure 6.5. To investigate the deformation behavior of the SiO_2 film during the nanoindentation process, the nearest neighbor *RDF* and bond angle distribution function (ADF) of the SiO_2 film with a truncation radius of 0.2 nm under a spherical indenter were analyzed. The results of these analyses at the maximum indentation depth of 3.2 nm are presented in Figure 6.6. The average distances between Si–O, O–O, and Si–Si atoms in the original SiO_2 are 0.163 nm, 0.269 nm, and 0.315 nm, respectively. After loading (when the indentation depth reaches its maximum, before unloading), the Si–O peaks slightly shift to the right. Simultaneously, the

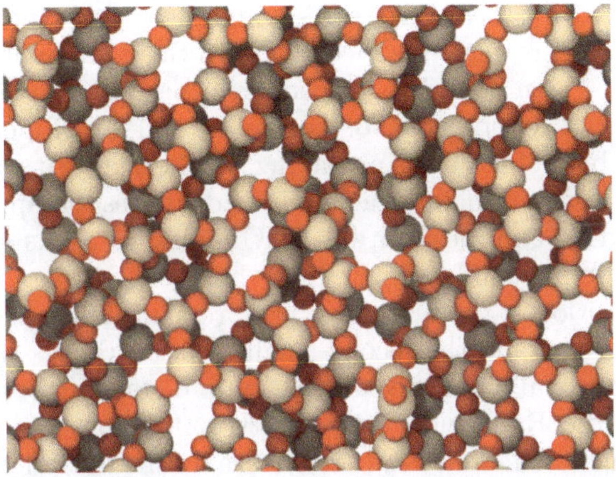

Figure 6.5. Amorphous SiO_2 membrane structure.

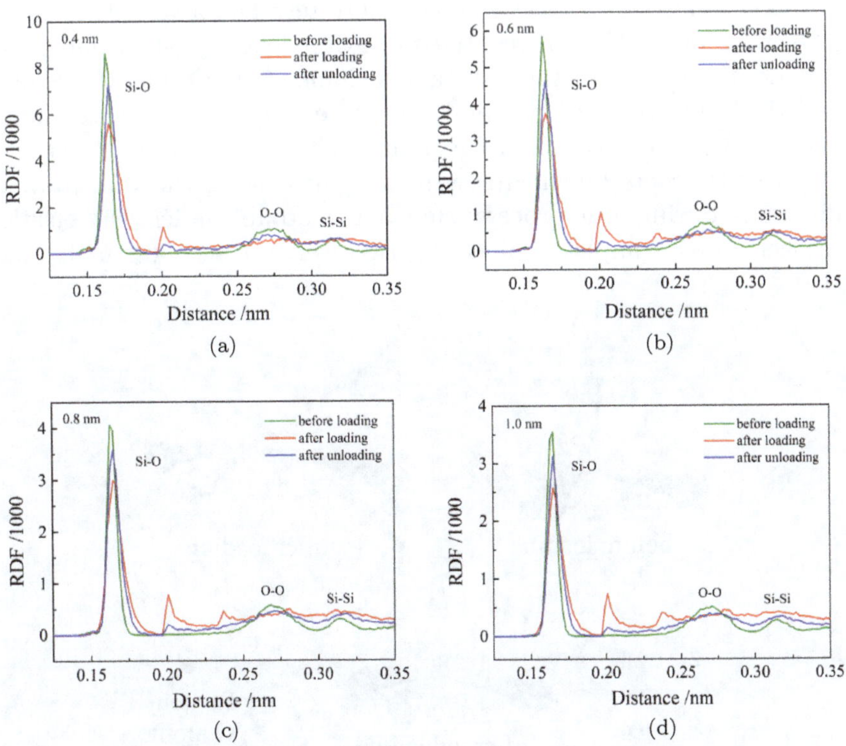

Figure 6.6. RDF analysis of three states of SiO_2 film with different thicknesses. (a) $L = 0.4\,\text{nm}$, (b) $L = 0.6\,\text{nm}$, (c) $L = 0.8\,\text{nm}$, (d) $L = 1.0\,\text{nm}$.

height of the Si–O peaks decreases while their width increases, indicating that the average distances between Si–O atoms increase during loading. Upon complete unloading, a partially opposite change is observed. The positions of the O–O and Si–Si peaks remain unchanged before and after unloading. Additionally, it was observed that small peaks emerged at atomic distances of approximately 0.20 nm and 0.24 nm between the Si–O and O–O peaks during loading. After complete unloading, due to the elastic recovery of the SiO_2 film, the position of the first small peak remains unchanged while its height decreases, and the second small peak nearly disappears.

The additional small peaks at atomic distances of 0.20 nm and 0.24 nm in the RDF plots can be attributed to the rotation and deformation of numerous SiO_4 tetrahedra during the indentation process, resulting in decreased distances between O–O and Si–Si atoms. As illustrated in Figure 6.7, the distances between oxygen atoms 1–2, 2–3, and 1–3 in the pristine SiO_4 tetrahedra are 0.258 nm, 0.284 nm, and 0.256 nm, respectively. After loading (at an indentation depth of 3.2 nm), these distances change to 0.200 nm, 0.228 nm, and 0.320 nm, respectively, and they revert to 0.248 nm, 0.259 nm, and 0.278 nm after complete unloading. This indicates that the original regular SiO_4 tetrahedra were flattened along the indentation direction during the process, and some atoms underwent elastic recovery after complete unloading. However, the second small peak

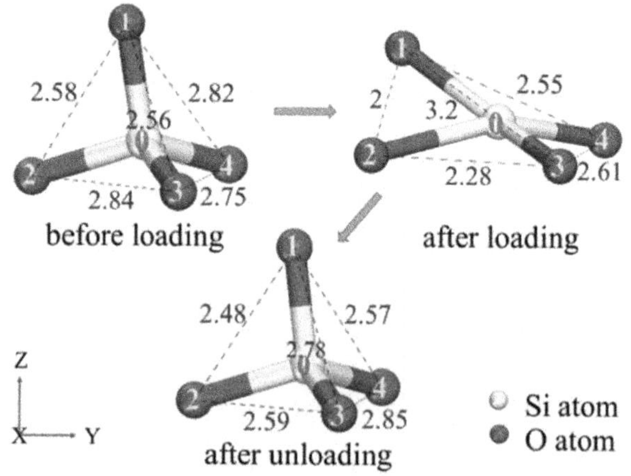

Figure 6.7. Configuration diagram of O–O atomic distance variation (Å).

in Figure 6.6(a) disappears. Compared to other thicker films, the SiO_2 film with an original thickness of 0.4 nm becomes the thinnest after loading. Unlike thicker films, the Si–Si atomic distances do not decrease, instead, they increase due to the substantial number of Si–O bond breaks.

The O–Si–O bond ADFs of amorphous SiO_2 membranes in three different states (pristine, loaded before unloading, and unloaded) are shown in Figures 6.8. The ADF of the pristine membranes exhibits a maximum bond angle distribution at 109.5° with a narrow peak width. After loading, the peak height decreases significantly while the width increases sharply, and an additional peak appears at 73°. Following complete unloading, the main peak undergoes a partially opposite change, with the height of the additional peak decreasing. The appearance of the additional peak is attributed to the rotation

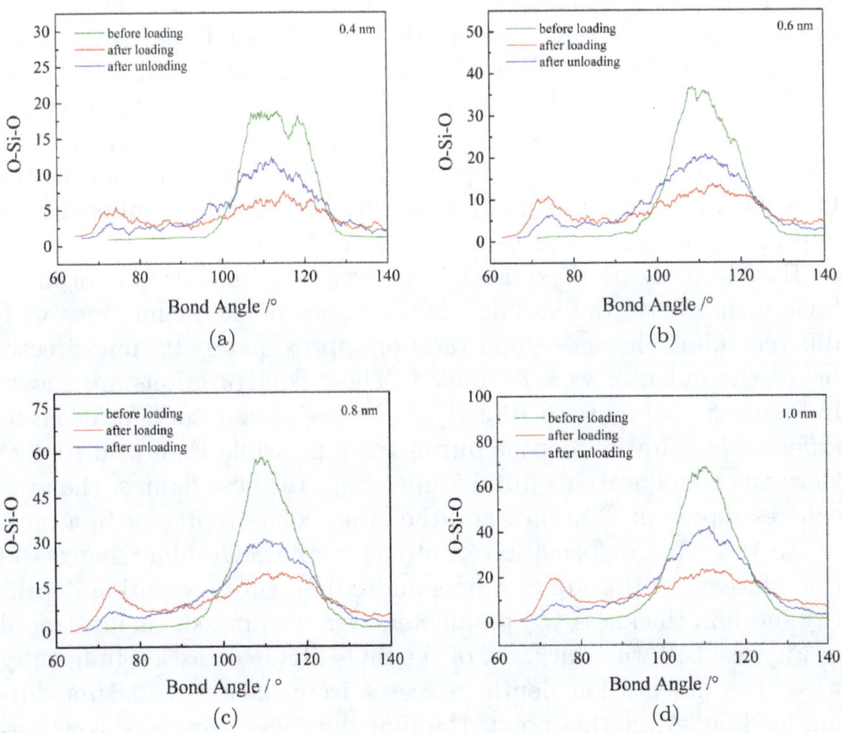

Figure 6.8. ADF analysis of three states of SiO_2 film with different thicknesses. (a) $L = 0.4$ nm, (b) $L = 0.6$ nm, (c) $L = 0.8$ nm, (d) $L = 1.0$ nm.

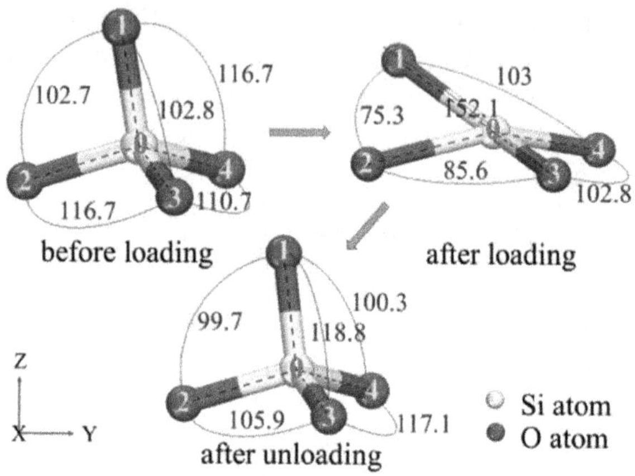

Figure 6.9. Configuration diagram of O–Si–O bond angle variation (Å).

and deformation of numerous SiO_4 tetrahedra during the loading process, leading to a decrease in the O–Si–O bond angle. As shown in Figure 6.9, the O–Si–O bond angles 1–0–2, 2–0–3, and 1–0–3 were 102.7°, 116.7°, and 102.8°, respectively, before loading, these angles changed to 75.3°, 85.6°, and 152.1° after loading, and finally to 99.7°, 105.9°, and 118.8° after complete unloading. The reduction of the O–Si–O bond angles in many such SiO_4 tetrahedra resulted in the appearance of additional peaks in the ADF.

Based on the RDF and ADF analysis results of the amorphous SiO_2 membranes, the atomic configurations of the membranes with different film thicknesses and radii of approximately 0.4 nm directly below the indenter were extracted. These configurations are shown in Figure 6.10. Figures 6.10(a–d) depict the atomic configurations at different indentation depths during loading, while Figures 6.10(e–f) show the configurations during unloading. In these figures, the gray spheres represent O atoms, and the other colors represent Si atoms.

To trace the deformation of atoms in the SiO_2 film, changes in the relative positions of Si atoms under different indentation depths (h) and film thickness (L) parameters were compared. As illustrated in Figures 6.10, the thickness of the SiO_2 film decreases significantly when the indentation depth increases from 0.0 nm to 0.8 nm during loading. After this point, the film thickness decreases slowly as the nanoindentation process continues, and then it increases slowly during unloading. However, the film does not return to its original

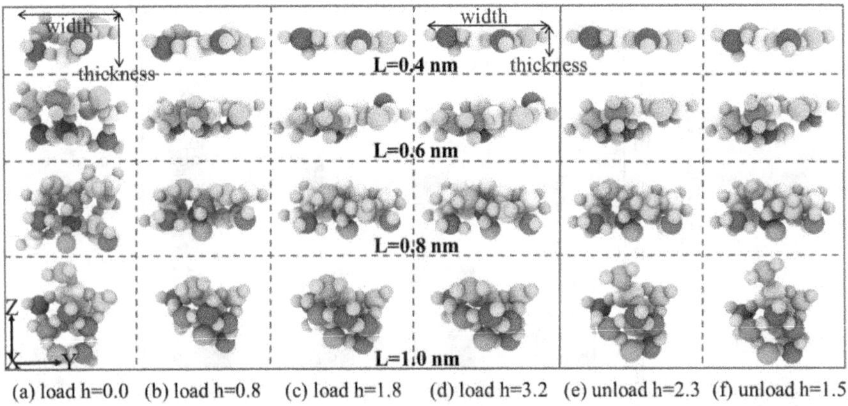

Figure 6.10. Atomic configuration diagrams of the deformation evolution of SiO_2 films with different indentation depths h and different thicknesses (h: nm).

thickness. This indicates that compaction and thinning of the SiO_2 membrane occurred in the vertical direction during loading, while extension occurred in the horizontal direction. Particularly for the membrane with a thickness of 0.4 nm, the film thickness reached its minimum at an indentation depth of 3.2 nm, approaching the thickness of a single atomic layer.

To analyze the compaction and expansion of amorphous SiO_2 membranes with different thicknesses, we examined the SiO_2 membrane directly below the indenter, as shown in Figure 6.10. We calculated the percentage of compaction along the direction of indentation (vertical) using the formula: ((membrane thickness at indent depth 0.0 nm—membrane thickness at indent dept 0.8 nm)/menbrane thickness at indent depth 0.0), and the percentage of expansion in the horizontal direction using the formula: (membrane width at indent depth 0.8 nm—membrane width at indent depth 0.0 nm)/membrane width at indent depth 0.0 nm). The width and thickness measurements of the deformed SiO_2 membrane are shown in Figure 6.10. It was found that as the thickness of the SiO_2 film increases, the percentage of compaction increases significantly, while the percentage of expansion decreases slightly, as shown in Figure 6.11. This, combined with Figure 6.10, indicates that thicker membranes have the potential for further compaction and extension. However, the 0.4 nm thick membrane reaches a critical compaction value at a maximum

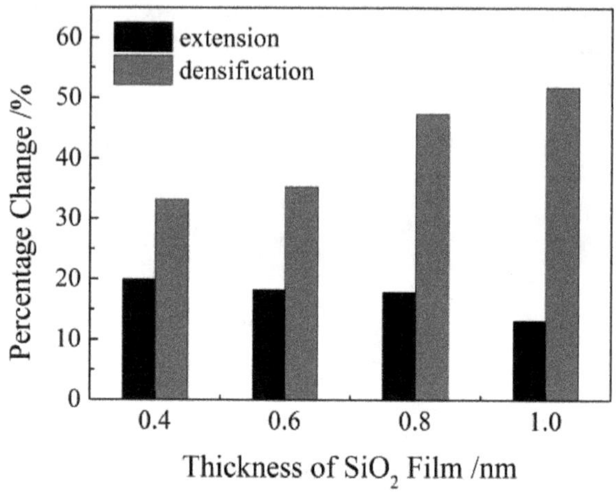

Figure 6.11. SiO_2 film extension and compaction percentage change.

indentation depth of 3.2 nm (with some Si–O bonds already broken). This membrane has a tendency to preferentially rupture under further loading, which reasonably explains the minimum hardness of the 0.4 nm thick SiO_2 membrane at an indentation depth of 3.2 nm.

6.2.2.4 *Plastic deformation of single-crystal silicon*

Experimental studies have revealed that phase change is the main plastic deformation mechanism of single-crystal silicon at room temperature [24, 26–28]. Diamond anvil indentation experiments on single-crystal silicon have shown that the transformation of pristine single-crystal silicon (Si-I phase) into the metallic Si-II phase during loading results in a volume shrinkage of nearly 20%. Nanoindentation experiments have found that not only is the Si-II phase generated in the indentation craters during loading, but also the presence of shear stresses in the indentation reduces the pressure required for the transformation of the Si-I phase to the Si-II phase. This leads to the generation of a new phase, the bct-5 phase. This process is irreversible, resulting in a mixture of crystalline phases (Si-XII and Si-III) from the Si-II phase generated during slow unloading, and amorphous α-Si during fast unloading [29–31]. Therefore, the following discussion focuses on the phase transition behavior of single-crystal silicon under different thicknesses of SiO_2 film.

When the spherical indenter is loaded along the indentation direction, the surface amorphous SiO_2 film is first deformed. Then, at a lower pressure, lattice torsion occurs on the surface of the single-crystal silicon substrate below the indenter, resulting in the generation of an amorphous layer of α-Si on the substrate surface. This is consistent with the results of numerous nanoindentation experiments [26, 32, 33]. As shown in Figure 6.12, panels (a–c) depict the phase transitions under different indentation depths during loading, and panels (d–e) show the phase transitions during unloading. The Si atoms are colored according to their coordination number, with pink, green, yellow, blue, and red representing Si-I, bct-5, β-Si, surface atoms, and other phase atoms, respectively.

At an indentation depth of 0.8 nm, an amorphous transformation occurs on the surface and sub-surface of the silicon substrate. As the indentation depth increases, the stress on the silicon substrate increases, resulting in a phase transformation of the silicon beneath the newly generated amorphous silicon. The transformation from the Si-I phase (diamond silicon) to the bct-5 (body-centered tetragonal) and β-Si phases (body-centered tetragonal β-tin) [34] continues to increase with loading, as shown in Figure 6.13. Additionally, at the same indentation depth, the phase transition region of the silicon substrate and the phase transition atoms of the newly generated phases

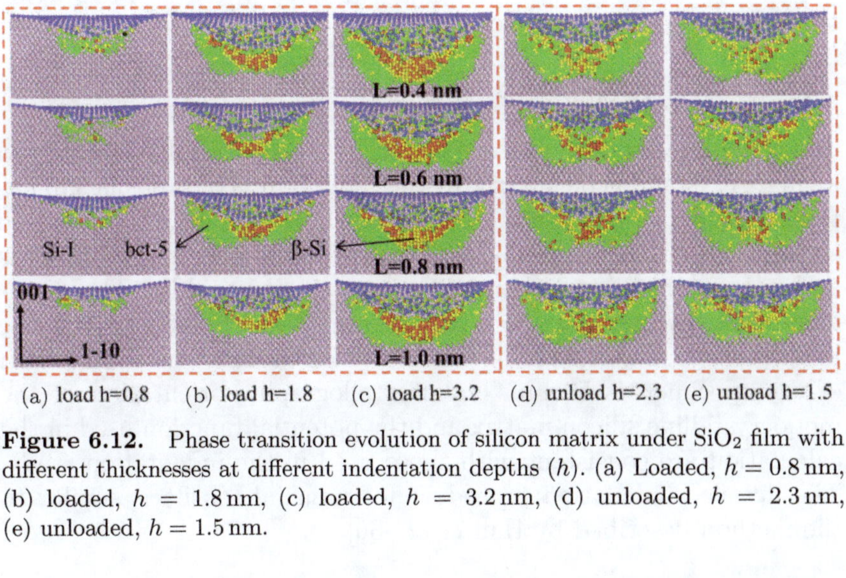

Figure 6.12. Phase transition evolution of silicon matrix under SiO_2 film with different thicknesses at different indentation depths (h). (a) Loaded, $h = 0.8$ nm, (b) loaded, $h = 1.8$ nm, (c) loaded, $h = 3.2$ nm, (d) unloaded, $h = 2.3$ nm, (e) unloaded, $h = 1.5$ nm.

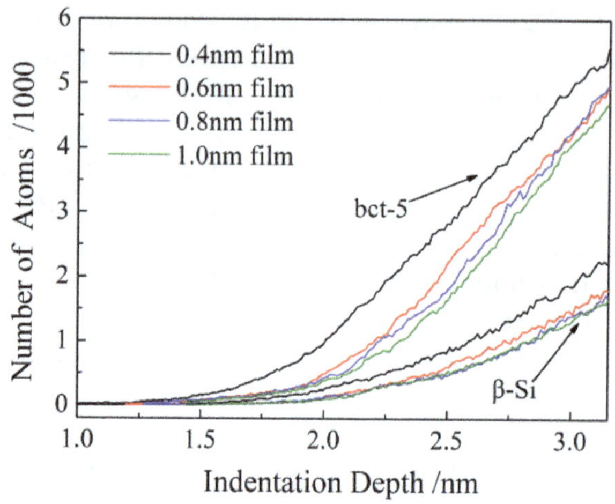

Figure 6.13. Variation curve of the number of Si phase transition atoms.

(amorphous silicon, bct-5, and β-Si) decrease with the increase in film thickness. This suggests that the amorphous SiO_2 film acts as a suppressor of the silicon phase transition.

The phase transition behavior of Si during unloading is not discussed in this section because the unloading velocities in the MD calculations are several orders of magnitude higher than those in actual nanoindentation, resulting in only amorphous Si appearing in the deformation craters after unloading [30, 34, 35].

6.2.2.5 *SiO_2/Si interfacial strength*

To investigate the interfacial strength between amorphous SiO_2 and a single-crystal silicon matrix in the model, uniaxial tensile simulations were performed using MD. As shown in Figure 6.14, the tensile specimen is a sandwich cylinder, with silicon sections at both ends and a 1 nm thick amorphous SiO_2 layer in the middle. The diameter of the cylindrical silicon at the upper end is 3.0 nm, and the length is 24.5 nm, containing 34,082 silicon atoms. The total length of the specimen is 50 nm. The crystallographic orientation of the monocrystalline silicon matrix and the potential function used in the calculation are consistent with those used in the indentation study. The detailed simulation procedure for uniaxial tensile is similar to the method described by Han *et al.* [36].

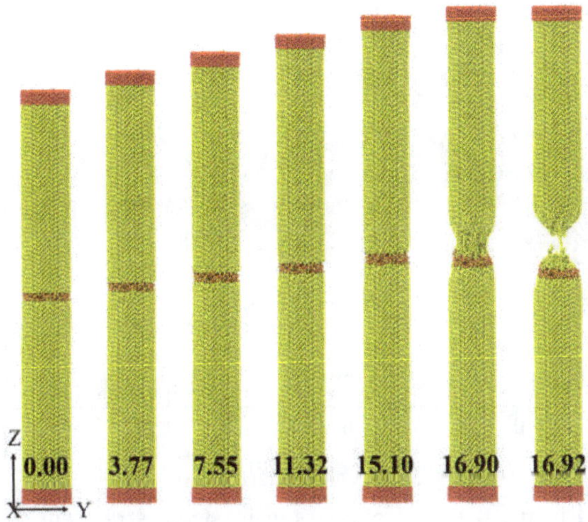

Figure 6.14. Uniaxial tensile deformation at different strains (%).

In the simulation, the cylindrical specimen was divided into three parts: deformed, fixed, and loaded layers. The fixed and loaded layers were defined as rigid bodies, with atoms in these layers at both ends of the specimen colored red. Prior to tensile loading, the simulation system was relaxed for 90 ps to equilibrate, using the same relaxation parameters as for nanoindentation. During subsequent stretching, the loaded layer was moved along the Z-axis at a constant strain rate of 2.5×10^7, while atoms in the fixed layer were frozen and did not participate in the tensile integration process. The stretching process was carried out in the NVT ensemble at 300 K.

A typical uniaxial tensile stress–strain curve is shown in Figure 6.15. The curve indicates that the specimen undergoes linear elastic deformation until fracture, with fracture stress and strain of 17.7 GPa and 16.9%, respectively. These results are close to Kang's [37] calculations (approximately 13.3 GPa, 16.2%) and Tang's [38] experimental results. Additionally, fitting the linear elastic portion of the stress–strain curve yields Young's modulus of 109.4 GPa, which is consistent with the indentation modulus obtained during nanoindentation.

The tensile deformation at different strains is depicted in Figure 6.14. The specimen elongates uniformly at strains from 0.00 to 15.10%, but when the strain increases beyond a certain value, an

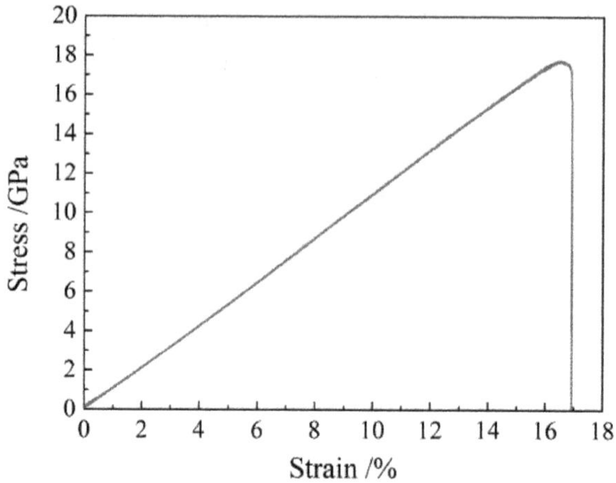

Figure 6.15. Uniaxial tensile stress–strain curve.

amorphous phase transition occurs on the surface of the silicon. This phase transition extends into the necking region of the specimen until it fractures. Outside of this necking region, the specimen maintains an ordered structure with no significant changes.

Since tensile fractures typically initiate from the weakest part of a material, the fact that the fracture site in our uniaxial tensile calculations is located in the region of monocrystalline silicon above the SiO_2/Si interface, rather than at the interface itself, directly proves that the interfacial bond between SiO_2 and Si is strong. The presence of the interface does not affect the tensile deformation mode of the monocrystalline silicon.

The structure of the SiO_2/Si interface at a strain of 16.9% is shown in Figure 6.16. In this figure, the red and yellow colors represent the oxygen and silicon atoms, respectively, while the short green lines represent the Si–O chemical bonds. The image demonstrates that the amorphous SiO_2 film and the Si matrix are tightly bonded through numerous chemical bonds. The Si–O bonding energy ($542\,\mathrm{kJ\,mol^{-1}}$) [39] is significantly higher than that of the Si–Si bond ($222\,\mathrm{kJ\,mol^{-1}}$) [40]. This difference in bonding energy explains why the tensile fracture occurs within the monocrystalline silicon rather than at the SiO_2/Si interface.

Through extensive examination of the von Mises stress distribution across the bilayer substrates at various indentation depths, it

Figure 6.16. Cross-sectional atomic configuration of SiO_2/Si interface (16.90%).

Figure 6.17. Cross-section of von Mises stress distribution for a film thickness of 2.0 nm SiO_2 sample.

has been observed that the stress level progressively rises with a deeper indentation, and pronounced stress concentration takes place within the amorphous SiO_2 layer. For instance, when evaluating a specimen with a film thickness of 2.0 nm at an indentation depth of 3.2 nm (as depicted in Figure 6.17), the average stress experienced by the SiO_2 film beneath the spherical indenter significantly surpasses that of the surrounding Si matrix. This phenomenon has also been reported by Yang *et al.* [41, 42], who discovered that the stress mutation at the interface intensifies as the indentation depth grows. However, the magnitude and rate of increase of the stress mutation in the current nanoindentation calculation are notably smaller than those obtained by Yang *et al.* through finite element simulation. This discrepancy may be attributed to the disparity in scale between the MD calculation and the finite element analysis, as well as the relatively weak interfacial bonding strength of the multilayer specimen in Yang *et al.*'s study. The occurrence of stress concentration within the SiO_2 film has important implications for the plastic deformation of the thin film material and the creep behavior of the composite specimen.

6.2.3 *Nanoindentation behavior under large indentation*

6.2.3.1 *Modeling*

To further examine the mechanical properties and deformation behavior of a single-crystal silicon coated with an amorphous SiO_2 film during nanoindentation, we set the indentation depth of the composite specimen to 8.2 nm. The simulation system included a spherical indenter and a double-layer composite matrix, with the indenter's radius set at 6.0 nm. The amorphous SiO_2 film was available in four thicknesses: 0.8, 1.0, 1.4, and 2.0 nm. To assess the effect of model size, we compared three different sizes of monocrystalline silicon substrates: $30.0 \times 30.0 \times 25.0\,\text{nm}^3$, $30.0 \times 30.0 \times 34.8\,\text{nm}^3$, and $34.2 \times 34.2 \times 34.8\,\text{nm}^3$. By comparing the indentation load–displacement curves at the maximum indentation depth of 8.2 nm, we found that the curves of the three models overlapped almost perfectly, with no significant difference. Therefore, we selected a single-crystal silicon model with simulated dimensions of $30.0 \times 30.0 \times 25.0\,\text{nm}^3$ and containing 1,130,685 atoms for further simulation. The details of the simulations, including the potential functions, are provided in Section 3.2.

6.2.3.2 *Mechanical property test*

To investigate the mechanical properties and deformation behavior of a bilayer matrix consisting of a surface amorphous SiO_2 film and a monocrystalline silicon matrix under large indentation, a simulation of nanoindentation at various depths was conducted. The load–displacement curves of different composite specimens were

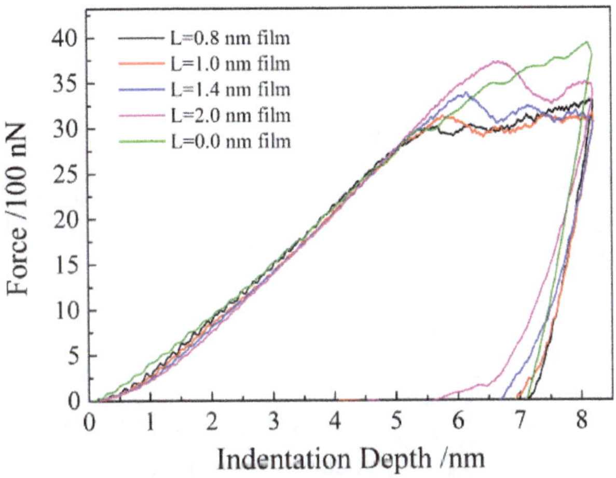

Figure 6.18. Load–displacement curve at an indentation depth of 8.2 nm.

obtained, and the curve at an indentation depth of 8.2 nm is illustrated in Figure 6.18. The results indicate that when the indentation depth is less than 5.5 nm, the indentation load increases with increasing indentation depth, however, when the indentation depth is greater than 5.5 nm, the load decreases with increasing indentation depth, depending on the size of the SiO_2 film thickness. After that, the indentation load increases again. All the specimens displayed significant plastic deformation as seen from Figure 6.18, and the SiO_2 film experienced compaction and thinning during loading until the indentation depth exceeded 5.5 nm. Additionally, the thinner the membrane, the earlier the load decreases, and the corresponding indentation depth is smaller when the load decreases compared to thicker membranes. This occurs because thinner membranes are more easily deformed and prone to early rupture.

The mechanical properties of the SiO_2/Si composites at various indentation depths were calculated using the Oliver and Pharr [23] method, and the resulting indentation modulus and hardness are shown in Figures 6.19 and 6.20. The results indicated that the indentation modulus of all specimens increased with increasing indentation depth. Additionally, at the same indentation depth, the modulus decreased with increasing film thickness. For example, the modulus of 2.0 nm specimens and pure monocrystalline silicon ranged

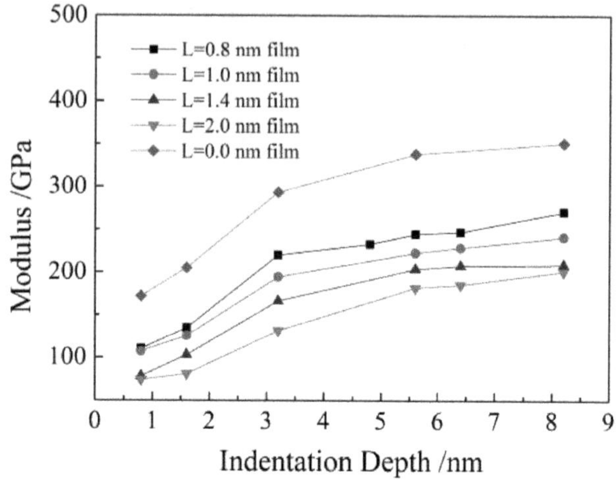

Figure 6.19. Variation of indentation modulus for different film thicknesses.

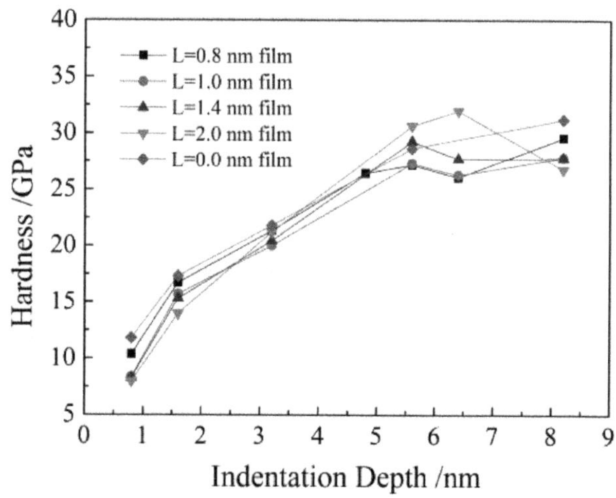

Figure 6.20. Variation of hardness for different film thicknesses.

from 73.9 GPa and 171.9 GPa at an indentation depth of 0.8 nm to 200.9 GPa and 350.7 GPa, respectively, at an indentation depth of 8.2 nm. The modulus of the pure monocrystalline silicon was found to be larger than that of the composites at the same indentation depth due to the monocrystalline silicon matrix. Notably, the indentation modulus and hardness at an indentation depth of 0.8 nm were consistent with experimental values [43, 44], while the values at other depths were larger. Possible explanations for this discrepancy are discussed later.

The hardness of pure monocrystalline silicon increases with greater indentation depth. As shown in Figure 6.20, for instance, at an indentation depth of 0.8 nm, the hardness is 11.8 GPa, and at an indentation depth of 8.2 nm, it increases to 31.1 GPa. In the case of composites, the hardness first increases with indentation depth (up to 5.6 nm), then decreases depending on film thickness, and eventually increases again with further indentation (from the depth 5.6 to 8.2 nm). The variation in hardness at the same indentation depth is quite complex. The hardness of the amorphous SiO_2 film varies with the indentation depth and film thickness. At small indentation depths of 0.8 nm and 1.6 nm, there is a slight decrease in hardness with increasing film thickness. However, when the indentation depth goes from 3.2 nm to 6.4 nm, the hardness increases with the film thickness. At the maximum indentation depth of 8.2 nm, the hardness decreases with the increase of the film thickness. This complex trend in hardness is due to the deformation behavior of the amorphous SiO_2 film during the indentation process. It undergoes stages of compaction, compaction–rupture transition, rupture, and elastic recovery during the indentation test loading process. The deformation behavior of SiO_2 is discussed in more detail in the following section. Additionally, the indentation modulus is more influenced by the substrate than the calculated hardness values due to the long-range nature of the elastic field. This means that the elastic deformation of the substrate under the indenter is not confined to the surface film itself but transfers the elastic field to the Si substrate [45].

Since the initial conditions of the simulation can influence the test results, we conducted nanoindentation simulations at depths of 0.8 nm and 2.0 nm. We varied the random numbers of the starting simulation states and calculated the average indentation modulus and hardness from five independent indentation simulations. We found that the values and the trends of the changes aligned with those depicted in Figures 6.19 and 6.20. This indicates that the mechanical properties and the corresponding plastic deformation results we obtained are trustworthy. Additionally, Chen *et al.* [46, 47] subjected single-crystal silicon surfaces coated with amorphous SiO_2 to impact by amorphous SiO_2 agglomerates at high speeds. The impact velocities (2500–4313 m/s) were much higher than our nanoindentation simulation velocities yet still provided significant experimental data when averaging the results from each independent MD simulation.

It's important to note that materials exhibit different mechanical properties when measured at different scales. For bulk homogeneous materials, the modulus and hardness obtained from sharp indenter experiments yield constant values, such as using Berkovich's and Vicker's indenter [45]. However, MD calculations for a small volume of material show that the modulus and hardness increase with the depth of the indentation when using a spherical indenter in the simulations. According to Tabor [48], the yield strength and hardness obtained from compression experiments increase with the equivalent indentation depth due to the geometrically dissimilar properties of the spherical indenter. On the other hand, for a conventional sharp indenter, the modulus and hardness obtained are independent of the load magnitude. It's worth mentioning that the mechanical property tests in this paper were conducted using Oliver and Pharr's method, which unfortunately does not account for the phenomenon of pile-up that occurs during the tests.

Pile-up tends to occur in specimens with soft films and hard substrates, causing the material contact area to be underestimated. This ultimately leads to high values of elastic modulus and hardness being measured [45, 49]. In our simulations, we observed the pile-up phenomenon in specimens with soft films and hard matrices. We also noted a slight pile-up phenomenon in both pure silicon and composites during our simulations, resulting in higher indentation modulus and hardness for both. The differences between properties obtained from indentation experiments and MD simulations are still under debate. These differences may be influenced by factors such as the scale of indentation [50, 51] and variations between ideal single-crystal silicon in MD simulations and single-crystal silicon containing multiple defects in experiments [52]. Regardless, spherical indenter nanoindentation offers a new perspective on studying mechanical properties of single-crystal silicon substrates coated with amorphous SiO_2 films and can be useful for engineering design and fabrication at the micro- and nanoscale.

6.2.3.3 *Plastic deformation behavior of amorphous SiO_2 films*

The atomic configuration of the 0.8 nm film changes as the spherical indenter applies pressure during the indentation process, as depicted in Figure 6.21. The film thins as the indentation depth increases, and the SiO_2 film breaks when the depth of the indentation reaches 6.0 nm, as shown in Figure 6.21(d). Additionally, greater indentation depth leads to more severe film rupture. It's observed that at the same indentation depth, the deformation of the thicker film specimen is smaller, as shown in Figure 6.22. At an indentation depth of 7.5 nm, the 0.8 nm film has completely ruptured, while the 2.0 nm film shows thinning and begins to rupture. The deformation behavior of the SiO_2 film is investigated using SiO_2 with a radius of 5.5 nm and a film thickness of 0.8 nm under a spherical indenter. In the following paragraph, the specific deformation characteristics are also studied using the *RDF* and the ADF of the bonds, with a truncation radius of 0.2 nm.

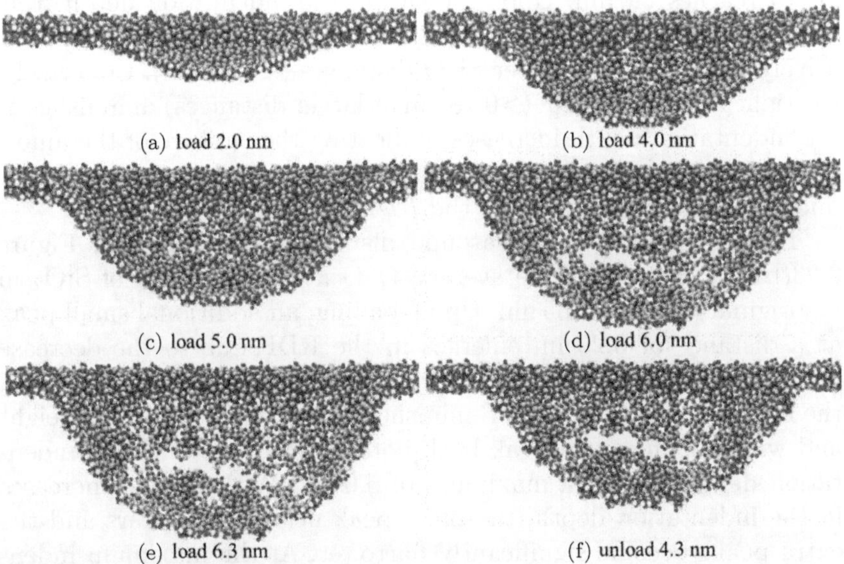

Figure 6.21. Deformation evolution of 0.8 nm SiO_2 film in the indentation crater during the indentation process. (a) Loaded, 2.0 nm, (b) loaded, 4.0 nm, (c) loaded, 5.0 nm, (d) loaded, 6.0 nm, (e) loaded, 6.3 nm, (f) unload, 4.3 nm.

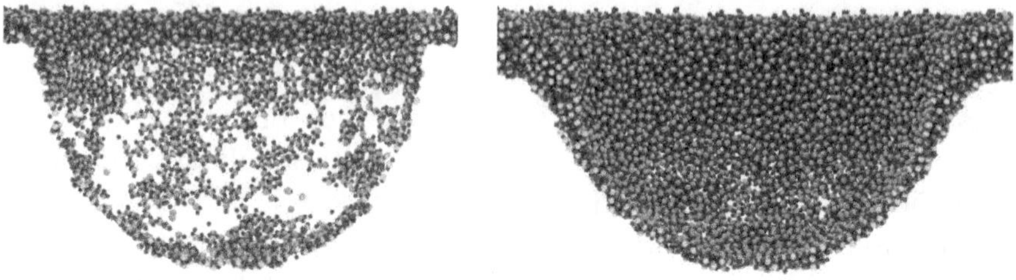

(a) (b)

Figure 6.22. Configuration of SiO_2 film with different thicknesses at indentation depth of 7.5 nm. (a) 0.8 nm film, (b) 2.0 nm film.

The *RDF* curves for atomic pairs in SiO_2 film at different depths of indentation are depicted in Figure 6.23. Figure 6.23(a) illustrates the correlation between the *RDF* of Si–O atoms and the distance of Si–O atoms. As the indentation depth increases, the peaks for Si–O atoms shift toward larger distances, resulting in a decrease in peak height and a considerable increase in peak width until the indentation depth reaches 6.0 nm. Upon increasing the indentation depth from 6.0 nm to 8.2 nm, the curves for the left half of the RDF remain relatively consistent at smaller Si–O distances (<0.18 nm). Conversely, the right half of the *RDF* (>0.18 nm at larger distances) diminishes as the indentation depth increases, indicating the rupture of the amorphous SiO_2 film. Upon unloading, there is a partially opposite change due to the elastic recovery in the *RDF*.

The *RDF* of the O–O atomic distance is displayed in Figure 6.23(b). It shows that the average O–O atomic distance of SiO_2 in its original state is 0.265 nm. Upon loading, an additional small peak at a distance of 0.20 nm emerges in the RDF due to the decrease in the O–O atomic distance. As the depth of indentation increases, the height of the O–O peak diminishes significantly, while the height and width of the extra peak both notably increase, until the indentation depth reaches a maximum of 5.0 nm. With further increases in the indentation depth, the main peak nearly disappears and the extra peaks become significantly narrower. At the maximum indentation depth, it is notable that the O–O atomic distances are broadly distributed from 0.20 nm to 0.32 nm. The atomic distances in their original state either decrease or increase, which is a result of the compaction of the indentation at small indentation depths during

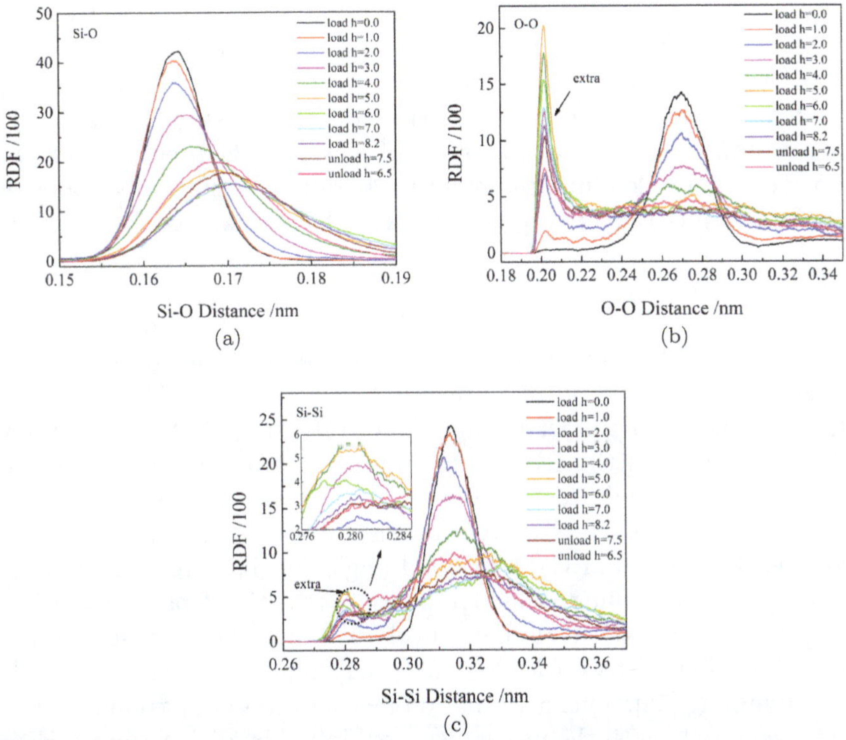

Figure 6.23. Atomic *RDF* variation curves at different indentation depths (h: nm). (a) Si–O bond, (b) O–O bond, (c) Si–Si bond.

the loading process (depths <5.0 nm) and the breaking of the chemical bonds of Si–O in the stretched state at larger indentation depths (depths >5.0 nm), respectively. When the material is unloaded, the main O–O peak still disappears, but its value slightly increases, and the additional peaks continue to diminish. It's worth mentioning that the sudden increase in the extra peak during indentation loading is a result of the combined effect of the Si–O bond length and the O–Si–O bond angle.

The *RDF* of Si–Si atoms in the SiO_2 material is illustrated in Figure 6.23(c). In the original state, the average distance between Si–Si atoms is 0.315 nm, with a relatively narrow peak width. Upon loading, an additional peak emerges at a distance of 0.28 nm. When the indentation depth reaches 5.0 nm, the main Si–Si peak height decreases, the width increases, and the extra peak becomes more prominent. With further indentation depth, from 6.0 nm to 8.2 nm, the main peak shifts to the right, and the right half of the RDF peak decreases, while the extra peak narrows. This indicates that the Si–Si distance in the pristine equilibrium state changes during loading, due to the compression of SiO_2 and the breakage of Si–O bonds. Upon unloading, the main peak undergoes partially opposite changes, while the extra peak continues to shrink, signifying a tendency of the system to return to equilibrium in order to minimize its potential energy [53].

In Figure 6.24, the O–Si–O ADF is shown at various indentation depths. The plot reveals that the highest concentration of bond angles occurs at the ideal tetrahedral angle of 110°, with an additional peak at 73° under loading. The ADF trend with indentation depth aligns with the RDF of Si–Si atoms. After complete unloading, the bond angle distribution spans a wider range of 70–150°. The analyses of RDF and ADF consistently indicate that the Si–O atomic distances and O–Si–O bond angles deviate from their equilibrium positions due to the high pressure resulting from indentation [53]. The deformation mechanism of this amorphous structure differs from that of crystal dislocations [54], phase transitions [55–57], and twinning [58]. This deformation mechanism in amorphous structures contrasts with the mechanisms exhibited by dislocations, phase transitions, and twinning in crystals.

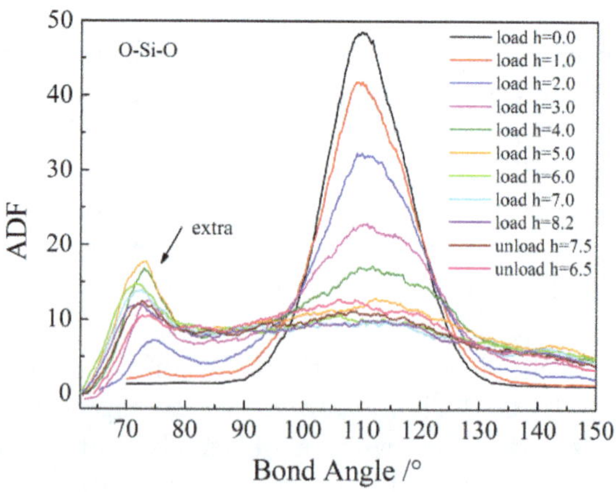

Figure 6.24. Variation curves of ADF of O–Si–O bond angles at different indentation depths (h: nm).

To further investigate the deformation characteristics of amorphous SiO_2 films during the indentation process, a parameter called percentage change of atom number of x-coordinated silicon (*PCNx* for short) was proposed. Here, *CNx* indicates the number of oxygen atoms connected to a silicon atom through Si–O bonding. The calculation of *PCNx* for an indentation depth of 6.0 nm is shown in Eq. (6.6). SiO_2 films with a radius of 5.5 nm and thicknesses of 0.8 nm and 2.0 nm under a spherical indenter were studied, as shown in Figure 6.25. The variation of *PCNx* with the depth of indentation illustrates that the plastic deformation of the SiO_2 film exhibits four phases: compaction, the transition phase from compaction to rupture, rupture, and elastic recovery in unloading.

During the compaction stage, *PCN5* increases significantly, while *PCN4* decreases significantly, and *PCN2* and *PCN3* remain near zero. In the transition stage, *PCN5* continues to increase, *PCN2* and *PCN3*

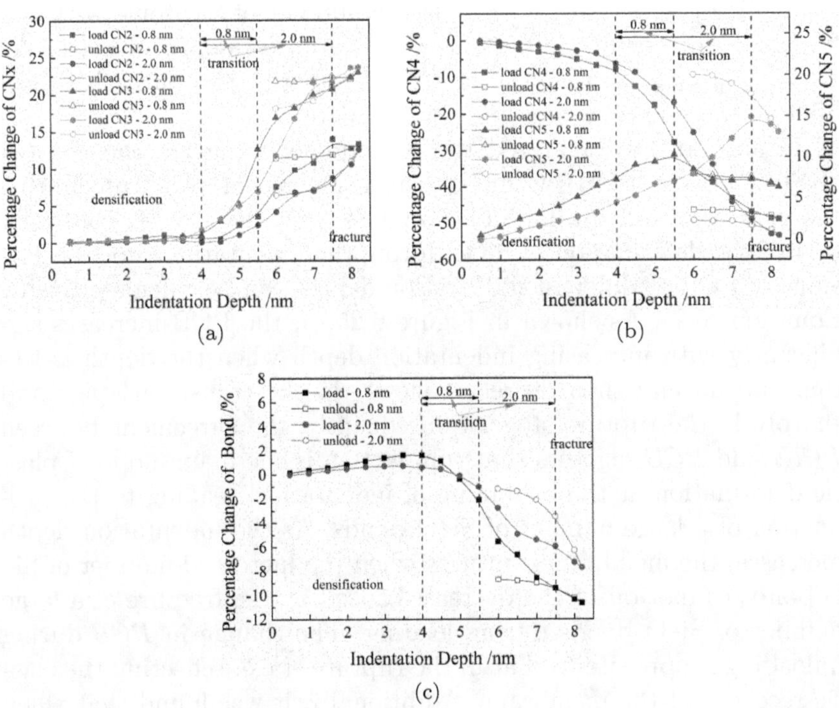

Figure 6.25. *PCNx* and *PCB* variation with indentation depth. (a) *PCN2* and *PCN3*, (b) *PCN4* and *PCN5*, (c) Si–O bonding.

begin to increase, and *PCN4* decreases sharply. At the rupture stage, *PCN4* and *PCN5* decrease sharply, while *PCN2* and *PCN3* increase significantly. In the final unloading stage, the opposite change to the rupture stage occurs. It was also found that at the same indentation depth, the *PCNx* values of the 0.8 nm film were significantly larger than those of the 2.0 nm film, except for cases when the indentation depth was larger. For the expression of *PCNx*, we have

$$PCNx = \frac{CNx \text{ atom number}_{\text{depth}=6.0\,\text{nm}} - CNx \text{ atom number}_{\text{depth}=0.0\,\text{nm}}}{CN4 \text{ atom number}_{\text{depth}=0.0\,\text{nm}}} \times 100\% \tag{6.6}$$

where *PCNx* is the percentage of Si atoms with different coordination numbers, $CNx_{\text{atom numberdept}=6.0\,\text{nm}}$ is the number of Si atoms with different coordination numbers at an indentation depth of 6 nm, $CNx_{\text{atom numberdept}=0.0\,\text{nm}}$ is the number of Si atoms with different coordination numbers at an indentation depth of zero, and $CN4_{\text{atom numberdept}=0.0\,\text{nm}}$ is the total number of Si atoms with coordination number 4 at zero indentation depth.

In addition, we analyzed the percentage change in the number of Si–O bonds (PCB) within a truncation radius of 0.2 nm during the process of indentation. This analysis provides further confirmation of the deformation of SiO_2 during the indentation process. The value of PCB at different indentation depths can be calculated using Equation (6.7). As shown in Figure 6.25(c), the PCB increases significantly with increasing indentation depth when the depth is less than 4.0 nm and then decreases slowly in the transition stage and sharply in the rupture stage. Consequently, the agreement between *PCNx* and *PCB* suggests that compaction is the main mode of plastic deformation at the beginning of indentation, leading to the production of a large number of Si–O bonds. As the indentation depth increases, the membrane ruptures when the increased number of Si–O bonds cannot offset the decrease caused by the fracture of a large number of Si–O bonds during loading. The change in *PCB* during unloading is opposite to that in the rupture stage, reflecting the elastic recovery of the membrane. Additionally, it was found that when the indentation depth was less than 4.0 nm, the *PCB* values of the 0.8 nm membranes were significantly higher than those of the 2.0 nm membranes. However, when the indentation depth was greater than 4.0 nm, the 0.8 nm membranes showed a larger change, and the *PCB* values were smaller than those of the 2.0 nm membranes, indicating that the 0.8 nm membranes were more plastically deformed at the same indentation depth. *PCB* can be calculated as follows:

$$PCB = \frac{\text{bond number}_{\text{depth}=6.0\,\text{nm}} - \text{bond number}_{\text{depth}=0.0\,\text{nm}}}{\text{bond number}_{\text{depth}=0.0\,\text{nm}}} \times 100\% \tag{6.7}$$

where *PCB* is the Si–O bond percentage change, bond $\text{number}_{\text{depth}=6.0\,\text{nm}}$ is the number of Si–O bonds at an indentation depth of 6 nm, and bond $\text{number}_{\text{depth}=0.0\,\text{nm}}$ is the number of Si–O bonds at an indentation depth of zero.

In the analysis of *PCNx* and *PCB* for both films, it was observed that the 0.8 nm film deformed more than the 2.0 nm film at a given indentation depth. After a brief transition phase, the 0.8 nm membrane ruptured at an indentation depth of about 5.5 nm, while the rupture of the 2.0 nm membrane occurred at a depth of 7.5 nm after sufficient deformation and thinning. This aligns with our previous conclusions that thicker membranes have the potential for further deformation and thinning.

During unloading, the elastic recovery of the SiO_2 film caused the Si–O atomic distance to decrease and new Si–O bonds to form. As a result, *PCN2* and *PCN3* decreased, while *PCN4*, *PCN5*, and *PCB* values increased. It is important to note that the *PCN5* and*PCB* values of the 2.0 nm membrane during unloading were higher than those during loading at the same indentation depth. This is because the 2.0 nm membrane experienced less plastic deformation and was more prone to elastic restitution.

In order to study the changes that occur in amorphous SiO_2 at the atomic level during deformation, a cluster of the material was extracted for analysis. The cluster is located directly beneath a spherical indenter. In the figure, oxygen atoms are represented in cyan, while the other atoms are silicon. Different Si–O rings are formed by the distribution of different-colored silicon atoms. The deformation evolution of SiO_2 during the indentation process is depicted in Figures 6.26 and 6.27. It was observed that as the indentation depth increased from 0.00 nm to 3.59 nm, the SiO_2 underwent compaction and thinning, even down to the thickness of a single atomic layer.

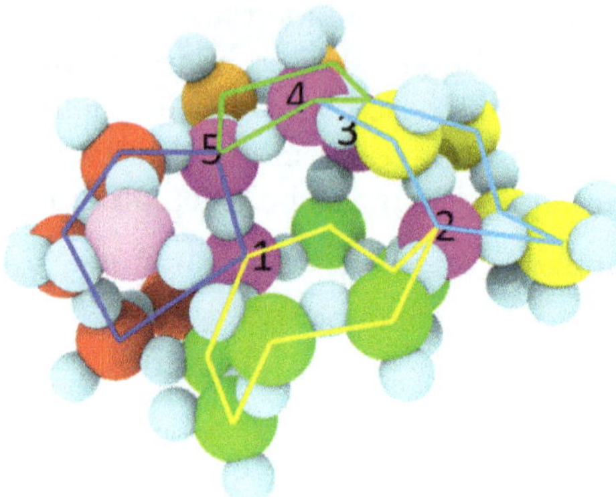

Figure 6.26. Diagram of Si–O ring in partial SiO_2.

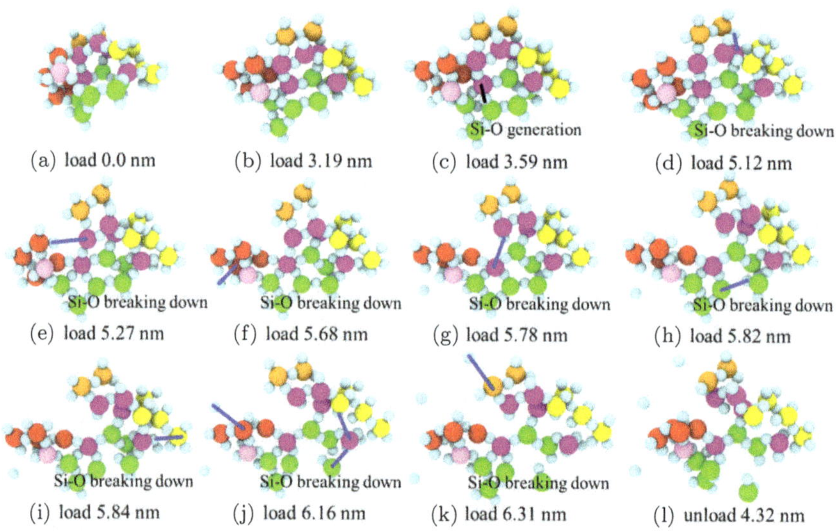

Figure 6.27. Partial SiO_2 deformation evolution (short black lines indicate bond generation, short blue lines indicate bond breakage). (a) Loaded, 0.0 nm, (b) loaded, 3.19 nm, (c) loaded, 3.59 nm, (d) loaded, 5.12 nm, (e) loaded, 5.27 nm, (f) loaded, 5.68 nm, (g) loaded, 5.78 nm, (h) loaded, 5.82 nm, (i) loaded, 5.84 nm, (j) loaded, 6.16 nm, (k) loaded, 4.32 nm, (l) unloaded, 4.32 nm.

The generation of Si–O bonds was observed during the indentation process. As the indentation continued, the Si–O bonds in different rings were broken. For example, at an indentation depth of 5.12 nm, the orange-colored Si–O ring was broken. Subsequently, at indentation depths of 5.27 nm and 5.78 nm, the red-colored Si–O ring was broken. The green-colored Si–O ring broke at depths of 5.82 nm and 6.16 nm, and the yellow Si–O ring broke at depths of 5.84 nm and 6.16 nm. This indicates that smaller Si–O rings broke and connected with surrounding Si–O rings to generate larger Si–O rings. As the loading continued, larger Si–O rings were generated, resulting in the formation of numerous nano-voids. Ultimately, this led to the rupture of the SiO_2 membrane. Therefore, the plastic deformation of amorphous SiO_2 under small indentation results in compaction and thinning, while larger indentations cause numerous Si–O bond breakages leading to the formation of larger Si–O rings and a decrease in the number of rings until the rupture of the SiO_2 membrane occurs. This is the characteristic behavior of the plastic deformation of the amorphous SiO_2 membrane.

The plastic deformation of non-crystalline materials can be explained by the free volume and shear transformation zones (STZ) models, which are based on the theory of non-crystalline plasticity [59–61]. When shear stress is applied, the STZ are the first to be easily initiated in areas with high free volume. This causes local twisting of the surrounding material and the spontaneous formation of shear bands, leading to the accumulation of free volume in the localized shear process. *In situ* experiments by Yavari *et al.* [62] found that the free volume in the deformed metallic glass bands was twice as high as in the as-cast state. However, theoretical studies indicate that the excess free volume is not stable. Therefore, thermodynamically, these excess free volumes merge to form nano-cavities in the shear band [63, 64]. On the other hand, voids, such as those formed by the rupture of SiO_2 in our simulations, reduce the stress required for further plastic deformation of the shear band, resulting in a reduction of material hardness [65], demonstrating the softening behavior of the non-crystalline material due to strain after compaction and thinning of the material. After the complete rupture of the SiO_2 film, the load starts to increase again due to the influence of the monocrystalline silicon substrate.

6.2.3.4 *Plastic deformation of monocrystalline silicon*

When a spherical indenter is pressed into the sample for testing, the amorphous SiO_2 film undergoes compaction, thinning, and even rupture. The detailed deformation behavior is described in the previous section. As the indentation continues, the high stress from the indentation is transmitted to the silicon matrix through the amorphous SiO_2 film, causing a phase transition in the silicon. This is the main mechanism of plastic deformation of single-crystal silicon under contact loading conditions. The SiO_2 film acts as a barrier to the phase transformation of single-crystal silicon, indicating that the silicon substrate under the thinner SiO_2 film is more susceptible to severe phase transformation, especially after the rupture of the SiO_2 film.

In order to compare the effect of an amorphous SiO_2 membrane on silicon phase transition with pure silicon, we analyzed the changes in silicon phase transition atomic numbers for both pure silicon and silicon substrate coated with a 0.8 nm SiO_2 membrane at different penetration depths. The penetration depth refers to the distance from the surface of pristine single-crystal silicon to the point of maximum depth of the indentation crater. In this study, the silicon substrate coated with the SiO_2 membrane is referred to as composite silicon.

As depicted in Figure 6.28, for both pure and composite silicon, the CN5 and CN6 silicon atom numbers increase with the increase in penetration depth. At the same penetration depth, when the indentation depth is less than 5.5 nm, the CN5 of pure silicon is slightly smaller than that of composite silicon. However, when the indentation depth equals 5.5 nm, the CN5 of pure silicon is significantly smaller than that of composite silicon. This difference is attributed to the fact that when the indentation depth exceeds 5.5 nm, the 0.8 nm

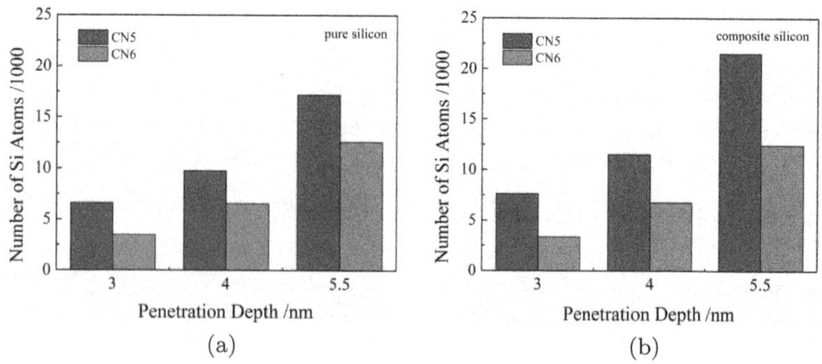

Figure 6.28. Variation of atomic number in phase transition of pure and composite silicon. (a) Pure silicon, (b) composite silicon.

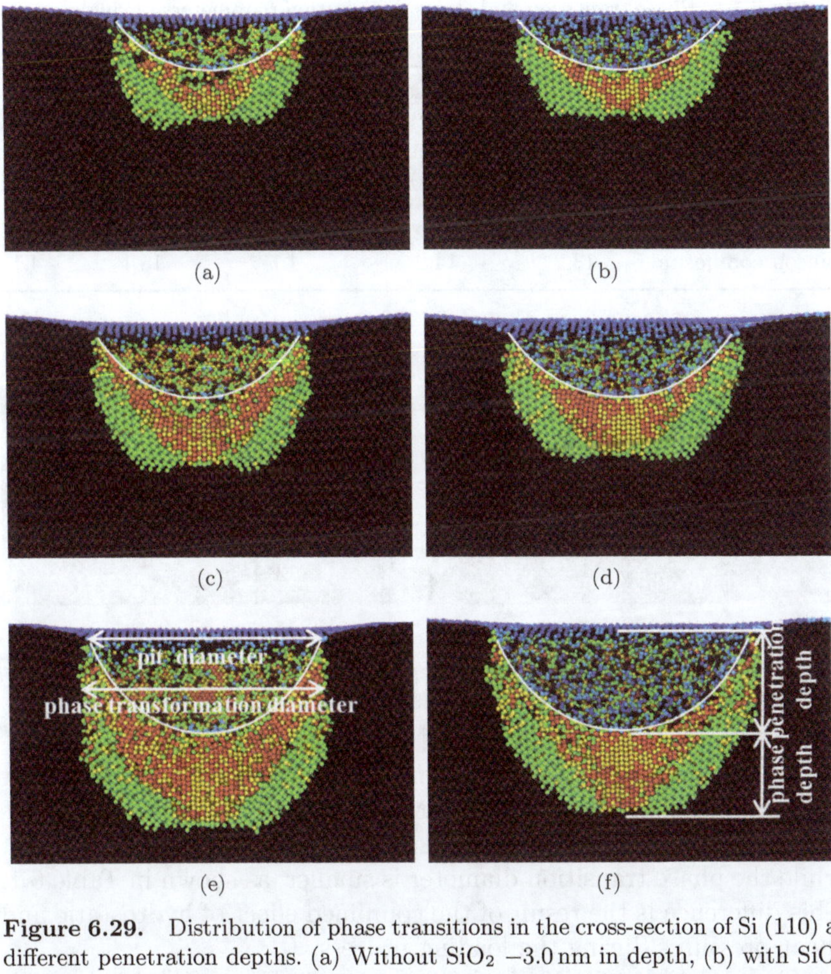

Figure 6.29. Distribution of phase transitions in the cross-section of Si (110) at different penetration depths. (a) Without SiO_2 −3.0 nm in depth, (b) with SiO_2 −3.0 nm in depth, (c) without SiO_2 −4.0 nm in depth, (d) with SiO_2 −4.0 nm in depth, (e) without SiO_2 −5.5 nm in depth, (f) with SiO_2 −5.5 nm in depth.

SiO_2 membrane ruptures, leading to an increase in force on the monocrystalline silicon and causing the CN6 value of pure silicon to be slightly larger than that of composite silicon.

The phase distribution diagrams of pure and composite silicon Si (110) cross-sections at different penetration depths are shown in Figure 6.29. In the figure, the burgundy, green, yellow, blue, and red atoms represent the Si-I, bct-5, and Si-II phases, surface atoms, and other phase atoms, respectively. It is observed that the diameter

Table 6.1. Phase transition and stress distribution parameters of single-crystal silicon at penetration depth of 5.5.

	Crater diameter (nm)	Phase transition diameter (nm)	Phase transition depth (nm)	Stress diameter (nm)	Stress depth (nm)
Pure silicon	12.2	13.2	5.3	12.8	4.9
Silicon composite	13.2	14.2	4.1	15.1	4.1

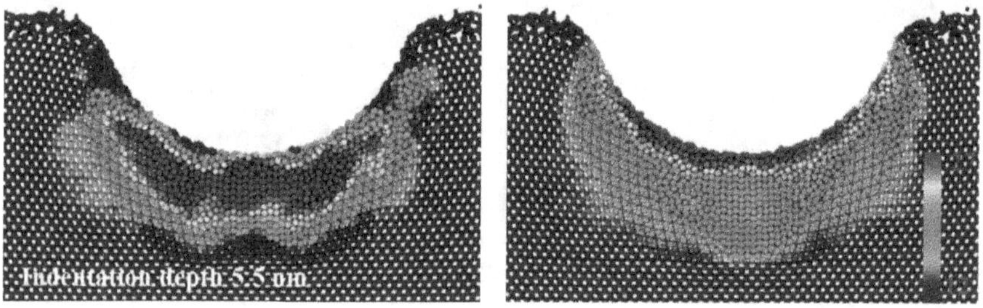

(a) (b)

Figure 6.30. Stress distribution in cross-section of 0.8 nm specimen (GPa). (a) Hydrostatic pressure, (b) von Mises stresses.

of the compression craters in composite silicon is larger than that in pure silicon at the same penetration depths. However, the phase change region of pure silicon is deeper than that of composite silicon, while the phase transition diameter is smaller, as shown in Table 6.1. This difference is the result of the combined effect of hydrostatic and shear pressures during the loading process.

The stress distribution of 0.8 nm specimens at an indentation depth of 5.5 nm is presented in Figure 6.30. The distribution indicates that the hydrostatic pressure is mainly concentrated directly below the spherical indenter, whereas the von Mises stresses are mainly distributed around the spherical indenter, with a broad range of distributions. Therefore, the following section focuses on the distribution of shear pressures. For the single-crystal silicon, the phase distribution is in line with the stress distribution. The Si-II phase is mainly generated under normal load, while bct-5 is generated under shear stress.

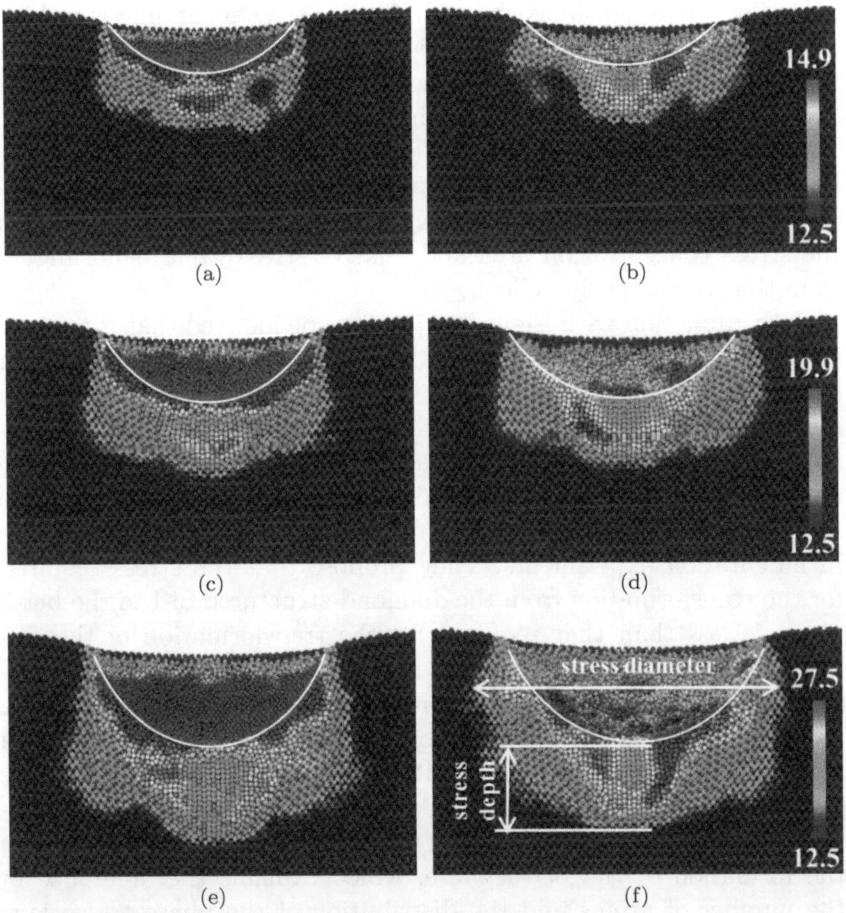

Figure 6.31. Von Mises stress distribution in S (110) cross-section (GPa). (a) Without SiO_2 −3.0 nm in depth, (b) with SiO_2 −3.0 nm in depth, (c) without SiO_2 −4.0 nm in depth, (d) with SiO_2 −4.0 nm in depth, (e) without SiO_2 −5.5 nm in depth, (f) with SiO_2 −5.5 nm in depth.

The von Mises stress distribution of single-crystal silicon at different penetration depths is shown in Figure 6.31. The shear stress increases with the increase of penetration depth. The stress of pure silicon extends mainly in the direction of the indentation (depth direction) and less in the horizontal direction (phase transition diameter). On the other hand, the stress of composite silicon is mainly around the abrasive, resulting in a larger stress diameter and a smaller stress depth, as shown in Table 6.1. This finding is in high agreement with the phase transition distribution of monocrystalline silicon, which is due to the higher hydrostatic pressure and lower shear stress directly below the spherical indenter.

The Si-II phase is generated under hydrostatic pressure, and the bct-5 phase is generated under shear stress. Additionally, stress concentration occurs in the area around the pure silicon indenter, with the stress concentration area and relative stress value being higher than that of composite silicon.

In conjunction with Figure 6.17, it can be inferred that the amorphous SiO_2 film helps dissipate the energy during the indentation process and transfers the stress from the indenter to the silicon matrix. It's important to note that the local stress in the specimen must exceed a specific threshold for the material to undergo a phase transition. Gerbig *et al.* [66, 67] provided sufficient evidence for the existence and possibility of the generation of the bct-5 phase in indentation experiments. They propose that the stress required for the transformation from the diamond-structured Si-I to the bct-5 phase is less than that required for the transformation of the Si-I phase to the Si-II phase. The theoretical stress required for the formation of the bct-5 phase in indentation experiments in the presence of shear stress is 8.5 GPa [68]. In our MD calculations, the minimum value of the von Mises stress is 12.5 GPa, and the maximum value of the stress increases from 14.9 to 27.5 GPa at different penetration depths. This is much higher than the theoretical pressure for the formation of the bct-5 phase, which explains the difference in the number of atoms and the distribution of the phase transitions between the pure and the composite silicon phases at different penetration depths. This is a reasonable explanation for the difference in the number and distribution of phase transition atoms between pure and composite silicon at different penetration depths [6].

6.2.4 *Effect of indentation speed on SiO_2/Si nano-materials*

6.2.4.1 *Modeling*

In order to study the impact of indentation speed on SiO_2/Si bilayer composites, we are focusing on the plastic deformation behavior of bilayer composites with a film thickness of 0.8 nm. The simulation uses displacement-controlled loading with identical loading and unloading speeds, and the indentation velocities are 10 m/s, 25 m/s, and 200 m/s. All other simulation details remain consistent with those previously described.

6.2.4.2 *Influence of indentation speed on the load–displacement curve*

The composite specimen load–displacement curves were obtained when different indentation speeds were used, as shown in Figure 6.32. It is found that when the indentation depth increases from 0.0 nm to about 5.5 nm, the indentation force is a monotonously increasing function of the indentation depth, and the curves at different indentation speeds almost overlap, indicating that the specimens are deformed in the same way, when the indentation depth increases from 5.5 nm to a maximum of 8.2 nm, the indentation force decreases to varying degrees according to the different indentation speeds, and then increases slowly. It can be seen that the unloading curves for different indentation speeds lagged far behind the loading curves, indicating that the specimens had undergone severe plastic deformation. Moreover, the load at which the indentation force decreases in the case of high-speed loading is smaller than that of low-speed

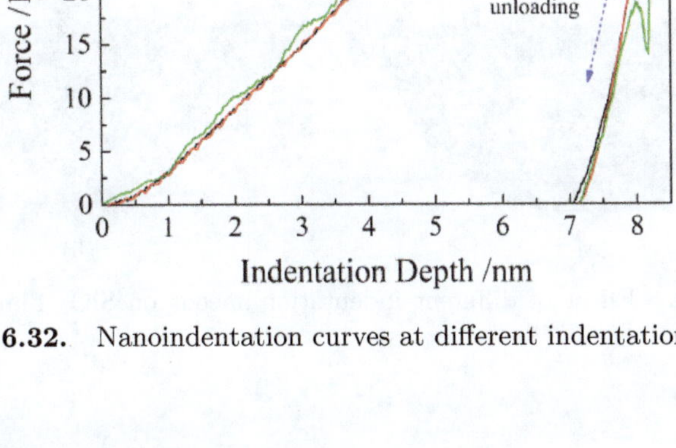

Figure 6.32. Nanoindentation curves at different indentation speeds.

indentation, which, combined with the findings of Section 6.2.3 in this chapter, indicates that early rupture behavior of the amorphous SiO_2 film occurs at higher indentation speeds and also indicates that the mechanical strength of the bilayer composites weakened with the increase of indentation speed, due to the fact that the strain energy increased with the increase of the indentation speed, which resulted in the increase of the defects of specimens in the process of nanoindentation that this in turn reduces the external force required for further deformation of the composite, so the specimen strength decreases. For the sake of simplicity and comparison, only the deformation behavior of amorphous SiO_2 film and silicon matrix at indentation velocities of 25 m/s and 200 m/s is analyzed and discussed.

6.2.4.3 *Effect of indentation speed on plastic deformation of amorphous SiO_2 film*

Figure 6.33 illustrates the cross-section of the SiO_2 membrane at different indentation speeds. It indicates that when the same indentation depth (5.4 nm) is applied, the SiO_2 membrane undergoes more deformation and ruptures under high-speed loading, while no rupture occurs under low-speed loading. This is due to the fact that the deformed atoms in the SiO_2 membrane do not have sufficient time for structural relaxation at higher loading speeds, leading to an increase in free volume and more severe material deformation.

In Sections 6.2.2 and 6.2.3, the plastic deformation analyses of the amorphous SiO_2 membranes led to the conclusion that the deformation during the indentation process showed four stages: compaction in loading, compaction–rupture transition, rupture, and elastic recovery in unloading. The compaction and thinning of the amorphous

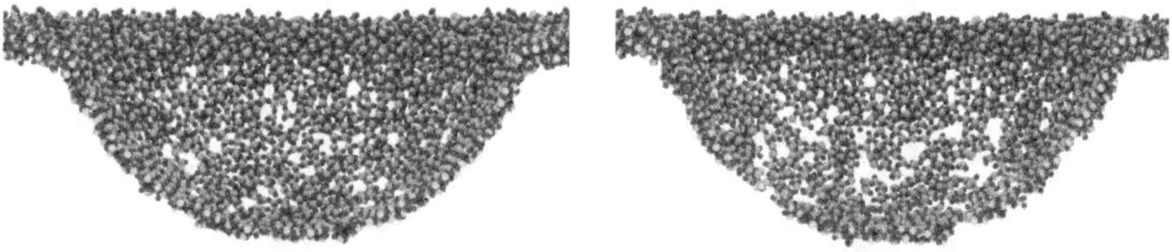

(a) (b)

Figure 6.33. Effect of different indentation speeds on SiO_2 film. (a) 25 m/s, (b) 200 m/s.

SiO_2 membrane occurred due to the deformation and rotation of a large number of SiO_4 tetrahedra at a small indentation depth. A large number of Si–O bonds in the stretched state were broken when the indentation depth increased, leading to the rupture of the deformed SiO_2 membrane. Additionally, the average distance of Si–O atoms became larger as the peak height decreased and the peak width increased during the compaction stage. During loading, an extra small peak appeared in the O–O and Si–Si RDFs, and the height and width of the extra peak increased with the increase of the indentation depth at shallow indentation depths, whereas the extra peak narrowed with the increase of the indentation depth at larger indentation depths. The deformation behavior of the amorphous SiO_2 film and silicon substrate at different indentation speeds was explored by analyzing the percentage change of RDF parameters (e.g., Si–O interatomic distance, Si–O peak height, Si–O half-peak full width, O–O, and Si–Si extra peak heights). The percentage change of parameters during the indentation process was obtained by the calculation of equation (6.8), and the results are shown in Figure 6.34:

$$PCD = \frac{D_{\text{depth}=6.0\,\text{nm}} - D_{\text{depth}=0.0\,\text{nm}}}{D_{\text{depth}=0.0\,\text{nm}}} \times 100\% \tag{6.8}$$

where PCD is the percentage average change in Si–O bond length, $D_{\text{dept}=6.0\,\text{nm}}$ is the average Si–O bond length when the indentation depth is 6.0 nm, and $D_{\text{dept}=0.0\,\text{nm}}$ is the average Si–O bond length when the indentation depth is zero.

The percentage change of Si–O bond length (PCD) and the full width at half maximum of the Si–O peak (PCW) showed slow increases during the compaction stage but significantly increased during the transition stage. Both parameters remained unchanged during the rupture stage. At the same indentation depth, the values of these parameters at 200 m/s were higher than those at 25 m/s. In contrast, the percentage change of height of the Si–O peak (PCH) decreased significantly during the compaction stage, decreased slowly during the transition stage, and remained almost constant during the rupture stage. Additionally, the peak height (PCH) values at 200 m/s were significantly lower than those at 25 m/s at the same depth.

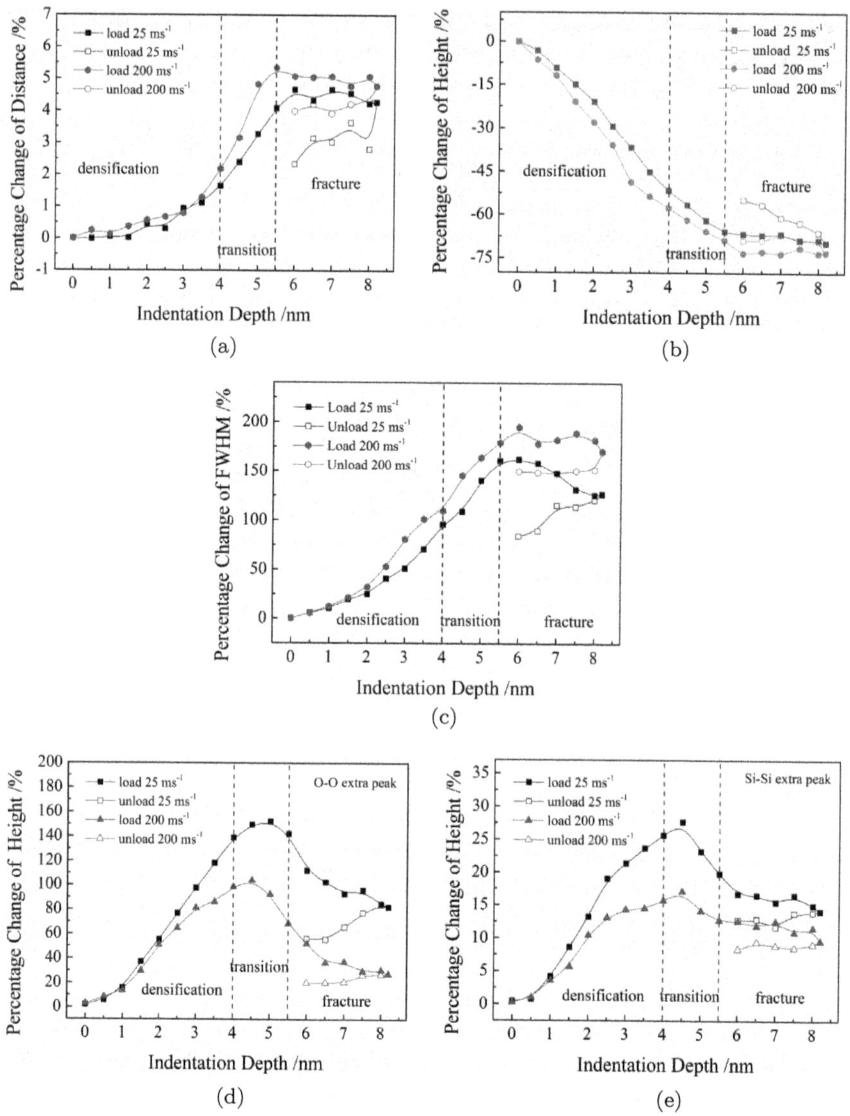

Figure 6.34. Percentage variation of SiO_2 parameters at different indentation speeds. (a) Mean Si–O atomic distance, *PCD*, (b) Si–O peak height, *PCH*, (c) Si–O peak half-height width, *PCW*, (d) O–O extra peak height, *PCHO*, (e) Si–Si extra peak height, *PCHSi*.

In the RDFs of O–O and Si–Si atoms, as shown in Figures 6.34(d–e), the percentage change of the height of O–O extra peak (*PCHO*) and the percentage change of the height of Si–Si extra peak (*PCHSi*) increase significantly with increasing indentation depth. The rate of increase of both peaks decreases slowly in the transition stage due to Si–O bond breaking. The percentage change of the height of Si–Si extra peak, *PCHSi*, increased significantly with the increase in indentation depth, while the rate of increase of both of them decreased slowly due to the breakage of the Si–O bond in the transition stage and even became negative at an indentation depth of 5.5 nm. After this, due to the rupture of the amorphous SiO_2 membrane, *PCHO* and *PCHSi* gradually decrease.

At an indentation speed of 25 m/s, the values of *PCHO* and *PCHSi* were higher than those at the same depth of 200 m/s, indicating that SiO_2 membranes experience a greater degree of compaction at low loading speeds. On unloading, due to the elastic recovery of the SiO_2 membrane, the above-mentioned parameters underwent partially opposite changes at slow unloading speeds, while the corresponding values remained essentially unchanged at high speeds.

To investigate the deformation characteristics of amorphous SiO_2 during nanoindentation at different indentation velocities, the atomic percentage change in the coordination number of silicon atoms within a truncation radius of 0.2 nm (*PCNx*) was analyzed. In the compaction stage, as depicted in Figure 6.35, *PCN2* and *PCN3* remain nearly zero as the indentation depth increases. Meanwhile, *PCN5* increases significantly, and *PCN4* decreases slowly. Interestingly, the value of *PCN4* at an indentation speed of 200 m/s is lower than that at 25 m/s for the same indentation depth, indicating compaction of SiO_2 and the generation of new Si–O bonds.

During the transitional stage,*PCN2* and *PCN3* start to increase suddenly, especially at an indentation speed of 200 m/s, with higher values under high-speed loading compared to low-speed loading. Additionally, the growth rate of *PCN5* differs at different indentation speeds, with a slower growth rate at 200 m/s and a faster rate at 25 m/s. At an indentation depth of 5.5 nm, the maximum value of *PCN5* at an indentation speed of 25 m/s is twice as high as that at 200 m/s. The sharp decrease of *PCN4*, with the value at 200 m/s being significantly lower than that at 25 m/s, suggests that a large number of bonds break during the transition phase, indicating that

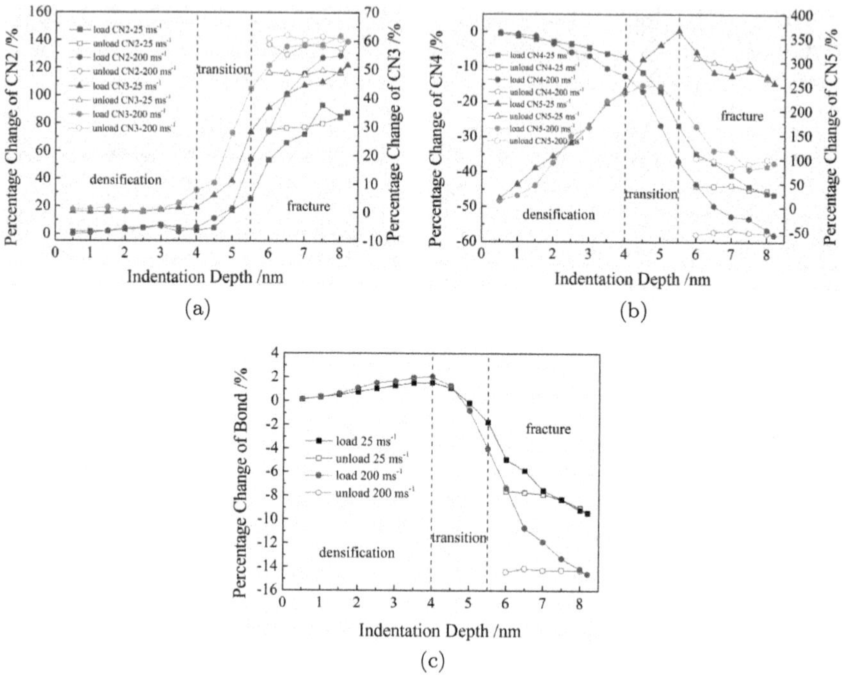

Figure 6.35. Percentage change in the number of Si coordination number atoms and Si–O bonds. (a) *PCN2* and *PCN3*, (b) *PCN4* and *PCN5*, (c) Si–O bonds.

a large number of chemical bonds are broken during this phase. Si–O bonds within the deformed SiO_4 tetrahedra break, resulting in a silicon atom being bonded to two or three oxygen atoms.

In the rupture phase, *PCN2* and *PCN3* increase significantly, with values at 200 m/s being significantly higher than the corresponding values at 25 m/s. Conversely, *PCN5* and *PCN4* decrease with increasing indentation depth, indicating severe plastic deformation of the SiO_2 film, especially at 200 m/s. During the final unloading, *PCNx* under a loading rate of 25 m/s undergoes local opposite deformation compared to the rupture stage, while the corresponding values under 200 m/s vibrate around a constant value.

In Figure 6.35(c), the curves for different indentation speeds almost overlap as the indentation depth increases during the compaction stage of the *PCB*. However, as the indentation depth continues to increase, some of the Si–O bonds break. When the increased number of Si–O bonds during the indentation process is unable to compensate for the reduced number of Si–O bonds, the total number of Si-Os decreases. As a result, the *PCB* value decreases slowly during the transition stage. During the rupture phase, the *PCB* decreases sharply, and the *PCB* values for the higher speed (200 m/s) are much smaller than those for 25 m/s at the same depth. This difference increases with the indentation depth, reaching up to 4.2% at the maximum indentation depth of 8.2 nm. During the final unloading, the *PCB* experiences a slight opposite change in elastic recovery when the indentation speed is 25 m/s. On the other hand, the *PCB* at 200 m/s maintains a nearly constant value. This happens because the amorphous SiO_2 undergoes significant deformation during high-speed indentation and loses its short-range ordered structural features. Basically, the original, more regular SiO_4 tetrahedra are flattened, and when the load keeps increasing, some of the Si–O bonds in the SiO_4 tetrahedra break and become SiO_3 or SiO_2. This is evidenced by the *RDF* results in Section 6.2.3, where the average distance of the Si–O bonds increases after the deformation. Additionally, the distribution of the O–Si–O bond angles becomes wider, the number of *CN4* atoms of silicon decreases, the number of *CN2* and *CN2* atoms increases, and the number of chemical bonds decreases. To sum it up, the changes in *PCNx* and *PCB* numbers indicate that the rupture of SiO_2 mainly occurs without sufficient compaction during high-speed indentation and that amorphous SiO_2 deforms much more extensively under high-speed loading compared to low-speed loading.

6.2.4.4 *Influence of indentation speed on the phase transition of single-crystal silicon*

When loading, the indenter is pressed downwards in the direction of the indentation into the single-crystal silicon specimen to be tested coated with amorphous SiO_2, which first induces plastic deformation of the SiO_2 film. As the indentation proceeds, the stress on the single-crystal silicon increases causing plastic deformation of the silicon. Since phase transformation is the main mechanism of plastic deformation of single-crystal silicon under contact loading conditions, we analyzed the phase transformation characteristics of silicon by coordination number (*CN*). In MD simulations, the Si-I phase of cubic diamond structure (Si-I phase with coordination number 4) transforms into the bct-5 (body-centered tetragonal, with coordination number 5) phase and Si-II (body-centered tetragonal β-tin, with coordination number 6) phase. Upon loading, when the indentation depth was greater than 1.5 nm, the phase transition atoms *CN5* and *CN6* of silicon began to increase and increased significantly with the increase of indentation depth, as shown in Figure 6.36, which indicates that not only the amorphous SiO_2 film played a role in preventing the phase transition of silicon but also the trends of the changes of *CN5* and *CN6* were slightly different, i.e., when the indentation depth was increased from 0.0 nm to 4.0 nm, the Si *CN5* change curves almost overlap, while when the indentation depth exceeds 4.0 nm, the amount of *CN5* at an indentation speed of 200 m/s significantly exceeds the value at the same depth at a lower speed. When the indentation depth exceeds 5.5 nm, the amount of *CN6* under high-speed loading is slightly lower than that under low speed.

The evolution of silicon's phase transition at different indentation depths is depicted in Figure 6.37. In the figure, burgundy, green, yellow, blue, and red atoms correspond to Si-I, bct-5, and Si-II phases, surface atoms, and other phase atoms, respectively. The phase transition region and the phase transition atoms (bct-5 and Si-II) visibly increase during the indentation process. The Si-II phase is observed

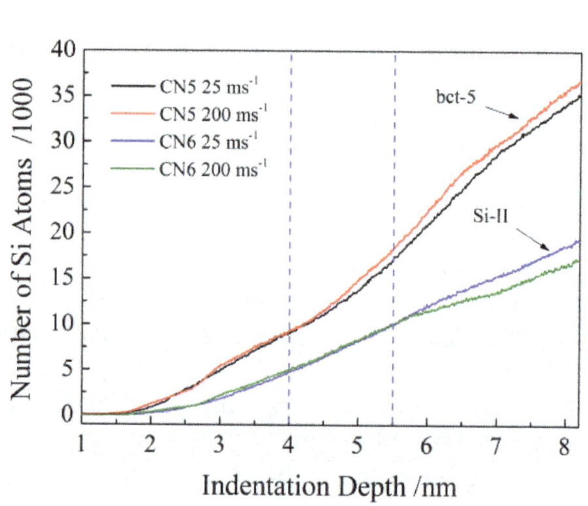

Figure 6.36. Atomic number curve for phase transition of Si.

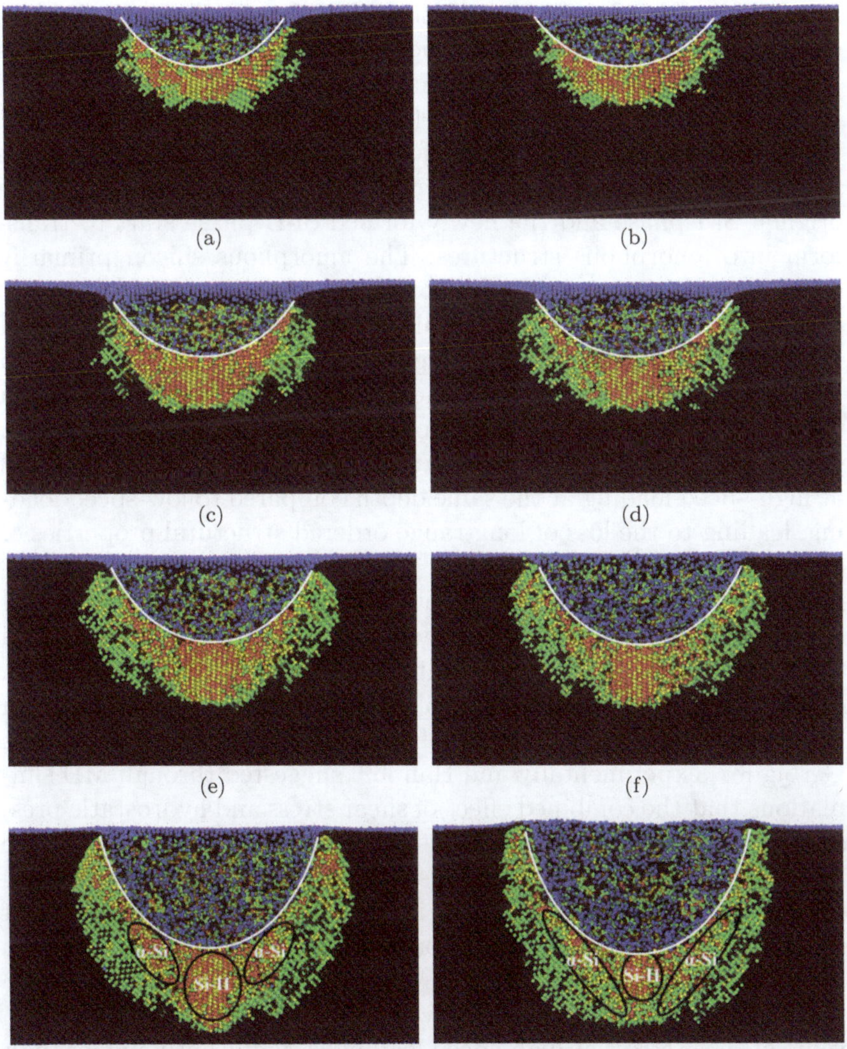

Figure 6.37. Distribution of phase transitions in the cross-section of Si (100) at different indentation depths. (a) 4.0 nm–25 m/s, (b) 4.0 nm–200 m/s, (c) 5.0 nm–25 m/s, (d) 5.0 nm–200 m/s, (e) 6.0 nm–25 m/s, (f) 6.0 nm–200 m/s, (g) 8.0 nm–25 m/s, (h) 8.0 nm–200 m/s.

near the surface of the silicon matrix at the center of the phase transition region, surrounded by the bct-5 phase. This observation aligns with the stress distribution characteristics of a spherical indenter.

At shallow indentation depths (from 0.0 nm to 4.0 nm), the phase transformation and amount of deformation are consistent across different indentation speeds. However, as the indentation continues, the original Si-I phase and the newly formed Si-II phase start to transform into amorphous structures. The amorphous silicon primarily consists of silicon atoms with coordination numbers of 4, 5, and 6, along with other atoms. This results in an increase in the number of *CN5*, explaining why there is a higher number of *CN5* in high-speed loading compared to low-speed loading at the same indentation depth when the depth reaches 4.0 nm to 8.2 nm. It can be observed from Figure 6.37 that the amorphous transition is much more pronounced at high-speed loading at the same depth compared to low-speed loading, leading to the loss of long-range ordered structural properties of crystalline silicon. This transition intensifies as the indentation progresses. For example, at the indentation depth of 8.0 nm, the area of the Si-II phase is significantly larger and more amorphous compared to the phase transition with a loading speed of 200 m/s versus 25 m/s.

Differences in the stresses applied to the silicon matrix at room temperature are the root cause of the different phase distributions. Gerbig [67] experimentally and Han [69] suggested through MD simulations that the combined effect of shear stress and hydrostatic pressure makes the crystal-amorphous transition easier. However, since the range of action of shear stress is greater than hydrostatic pressure in this simulation, the focus is on analyzing the distribution of shear stress. The stress distribution of Si (100) von Mises at different indentation depths is shown in Figure 6.38, which shows that the shear stress increases significantly with indentation depth, and the value of shear stress at high-speed loading is significantly higher than that at low speed for the same indentation depth. Interestingly, the sudden increase in shear stress at an indentation depth of 5.0 nm under high-speed loading conditions is caused by the fracture of the heavily stretched Si–O bonds. As mentioned earlier, the SiO_2 film undergoes a compaction–breakage transition when the indentation depth increases from 4.0 nm to 5.5 nm, and the SiO_2 film acts as a medium for dissipating energy and transferring stress, which impedes

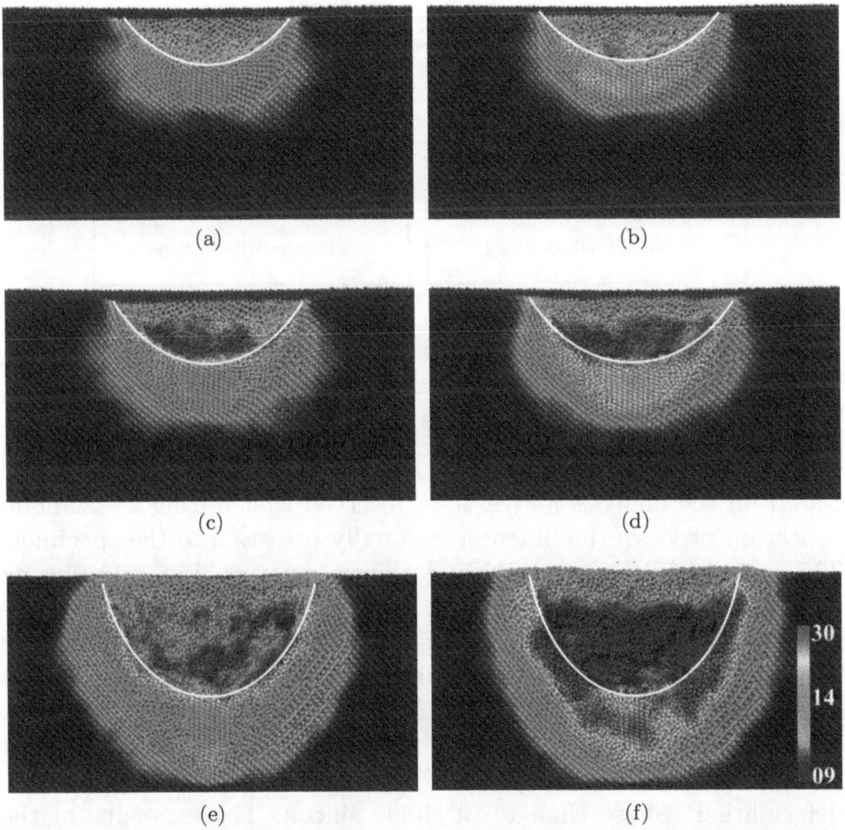

Figure 6.38. Von Mises stress distribution in Si (100) cross-section (GPa). (a) 4.0 nm–25 m/s, (b) 4.0 nm–200 m/s, (c) 5.0 nm–25 m/s, (d) 5.0 nm–200 m/s, (e) 8.0 nm–25 m/s, (f) 8.0 nm–200 m/s.

the phase transformation of silicon, so once the SiO_2 film starts to break up, the stress suffered by the silicon substrate increases suddenly, which rationally explains the reason for the difference in the distribution of silicon phases at different indentation speeds. In addition, the value of silicon shear stress under high-speed loading is around 30 GPa, while the local shear stress at low speed also exceeds 24 GPa, which is the theoretical transition stress of diamond-type Si-I phase to amorphous silicon [8] that reveals the reason for the generation of more amorphous silicon under high-speed loading conditions.

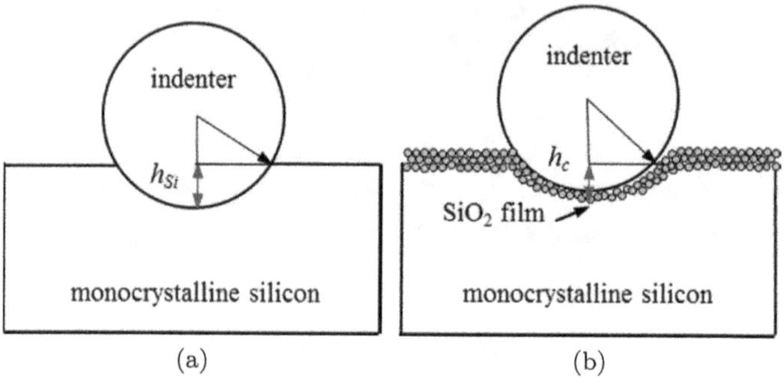

Figure 6.39. Nanoindentation model diagram. (a) Without SiO_2 membrane, (b) with SiO_2 membrane.

6.2.4.5 *Indentation models*

Based on the analysis above, it is observed that during the nanoindentation process, the indenter gradually presses into the specimen. The surface of the single-crystal silicon deforms first, forming an amorphous SiO_2 membrane. As a result of the dissipation of energy and stress transfer in the process of indentation, the indentation depth (h_c) of a composite specimen of monocrystalline silicon is smaller than the depth of indentation (h_{Si}) of the pure monocrystalline silicon substrate, as shown in Figure 6.39. This indicates that the presence of the amorphous SiO_2 membrane can inhibit plastic deformation (phase change) of single silicon. The existence of the amorphous SiO_2 film can inhibit the occurrence of plastic deformation (phase transition) of single-crystal silicon, which provides a physical basis for chemical mechanical polishing to obtain a smooth surface of single-crystal silicon with few or no defects.

6.2.5 *Summary*

We utilized a MD simulation method to investigate the nanoindentation behavior of composite specimens consisting of an amorphous SiO_2 film and monocrystalline silicon bilayer. We calculated the mechanical properties of these composite specimens with varying indentation depths and film thicknesses and analyzed the evolution of plastic deformation during the nanoindentation process. The primary findings of the study are as follows:

(1) The modulus of nanoindentation increases as the depth of indentation increases. However, when the indentation depth is the same, the modulus of the specimen decreases as the thickness of the amorphous SiO_2 film increases. For indentation depths less than 5.6 nm, the hardness of the composites increases with increasing depth, but when the depth is increased from 5.6 nm to 8.2 nm, the hardness first decreases and then increases. Additionally, when the indentation depth is constant, the plastic deformation of amorphous SiO_2 has a greater impact on the hardness of the composite specimens.
(2) Furthermore, changes in the coordination number (*CN*) and the number of Si–O bonds within the amorphous SiO_2 during the indentation process confirm four stages of deformation: compaction in loading, compaction–breakage transition, rupture, and elastic recovery in unloading. When the indentation is small, the SiO_2 membrane plastically deforms into compaction and thinning. As the indentation depth increases, a large number of Si–O bonds break, leading to the enlargement of the Si–O ring and the generation of numerous nanoholes. Eventually, the nanoholes merge, resulting in the rupture of the SiO_2 membrane.
(3) When there is a SiO_2 film on the surface of single-crystal silicon, it prevents the phase transition of the silicon. The thinner the film, the easier the phase transition of silicon occurs. Monocrystalline silicon in the pristine Si-I phase during loading changes into bct-5 and β-Si phases, and the number of phase transition atoms generated decreases as the film thickness increases at the same indentation depth.
(4) A higher nanoindentation speed (200 m/s) decreases the mechanical properties of the composites. The percentage change in the coordination number (*CN*) of Si atoms and the number of Si–O bonds of the SiO_2 film shows that the rupture of the amorphous SiO_2 film occurs more readily before sufficient compaction, and the degree of deformation is much larger compared to low-speed indentation at the same depth.
(5) The strength of the SiO_2/Si interface was analyzed using sandwich-type uniaxial tensile calculations, and it showed that the interfacial strength was high. The interface did not significantly affect the fracture mode of the silicon specimens, as the

interface between amorphous SiO_2 and Si was tightly bound together by a large number of chemical bonds.

6.3 Creep and Stress Relaxation Behavior of SiO_2/Si Bilayer Nanomaterials

6.3.1 *Introduction*

When nanoindentation was used to test the mechanical properties of thin film specimens, it was observed that under constant load from the indenter, the specimen would slowly deform and the depth of the nanoindentation would increase. This long-time loading had an impact on the material's mechanical properties and led to plastic deformation. In MEMS devices, amorphous SiO_2 is used as a dielectric material covering the surface of single-crystal silicon, causing creep in the device material after prolonged contact loading.

To accurately predict material properties and understand the deformation behavior of materials, it is crucial to investigate the time-dependent changes in properties such as creep and stress relaxation in bilayer composites. Currently, the creep properties of monocrystalline silicon materials at high temperatures and low stresses are obtained using uniaxial tensile and bending tests in the temperature range of 800–1300°C and stresses of 2–150 MPa. The creep mechanism found in these conditions is the slip of dislocations [70]. However, at room temperature, nanoindentation creep experiments revealed that the main creep mechanisms are phase transition and dislocation mechanisms. In their research, Gerbig *et al.* [71] studied the creep behavior of single-crystal silicon at room temperature using nanoindentation. They observed that the Si-II phase, formed during loading, transforms into Si-III and Si-XII phases during the creep phase. When it comes to amorphous materials, the creep mechanism involves the diffusion of atoms when the indentation is small. The deformation mechanism can be explained by the free volume and *STZ* models, based on the amorphous plasticity theory of Spaepen and Argon *et al.* Cao [72] investigated the nanoindentation creep behavior of laser-deposited SiO_2 films. They found that the stress index versus indentation size values were more influenced in the pristine deposited specimens with larger defect densities and lower *STZ* shear resistances compared to less defective specimens in the annealed state. Descratere extensive research on amorphous materials and crystalline silicon, the nanoindentation creep behavior of confined thin films and bilayer composites, especially the atomic scale mechanism, has been little investigated due to experimental constraints.

This section focuses on discussing the creep and stress relaxation behavior of nanoindentation of amorphous SiO_2/monocrystalline silicon bilayer composites in the context of the *CMP* process. Both creep and stress relaxation studies examine the time-dependent plastic deformation of materials. In creep tests, the load is held constant, and the indenter displacement is monitored when the load reaches a target value. In stress relaxation tests, the indenter position is kept constant, and the indentation force applied to the indenter is monitored. However, it is often challenging to keep the indenter position constant, so creep testing is more commonly used than stress relaxation. This section aims to investigate the creep of conical nanoindentation with a self-similar shape and analyze the creep properties and deformation behavior of the specimens.

Additionally, the present authors qualitatively analyze the stress relaxation behavior of materials with spherical indenters, relevant to chemical mechanical polishing and MEMS microcontacts, by utilizing MD methods. The focus of the work is to explore the effects of different loads on the creep behavior of bilayer composites and different indentation depths on the stress relaxation behavior. It also aims to analyze the time-dependent plastic deformation of nanoscale amorphous SiO_2 films and single-crystal silicon, providing a new perspective for the creep analysis of amorphous/crystalline composites.

6.3.2 *Creep behavior*

6.3.2.1 *Modeling*

The nanoindentation creep simulation system includes a conical indenter and a bilayer matrix consisting of an amorphous SiO_2 film and a monocrystalline silicon matrix, as shown in Figure 6.40. The monocrystalline silicon matrix has crystallographic orientations of [100], [010], and [001] on the X, Y, and Z axes, respectively. The amorphous SiO_2 was obtained by quenching and melting crystalline SiO_2, which was used as an input file in the MD simulations.

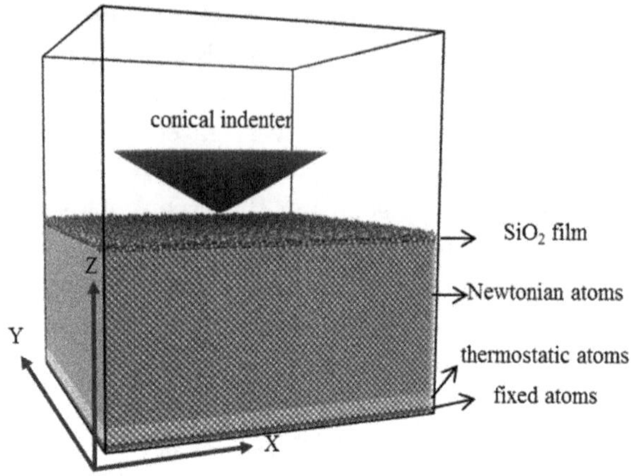

Figure 6.40. Schematic diagram of indentation creep model.

The composite specimen was divided into three parts based on the type of atoms: fixed layer, hot bath layer, and Newtonian layer atoms. The model has periodic boundaries in the X and Y directions and free boundaries in the Z direction. The specific parameters of the model and the potential functions used during the simulation are detailed in Table 6.2.

The creep simulation involved four stages: relaxation, loading, holding pressure, and unloading. Initially, the simulated system underwent local energy minimization using the steepest descent algorithm. This was followed by a relaxation period of 60 picoseconds under an NVT ensemble with a Nosé–Hoover heat bath to reach equilibrium. Once equilibrium was achieved, the conical indenter was loaded to the target load in the Z direction at a rate of 8.125 nN/ps. It was then held at this load for 200 picoseconds before being unloaded at the same rate in the Z direction. The indentation simulation utilized the NVE ensemble with a Langevin heat bath throughout.

6.3.2.2 *Creep property analysis and stress relaxation behavior of SiO_2/Si bilayer nanomaterials*

The conical indenter for each specimen was loaded vertically to three target creep loads (480 nN, 2430 nN, and 4080 nN) at a loading rate of 8.125 nN/ps. The load was held at this level for 200 ps and then unloaded. The specific creep simulation methodology can be

Table 6.2. Simulation parameters.

Element	Parameters
Amorphous SiO_2 film (nm)	$30 \times 30 \times 0.8$ (47040) $30 \times 30 \times 1.0$ (58996) $30 \times 30 \times 1.4$ (83104) $30 \times 30 \times 2.0$ (119168)
Monocrystalline silicon (nm)	$30 \times 30 \times 25$ (1130685)
Conical indenter (nm)	Half angle of 60° (121, 300)
SiO_2/Si potential function	Extended Si–O Tersoff potential function
C–Si potential function	Morse potential function ($D_0 = 0.435\,\text{eV}$, $\alpha = 4.6487\,\text{Å}^{-1}$, $r_0 = 1.9475\,\text{Å}$)
C–O potential function	L-J potential function ($\varepsilon = 0.1\,\text{eV}$, $\sigma = 3.275\,\text{Å}$)
Operating temperature (K)	300
Load-offload rate (nN–$(\text{ps})^{-1}$)	8.125
Time step (fs)	0.5
Fixed layer thickness (nm)	1
Heat bath thickness (nm)	1.5

found in Figure 6.41, which illustrates the specimen undergoing creep while the indenter continues to press into it, causing an increase in the depth of the indentation. The corresponding creep data were recorded. For instance, Figure 6.42(a) shows the change in indentation depth with creep time for a 0.8 nm specimen, indicating a sharp increase in indentation depth initially, followed by a steady increase in the later stages of creep.

To investigate the creep behavior of the specimen at the nanoscale, an empirical formula was used. For a self-similar indenter, such as the conical indenter described in this chapter, the indentation strain rate can be expressed as follows:

$$\dot{\varepsilon}_I = \frac{1}{h}\frac{dh}{dt} \tag{6.9}$$

where h is the indentation depth/nm and t is the creep time/ps. dh/dt is obtained by fitting the empirical equation to the creep displacement, using the nano-scale creep displacement equation of Wang *et al.* [73], as shown in following equation:

$$h(t) = h_0 + a \times (t - t_0)^{0.5} + b \times (t - t_0)^{0.25} + c \times (t - t_0)^{0.125} \tag{6.10}$$

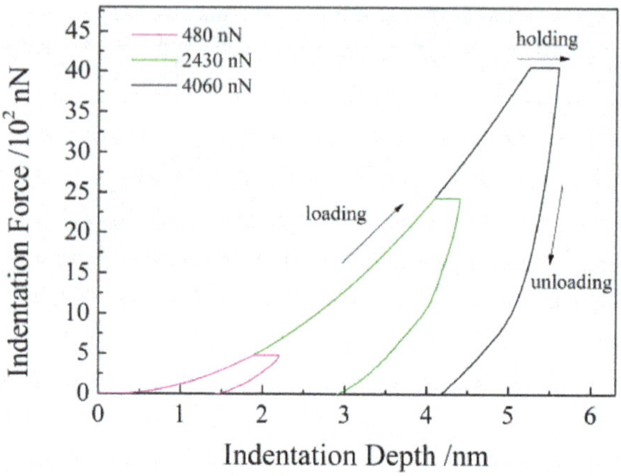

Figure 6.41. Load–displacement curve under different creep loads.

where h_0 is the starting indentation depth/nm, a, b, c are fitting constants, and t_0 is the creep onset time/ps. As shown in Figures 6.42(a–b), Equation (6.11) can fit most of the simulated data very well and capture all trends in the data. Therefore, equation (6.11) can be used to fit the creep curve of the specimen.

After analyzing the data, we found that the creep displacement of the indentation increases rapidly with creep time and then rises steadily, with larger final creep displacement under higher creep loads. For instance, in Figure 6.42(b), the creep displacements are 0.27 Å and 1.06 Å under 480 nN and 4060 nN, respectively, for the 0.8 nm specimen. The final creep displacements of specimens under different creep loads are shown in Figure 6.42(c), revealing a significant increase with higher creep loads. Additionally, we observed a slight decrease in creep displacement with the increase in SiO_2 film thickness of the specimen at the same creep load.

The stress on the specimen under different creep loads can be determined using the formula $H = P/A$, where P represents the load during creep and A is the instantaneous projected contact area. It is evident that the stress decreases rapidly at first and then stabilizes as the creep time increases. Additionally, the stress on the specimen rises with the creep load. For instance, the stress change of a 0.8 nm specimen under various creep loads is illustrated in Figure 6.43(a). This change in stress is also evident in

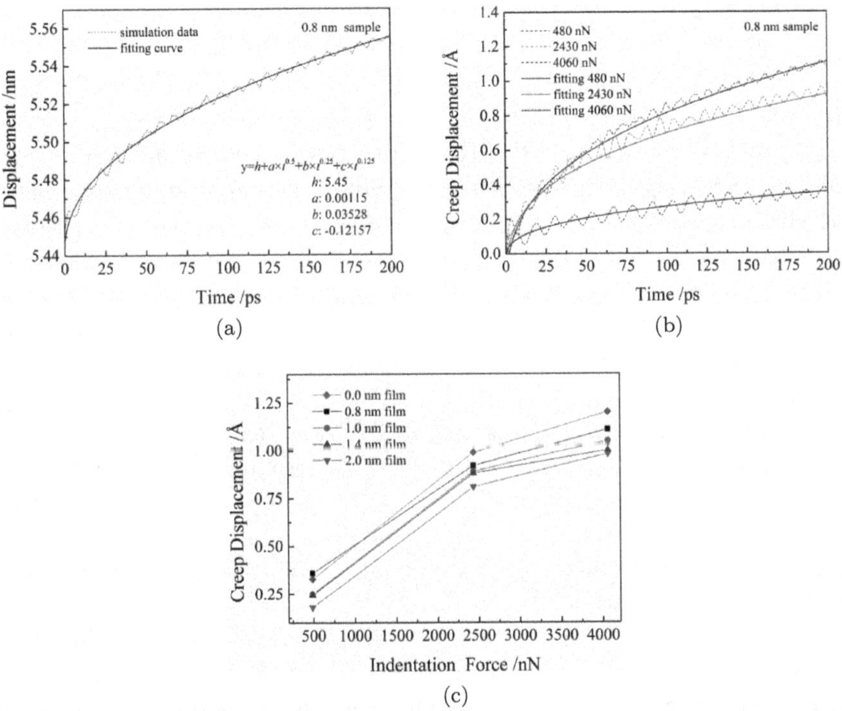

Figure 6.42. Displacement change during creep process for 0.8 nm SiO_2 film thickness. (a) Change in indentation depth, (b) change in creep displacement, (c) change in creep displacement.

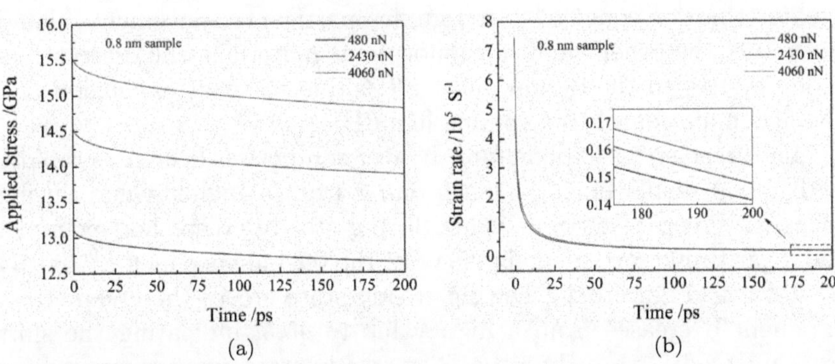

Figure 6.43. Applied stress and strain rate changes during creep of 0.8 nm specimens. (a) Applied stress, (b) strain rate.

nanoindentations of metallic glasses with indentation depths of less than 50 nm, which is attributed to stronger surface effects [74].

The rate of strain during creep can be determined using equation (6.10). Figure 6.43(b) illustrates the strain rate of the 0.8 nm composite specimen under various creep loads. It shows that the strain rate at the beginning of the creep is very high, but as creep progresses, the strain rate in the later stages rapidly decreases to a low value. Moreover, with an increase in creep time, the strain rate remains relatively stable, indicating that the creep has nearly entered the steady-state stage. The inserted magnified figure demonstrates that the specimen exhibits a higher creep rate under higher creep loads, consistent with the trend of creep displacement. Similar changes are observed in the nanoindentation creep of amorphous Ta film. It is worth noting that the strain rates in the simulations are approximately 10^5, which is several orders of magnitude larger than the values around 10^{-2} in the indentation experiments [75]. This discrepancy is due to the intrinsic nature of MD simulations, involving small deformation volumes and very fast loading rates.

The creep stress index can be obtained by determining the slope of the double logarithmic stress–strain curve, i.e., $n = \partial(\log \dot{\varepsilon}_I)/\partial(\log \bar{\sigma})$, providing information on time-dependent plastic deformation. For crystalline materials at the steady state stage, the creep stress index can reveal the creep mechanism. When $n = 1$, it indicates Cobal creep, when $n = 2$, it points to grain boundary slip, when n ranges from 3 to 7, it denotes dislocation slip, and when n is greater than or equal to 8, it signifies particle enhancement-related creep. On the other hand, for amorphous materials, the creep stress index relates to the operation of *STZs* due to shear expansion and the accumulation of free volume [60, 61].

The creep stress exponent for bilayer composites decreases rapidly with creep displacement, transitioning from a higher value during transient creep to a lower value during steady creep. For example, the logarithmic stress–strain curve of the 0.8 nm specimen at 480 nN is shown in Figure 6.44. During steady-state creep, the creep stress exponent decreases from a high value to 45.5, indicating the shift from unsteady to steady creep. The steady-state creep stress indices of different specimens are presented in Figure 6.45, showing that the general trend is for n to decrease with an increase in maximum indentation force. In addition, n decreases slightly with increasing SiO_2

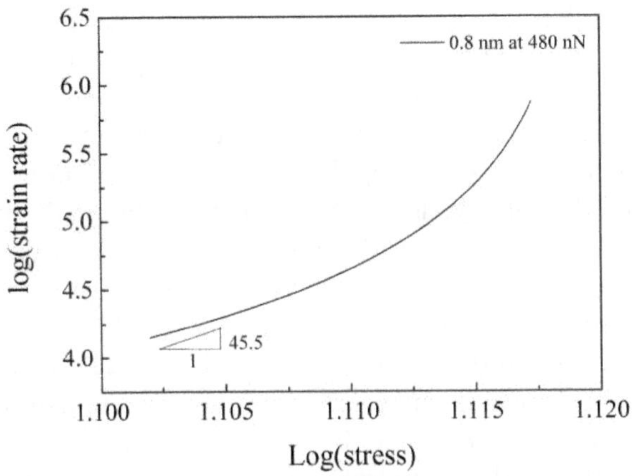

Figure 6.44. Curve of logarithmic variation of creep stress and strain rate.

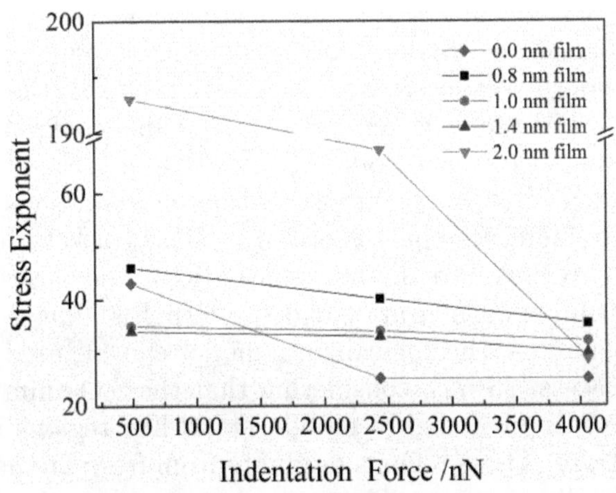

Figure 6.45. Variation of creep stress index under different creep loads.

film thickness at the same indentation force, except for the 2.0 nm specimen at 480 nN. Furthermore, the physical strain rate sensitivity index ($m = 1/n$) reflects the material's resistance to localized deformation during stable creep. Higher values of m (and smaller values of n) indicate relatively uniform creep, suggesting more uniform material deformation at higher creep loads for a specific specimen.

For amorphous materials, Argon suggests that the STZ operates in the elastic confines of the surrounding material. It moves from a position of relatively low energy to another position of low energy, and the shear expansion due to deformation redistributes stress and strain in the region around the *STZ* [61]. For amorphous SiO_2 films, the first *STZ* is initiated from defects already present in the original material, such as weaker Si–O bonding, dangling bonds, and broken SiO_3 bonds. The yield strength of the material can easily be exceeded by the application of stress, and the initiation of the first *STZ* will result in a free-volume expansion of the surrounding area. The accumulation of local strain in the material and an increase in free volume can compensate for shear strain at large scales. This can lead to plastic deformation in metallic glass or amorphous SiO_2. In a composite specimen that has been loaded, the plastic deformation of the SiO_2 film and the increase in free volume both increase as the load increases. This is in line with Spaepen's work [60], which suggests that the rate of free volume generation is proportional to the product of stress and strain. Therefore, before creep, the free volume density generated under a load of 4060 nN is significantly higher than that under lower loads. So, the higher the load, the greater the increase in free volume. When the load remains constant, a thicker film composite specimen is softer and leads to a greater indentation depth, which involves more atoms in the deformation process and, consequently, increases the free volume. The thicker the film, the more atoms are involved in the deformation. The plastic deformation of a 2.0 nm specimen at a load of 480 nN is almost exclusively in the SiO_2 film. A rheological unit model can be used to explain the creep process [76, 77], considering that the SiO_2 film consists of numerous low-density, highly mobile liquid-like regions distributed on the substrate. These regions originate from free volume generated during the loading process. The free volume generated at high loads is uniformly distributed over the deformed region, allowing a large number of regions to start and operate simultaneously at different locations. This results in relatively homogeneous rheology during the subsequent creep, reducing the creep stress index.

The increased net free volume and loading distribution can alter the activation energy and activation volume of *STZ* in SiO_2 membranes. Based on Johnson–Samwer's Collaborative Shear Model (*CSM*), Pan *et al.* [78] systematically characterized the *STZ* of bulk metallic glass using the rate-jump method. The activation volume of the *STZ* can be expressed as a function of the

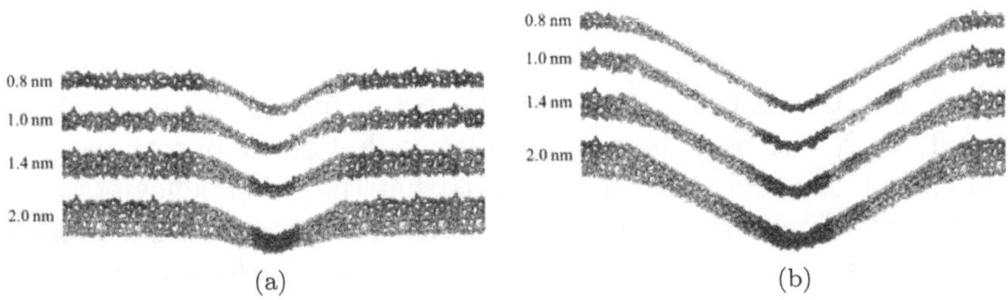

Figure 6.46. Shear stress distribution of SiO_2 membrane under different creep loads. (a) 480 nN, (b) 4060 nN.

Boltzmann constant (k), temperature (T), shear stress (σ), and the creep strain rate sensitivity index (m), i.e., $\Delta V^* = \frac{kT}{m\tau}$. In this chapter, the shear stress of the compressed SiO_2 film is calculated using von Mises and shown in Figure 6.46. It is observed that the Mises stress and τ increase with the creep load. Additionally, at the same creep load, the stress concentration area and τ increase with the thickness of the film. The creep stress index decreases as the creep load increases. When the creep load is constant, the creep stress index of the thick film composite specimen is slightly lower. Based on the *STZ* activation volume calculation formula, it is concluded that the *STZ* start activation volume of the SiO_2 film is lower when the creep load is higher, and the corresponding activation energy is also lower. Therefore, the *STZ* in the deformed SiO_2 under a high creep load and thick membrane is more likely to activate and operate during creep. In summary, it can be inferred from the analysis above that the plastic deformation of SiO_2 film during creep initiates from numerous potentials with lower potential energy due to the increase of free volume and uniform distribution of the free volume in the deformation craters, leading to more homogeneous rheology and a lower n-value of the SiO_2 film with the increase of creep load or SiO_2 film thickness. Additionally, by observing the creep displacement curves in Figure 6.42(b), the curves of the 0.8 nm specimens appear smoother than those of 480 nN at a load of 4080 nN, displaying less serrated behavior and more uniform deformation.

The n value for pure monocrystalline silicon specimens decreases as the creep load increases. The n value stabilizes at a higher load of 2430 nN, as shown in Figure 6.45, indicating that the creep mechanism remains consistent at different creep loads. This aligns with the phase change of silicon being the primary plastic deformation mechanism at room temperature. Furthermore, as the creep load increases, the creep stress index of the composite specimen decreases significantly and becomes more similar to that of the pure monocrystalline silicon specimen. This suggests that the creep behavior of the monocrystalline silicon matrix becomes dominant at higher creep loads. However, at low creep loads, the plastic deformation of the amorphous SiO_2 film is the primary creep mechanism, for example, at 480 nN for the composite specimen with 2.0 nm film thickness.

6.3.2.3 *Plastic deformation behavior in creep*

To study the behavior of amorphous SiO_2 film under different loads, the SiO_2 film with a radius of 8.0 nm and a thickness of 0.8 nm directly under the indenter is analyzed to explore its deformation characteristics. The nearest neighbor RDF with a truncated radius of 0.2 nm is used for analysis. As depicted in Figure 6.47(a), when the creep load is increased, the Si–O peak shifts to the right. The peak height decreases but the peak width significantly increases, indicating that the average distance between Si–O atoms becomes larger at higher loads. Additionally, with increasing loads, extra peaks appear on the *RDF* curves of O–O and Si–Si atom pairs at distances of about 0.20 and 0.28 nm, respectively, as shown in Figures 6.47(b–c). The width and height of the extra peak increase with the load, while the O–O and Si–Si main peaks decrease accordingly, suggesting that some of the O–O and Si–Si atomic distances decrease due to the compaction and thinning of the amorphous SiO_2 film. Based on the analyses, it can be inferred that the SiO_2 film becomes more disordered after deformation under high loads, leading to larger plastic deformation. According to the results in Section 6.2.3, the SiO_2 film with a thickness of 2.0 nm also displays similar deformation behavior, albeit to a slightly smaller degree.

In previous studies involving spherical nanoindentations of SiO_2/Si bilayer composites, it was observed that the deformation of the amorphous SiO_2 film during the loading process can be divided into three phases: compaction, transition, and fracture. Initially, compaction occurs due to the rotation and deformation of the numerous SiO_4 tetrahedra when the indentation is shallow. Subsequently, during further loading, the SiO_2 membrane fractures as a result of the

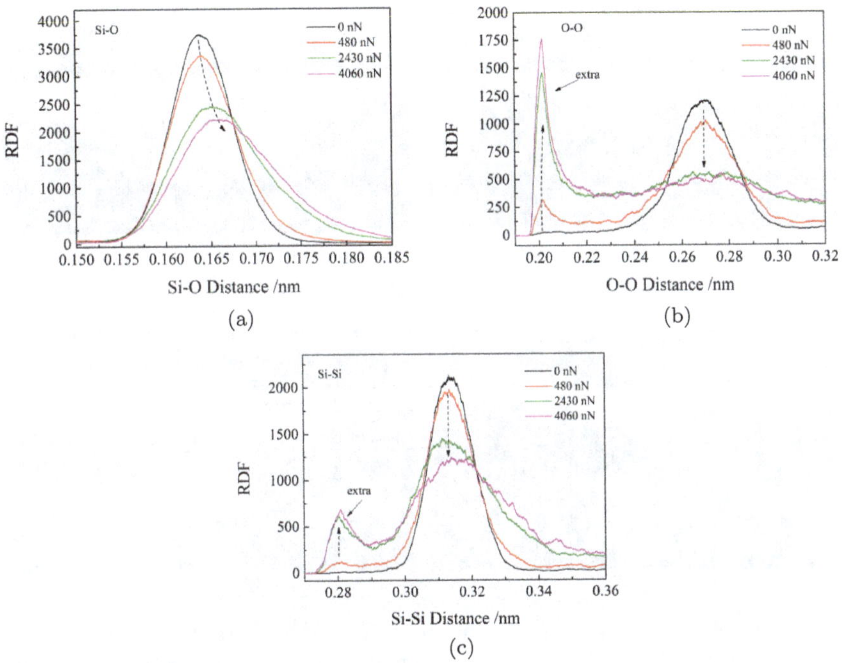

Figure 6.47. RDF of 0.8 nm SiO_2 film at different loads. (a) Si–O, (b) O–O, (c) Si–Si.

heavily stretched Si–O bonds breaking. In the case of conical nanoindentations, the specific evolution of the composites before and after deformation at different creep loads is depicted in Figures 6.48. It was noted that for a particular SiO_2 film, the ratio of before/after was lower at a high creep load (4080 nN) compared to a low load (480 nN). Additionally, at the same creep load, the before/after ratio for the 2.0 nm specimen was lower than that for the 0.8 nm specimen, suggesting that the deformation of amorphous SiO_2 films increases with both creep load and decreasing film thickness.

To analyze the time-dependent deformation characteristics of the SiO_2 film during creep, we can calculate the atomic percentage change in the coordination number of Si (abbreviated as *PCNx*) using Eq. (6.11). Additionally, the percentage change in the number of Si–O bonds (abbreviated as *PCB*) can be used to further confirm the deformation behavior of the SiO_2. The value of *PCB* can be obtained by using Eq. (6.12). In the equations, *CN4* atom number$_{\text{as-prepared}}$

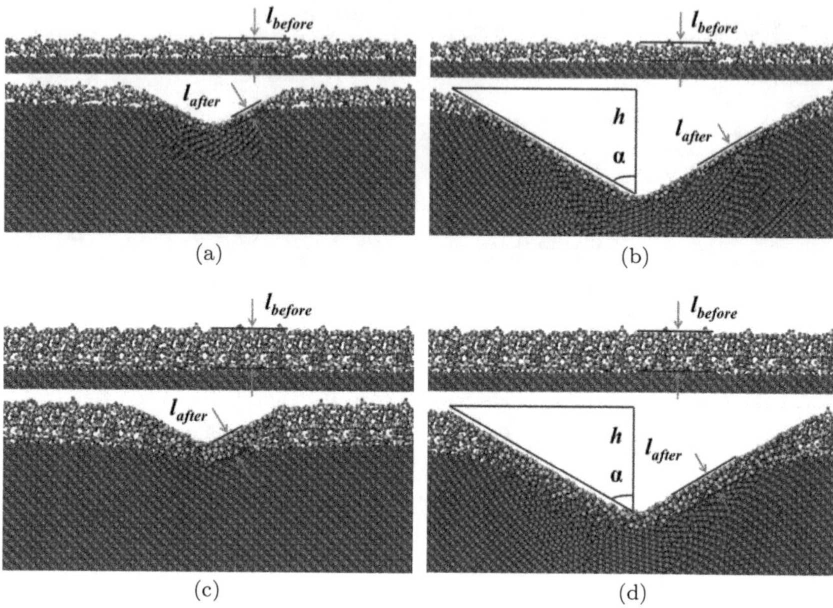

Figure 6.48. Cross-section of the specimen before creep. (a) 0.8 nm specimen at a creep load of 480 nN, (b) 0.8 nm specimen at a creep load of 4060 nN, (c) 2.0 nm specimen at a creep load of 480 nN, (d) 2.0 nm specimen at a creep load of 4060 nN.

and bond number$_{\text{as-prepared}}$ represent the number of atoms with Si coordination number 4 and the number of Si–O chemical bonds in the pristine SiO_2 film, respectively. CNx atom number$_{\text{time=6.0 ps}}$ represents the number of atoms with a Si coordination number of x at a creep time of 6 ps:

$$PCNx = \frac{CNx \text{ atom number}_{\text{time=6.0 ps}} - CNx \text{ atom number}_{\text{time=0.0 ps}}}{CN4 \text{ atom number}_{\text{as-prepared}}} \times 100\% \tag{6.11}$$

$$PCB = \frac{\text{bond number}_{\text{time =6.0 ps}} - \text{bond number}_{\text{time=0.0 ps}}}{\text{bond number}_{\text{as-prepared}}} \times 100\% \tag{6.12}$$

The properties of amorphous SiO_2 membranes with thicknesses of 0.8 nm and 2.0 nm under different levels of stress were examined.

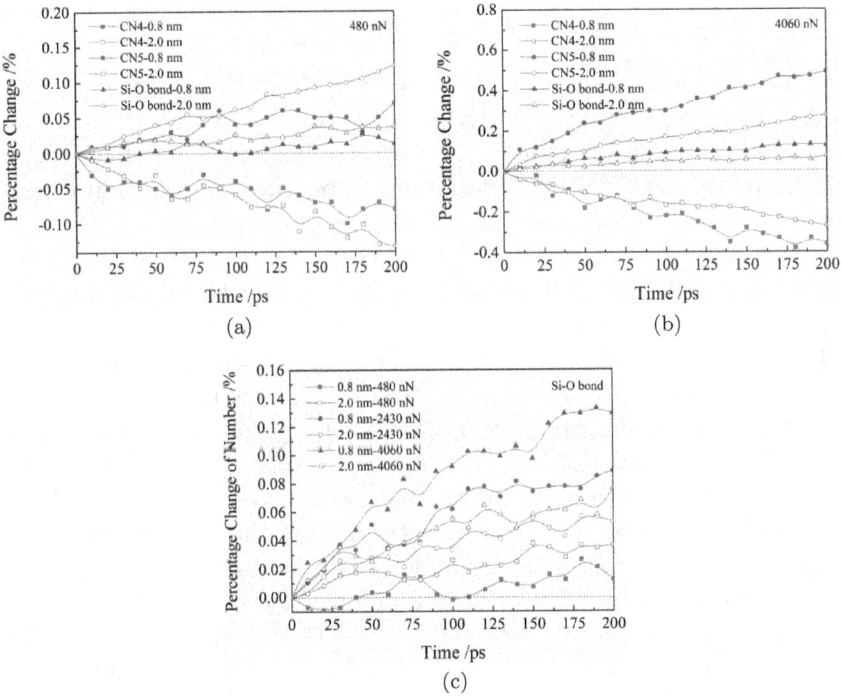

Figure 6.49. Variation of coordination number of Si atoms and number of Si–O bonds in SiO_2 film under different creep loads. (a) 480 nN, (b) 4060 nN, (c) variation in the number of Si–O bonds.

As depicted in Figure 6.49, with increasing load, both membranes displayed a significant increase in *PCN5* and *PCB*, while the value of *PCN4* decreased significantly. This resulted in additional oxygen atoms being added to a deformed tetrahedral SiO_4, causing the structure to transform into SiO_5. The introduction of a large number of oxygen atoms led to an increase in *CN5* and a decrease in *CN4*, indicating compaction of the SiO_2 film. At a lower load (480 nN), the *PCN5* and *PCN* of the 2.0 nm membrane were significantly larger than those of the 0.8 nm membrane at the same creep time, while the *PCN4* of the 2.0 nm membrane was lower than that of the 0.8 nm membrane, indicating more serious compaction deformation in the 2.0 nm membrane. However, the specific plastic deformation of the 2.0 nm membrane at 480 nN contradicted the results shown in Figure 6.48, providing further evidence of the accuracy of the quantitative analysis. At higher creep loads, as shown in Figure 6.49(b), the absolute values of *PCNx* and *PCB* for the 2.0 nm membrane were significantly lower than those for the 0.8 nm membrane, indicating that the film deforms more at larger creep loads and the compaction increases significantly with creep time. In summary, all SiO_2 films undergo compaction deformation during creep.

The graph in Figure 6.49(c) shows that the values of *PCB* increase in a linear fashion as the creep time increases for two films of 0.8 nm and 2.0 nm. This indicates that plastic deformation occurs in both films, in the form of compaction rather than rupture. It's interesting to note that the *PCB* values for 0.8 nm are generally greater than those for 2.0 nm at different creep loads for a given creep time, except at 480 nN. For instance, at a load of 4060 nN, the *PCB* value for 0.8 nm is about 87% higher than that for 2.0 nm.

These differences in values are attributed to the bilayer composites and the state of stress. At a smaller creep load of 480 nN (indentation depth ~2.0 nm), the indenter deformed the thin film and the monocrystalline silicon of the 0.8 nm specimen, while for the 2.0 nm specimen, the plastic deformation was confined to the SiO_2 itself, with only partial distortion of the monocrystalline silicon and no plastic deformation occurring.

As the creep load increased to 4080 nN, plastic deformation occurred in both the SiO_2 film and the single-crystal silicon. The higher the creep load, the more monocrystalline silicon was involved in the plastic deformation, causing the n values of the composites to get closer to those of the monocrystalline silicon specimens.

During the initial stages of creep in single-crystal silicon, the number of Si atoms in *CN5* and *CN6* increases rapidly. Subsequently, during the creep stabilization stage, the increase becomes linear, as depicted in Figure 6.50. This pattern aligns with the trend of creep displacement change. When subjected to the same creep load, the *CN5* and *CN6* of pure monocrystalline silicon specimens are notably larger than those of the composites. This disparity arises from the hindering effect of SiO_2 on the monocrystalline silicon matrix, which prevents phase transformation. Of particular note is that, under a creep load of 480 nN and within a 2.0 nm film, the *CN5* and *CN6* values of single-crystal silicon are the smallest. At this point, the atoms only form clusters with different coordination numbers rather than

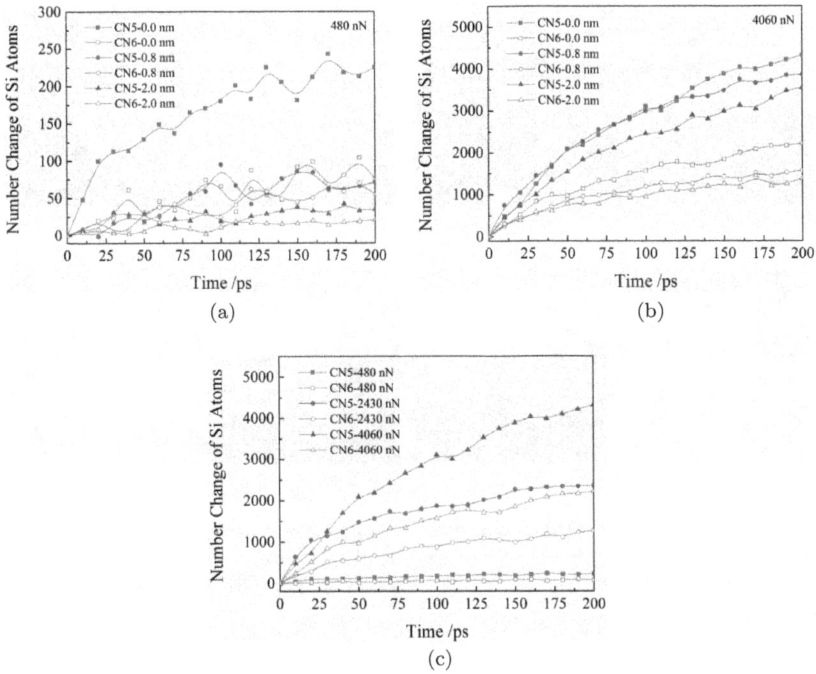

Figure 6.50. Variation of atomic changes in phase transition of single-crystal silicon. (a) 480 nN, (b) 4060 nN, (c) effect of different creep loads on phase change atoms.

a stable phase. However, at a creep load of 4060 nN, the *CN5* value of the 0.8 nm monocrystalline silicon matrix is similar to that of the pure monocrystalline silicon matrix.

We discuss the analysis of phase transitions in single-crystal silicon substrates under different creep loads. The increase in *CN5* and *CN6* atoms of single-crystal silicon under a 2.0 nm film is shown in Figure 6.50(c). It is observed that *CN5* and *CN6* atoms increase sharply in the early stage of creep and then increase slowly and steadily in the stable stage or near the late stage of creep. It is also noted that the number of *CN5* atoms is much larger than that of *CN6*, which is attributed to the phase transition characteristics and stress distribution of single-crystal silicon.

In Figure 6.51, the phase distribution on the Si (110) surface under different creep loads is shown, revealing that the Si-II phase is surrounded by the bct-5 phase, which is distributed in the middle of

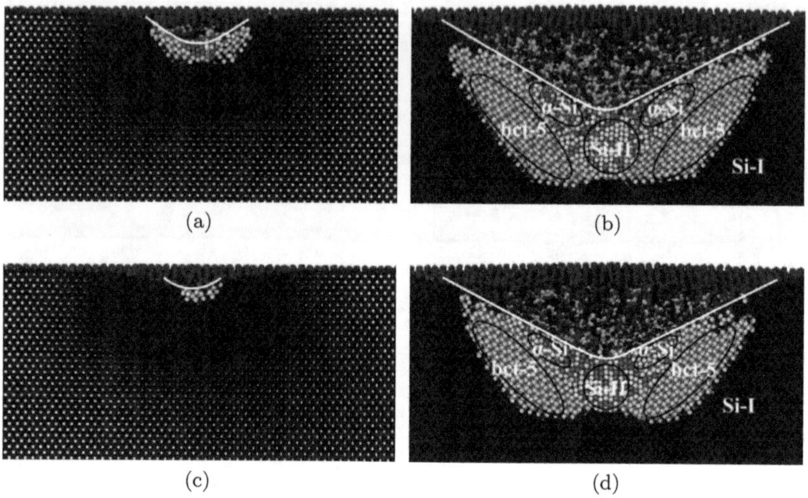

Figure 6.51. Phase transition distribution of single-crystal silicon Si (110) under thin film. (a) 0.8 nm at a creep load of 480 nN, (b) 0.8 nm at a creep load of 4060 nN, (c) 2.0 nm at a creep load of 480 nN, (d) 2.0 nm at a creep load of 4060 nN.

the phase transition region near the surface of the silicon substrate. The phase transition of single-crystal silicon at 0.8 nm film thickness is significantly larger than that at 2.0 nm film thickness.

Furthermore, it is observed that the deformed atoms of 2.0 nm single-crystal silicon are merely clusters of atoms with different coordination numbers, making it difficult to form a stable phase, and these atoms are transformed into the original substrate atoms, i.e., the Si-I phase, when unloaded. Thus, it is concluded that no significant phase transformation occurs in the monocrystalline silicon under 2.0 nm at a creep load of 480 nN, providing an explanation for the difference in the amount of plastic deformation between the 0.8 and 2.0 nm films under 480 nN.

6.3.3 *Stress relaxation behavior*

6.3.3.1 *Modeling*

The stress relaxation simulation system includes a spherical diamond indenter and a bilayer composite consisting of an amorphous SiO_2 film and a monocrystalline silicon matrix. The diamond sphere has a radius of 6 nm and contains 158921 atoms. The single-crystal silicon matrix shares the same dimensional size and crystallographic orientation as in the creep behavior simulation calculations. The thickness of the amorphous SiO_2 film varies, with options of 0.8 nm, 1.0 nm, 1.4 nm, and 2.0 nm. The Si–O Tersoff potential function used in the simulations remains constant.

The stress relaxation MD simulation involves four processes: relaxation, loading, stress relaxation, and unloading. Before the start of nanoindentation, the simulated system was initially relaxed under the steepest descent algorithm to achieve local energy minimization. This was followed by 90 ps of relaxation under the NVT system with a Nosé–Hoover thermal bath. After the system was stabilized, the spherical indenter was loaded along the negative Z-axis to the target indentation depth at a speed of 25 m/s. The indenter position was then held constant for 200 ps before being unloaded at a speed of 25 m/s. The entire indentation process was carried out in an NVE ensemble with a Langevin heat bath, using 0.5 fs time steps.

6.3.3.2 *Stress relaxation curve*

The stress relaxation behavior of SiO_2/Si composites was studied using nanoindentation simulation, as depicted in Figure 6.52. In this study, a spherical indenter was used to indent five composite specimens, including pure monocrystalline silicon, to various maximum indentation depths (0.8, 3.2, and 5.6 nm). Subsequently, the indenter was held stationary for a specific period (200 ps) before being unloaded. The results indicated that the applied indentation force decreased over time, and stress relaxation increased with the maximum indentation depth, as illustrated in Figure 6.52(b).

The graph in Figure 6.53 displays the changes in the indentation force of different specimens during stress relaxation. It indicates that the indentation force decreases rapidly at the early stage of stress relaxation and then decreases slowly and steadily at the later stage. Additionally, it is observed that the indentation force varies for different specimens at the early stage of stress relaxation. For instance, at an indentation depth of 3.2 nm, the pure silicon specimen and the 0.8 nm specimen exhibit the highest indentation force, while the 2.0 nm specimen demonstrates the lowest. However, at an indentation depth of 5.6 nm, the situation changes, and the indentation

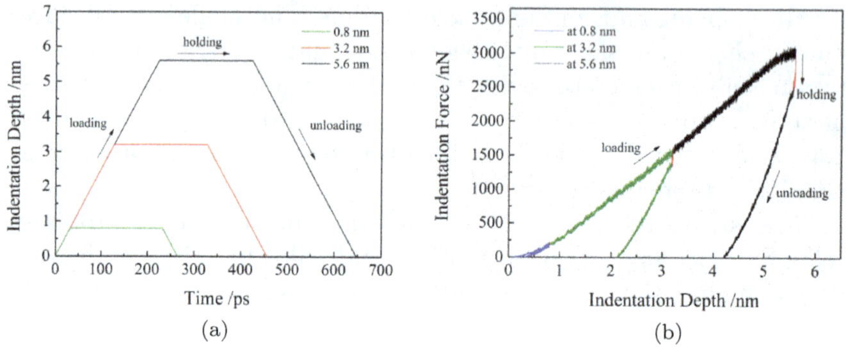

Figure 6.52. Schematic diagram of stress relaxation methods. (a) Indentation displacement–time, (b) load–displacement.

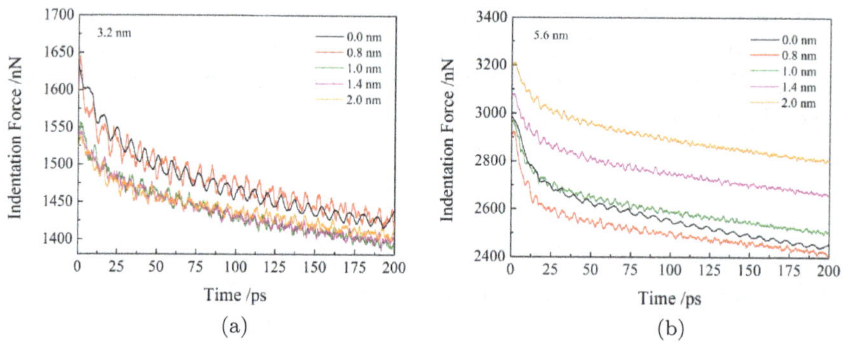

Figure 6.53. Variation of indentation force with time at different indentation depths. (a) 3.2 nm, (b) 5.6 nm.

force increases with the thickness of the SiO_2 thinness, as depicted in Figure 6.53(b). This behavior is attributed to the plastic deformation of the amorphous SiO_2 membrane during the loading process. When the indentation depth is shallow, the plastic deformation of the SiO_2 membrane involves compaction and thinning. Conversely, an indentation depth of 5.5 nm leads to the rupture of the 0.8 nm SiO_2 membrane. The strain-softening behavior of the amorphous SiO_2 membrane results in a reduced external force required for further deformation of the 0.8 nm membrane. Hence, the 0.8 nm specimen at the early stage of stress relaxation exhibits a smaller indentation force than the thick film specimen. Analysis in Section 6.2.2 suggests that thicker SiO_2 membranes have the potential for further compaction and likely

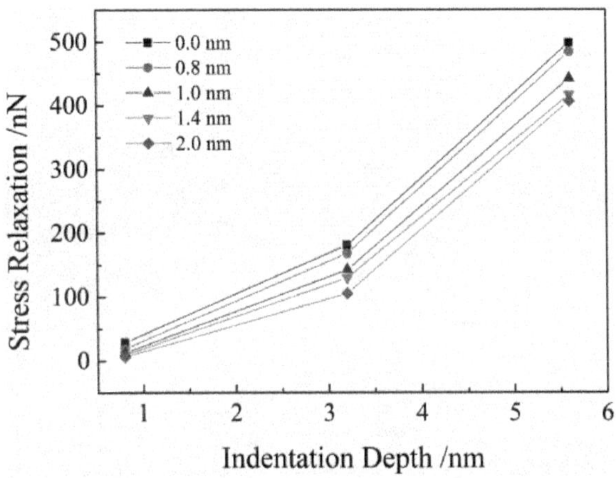

Figure 6.54. Variation of stress relaxation at different indentation depths.

rupture at larger indentation depths, for example, 2.0 nm SiO_2 membranes rupture at an indentation depth of approximately 7.5 nm. Consequently, at an indentation depth of 5.6 nm, the thick film specimen demonstrates a greater indentation force than the pure single-crystal silicon and 0.8 nm specimens, as discussed in the analysis results in Section 6.2.2.

The stress relaxation of different specimens at the end of stress shows in Figure 6.54 that the stress relaxation increases significantly with the increase in the indentation depth. For example, for the specimen with an indentation depth of 0.8 nm, the stress relaxation is 28 nN, while at an indentation depth of 5.6 nm, the relaxation is 483 nN. For the same indentation depth, the stress relaxation of composite specimens decreases slightly with the decrease of the SiO_2 film thickness. The stress relaxation of the specimen to be tested was found to have the same deformation trend as creep, both describing the relationship between material changes with time.

6.3.3.3 *Plastic deformation behavior during stress relaxation*

To investigate how a 0.8 nm-thick SiO_2 film with a radius of 6.0 nm deforms during the loading process and its effect on stress relaxation, researchers analyzed the specific deformation characteristics of

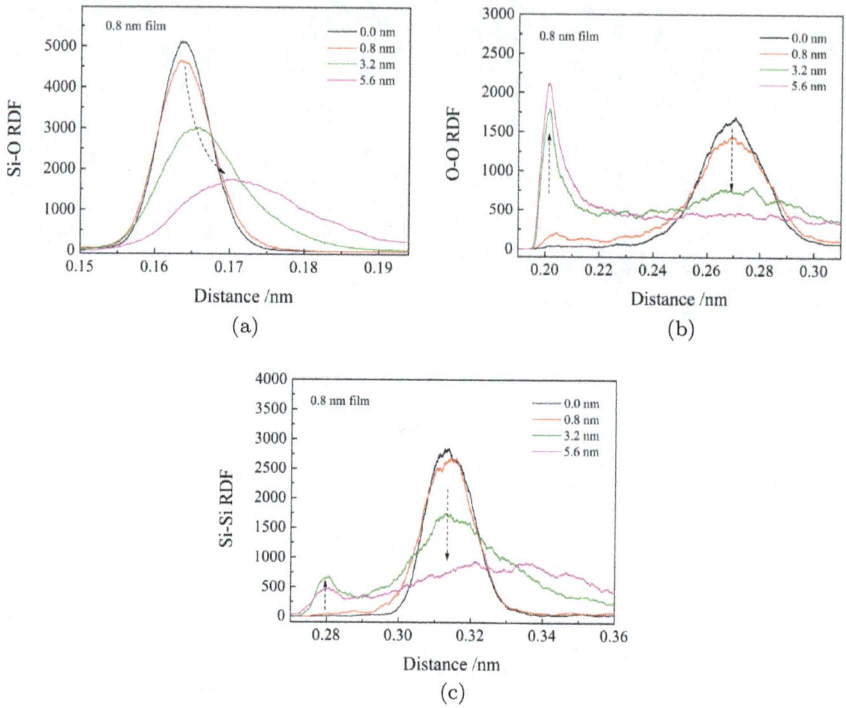

Figure 6.55. Variation of RDF with indentation depth for 0.8 nm thick SiO_2 film. (a) Si–O atoms, (b) O–O atoms, (c) Si–Si atoms.

multiple atom pairs using nearest neighbor RDFs. We found that the Si–O peak shifted to the right and the peak width increased, as shown in Figure 6.55, while the peak height decreased as the indentation depth increased. This indicated that the average distance between the Si–O bonds increased from 0.165 nm to 0.172 nm at an indentation depth of 5.6 nm. Additionally, an extra peak appeared in the *RDF* of the O–O and Si–Si atomic pairs with distances of about 0.2 nm and 0.28 nm, respectively. As the indentation depth increased, the main peak in the *RDF* of O–O and Si–Si decreased in height, while the two extra peaks increased in height and width. However, at an indentation depth of 5.6 nm, the extra peak of Si–Si atoms narrowed, indicating a breakup of 0.8 nm of SiO_2. This analysis showed that the original SiO_2 film consisting of numerous regular SiO_4 structures became more disordered during indentation loading, and the plastic deformation of SiO_2 gradually increased with the increase of the indentation depth. The 0.8 nm film changed from compaction to rupture, while the thicker film only underwent compaction deformation.

The plastic deformation analysis of SiO_2 membranes reveals that the deformation during loading occurs in three stages: compaction, transition, and rupture. During the compaction stage, *PCN5* and *PCB* increase while *PCN4* decreases. As the membranes transition from compaction to rupture,*PCB* and *PCN4* decrease while *PCN5* increases. Finally, in the rupture stage, *PCN4*, *PCN5*, and *PCB* all decrease.

The in-load RDF analysis shows that the SiO_2 films experienced compaction at different indentation depths during loading, except for the SiO_2 film at 0.8 nm under an indentation depth of 5.6 nm, which underwent rupture.

The time-dependent deformation behavior of SiO_2 films during stress relaxation is described by the previously defined atomic percentage change in the coordination number of silicon (*PCNx*) and percentage change in the number of Si–O bonds (*PCB*), as shown in Eqs. (6.11) and (6.12). The trends of *PCNx* and *PCB* during stress relaxation are illustrated in Figure 6.56. It is observed that *PCN4*, *PCN5*, and*PCB* change faster at the beginning of the stress relaxation period and become more stable later on. As the stress relaxation time increases, the values of *PCN5* and *PCB* increase while *PCN4* decreases, indicating compaction of SiO_2 membranes during stress relaxation, except for the 0.8 nm membranes at a larger indentation depth (5.6 nm). For the 0.8 nm film, at an indentation depth of 3.2 nm, the *PCB* value is smaller than the *PCB* value at an indentation depth of 0.8 nm, as shown in Figure 6.56(d). This, along with the results in Figure 6.56(b), indicates that the SiO_2 film still undergoes compaction, but the degree of compaction is relatively low because the 0.8 nm film at an indentation depth of 3.2 nm shows a change toward the transition stage of the tendency. At an indentation depth of 5.6 nm, *PCN4* vibrates around the zero value, while *PCN5* and *PCB* increase slowly with time, implying that the SiO_2 film undergoes elastic recovery.

The SiO_2 film, with a thickness of 2.0 nm, showed increased values of both *PCN5* and *PCB* as the indentation depth during loading increased. Meanwhile, *PCN4* decreased, indicating compaction of the films at different indentation depths during stress relaxation. Furthermore, the *PCB* values of the 0.8 nm films were significantly

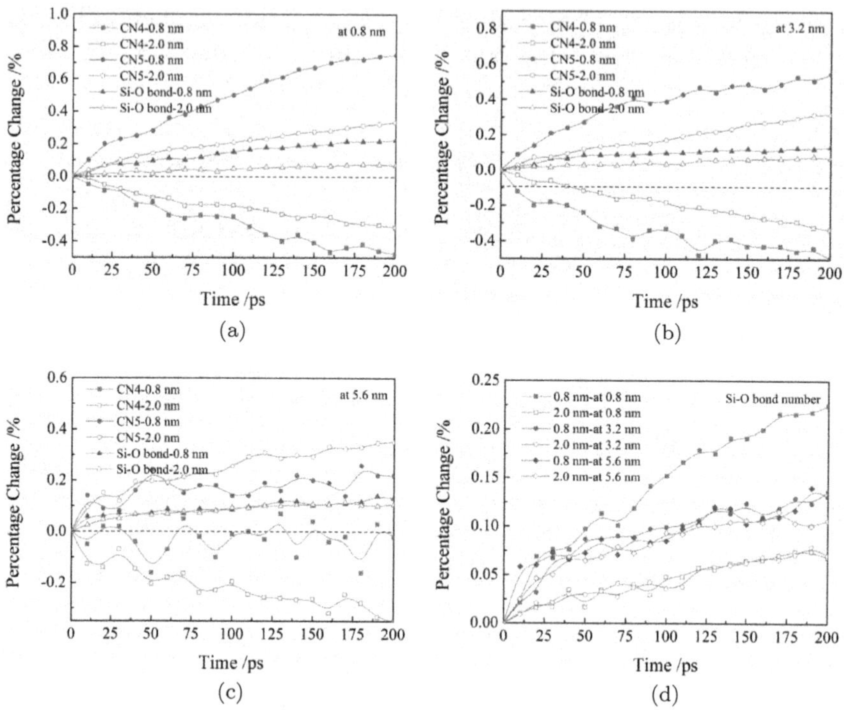

Figure 6.56. Variation of coordination number of silicon atoms and number of Si–O bonds in different membranes during stress relaxation process. (a) Indentation depth 0.8 nm, (b) indentation depth 3.2 nm, (c) indentation depth 5.6 nm, (d) Si–O quantity.

larger than those of the 2.0 nm films at the same indentation depth, suggesting that the 0.8 nm films deformed more during stress relaxation due to greater plastic deformation, which occurs in a similar manner during nanoindentation loading.

The phase transformation of single-crystal silicon during stress relaxation was studied by analyzing the changes in the number of coordination number atoms of silicon, specifically *CN5* and *CN6*. The graph in Figure 6.57 shows that *CN5* and *CN6* increase rapidly with the extension of the stress relaxation time and then stabilize, in line with the trend of the change in the indentation force. It was observed that *CN5* and *CN6* increased significantly with increasing indentation depth. For pure single-crystal silicon, the *CN5* atomic increment increased from 20 at an indentation depth of 0.8 nm to 1600 at an

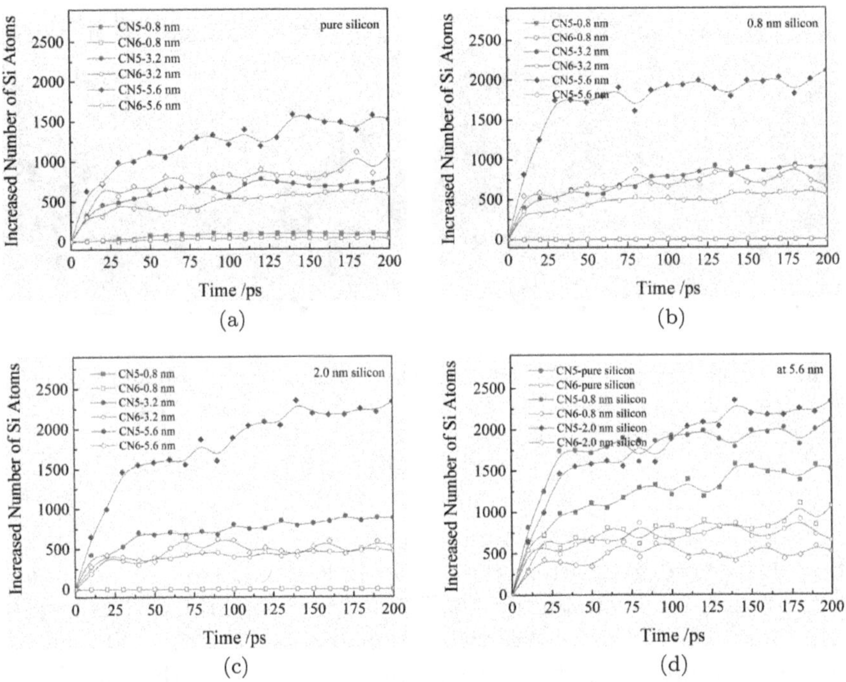

Figure 6.57. Phase transformation of single-crystal silicon during stress relaxation. (a) Pure silicon, (b) 0.8 nm silicon, (c) 2.0 nm silicon, (d) phase variable under indentation depth 5.6 nm.

indentation depth of 5.6 nm. It was noted that the number of atoms in *CN5* was much larger than that in *CN6* under the same conditions, which was attributed to the nature of the single-crystal silicon phase transition and the stress distribution. In nanoindentation experiments, Gerbig *et al.* found that the pressure of the Si-I phase transforming into the bct-5 phase is smaller than the pressure of the Si-I phase transforming into the Si-II phase. The atomic configuration diagram of single-crystal silicon phase transformation before stress relaxation (after the end of loading) shown in Figure 6.58 revealed that Si-II is located at the center of the whole deformation region, with the phase transformation region being smaller and surrounded by the bct-5 phase. This results in there being more silicon atoms in the bct-5 phase than in the Si-II phase, leading to a higher number of atoms in the Si-I phase than in the Si-II phase, resulting in more *CN5* atoms appearing in the stress relaxation process.

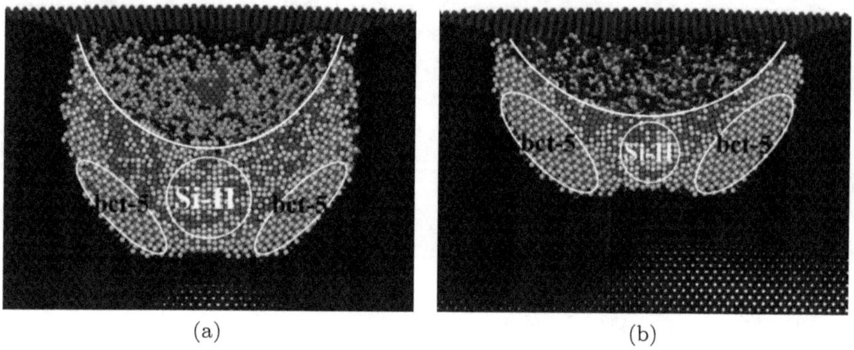

Figure 6.58. Phase distribution of single-crystal silicon Si (110) under different SiO_2 films at an indentation depth of 5.6 nm. (a) Pure monocrystalline silicon, (b) 2.0 nm monocrystalline silicon.

When the indentation depth is 5.6 nm, the change in the number of atoms during stress relaxation in single-crystal silicon phase transition is illustrated in Fig. 6.57(d). It is observed that the atomic increment of *CN5* increases with the increase of the SiO_2 film thickness within a certain stress relaxation time. On the other hand, the increment of *CN6* decreases. This phenomenon can be explained by the atomic configuration diagrams of the single-crystal silicon phase transition before and after stress relaxation, as shown in Figure 6.58. At an indentation depth of 5.6 nm, the Si-II phase region of the pure single-crystal silicon specimen is significantly larger than that of the composite silicon at 2.0 nm, while the bct-5 phase is less than that of the composite silicon. Similar variations are depicted in the atomic variations of phase transitions of different monocrystalline silicon shown in Figures 6.28 and 6.29, where the crater diameter of the monocrystalline silicon below the SiO_2 film is larger and the indentation depth is shallower than that of the pure monocrystalline silicon when the indentation depth is the same.

As the SiO_2 membrane has the role of transferring stress and dissipating energy, the greater the thickness of the membrane, the smaller the composite single-crystal silicon is subjected to stress, as shown in Figure 6.59 for the von Mises stress distribution diagram. The white circle in the figure indicates that the shear stress below the indenter is smaller and the hydrostatic pressure is in a larger area. It can be seen that in the same depth of the indentation, pure single-crystal

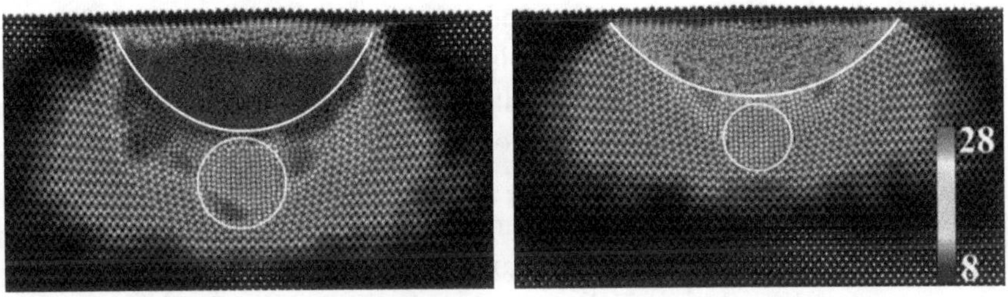

(a) (b)

Figure 6.59. Shear stress distribution (GPa) of single-crystal silicon with different SiO_2 film downwards at indentation depth 5.6 nm. (a) Pure monocrystalline silicon, (b) 2.0 nm monocrystalline silicon.

silicon specimen shear stress is greater than the 2.0 nm composite silicon, and the stress concentration area and stress value are also larger.

Therefore, the indentation loading process generates significantly more Si-II from the Si-I phase transition and less bct-5 phase in pure monocrystalline silicon specimens compared to the composite silicon. From Figure 6.56(d), it can be seen that the phase transformation during stress relaxation of pure monocrystalline silicon specimens during loading has a great influence on the subsequent stress relaxation, and the larger plastic deformation during loading leads to larger time-dependent plastic deformation during relaxation.

6.3.4 *Summary*

The study used MD simulations to analyze the creep and stress relaxation behavior of SiO_2/Si composite specimens with different SiO_2 film thicknesses. The findings are as follows:

(1) During the creep stage, the creep displacement increases rapidly at first and then increases linearly in the stable creep stage. The final creep displacement of the specimen increases with the creep load due to the occurrence of a large amount of plastic deformation.
(2) The creep stress index decreases with increasing creep load due to the generation and distribution of free volume in the SiO_2 membrane during nanoindentation loading. Higher creep loads lead to

greater plastic deformation and a larger number of free volumes, resulting in more uniform plastic deformation. The specimen stress index also decreases slightly with increasing film thickness of amorphous SiO_2 at the same creep load, except for thick film specimens at small loads.

(3) The original short-range ordered SiO_2 film in the nanoindentation undergoes compaction during indentation loading, leading to more disordered SiO_4 tetrahedra. This plastic deformation continues during creep, especially at higher creep loads.

(4) In the stress relaxation process of spherical nanoindentation, two stages were observed in the stress relaxation curve. In the first stage, the indentation force decreases rapidly and the curve is nearly parabolic. In the second stage, the indentation force decreases slowly and the curve is nearly straight. The stress relaxation of the composite specimen SiO_2/Si increases with the increase of the indentation depth. Additionally, the stress relaxation of the thick film specimen is relatively small under the same indentation depth, which aligns with the creep results.

(5) During spherical nanoindentation loading with a maximum indentation depth of 5.6 nm, the 0.8 nm film underwent deformation from compaction to rupture, while the thicker film specimen only underwent compaction. Throughout stress relaxation, the thicker SiO_2 film continued to undergo compaction, while the 0.8 nm thin film that ruptured during loading underwent elastic recovery. The single-crystal silicon under the amorphous film in the nanoindentation underwent a phase transition, with more phase transition atoms observed during stress relaxation as the depth of the indentation increased.

6.4 Wear Behavior of SiO_2/Si Bilayer Nanomaterials

6.4.1 *Introduction*

The abrasion method, known as two-body abrasion, is commonly used to investigate surface properties of materials and to study wear and material removal. This is typically achieved through nanoscratching experiments using a single needle tip in specific tests.

Sun [55, 79] conducted studies on two-body and three-body abrasive wear of monocrystalline silicon. Additionally, Fang and Shi *et al.* [16, 80] investigated how the shape of abrasives impacts the phase transition of single-crystal silicon. Si Lina *et al.* [81, 82] employed MD to study the scratch removal of amorphous SiO_2 abrasive on single-crystal silicon substrates, discovering that both rolling and sliding of the abrasive contribute to material removal.

Chen [83] investigated the high-speed impact removal of colloidal SiO_2 particles from single-crystal silicon substrates, finding that the removal mechanism is adhesion. Linmao Qian's group [84] performed nano-scratching experiments using SiO_2 glass sphere tips on monocrystalline silicon with three different hydrophilic surfaces in four different environmental atmospheres, revealing that friction chemical reaction becomes apparent with lower applied pressure. Furthermore, it was found that an increase in the hydrophilicity of the silicon surface leads to a deeper wear depth in monocrystalline silicon surfaces.

Barnette [85], Yu [86], and Wang [87] found that in experiments involving ball-and-disc wear and nano-friction on SiO_2 balls and monocrystalline silicon, the friction chemistry intensified and the depth of wear increased when the humidity was increased from 0% to 50%. While these experiments have discovered the frictional chemical reaction during scratching, the specific material deformation behavior in scratching, as well as the relationship between friction and material deformation and the relationship between friction and material removal, remains unresolved. On the other hand, amorphous colloidal SiO_2 abrasive is used as the main polishing abrasive for silicon wafers, and the wafers obtain an atomically flat, defect-free surface after chemical mechanical polishing. However, due to experimental conditions, the polishing behavior is still unclear. Therefore, it is necessary to study the tribological properties of materials at the atomic scale and the material removal mechanism.

We used MD simulations to investigate the process of scratching single-crystal silicon substrates with amorphous colloidal SiO_2 abrasives coated with amorphous SiO_2 films. The goal was to analyze the frictional properties and material deformation behavior of SiO_2/Si composite specimens with varying film thicknesses under different positive pressures and to evaluate material removal efficiency. We also analyzed the impact of abrasive size on frictional properties and removal efficiency. Lastly, diamond abrasives were used in a comparative experiment to explore different material removal mechanisms during the scratching process.

6.4.2 *Modeling*

The nano-scratch simulation system is comprised of an amorphous SiO_2 abrasive and a single-crystal silicon substrate coated with amorphous SiO_2 films of varying thicknesses on the surface. To more accurately simulate the chemical mechanical planarization (*CMP*) process, a layer of water molecules with a thickness of 0.3 nm is present on the specimen's surface, as depicted in Figure 6.60. The amorphous SiO_2 abrasive has a cylindrical shape at the upper part and a hemisphere with a radius of 3.0 nm at the bottom part, comprising a total of 8844 atoms. In the simulation of the abrasive's motion during the CMP process, the uppermost part of the abrasive was designated as a movable stationary layer, represented by green atoms, while the middle part was a hot bath layer, represented by yellow atoms. The hot bath and hemispherical parts of the abrasive were available for the Newtonian integrations.

The thicknesses of the SiO_2 films were 0.2, 0.4, 0.6, 0.8, and 1.0 nm, containing different numbers of atoms. Both the SiO_2 abrasive and the films were obtained by quenching the molten SiO_2 crystals. The size of the single-crystal silicon matrix is $34.2 \times 23.5 \times 15.0\,nm^3$, containing a certain number of atoms. In the simulation,

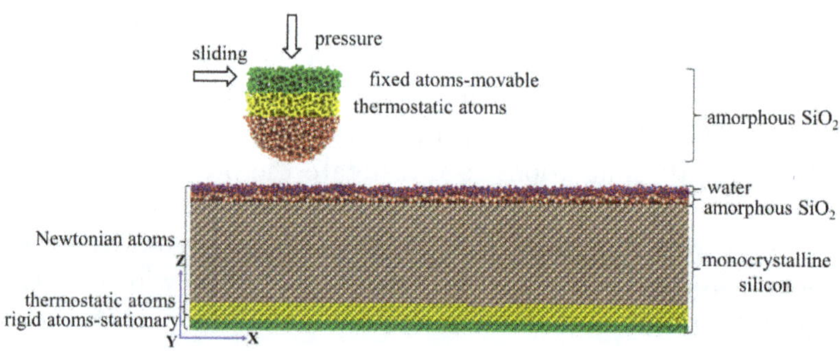

Figure 6.60. Schematic of the two-body scratch model.

the bilayer SiO_2/Si composite matrix can be divided into three layers: a fixed layer, a hot bath layer, and a Newtonian layer. The thickness of each layer is set to be the same as that of nanoindentation. The SiO_2/Si matrix is subjected to periodic boundary conditions in the X- and Y-axis directions and non-periodic boundary conditions in the Z-axis direction.

A Tersoff three-body potential function based on first-principles calculations was used to describe the interaction between amorphous SiO_2 and Si in the scratch system. The interaction between water molecules was modeled using the TIP4T model [88], which contains three fixed point charges and a Lennard–Jones center. The effect of hydrogen atoms on the thermodynamic properties of the system is very small, so the interactions between hydrogen atoms are neglected, and the Lennard–Jones potential [21, 88, 89] is used between Si and O atom in the water molecule, O and O atoms, and the relevant parameters are shown in Table 6.3.

Before performing the nano-scratch calculations, the simulated system underwent local energy minimization using the steepest descent algorithm and structural relaxation at 300 K. The system was then subjected to local energy minimization again using the steepest descent algorithm. After the system structure was stabilized, the scratch calculation was carried out.

In the scratch calculation, the amorphous SiO_2 abrasive was pressed into the specimen surface along the Z direction under a certain positive load in a constant-load manner after the system was equilibrated. Then, the abrasive slipped along the X-axis in the positive direction for a certain distance (18 nm) under the action of this load. The abrasive's movement speed during the simulation of the scratch process was 25 m/s, and finally, it was unloaded in a constant-load manner.

Table 6.3. TIP4P and Lennard-Jones potential function parameters.

Atomic pair	E (kcal/mol)	δ (nm)	Cut-off radius (nm)
O–O	0.1554	0.31655	0.6
Si–O	0.0520594	0.349575	0.5

Throughout the simulation, an NVE ensemble with a Langevin thermal bath and a time step of 0.5 fs was used.

6.4.3 *Effect of load and Film Thickness on friction*

In this study, we conducted scratch tests using various positive loads (10, 20, 40, 60, 80, 100 nN) on monocrystalline silicon samples with different thicknesses of amorphous SiO_2 films. We obtained friction change curves during the scratching process, which are shown in Figures 6.61 and 6.62. We found that as scratching begins, the friction

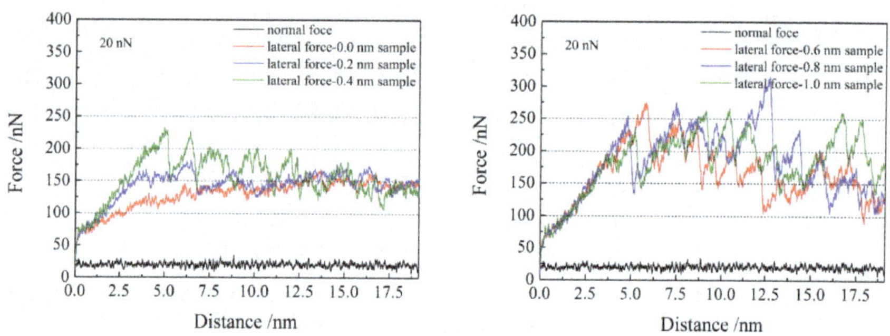

(a) (b)

Figure 6.61. Variation of friction of 20 nN specimens with different film thicknesses under positive loading. (a) Film thickness 0.0–0.4 nm, (b) film thickness 0.6–1.0 nm.

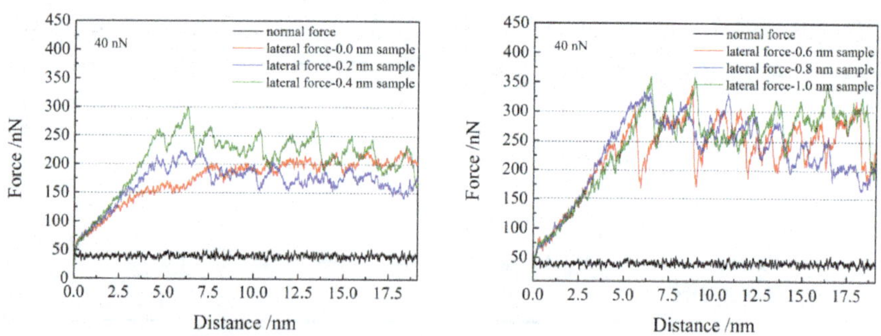

(a) (b)

Figure 6.62. Variation of friction of 40 nN specimens with different film thicknesses under positive loading. (a) Film thickness 0.0–0.4 nm, (b) film thickness 0.6–1.0 nm.

of the silicon sample gradually increases from zero. For the pure monocrystalline silicon specimen, the friction slowly increases to a steady state, while the friction of the composite specimen first increases and then decreases, showing a peak in the curve.

From the friction curves, we can see that the friction during the scratching process increases gradually with the thickness of the amorphous SiO_2 film. For example, when the positive load is 20 nN, the maximum friction force ranges from 150 nN for the pure monocrystalline silicon to 250 nN for the 0.6 nm composite specimen. However, there is no significant difference in the friction curves of the specimens when the film thickness is greater than 0.6 nm.

We also observed that the friction curves of pure monocrystalline silicon are relatively smooth, while the curves of the composite specimens all show obvious serrated features, especially the friction curves of the 0.6 nm composite specimens at a positive load of 40 nN. This serrated pattern is a characteristic of the inhomogeneous plastic deformation of the amorphous material and is the result of the interaction between the abrasive material and the SiO_2 film during the scratching process.

The friction curves of 0.4 nm specimens scratched under different positive loads are depicted in Figure 6.63. The curves reveal that friction rapidly increases with scratching distance and then decreases. Additionally, friction rises with the increase in positive load at the same scratching distance. At a positive load of 10 nN, the maximal friction is about 200 nN, while at a load of 100 nN, the maximal friction of the specimen reaches about 400 nN. This phenomenon occurs because the increased positive load leads to significant plastic deformation of the amorphous SiO_2 abrasive, thus increasing the contact area between the abrasive and the specimen.

The potential energy is affected by the displacement of atoms during the scratching process due to the interaction between the abrasive and the specimen surface. This causes some of the atoms on the surface to move from their original equilibrium position. As a result, the potential energy of the system increases with the scratching going on. Figure 6.64 illustrates the potential energy and friction during the scratching process of a 0.6 nm specimen. It shows that the trends in potential energy and friction changes are similar. For instance, in the zone I interval of Figure 6.64, when the atoms are in their original position, the potential energy is low. As the scratching

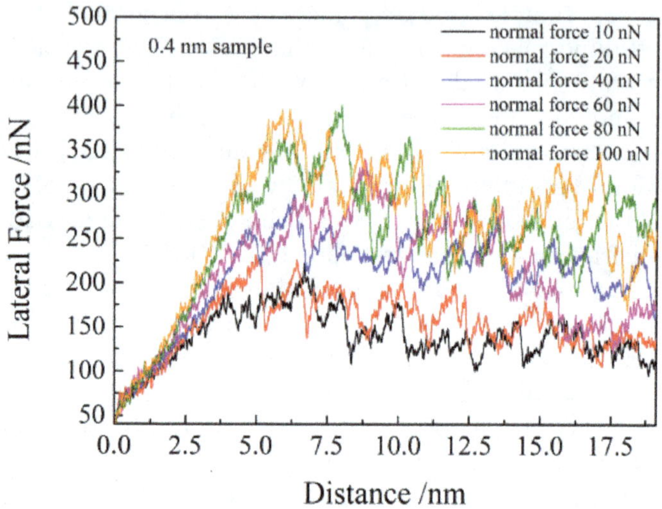

Figure 6.63. Variation of friction of 0.4 nm specimen under different positive loads.

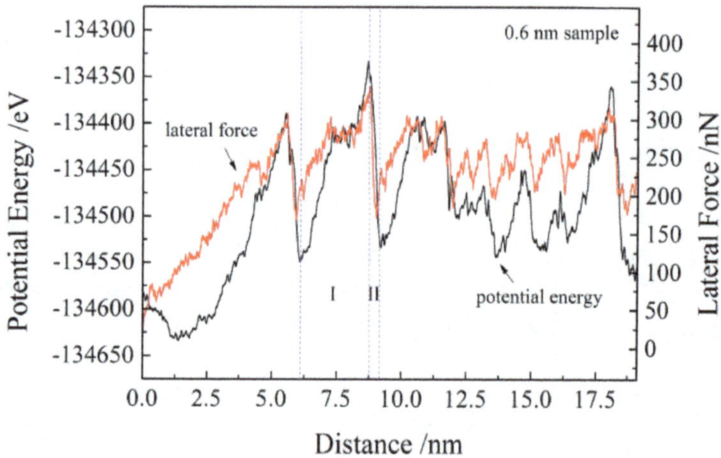

Figure 6.64. Relationship between frictional force and potential energy.

occurs, some of the atoms on the sample surface adhere to the abrasive, causing the potential energy of the system to rise. This increase in potential energy requires a higher external force to overcome the displacement of the atoms, leading to an increase in friction. When the displacement of atoms exceeds the bond length between them, these surface atoms may leave the matrix, remain on the specimen's surface, or adhere to the abrasive. Subsequently, these atoms undergo structural relaxation to reduce the potential energy, naturally lowering the potential energy of the system and, in turn, reducing friction, as shown in the zone II interval of Figure 6.64.

In order to explore the relationship between friction and the movement of atoms on the surface of a specimen, the change in the number of displaced atoms during the scratching of the 0.6 nm specimen was analyzed and is illustrated in Figure 6.65. The actual count of the displaced atoms is found to be greater than 0.2 nm for the amorphous SiO_2 film, where the atoms are removed from their original positions on the substrate. For monocrystalline silicon, the atoms are displaced more than 0.26 nm, resulting in their detachment from the original position of the substrate (0.2 nm and 0.26 nm are the fracture distances of the Si–O and Si–Si bonds, respectively). It is considered that these atoms are removed because, in the actual CMP process, the atoms detached from the substrate are removed when the polishing solution is flushed. When the displacement of the atoms is less than the fracture distance, the atoms undergo structural relaxation after abrasive scratches and spontaneously rearrange to reduce the potential energy, returning to a relative equilibrium state. From Figure 6.65, it can be observed that the trend between the friction force and the removal of atoms is consistent, indicating that the

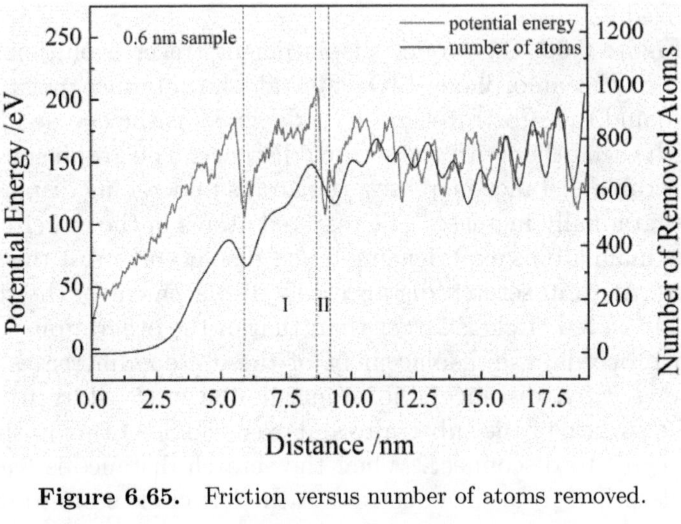

Figure 6.65. Friction versus number of atoms removed.

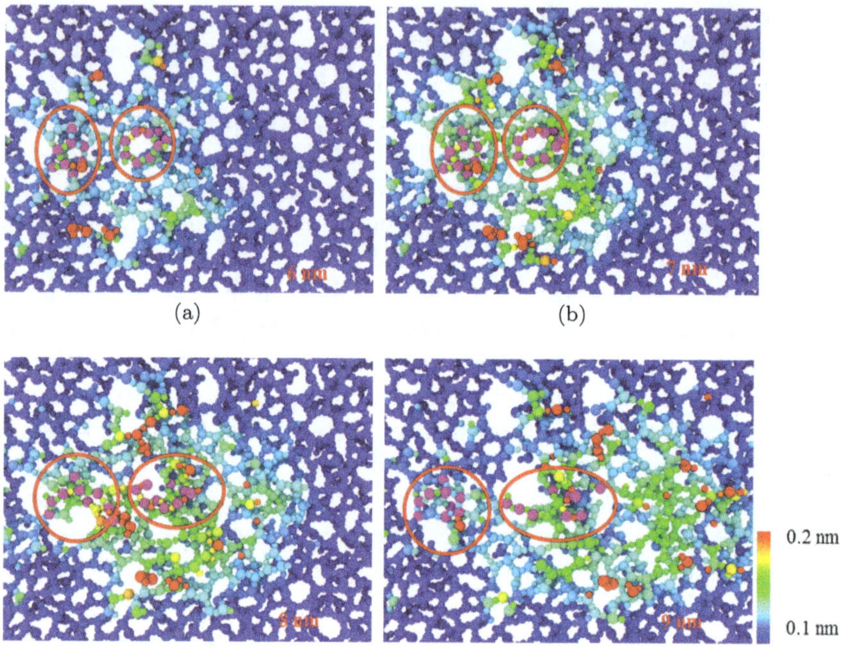

(a) (b) (c) (d)

Figure 6.66. Evolution of atomic displacement deformation on SiO_2 surface for different scratch distances. (a) 6 nm, (b) 7 nm, (c) 8 nm, (d) 9 nm.

removal of more atoms requires overcoming a larger friction force to be realized.

In Figure 6.66, the atomic displacement tracer evolution configuration of the amorphous SiO_2 film at the interface between the abrasive and the substrate surface is depicted as they come into contact with each other with the scratch distance. The graph shows that as the scratch distance increases, the atoms undergoing displacement changes gradually increase. The rosy-red atoms in the red circle indicate the main atoms experiencing larger displacements at the scratch distance. At 6 nm scratch distance, the displacement of the rosy-red atoms in the red circle is larger than that of the other atoms. As the scratch proceeds, the displacement of these atoms increases slowly. At 8 nm scratch distance, the atoms in the circle start to detach from the surface of the substrate, and even the Si–O ring in the right circle begins to disconnect. When the scratch distance is 9 nm, the atoms detached from the substrate in the left circle start to undergo structural relaxation and attempt to return to a relative equilibrium position. The Si–O ring in the right circle is completely disconnected and undergoes structural relaxation, completing a serrated deformation, as shown in zones I and II in Figures 6.64 and 6.65.

6.4.4 *Material removal behavior and surface quality analysis*

6.4.4.1 *Effect of load, film thickness on atom removal efficiency*

The statistical analysis showed that as the thickness of SiO_2 film increases, the rate of atomic removal from the single-crystal silicon matrix decreases and becomes almost negligible when the film thickness exceeds 0.6 nm, as shown in Figure 6.67. This is because the SiO_2 film restricts the plastic deformation of the single-crystal silicon. When the applied load is 20 nN, the atomic removal of SiO_2 film remains at around 200, but at 40 nN, the atomic removal curve reaches peaks, with the 0.6 nm film exhibiting the highest removal rate. Consequently, the overall atomic removal of the composite specimens (combination of single-crystal silicon and SiO_2 film) also peaks, with the 0.6 nm specimen showing the highest removal rate. This information is illustrated in Figure 6.67(b).

The number of atoms removed at different positive loads was counted, and it was found that the removal of atoms increased almost linearly with the increase in positive load. The overall number of

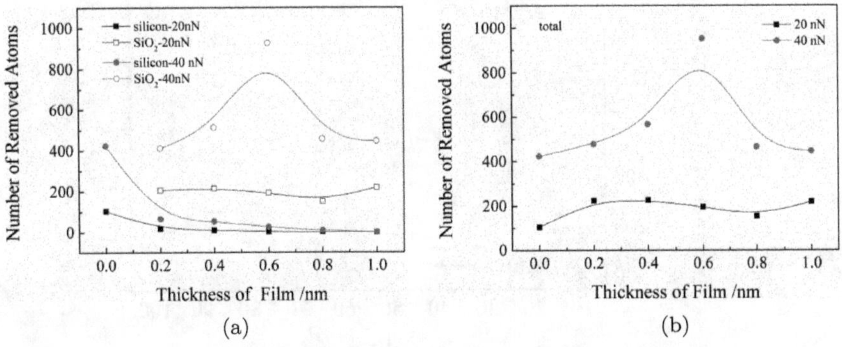

Figure 6.67. Variation of the number of atoms removed from the specimen under different positive loads. (a) SiO_2 and monocrystalline silicon, (b) specimens.

atoms increased from 140 at 10 nN to 1,130 at 100 nN, with a growth rate of about 707%. This increase was attributed to the fact that the contact area between the abrasive and the specimen increased with the elevation of the positive load. The increase in contact area is caused by the plastic deformation of the soft SiO_2 abrasive and the film. This plastic deformation intensifies with the scratch, leading to a significant increase in the number of contact atoms between the abrasive and the specimen as the scratch distance increases. For example, the number of contact atoms under the positive load of 10 nN and 100 nN at the scratch distance of 0 nm is about 6 and 24, respectively, with a growth rate of about 283%. Once the scratch is completed, the number of contact atoms at positive loads of 10 nN and 100 nN are about 46 and 180, respectively, with a growth rate of about 291%. The number of removal atoms and the number of contact atoms change in the same trend for different positive loads, but the growth rate is not consistent, which is determined by the removal mechanism of the material (Figures 6.68 and 6.69). When counting the number of atoms touching between the abrasive and the specimen, we consider that atoms are in contact if the distance between them is less than 0.2 nm. This can be observed from the *RDF* of Si–O atoms in Section 6.3.3, where the average bond length between the Si–O atoms is 0.165 nm, and the Si–O bond length when in the tensile state is less than 0.2 nm. Therefore, we have chosen 0.2 nm

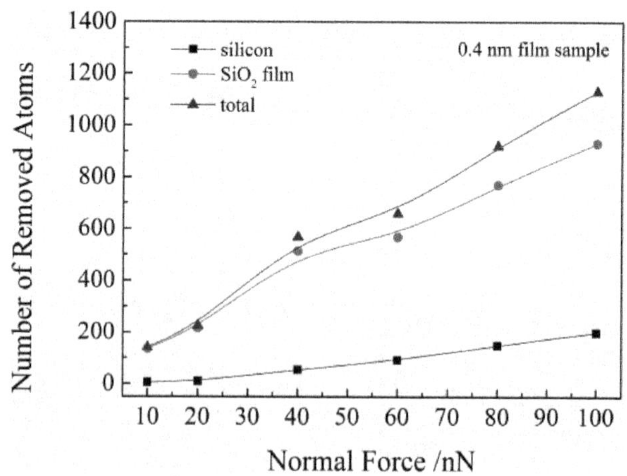

Figure 6.68. Effect of load on specimen atom removal.

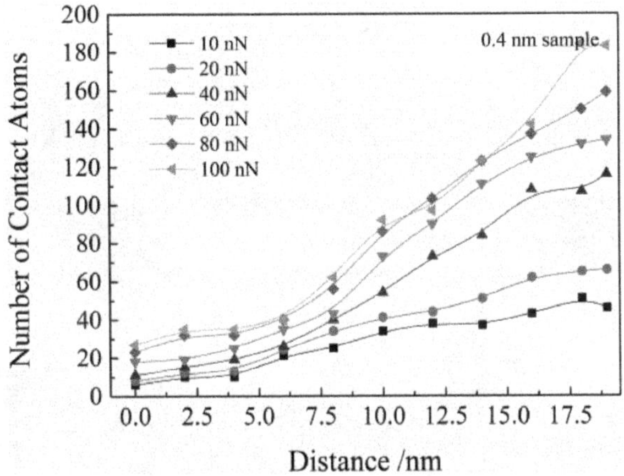

Figure 6.69. Effect of load on contact atomic weight during scratching.

as the maximum Si–O bond length for considering atoms to be in contact.

6.4.4.2 *Plastic deformation behavior of SiO_2 abrasives in scratches*

The interfacial cross-sectional configuration of the 0.6 nm specimen was examined to understand the contact between the abrasive and the specimen during the scratching process. The examination was conducted at different positive loads and scratch distances, as depicted in Figure 6.70. In the figure, the green color signifies the presence of a SiO_2 thin film with water molecules on it. The results indicate that as the scratch distance increases, the plastic deformation of the amorphous SiO_2 abrasive also increases. Additionally, the deformation of the abrasive becomes more pronounced when the load is increased to 40 nN. Interestingly, the cross-sectional changes in the SiO_2/Si composite specimens were found to be insignificant, requiring further detailed analysis.

In order to better understand the plastic deformation behavior of the amorphous SiO_2 abrasive during scratching, we analyzed the *RDF* of the abrasive for various states of the film thickness, as shown in Figures 6.71 and 6.72. We observed that the peaks corresponding to Si–O bonds in the undeformed SiO_2 are initially high and narrow. As the scratch displacement increases, the peak height decreases

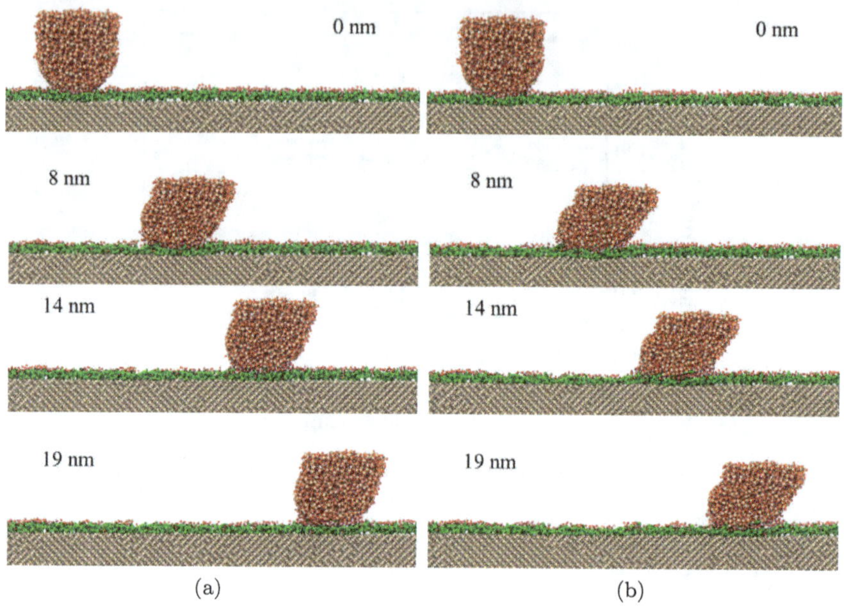

Figure 6.70. Cross-section of abrasive and specimen interface for different scratching distances. (a) Positive load 20 nN, (b) positive load 40 nN.

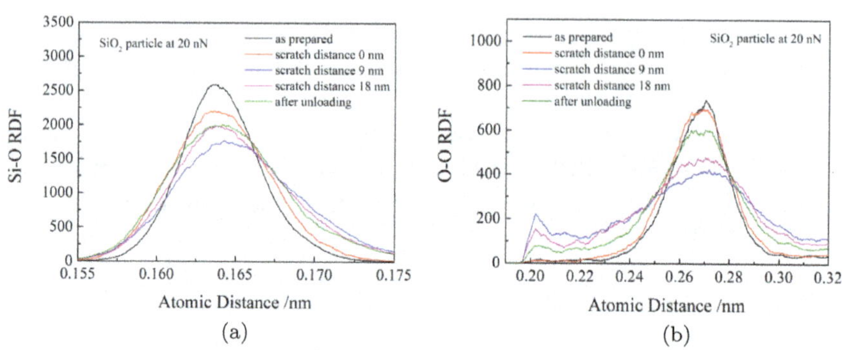

Figure 6.71. RDF of SiO_2 abrasive in different states at a positive load of 20 nN. (a) Si–O RDF, (b) O–O RDF.

while the peak width increases. Additionally, the Si–O peaks shift slightly to the right, indicating a widening of the average distance between the Si–O atoms.

Upon unloading, as the abrasive separates from the substrate, there is partial elastic recovery, causing the average distance between

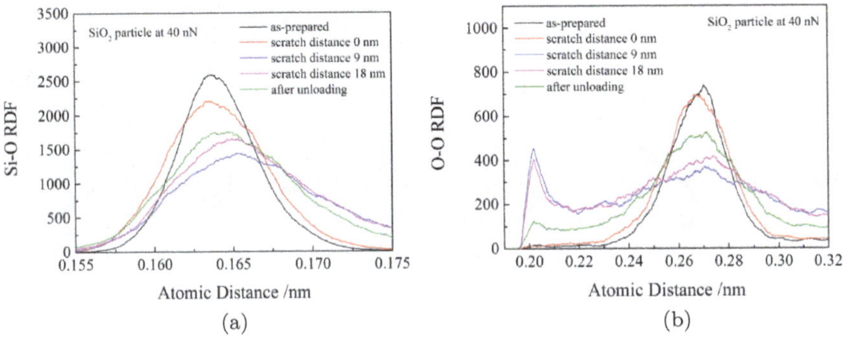

Figure 6.72. RDF of SiO_2 abrasive in different states at a positive load of 40 nN. (a) Si–O RDF, (b) O–O RDF.

the Si–O atoms to decrease again. The *RDF* for O–O atoms also shows changes: The height of the main peak decreases as the scratch progresses, resulting in a wider distribution of O–O atomic distances, ranging from 0.198 nm to 0.32 nm. An additional peak at a distance of about 0.2 nm becomes evident, and its height and width increase as the scratch advances. This additional peak indicates a partial decrease in the distance of the O–O atoms.

Upon unloading and elastic recovery, the height of the main O–O peak increases, the width decreases, and the extra peak shrinks. These observations collectively suggest that the abrasive undergoes compaction plastic deformation during the scratching process. The Si–O, O–O atomic *RDF* at a scratching distance of 9 nm shows more intense abrasive deformation compared to that at 18 nm. This is because of the different forces at the friction interface at these two displacement points. The interaction force between the abrasive and the specimen's surface is stronger at the 9 nm scratch distance, which is the main stage of material adhesion removal. Conversely, the interaction force weakens at the 18 nm scratch distance as indicated by the friction force in Figure 6.63.

In Figure 6.73, the *RDF* characteristics of SiO_2 abrasive under different loads after the end of scratching (before unloading) show a similar deformation trend of Si–O and O–O atomic *RDF* to that of Figure 6.72. The primary plasticity of amorphous SiO_2 abrasive is compaction, and the deformation of abrasive compaction becomes more severe with increasing load.

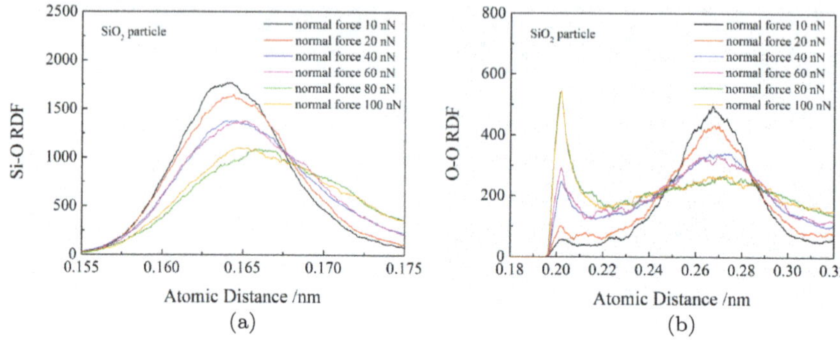

Figure 6.73. Effect of loading on RDF of SiO_2 abrasives. (a) Si–O RDF, (b) O–O RDF.

6.4.4.3 *Mechanism of amorphous SiO_2 film removal and surface quality analysis*

In order to study the efficiency of material removal and the quality of the surface during scratching, we analyzed the displacement of atoms in an amorphous SiO_2 film. The analysis is shown in Figures 6.74 and 6.75. We found that as the load increases, more Si and O atoms in the SiO_2 film are displaced. A wider abrasion mark corresponds to a greater number of displaced atoms. In some local areas, particularly in the 0.2–0.6 nm film, atoms colored in rose red are detached from the substrate surface, leading to film rupture. When the load is low (20 nN), localized regions of the film undergo stripping, while at a high load (40 nN), atoms peel off throughout the groove. This peeling phenomenon occurs in films of 0.2 nm and 0.4 nm thickness but not in films thicker than 0.4 nm. It is also worth noting that at a load of 20 nN, the displacement of atoms in films of different thicknesses does not vary much, while at 40 nN, the 0.6 nm film displaces the most atoms, followed by the 0.4 nm film, and the 0.2 nm film displaces the least.

The graph in Figure 6.75 displays the number of atoms displaced (i.e., removed) by a 0.4 nm thick SiO_2 film under varying positive loads. It indicates that the number of atoms removed increases rapidly and then levels off as the scratching distance increases. Furthermore, at higher positive loads, such as 80 nN and 100 nN, the curves representing atom removal almost overlap. This is due to the fact that when the load exceeds 80 nN, the SiO_2 film within the

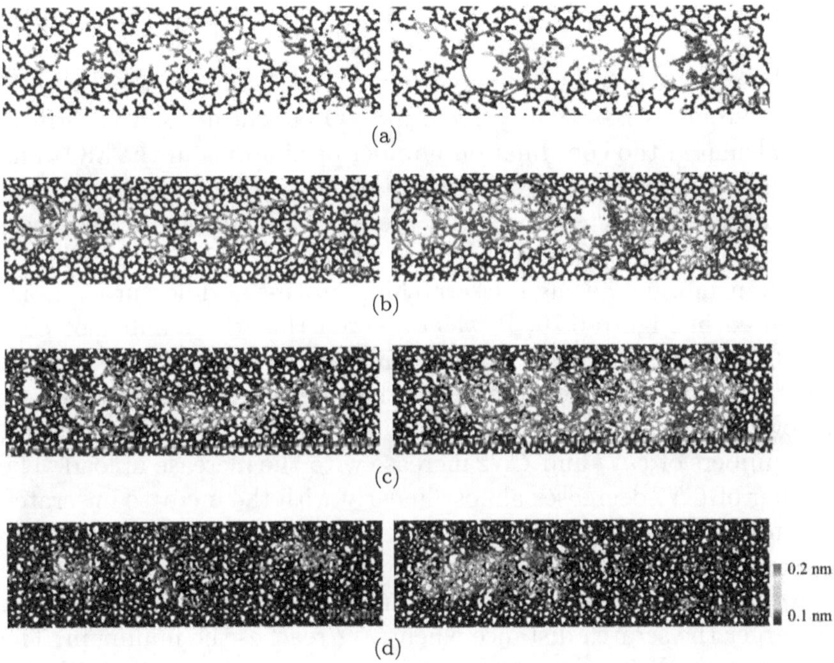

Figure 6.74. Atomic displacement diagrams of SiO_2 films under different positive loads (left: 20 nN; right 40 nN). (a) 0.2 nm film, (b) 0.4 nm film, (c) 0.6 nm film, (d) 0.8 nm film.

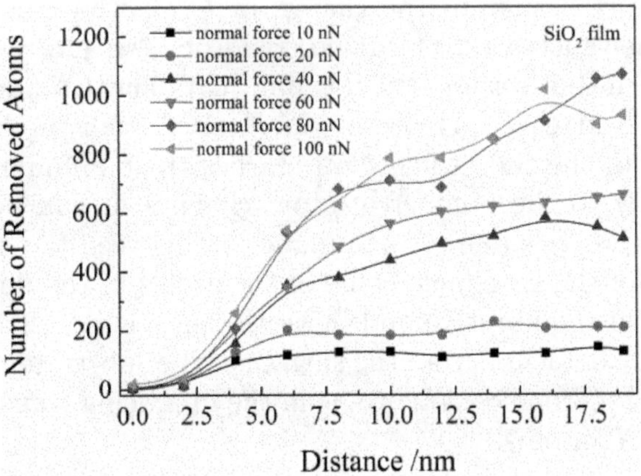

Figure 6.75. Effect of load on atom removal by SiO_2 membrane.

scratched area is nearly completely removed, resulting in the number of atoms removed approaching saturation.

The present researchers found that when the distance between Si and O atoms is less than 0.2 nm, a Si–O chemical bond is formed. We calculated the coordination number of Si atoms in the SiO_2 film and the number of Si–O chemical bonds by counting the number of O atoms connected to the Si atoms. The changes in the coordination number of Si atoms and the number of Si–O bonds in the 0.4 nm membrane during the scratching process at different loads are presented in Figure 6.76. It was observed that the number of *CN1* and *CN2* atoms increases rapidly with the increase of load at first, and the higher the load, the faster the rate of increase, followed by a slower rate of increase. Moreover, at the same scratch distance, the numbers of *CN1* and *CN2* increase with the increase in load. The number of *CN3* decreases almost linearly with the increase in scratch distance. The larger the load, the larger the rate of *CN3* decrease. The change in the number of *CN4* shows a tendency to decrease and then increase, as shown in Figure 6.76(d). The larger the load, the larger the scratch distance when *CN4* reaches the minimum. The number of *CN5* increases monotonically and also increase with the increase in the positive load.

The change in the number of Si–O bonds is the combined result of considering the number of atoms of *CNx* (x is from 1 to 5), as shown in Figure 6.76(f). The trend of the change in the number of Si–O bonds is similar to the change of *CN4*, which also decreases first and then increases. From the trend of *CNx* and Si–O changes, it can be seen that for a particular film, *CN3* and *CN4* decrease, and *CN1*, *CN2*, and *CN5* increase in the interval from the beginning of the scratch to the time when *CN4* reaches the minimum value.

According to the number of atom changes, it can be concluded that in this process, due to the interaction between the abrasive and the surface of the specimen, some of the atoms in the film adhere to the surface of the abrasive and, with the movement of the abrasive, are gradually detached from the surface of the substrate. This results in part of the *CN3* and *CN4* atoms in the film being transformed into *CN1* and *CN2* atoms.

Due to the abrasive's role of the vertical load, the distance between some of the atoms in the film is reduced, compaction deformation occurs, and part of the *CN4* atoms in the film is transformed

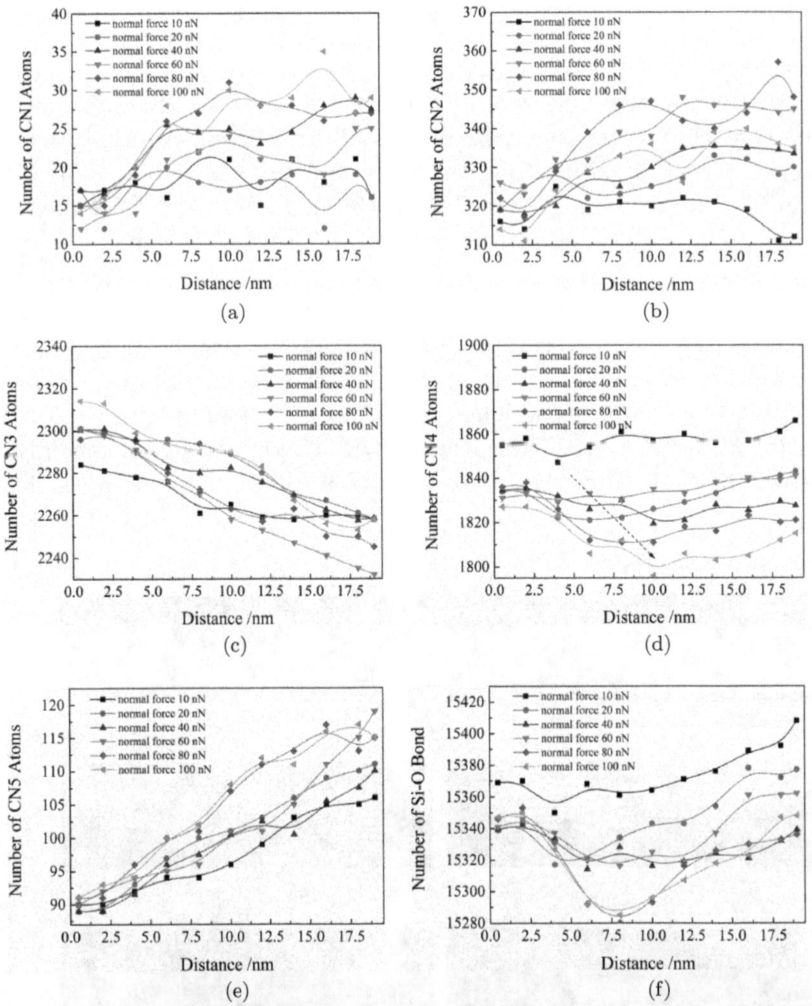

Figure 6.76. Effect of loading in scratches on the coordination number of Si atoms and the number of Si–O bonds in 0.4 nm films. (a) *CN1*, (b) *CN2*, (c) *CN3*, (d) *CN4*, (e) *CN5*, (f) Si–O bonding.

into *CN5* atoms. In the later stage of the scratch, which is mainly a compaction stage accompanied by a small amount of atom removal, *CN1* and *CN2* increase slightly, *CN3* decreases significantly, the number of *CN4* and *CN5* increases, and *CN3* transforms into *CN4* and *CN5* atoms.

6.4.4.4 *Single-crystal silicon removal mechanism and surface quality analysis*

In this study, we looked at how the thickness of a SiO_2 film affects the scratching process on monocrystalline silicon. We analyzed the displacement of silicon atoms in composite specimens with SiO_2 film thickness ranging from 0.0 to 0.6 nm under loads of 20 nN and 40 nN. The results, shown in Figure 6.77, revealed that as the SiO_2 film thickness increased, the displacement of monocrystalline silicon atoms underneath the film under the same load significantly decreased. We also observed differences in the distribution of atoms within the groove. For pure single-crystal silicon specimens, a large number of Si atoms experienced greater displacement throughout the entire abrasion range. However, for the single-crystal silicon underneath the SiO_2 film, the Si atoms with a greater change in displacement were mainly concentrated in the early stage of the scratch.

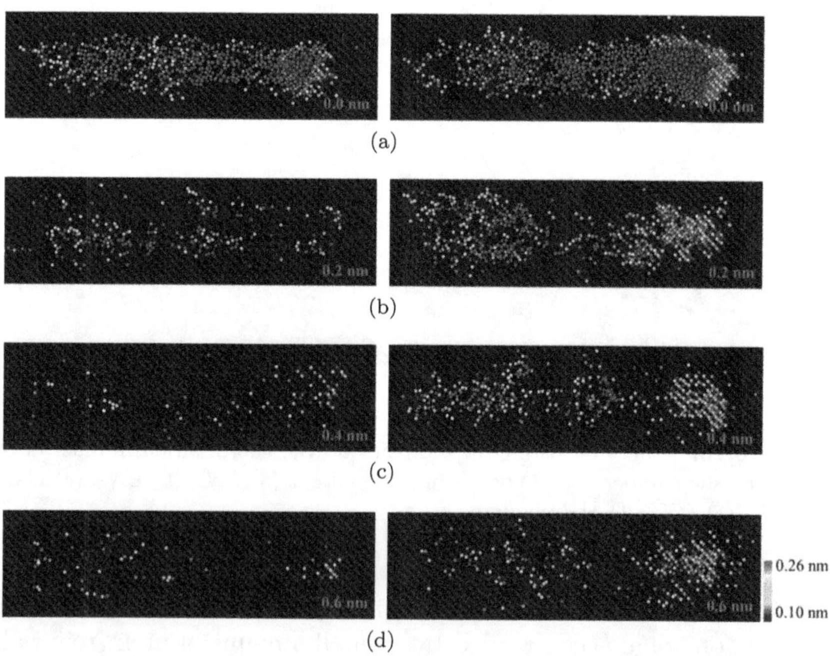

Figure 6.77. Atomic displacement diagram of single-crystal silicon under different loads (left: 20 nN; right 40 nN). (a) 0.0 nm silicon, (b) 0.2 nm silicon, (c) 0.4 nm silicon, (d) 0.6 nm silicon.

Additionally, we observed that the number of atoms undergoing displacement increased with the increase in loading.

Furthermore, we investigated the effect of positive loading on the removal of single-crystal silicon by counting the number of Si atoms with atomic displacements greater than 0.26 nm at different loads at the end of the scratch (before unloading). As depicted in Figure 6.78(a), we found a distinct peak in the number of removed Si atoms at a scratch distance of about 7 nm, and the peak became larger with increasing load. We also observed an increase in the number of removed Si atoms with increasing load at the same scratch distance. It was observed that the number of removed Si atoms increased as the load increased at the same scratch distance. The relationship between the number of atoms removed from the specimen (the sum of atoms removed from the SiO_2 film and the monocrystalline silicon surface) as a function of the scratch distance is shown in Figure 6.78(b). It was found that the number of atoms removed from the specimen (including SiO_2 file and single-crystal silicon together) at different loads increases rapidly and then slows down as the scratch proceeds. This suggests that the interaction of the abrasive with the atoms on the surface of the composite material during indentation leads to the formation of bonding bridges between the SiO_2 abrasive and the film.

During the subsequent scratching process, due to the movement of the abrasive, the surface atoms adhering to the abrasive undergo tensile deformation under shear and finally leave the specimen surface.

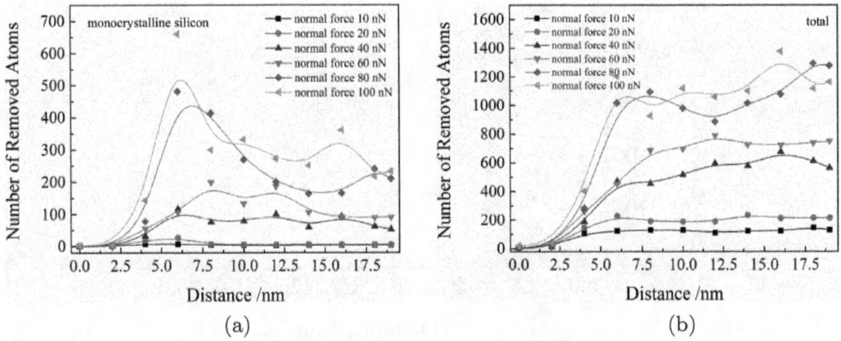

Figure 6.78. Effect of positive loading on the removal of atoms from a material. (a) Monocrystalline silicon, (b) composite specimens.

At lower loads, there are fewer contact atoms between the abrasive and the specimen surface, and only the atoms on the surface of the SiO_2 film are detached from the specimen surface and move with the abrasive during the scratching. At higher loads, the number of contact atoms on the surface of the abrasive and the film is higher, and under the action of the shear force, the backing bonds of the bond bridge break first because the bonding energy of the Si–O bond is larger than the bonding energy of the Si–Si bond. This results in the stripping of a large number of atoms from the surface of the specimen, leading to an increase in friction.

In the subsequent scratching process, due to the deformation of the amorphous SiO_2 abrasive, the contact area between the abrasive and the surface of the specimen increases, and the pressure between the contact interfaces decreases under the same positive load. Additionally, more water molecules are present between the friction interfaces, as shown in Figure 6.79, decreasing the contact between the abrasive and the composite sample. This also has a lubricating effect, resulting in a reduction in the removal rate of atoms.

The main concentration of atom removal is in the early stage of scratching due to the following reasons: The deformation of the abrasive is small, the contact area between the abrasive and the specimen is small, and the stress on the specimen is increased. The contact

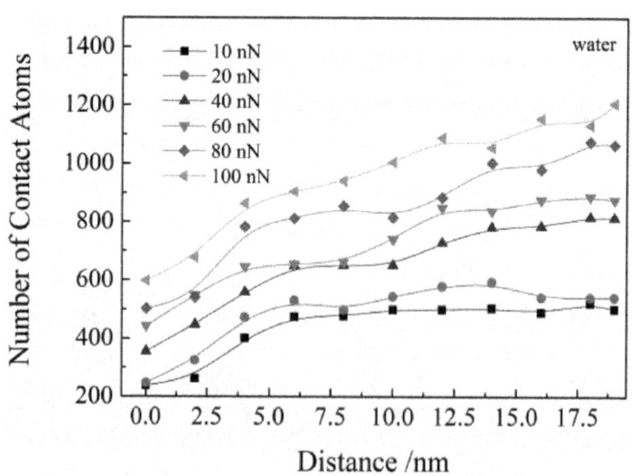

Figure 6.79. Effect of positive load on the number of contact atoms between abrasive and water film.

time between the abrasive and the specimen is long in nanoindentation loading, and more Si–O–Si bonding bridges will form between the abrasive and the specimen. With the movement of the abrasive, a large number of atoms on the surface of the specimen and the atoms of the monocrystalline silicon are detached, and atom removal takes place. The number of contacts between the abrasive and water molecules is small, weakening the lubrication ability of water.

In Figure 6.80, the cross-section of the atomic displacement of single-crystal silicon under a 0.4 nm film under different loads is shown. It's observed that a small number of atoms are removed from the surface of single-crystal silicon when the load is low. However, the number of displaced atoms significantly increases when the load is raised to 100 nN. These displaced atoms can be easily removed under the action of the polishing solution and abrasive material, and the main area of removal is concentrated in the early stage of the scratch.

As phase transformation is the primary mode of plastic deformation of single-crystal silicon, the phase transformation of single-crystal silicon under 0.4 nm film at different loads after the end of scratching is analyzed in Figure 6.81. In this figure, light blue, blue, green, and orange colors represent the Si atoms with coordination numbers of 2, 3, 4, and 5, respectively. It's important to note that the Si atoms with coordination numbers of 2 and 3 are the surface atoms. It has been observed that at a force of 20 nN, the surface of single-crystal silicon remains perfect in structure with no phase transition taking place. However, at 100 nN, there is a significant increase in atoms with a coordination number of 5. These atoms are dispersed and do not form the expected bct-5 phase. Instead, they create a thin layer of amorphous silicon along with other atoms with

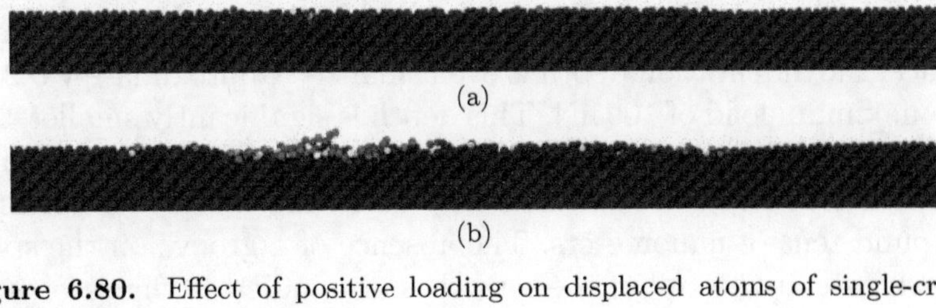

(a)

(b)

Figure 6.80. Effect of positive loading on displaced atoms of single-crystal silicon. (a) 20 nN, (b) 100 nN.

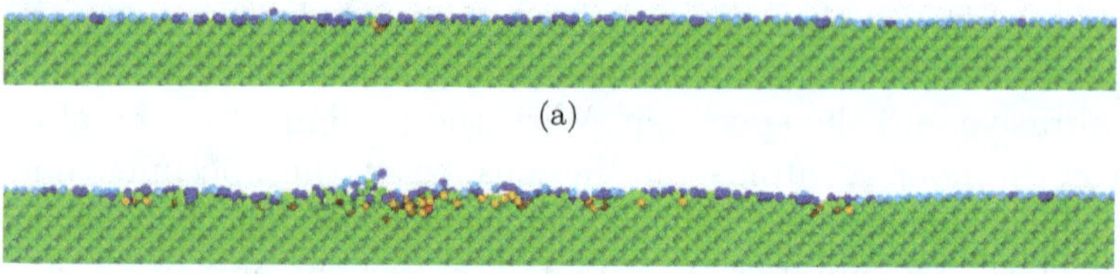

Figure 6.81. Effect of positive loading on phase transition of single-crystal silicon. (a) 20 nN, (b) 100 nN.

different coordination numbers. The amorphous silicon is mainly distributed in the scratches at the early stage of the process, while in other areas, its thickness is reduced and the amount is smaller. Even at 100 nN, it was observed that the single-crystal silicon under the amorphous silicon showed almost no lattice distortion and did not produce any other defects.

In order to investigate the mechanism of atom removal, the contact radii of the abrasive and the specimen surface were calculated to be about 1.3 nm and 2.0 nm, respectively. This was under the positive loads of 10 nN and 80 nN. The contact pressures were initially obtained as 4.8 and 6.3 GPa. However, it is worth noting that these values are actually less due to the soft materials—the abrasive and the SiO_2 film—deforming plastically. Thus, the contact pressures between the abrasive and the substrate should be lower than 4.8 and 6.3 GPa. The hardness of the amorphous SiO_2 film is approximately 7.4 GPa, which means that it is not feasible to remove atoms from the specimen's surface through abrasive mechanical scratching with a positive load below this value.

During the *CMP* process, the individual abrasive's indentation depth into the substrate is less than 1 nm. Furthermore, through simulations, it was found that due to the plastic deformation of the soft amorphous SiO_2 abrasive and the SiO_2 film on the substrate's surface, the indentation depth is even smaller—approximately 0.2 nm at a maximum load of 100 nN. This depth is significantly smaller than reported in previous studies by Son *et al.* [90]. The suggested minimum cutting depth for mechanically removing individual abrasives is around tens of nanometers. The absence of a groove on the specimen surface at the end of the scratches also indicates that atoms are removed from the specimen surface through adhesive removal.

6.4.5 *Effect of abrasive size on friction and wear*

In order to study how different sizes of abrasives affect the friction and wear behavior of SiO_2/Si composites, a simulation was conducted to analyze the scratching behavior of a specimen with a film thickness of 0.4 nm. The simulation involved a larger amorphous SiO_2 abrasive with a cylindrical radius at the upper end and a lower hemisphere with a radius of 4 nm, following the same abrasive shape and simulation details, as described in Section 6.4.2. The friction curves at different loads are presented in Figure 6.82. The analysis revealed that at lower loads (20 nN), the friction trend was consistent with the findings of the previous Section 6.4.3, with peaks occurring. However, for larger abrasives, the friction initially increased and then remained relatively constant at 60 nN. Furthermore, it was observed that the friction value increased with the radius of the abrasive for the same scratching distance.

During the experiment, we counted the number of atoms removed from both the amorphous SiO_2 film and the single-crystal silicon at the end of the scratch, using the method illustrated in Figure 6.83. At a load of 20 nN, we observed that more atoms were removed from the SiO_2 film when scratched by a 4 nm abrasive compared to a 3 nm abrasive. Similarly, more Si atoms were removed from the surface of single-crystal silicon when using the 4 nm abrasive. When the load was increased to 60 nN, we found that the amount of atoms removed from the SiO_2 film by the 4 nm abrasive was significantly higher than that removed by the 3 nm abrasive. This also resulted

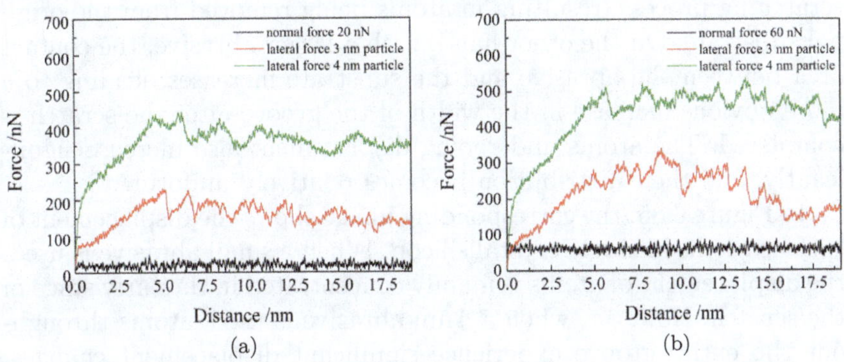

Figure 6.82. Effect of abrasive size on friction under different positive loads. (a) 20 nN, (b) 60 nN.

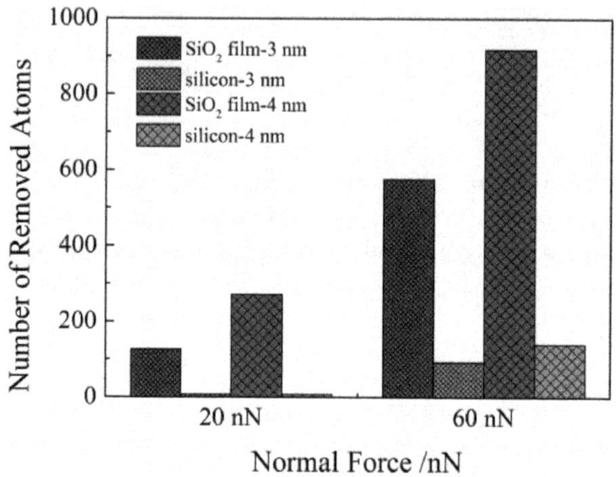

Figure 6.83. Plot of the effect of different abrasive sizes on the number of atoms removed.

in a higher number of single-crystal silicon atoms being removed. The higher total number of atoms removed using the larger diameter abrasive led to an increase in the external force needed to overcome the atoms on the original substrate surface, thus increasing friction during scratching with the larger abrasive.

The graph in Figure 6.84 illustrates the changes in atomic displacement of SiO_2 of 0.4 nm after being scratched with two different abrasives under a 60 nN force. It is evident that with a smaller SiO_2 abrasive, the primary stage of atomic displacement occurs early in the scratching process, resulting in atoms being removed from the original substrate. On the other hand, with a larger abrasive, the contact area between the abrasive and the substrate increases, leading to a more obvious increase in the width of the groove after the scratch is completed. The atoms undergoing displacement also increase significantly, and their distribution becomes relatively uniform.

In Figure 6.85, the corresponding image shows the displacement of surface atoms in single-crystal silicon. When a small abrasive is used, the displacement of atoms is mainly concentrated in the early stage of the scratch. However, when a 4 nm abrasive is used, atoms throughout the entire groove experience significant displacement changes. This suggests that with larger abrasives and higher loads, the number of atoms removed increases, resulting in higher friction.

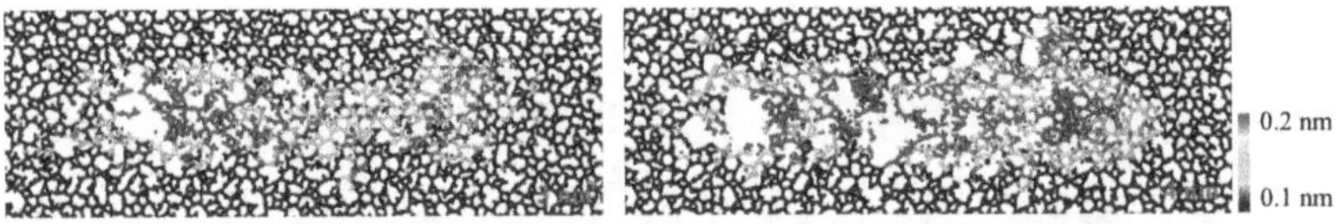

Figure 6.84. Effect of different sizes of abrasives on the atomic displacement of 0.4 nm membrane. (a) 3 nm, (b) 4 nm.

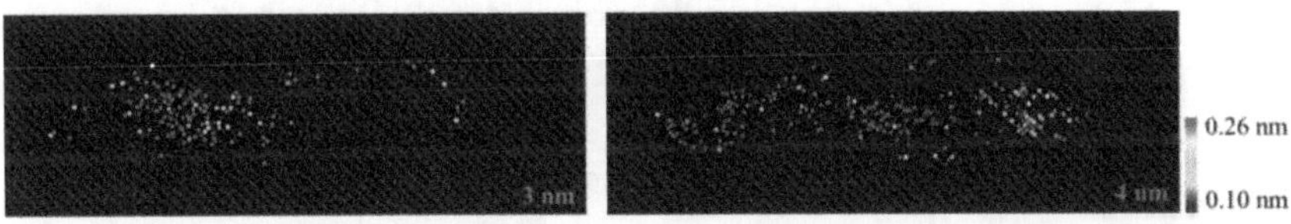

Figure 6.85. Effect of different sizes of abrasives on the atomic displacement of single-crystal silicon. (a) 3 nm, (b) 4 nm.

6.4.6 *Effect of abrasive type on wear*

In order to compare how different abrasives affect the removal of SiO_2/Si materials, we conducted a two-body abrasion using diamond abrasives with a radius of 3 nm. The load and friction curves for scratching a specimen with a film thickness of 0.4 nm using diamond abrasives, as well as the comparison curves of friction between the two types of abrasives, are illustrated in Figure 6.86. The results indicate that at a load of 60 nN, the friction force remains around 20 nN throughout the scratching process using diamond abrasives, which is significantly different from the results obtained using amorphous SiO_2 abrasives. This difference is attributed to the interaction between the SiO_2/Si matrix and the abrasive during the scratching process. With diamond abrasives, there is mechanical contact between the abrasive and the substrate, resulting in minimal interaction between the two contact surfaces. On the other hand, when using SiO_2 abrasives, the contact between the abrasive and the substrate occurs through adhesion, leading to higher interaction forces.

The cross-section of the specimen at the end of the scratch is shown in Figure 6.87. It was discovered that compaction of the SiO_2 film on the substrate surface had occurred, and the scratch was a groove with no accumulation of atoms around it. The analysis of the

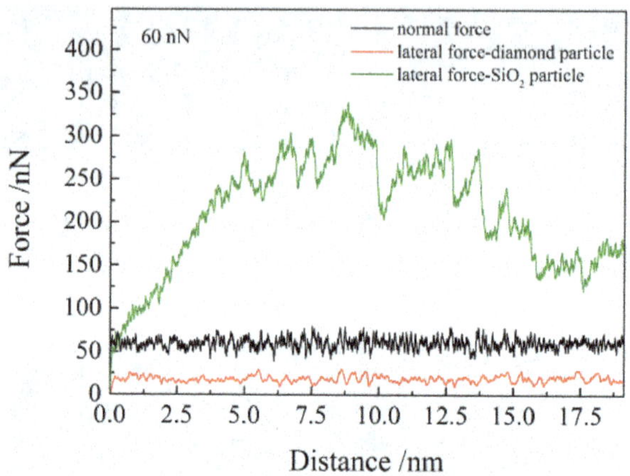

Figure 6.86. Effect of different types of abrasives on friction.

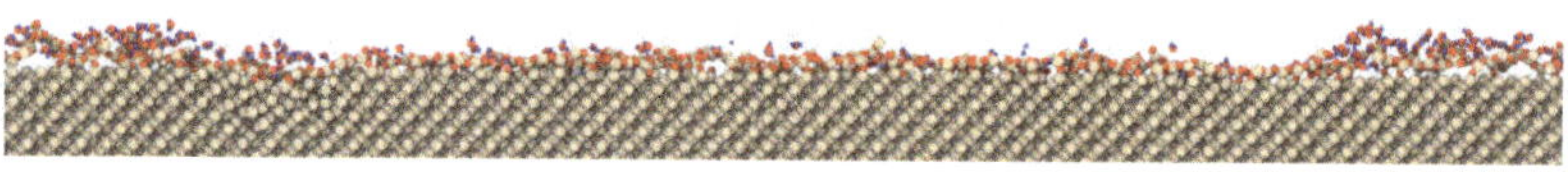

Figure 6.87. Cross-section of 0.4 nm specimen after scratching.

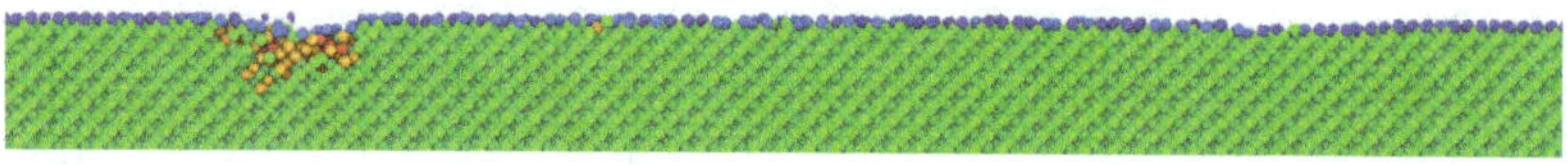

Figure 6.88. Phase transition cross-section of 0.4 nm single-crystal silicon after scratching.

phase transition of the single-crystal silicon beneath the film revealed a significant phase transition within the groove at the beginning of the scratch, forming the smaller bct-5 phase. No considerable phase transition was observed within the rest of the groove, as depicted in Figure 6.88. When the diamond abrasive was unloaded, it was found that statistically, no atoms were removed. This is due to the weaker van der Waals forces between the diamond and the abrasive, which prevents adhesion and removal of a large number of atoms. On the other hand, smaller positive loads resulted in a minimal depth of diamond abrasive pressed into the substrate, which was less than the proposed minimum removal depth by Son *et al.* [90]. Therefore, the removal efficiency of diamond abrasives was almost zero. It was observed that only when the abrasive indentation depth exceeded the proposed minimum mechanical removal depth did the post-scratch morphology reveal the appearance of mechanical grooves and atomic build-up. Sun *et al.* [55] confirmed this through molecular simulations of diamond abrasive nano-scratched single-crystal silicon specimens. They found that atom removal occurs at larger indentation depths, resulting in obvious grooves on the specimen surface, and the scratched single-crystal silicon specimen undergoes large plastic deformation, with a layer of amorphous silicon on the surface and a phase transition of atoms underneath. Zhang *et al.* [91] observed dislocations and stacking faults on the specimen after the single-crystal silicon nano-scratch experiment. Hence, soft abrasives (colloidal SiO_2) are more suitable for the removal of monocrystalline silicon and SiO_2 media through chemical mechanical polishing.

6.4.7 *Effect of SiO_2 abrasive properties on friction wear*

6.4.7.1 *Friction changes*

To study the impact of different types of amorphous SiO_2 abrasives on the removal of atoms from SiO_2/Si bilayer composite specimens, we conducted calculations on specimens with a film thickness of 0.4 nm using two types of abrasives (rigid and non-rigid SiO_2) and three different scratch cases. The specific simulation system consisted of rigid SiO_2 abrasives and SiO_2/Si composite specimens, rigid SiO_2 abrasives and water-containing film SiO_2/Si composite specimens, and non-rigid SiO_2 abrasives and aqueous film SiO_2/Si composite specimens.

The bilayer composite specimen grew up to 51 nm in the scratch direction, maintaining constant width and thickness scales and using an abrasive radius of 3 nm. We applied a constant load of 40 nN for both nanoindentation and scratching, with a water film thickness of 0.3 nm, and an abrasive scratching speed of 25 m/s. The other details of the simulations remained the same as in Section 6.4.2.

The results displayed in Figure 6.89 indicate that the friction force during the scratching process using a non-rigid SiO_2 abrasive is the highest, while the friction force of the rigid abrasive scratching the composite specimen with an aqueous film is the lowest for the same positive load of 40 nN. It is also observed that the friction force

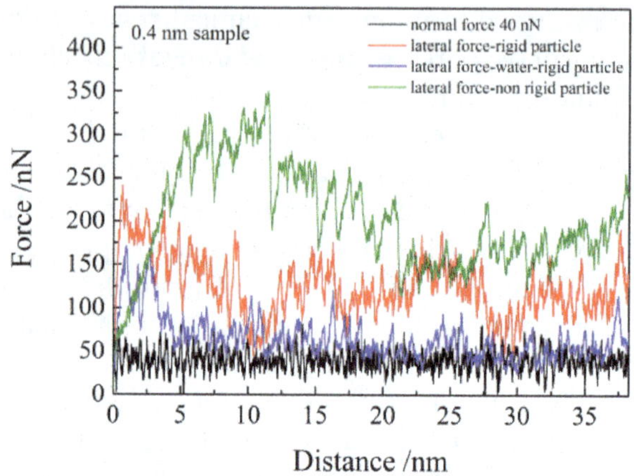

Figure 6.89. Effect of different SiO_2 abrasives on the friction of 0.4 nm specimens.

during the scratching process initially increases, then decreases, and ultimately reaches a state of equilibrium.

The graph in Figure 6.90 shows the overall atomic removal (the sum of the atomic removal from the film and monocrystalline silicon) of the amorphous SiO_2 film and monocrystalline silicon during the scratching process for three different conditions. The results indicated that the maximum number of atoms removed occurred when a rigid abrasive was used to scratch the specimen without an aqueous film. Conversely, the minimum number of atoms removed was observed when a rigid abrasive was used to scratch the specimen with an aqueous film. Additionally, the number of atoms removed during the scratching process was found to be in between these extremes for the non-rigid abrasive. It was also observed that the number of atoms removed increased rapidly and then leveled off, with the non-rigid abrasives demonstrating a moderate rate in scratching.

6.4.7.2 *Change in the number of atoms in contact between the abrasive and the specimen*

The diagrams in Figure 6.91 illustrate the change in the number of contact atoms between the SiO_2 abrasive and the SiO_2 film and water film of the composite specimen. It reveals that when using a non-rigid abrasive, the number of contact atoms between the abrasive

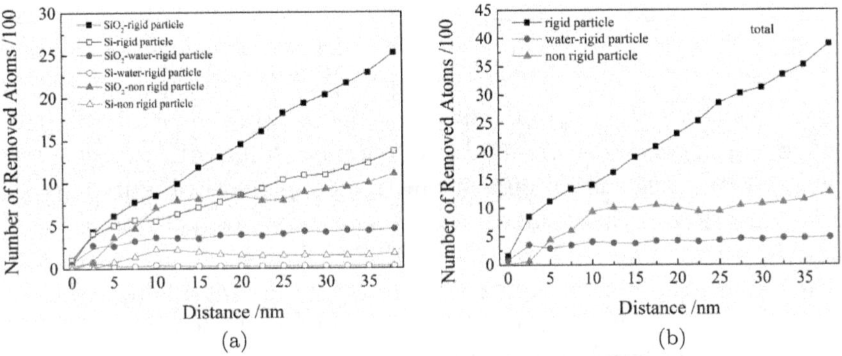

Figure 6.90. Effect of different SiO_2 abrasives on the removal of atoms from 0.4 nm specimens. (a) SiO_2 film and single-crystal silicon atom removal, (b) total atom removal.

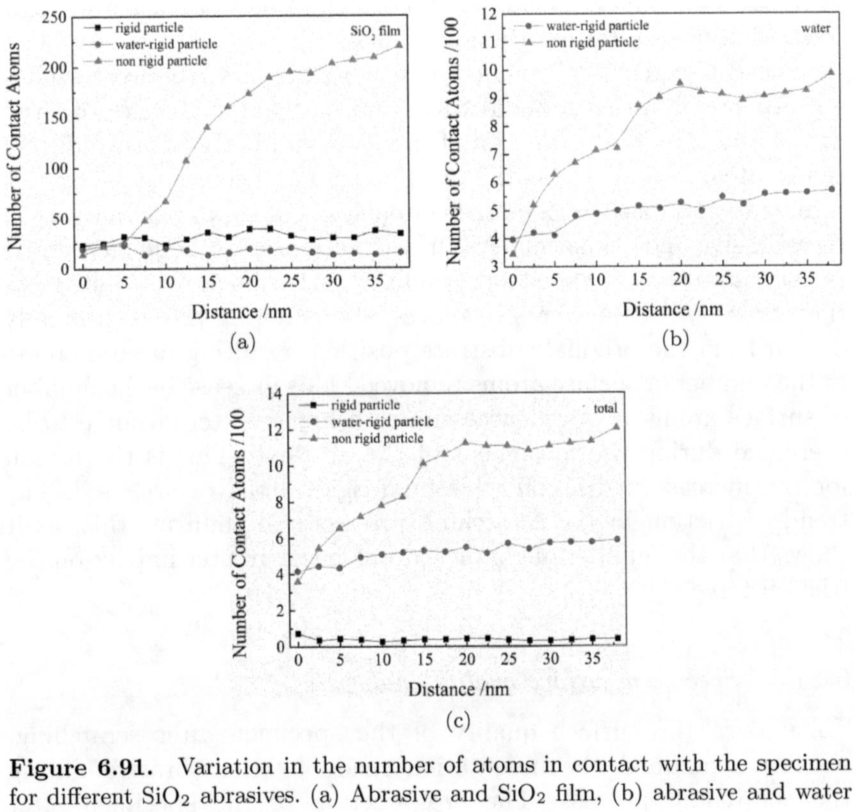

Figure 6.91. Variation in the number of atoms in contact with the specimen for different SiO_2 abrasives. (a) Abrasive and SiO_2 film, (b) abrasive and water film, (c) abrasive and specimen.

and the SiO_2 film is significantly higher compared to the other two cases. Additionally, the number of contact atoms exhibits a pattern of rapid increase followed by a gradual rise. When a rigid body abrasive is used, the number of contact atoms oscillates around a specific value, and it has the least number of contact atoms with the SiO_2 film of the water film-containing specimen, as depicted in Figure 6.91(a). The variation in the number of contact atoms between the abrasive and the water film is illustrated in Figure 6.91(b). It is evident that when a non-rigid abrasive is utilized, the contact atoms between the abrasive and the water film increase rapidly and then plateau, with a much greater number of contact atoms compared to the rigid abrasive. Overall trends in contact atom changes between the abrasive and the specimen during the scratching process are demonstrated in Figure 6.91(c), with the use of a non-rigid body abrasive, resulting in the most significant change, and the rigid body abrasive without a water film, resulting in the least change.

The difference in the number of contact atoms between different abrasives and amorphous SiO_2 film and water film is primarily due to the plastic deformation of the non-rigid body abrasive during the scratching process. As the positive load and composite specimen come into play (as described in Section 6.3), the contact area between the abrasive and the specimen surface increases. Consequently, more surface atoms of the specimen (mainly SiO_2 film atoms) adhere to the abrasive. As the abrasive moves, the adhering atoms gradually detach from the original substrate position, resulting in an increase in the number of surface atoms removed. This increase in the number of surface atoms removed necessitates a higher external force to be overcome during the movement of the abrasive. This is the reason for the increase in friction when non-rigid abrasives are used. The trend of friction in the scratching process also confirms this, as it shows that the number of contact atoms has a greater impact on the friction force.

6.4.7.3 *Specimen surface quality analysis*

To analyze the surface quality of the specimen after scratching, we first examined the indentation made by the abrasive during the scratching process. The cross-section of the specimen after the scratching, with a film thickness of 0.4 nm before scratching,

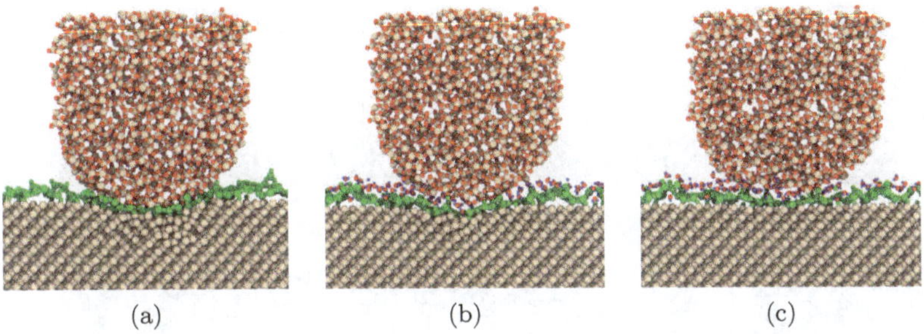

Figure 6.92. Cross-section at the end of the indentation for different SiO_2 abrasives. (a) Rigid particles without water film, (b) rigid particles with water film, (c) non-rigid particles with water film.

is depicted in Figure 6.92. In the figures, the green atoms represent the amorphous SiO_2 film, and the blue represents the hydrogen atoms in the aqueous film. We found that under the same positive load (40 nN), the rigid body abrasive caused the greatest depth of penetration into the surface of the composite specimen, resulting in significant deformation of the SiO_2 film and the single-crystal silicon substrate beneath the indenter, as shown in Figure 6.92(a). However, the SiO_2 film and monocrystalline silicon under the non-rigid body abrasive did not change significantly. Even the lattice distortion of the monocrystalline silicon matrix could not be observed, as shown in Figure 6.92(c). This demonstrates that the specimen of rigid abrasive is between the two cases with the water-containing film.

The atomic displacement changes of a 0.4 nm amorphous SiO_2 membrane under three scratching states after the end of scratching are depicted in Figure 6.93. It is observed that when a rigid abrasive scratches the surface of the anhydrous membrane specimen, the number of displaced atoms in the SiO_2 membrane is the highest, leading to ruptures in many places (Figure 6.93(a)). This is because of the greater depth of the rigid abrasive pressed into the specimen, resulting in more atoms being contacted, and hence more displacement in the surface. In the presence of a water film, the wear scar is narrower and the number of displaced atoms is lower (Figure 6.93(b)), as the water film dissipates some of the energy, making the abrasive indentation depth shallower and hindering direct contact between the abrasive and the SiO_2 film. When a non-rigid abrasive is used, the wear scar becomes wider and the film ruptures at the beginning

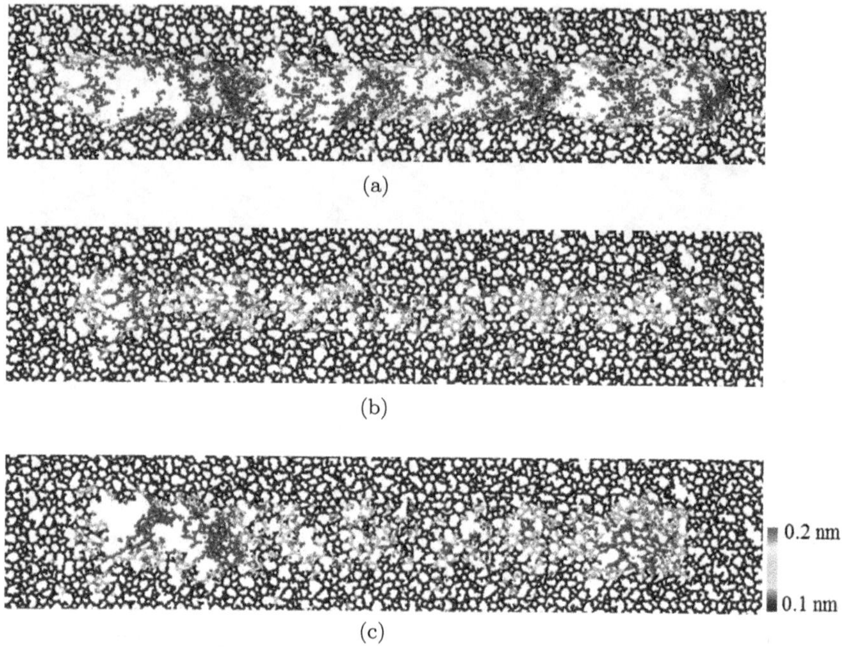

Figure 6.93. Configuration of atomic displacement changes of 0.4 nm SiO_2 film scratches by different SiO_2 abrasives. (a) Rigid particles without water film, (b) rigid particles with water film, (c) non-rigid particles with water film.

of the scratch, consequently increasing the number of atoms undergoing displacement (Figure 6.93(c)). This is because of the plastic deformation of the non-rigid abrasive during the scratching process, leading to increased contact area and detachment of a large number of atoms from the original position of the substrate.

Figure 6.94 illustrates the phase transition analysis of a single-crystal silicon substrate under a SiO_2 film with a 0.4 nm thickness in three scratching cases. In the figure, blue, light blue, green, yellow, and red represent silicon atoms with coordination numbers of 1, 2, 3, 4, and 5, respectively. It is observed that when using a rigid abrasive to scratch the anhydrous film, a significant number of silicon atoms on the surface of the single-crystal silicon are shown by the blue and light blue colors in the figure. This results in the removal of the highest number of silicon atoms. However, after the scratching process, a thicker layer of amorphous silicon, composed of different coordination numbers of silicon atoms, is generated on the single-crystal

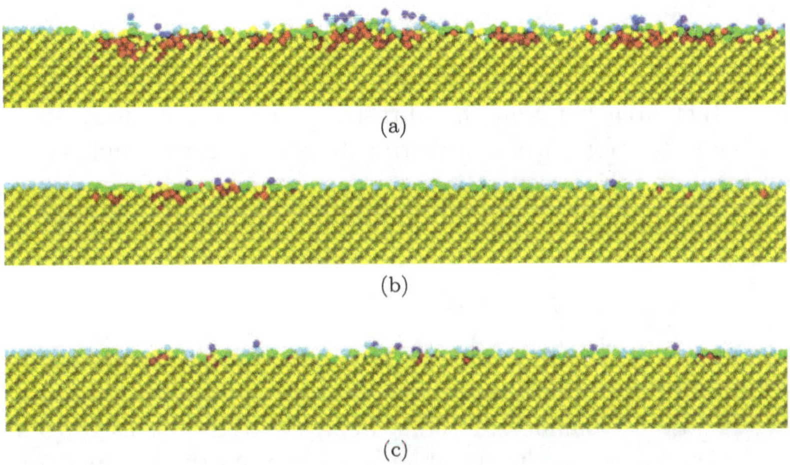

Figure 6.94. Configuration of the phase transition of single-crystal silicon scratches by different SiO_2 abrasives. (a) Rigid particles without water film, (b) rigid particles with water film, (c) non-rigid particles with water film.

silicon surface, causing a more serious amorphous phase transition, as indicated in Figure 6.94(a).

When using a non-rigid abrasive to scratch a sample with a water film, the contact area between the abrasive and the sample increases, resulting in relatively low stress on the single-crystal silicon surface. Due to the adhesive removal mechanism between the abrasive and the specimen, some silicon atoms are removed from the surface of the single-crystal silicon, leading to increased removal of atoms. After scratching, a small number of amorphous atoms are generated on the single-crystal silicon surface, but the substrate does not have any crystalline distortion, as shown in Figure 6.94(c). Conversely, when using a rigid abrasive in a water-containing film to scratch the composite specimen surface, the amount of atom removal at the end is less, leading to the generation of more phase change atoms on the single-crystal silicon surface. This is depicted in Figure 6.94(b).

Based on the analysis above, it can be concluded that when the surface of a SiO_2/Si composite specimen is scratched with a non-rigid SiO_2 abrasive under the same positive load (40 nN), the friction force experienced by the abrasive is the greatest. As a result, more atoms are removed from the amorphous SiO_2 film and the monocrystalline silicon. The adhesion removal of single atoms occurs on the monocrystalline silicon surface, leaving the surface with good quality and no substrate defects. This is attributed to the occurrence of plastic deformation during the scratching process, which increases the contact area between the abrasive and the specimen's surface. This leads to more atoms adhering to the abrasive, and with the movement of the abrasive, these atoms are removed from the original position on the substrate.

6.4.8 *Removal of models*

The friction and wear processes involve placing abrasive material under a constant positive load near the surface of the specimen. After creating a balance in the system, allow the abrasive to slide at a certain speed on the specimen's surface. As the hardness of the amorphous SiO_2 abrasive is less than that of single-crystal silicon, the abrasive causes slight plastic deformation on the surface of composite specimens. However, single-crystal silicon does not produce lattice distortion, phase transition, dislocation defects, etc. As indicated in Figure 6.95, with scratching, this results in the amorphous SiO_2 atoms adhering to the surface of the abrasive which are detached and then removed from the single-crystal silicon substrate. The presence of the amorphous SiO_2 film on the surface of the single-crystal silicon not only increases the number of atoms removed during wear but also protects the substrate, reducing the degree of damage.

6.4.9 *Summary*

We investigated the friction characteristics between interfaces by simulating the scratching of colloidal SiO_2 abrasive on a single-crystal silicon substrate coated with an amorphous SiO_2 film. The aim was to understand the removal efficiency and removal mechanism of SiO_2/Si composites. The main findings are as follows:

(1) The friction of composite specimens with different SiO_2 film thicknesses varies under the same positive load. Friction increases with increasing SiO_2 film thickness, but the friction does not change significantly when the film thickness is above 0.6 nm. With increasing scratch distance, the friction force initially rises, then decreases, and finally reaches a peak. For the

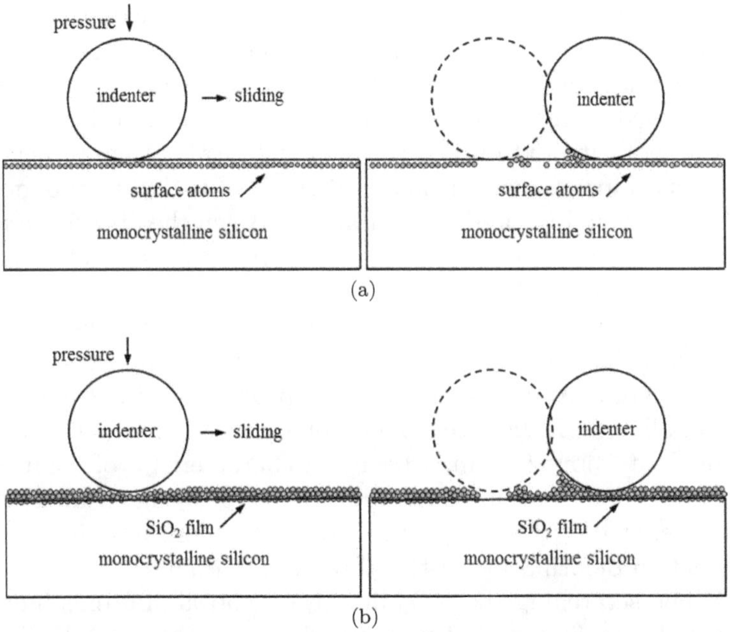

Figure 6.95. Friction wear model diagram. (a) Without SiO_2 membrane, (b) with SiO_2 membrane.

same specimen (0.4 nm specimen), the friction force at the same scratch distance increases with increasing positive load. This is caused by the rise of contact area between the SiO_2 abrasive and the specimen, the increase of interfacial interaction force, and the sliding of the abrasive, which results in more atoms on the surface of the specimen being detached from the original matrix.

(2) The friction curve of the composite specimen exhibits irregular serrated characteristics, which is attributed to the amorphous SiO_2 exhibiting inhomogeneous plastic deformation characteristics. The interaction between the abrasive and the specimen causes some atoms on the specimen surface to adhere to the abrasive. As the scratching occurs, these atoms and the abrasive material deviate from their original positions, requiring an increased external force to overcome this displacement of atoms, resulting in increased friction. When some of the atoms are detached from the substrate, the friction force decreases.

(3) When a small positive load (20 nN) is applied, the removal of material is not significantly related to the thickness of the SiO_2 film. However, at a load of 40 nN, the amount of material removed increases and then decreases with film thickness, with the most amount of removal occurring in the 0.6 nm thick specimen. This increase in material removal is due to the growing interaction force and contact area between the abrasive and the specimen surface caused by the increasing load. The removal mechanism is adhesion.
(4) During the scratching process, it was observed that the Si–O peak height of the *RDF* of the Si–O atoms in the amorphous SiO_2 abrasive decreased, and the peak width increased. Additionally, the *RDF* of the O–O atoms showed an additional small peak at about 0.2 nm, which is a characteristic of compaction plastic deformation of the amorphous SiO_2 material. Furthermore, the higher the positive load, the more intense the compaction deformation of the abrasive material.
(5) When scratching the SiO_2 film, the removal of atoms increases rapidly at first and then more slowly as the scratch distance increases. A higher load leads to the removal of more atoms for the same slip distance. The diagrams showing atomic displacement during scratching reveal that a wider wear scar and more displaced O and Si atoms occur with a larger positive load. Thinner SiO_2 films experience detachment from the substrate across a larger range when scratched, whereas thicker films, above 0.4 nm, mainly undergo atom detachment in the early stage of scratching. Thicker films show adhesion removal of individual atoms without any group detachments.
(6) As the scratching distance increases, the *CN* of Si atoms in the film changes. *CN1* and *CN2* increase, *CN3* decreases, *CN4* decreases and then increases, and *CN5* increases. The number of Si–O bonds decreases and then increases with the increase of the slip distance. Based on the changes in the coordination number of atoms, it can be concluded that the early stage of the scratch, up to the point where *CN4* is at a minimum, is the main stage of atom removal. Some *CN3* and *CN4* atoms in the film transform into *CN1* and *CN2* atoms, and some *CN4* atoms transform into *CN5* atoms due to the compaction deformation

caused by the reduction of distance between atoms under positive loading. In the later stage of the scratch, which is mainly a compaction stage with a small amount of atom removal, *CN1* and *CN2* increase slightly, *CN3* decreases significantly, and the number of *CN4* and *CN5* increases. *CN3* transforms into *CN4* and *CN5* atoms.

(7) The removal of single-crystal silicon atoms under the film increases with the scratch distance. There is a distinct peak at a scratch distance of about 7 nm, which becomes larger with increasing load. The overall atom removal from the specimen (the sum of atoms removed from the SiO_2 film and the single-crystal silicon) increases rapidly and slowly with slip distance. This is due to the interaction between the abrasive and the SiO_2/Si substrate. The friction interface between the formation of Si–O–Si bond bridges occurs as the abrasive slides under the action of shear. This happens because the bond energy of the Si–O bond is greater than the Si–Si bond energy. So, the backward bonding of the bridge is broken, and part of the SiO_2 atoms and monocrystalline Si atoms are detached from the substrate. They are then removed in the polishing fluid, explaining the high friction at the beginning of the scratch.

(8) When larger colloidal SiO_2 abrasives were used to scratch the SiO_2/Si substrate, it was found that the friction during the scratching process was greater and that the number of atoms removed from the SiO_2 film and the monocrystalline silicon was higher. This was caused by a larger area of interaction between the larger abrasive and the substrate, leading to more atoms on the sample surface being detached from the substrate.

(9) When using diamond abrasives, the friction load during the scratching process is much smaller compared to colloidal SiO_2 abrasives of the same size. After scratching the surface of the specimen, only a SiO_2 film compaction occurs, and there is zero atom removal. However, after scratching the surface of monocrystalline silicon, phase transformation is more serious, leading to the generation of the bct-5 phases. When a soft amorphous SiO_2 film scratches the monocrystalline silicon surface, only a small amount of non-crystalline atoms are removed, and no substrate lattice distortion occurs.

(10) The calculation of scratches using SiO_2 abrasives in three different states shows that the material removal efficiency is higher when non-rigid abrasives are used. The removal of the monocrystalline silicon surface is of good quality, with no defects in the substrate.

References

[1] Huang QH. International technology roadmap for semiconductors (ITRS) 2013 edition overview (3). *China Integr Circuit*, 2014, 23: 14–26.

[2] Ayazi F, Najafi K. High aspect-ratio combined poly and single-crystal silicon (HARPSS) MEMS technology. *J Microelectromech Syst*, 2000, 9: 288–294.

[3] Zhai Yujing, Yang Kaiyong, Pan Yao. History, current status and prospect of gyroscope. *Fly Missiles*, 2018, 12: 84–88.

[4] Xu J, Luo JB, Wang LL, *et al.* The crystallographic change in subsurface layer of the silicon single crystal polished by chemical mechanical polishing. *Tribol Int*, 2007, 40: 285–289.

[5] Estragnat E, Tang G, Liang H, *et al.* Experimental investigation on mechanisms of silicon chemical mechanical polishing. *J Electron Mater*, 2004, 33: 334–339.

[6] Zarudi I, Zhang LC, Cheong WCD, *et al.* The difference of phase distributions in silicon after indentation with Berkovich and spherical indenters. *Acta Mater*, 2005, 53: 4795–4800.

[7] Mylvaganam K, Zhang LC. Effect of residual stresses on the stability of bct-5 silicon. *Comput Mater Sci*, 2014, 81: 10–14.

[8] Jang JI, Lance MJ, Wen SQ, *et al.* Indentation-induced phase transformations in silicon: influences of load, rate and indenter angle on the transformation behavior. *Acta Mater*, 2005, 53: 1759–1770.

[9] Wu M, Liang YF, Jiang JZ, *et al.* Structure and properties of dense silica glass. *Sci Rep*, 2012, 2: 398–404.

[10] Mott NF. The viscosity of vitreous silicon dioxide. *Philos Mag B*, 1987, 56: 257–262.

[11] Chowdhury SC, Haque BZ, Gillespie JW, Jr. Molecular dynamics simulations of the structure and mechanical properties of silica glass using ReaxFF. *J Mater Sci*, 2016, 51: 10139–10159.

[12] Jin W, Kalia RK, Vashishta P, *et al.* Structural transformation, intermediate-range order, and dynamical behavior of SiO_2 glass at high-pressures. *Phys Rev Lett*, 1993, 71: 3146–3149.

[13] Munetoh S, Motooka T, Moriguchi K, *et al.* Interatomic potential for Si–O systems using Tersoff parameterization. *Comput Mater Sci*, 2007, 39: 334–339.

[14] Zhao S, Xue J. Modification of graphene supported on SiO_2 substrate with swift heavy ions from atomistic simulation point. *Carbon*, 2015, 93: 169–179.

[15] Chen J, Zhang G, Li BW. Thermal contact resistance across nanoscale silicon dioxide and silicon interface. *J Appl Phys*, 2012, 112: 064319.

[16] Shi J, Chen J, Wei X, *et al.* Influence of normal load on the three-body abrasion behaviour of monocrystalline silicon with ellipsoidal particle. *RSC Adv*, 2017, 7: 30929–30940.

[17] Plimpton S. Fast parallel algorithms for short-range molecular-dynamics. *J Comput Phys*, 1995, 117: 1–19.

[18] Stukowski A. Visualization and analysis of atomistic simulation data with OVITO-the open visualization tool. *Model Simul Mater Sci Eng*, 2010, 18: 015012.

[19] Srikanth K, Ananthakrishna G. Dynamical approach to displacement jumps in nanoindentation experiments. *Phys Rev B*, 2017, 95: 014107.

[20] Gouldstone A, Koh HJ, Zeng KY, *et al.* Discrete and continuous deformation during nanoindentation of thin films. *Acta Mater*, 2000, 48: 2277–2295.

[21] Peng P, Liao G, Shi T, *et al.* Molecular dynamic simulations of nanoindentation in aluminum thin film on silicon substrate. *Appl Surf Sci*, 2010, 256: 6284–6290.

[22] Golovin YI, Ivolgin VI, Khonik VA, *et al.* Serrated plastic flow during nanoindentation of a bulk metallic glass. *Scr Mater*, 2001, 45: 947–952.

[23] Oliver WC, Pharr GM. Measurement of hardness and elastic modulus by instrumented indentation: Advances in understanding and refinements to methodology. *J Mater Res*, 2004, 19: 3–20.

[24] Goel S, Kovalchenko A, Stukowski A, *et al.* Influence of microstructure on the cutting behaviour of silicon. *Acta Mater*, 2016, 105: 464–478.

[25] Xu KW, Hou GL, Hendrix BC, *et al.* Prediction of nanoindentation hardness profile from a load-displacement curve. *J Mater Res*, 1998, 13: 3519–3526.

[26] Zarudi I, Zhang LC. Structure changes in mono-crystalline silicon subjected to indentation—experimental findings. *Tribol Int*, 1999, 32: 701–712.

[27] Goel S, Faisal NH, Luo X, *et al.* Nanoindentation of polysilicon and single crystal silicon: Molecular dynamics simulation and experimental validation. *J Phys D Appl Phys*, 2014, 47: 275304.
[28] Sun J, Li C, Jing H, *et al.* Nanoindentation induced deformation and pop-in events in a silicon crystal: Molecular dynamics simulation and experiment. *Sci Rep*, 2017, 7: 10282.
[29] Budnitzki M, Kuna M. Stress induced phase transitions in silicon. *J Mech Phys Solids*, 2016, 95: 64–91.
[30] Goel S, Luo XC, Agrawal A, *et al.* Diamond machining of silicon: A review of advances in molecular dynamics simulation. *Int J Mach Tools Manuf*, 2015, 88: 131–164.
[31] Lin Y-H, Chen T-C, Yang P-F, *et al.* Atomic-level simulations of nanoindentation-induced phase transformation in mono-crystalline silicon. *Appl Surf Sci*, 2007, 254: 1415–1422.
[32] Wu YQ, Yang XY, Xu YB. Cross-sectional electron microscopy observation on the amorphized indentation region in 001 single-crystal silicon. *Acta Mater*, 1999, 47: 2431–2436.
[33] Zarudi I, Cheong WCD, Zou J, *et al.* Atomistic structure of monocrystalline silicon in surface nano-modification. *Nanotechnology*, 2004, 15: 104–107.
[34] Sun J, Ma A, Jiang J, *et al.* Orientation-dependent mechanical behavior and phase transformation of mono-crystalline silicon. *J Appl Phys*, 2016, 119: 095904.
[35] Han J, Xu S, Sun J, *et al.* Pressure-induced amorphization in the nanoindentation of single crystalline silicon. *RSC Adv*, 2017, 7: 1357–1362.
[36] Han J, Fang L, Sun J, *et al.* Length-dependent mechanical properties of gold nanowires. *J Appl Phys*, 2012, 112: 114314.
[37] Kang KW, Cai W. Size and temperature effects on the fracture mechanisms of silicon nanowires: Molecular dynamics simulations. *Int J Plast*, 2010, 26: 1387–1401.
[38] Tang DM, Ren CL, Wang MS, *et al.* Mechanical properties of Si nanowires as revealed by in situ transmission electron microscopy and molecular dynamics simulations. *Nano Lett*, 2012, 12: 1898–1904.
[39] Wang XD, Guo J, Chen C, *et al.* A simple method to control nanotribology behaviors of monocrystalline silicon. *J Appl Phys*, 2016, 119: 044304.
[40] Yajima S, Hasegawa Y, Hayashi J, *et al.* Synthesis of continuous silicon-carbide fiber with high-tensile strength and high Youngs modulus. 1. Synthesis of polycarbosilane as precursor. *J Mater Sci*, 1978, 13: 2569–2576.

[41] Yang Y, Liao N, Zhang M, *et al.* Numerical investigation on the bond strength of a SiCN-based multi-layer coating system. *J Alloys Compd*, 2017, 710: 468–471.
[42] Yang Y, Liao N, Zhang M, *et al.* Evaluation of the elastic-plastic properties of SiCN coating system by finite element simulations. *J Eur Ceram Soc*, 2017, 37: 3891–3897.
[43] Schiffmann KI. Determination of fracture toughness of bulk materials and thin films by nanoindentation: comparison of different models. *Philos Mag*, 2011, 91: 1163–1178.
[44] Huang H, Zhao HW, Shi CL, *et al.* Randomness and statistical laws of indentation-induced pop-out in single crystal silicon. *Materials*, 2013, 6: 1496–1505.
[45] Saha R, Nix WD. Effects of the substrate on the determination of thin film mechanical properties by nanoindentation. *Acta Mater*, 2002, 50: 23–38.
[46] Chen RL, Wu YH, Lei H, *et al.* Study of material removal processes of the crystal silicon substrate covered by an oxide film under a silica cluster impact: Molecular dynamics simulation. *Appl Surf Sci*, 2014, 305: 609–616.
[47] Chen R, Luo J, Guo D, *et al.* Extrusion formation mechanism on silicon surface under the silica cluster impact studied by molecular dynamics simulation. *J Appl Phys*, 2008, 104: 104907.
[48] Tabor D. *The hardness of metals. Meas Tech*, 1951, 5: 281.
[49] Pharr GM. Measurement of mechanical properties by ultra-low load indentation. *Mater Sci Eng A*, 1998, 253: 151–159.
[50] Komanduri R, Chandrasekaran N, Raff LM. MD simulation of indentation and scratching of single crystal aluminum. *Wear*, 2000, 240: 113–143.
[51] Mulliah D, D Kenny S, McGee E, *et al.* Atomistic modelling of ploughing friction in silver, iron and silicon. *Nanotechnology*, 2006, 17: 1807–1818.
[52] Fang T-H, Chang W-J, Weng C-I. Nanoindentation and nanomachining characteristics of gold and platinum thin films. *Mater Sci Eng A*, 2006, 430: 332–340.
[53] Rimsza JM, Yeon J, van Duin ACT, *et al.* Water Interactions with Nanoporous Silica: Comparison of ReaxFF and ab Initio based Molecular Dynamics Simulations. *J Phys Chem C*, 2016, 120: 24803–24816.
[54] Shi J, Chen J, Fang L, *et al.* Atomistic scale nanoscratching behavior of monocrystalline Cu influenced by water film in CMP process. *Appl Surf Sci*, 2018, 435: 983–992.
[55] Sun J, Fang L, Han J, *et al.* Abrasive wear of nanoscale single crystal silicon. *Wear*, 2013, 307: 119–125.

[56] Sun JP, Li C, Jing H, *et al.* Nanoindentation induced deformation and pop-in events in a silicon crystal: Molecular dynamics simulation and experiment. *Sci Rep*, 2017, 7: 10282.

[57] Liu Q, Wang L, Shen S. Effect of surface roughness on elastic limit of silicon nanowires. *Comput Mater Sci*, 2015, 101: 267–274.

[58] Bradby JE, Williams JS, Wong-Leung J, *et al.* Nanoindentation-induced deformation of Ge. *Appl Phys Lett*, 2002, 80: 2651–2653.

[59] Spaepen F. Homogeneous flow of metallic glasses: A free volume perspective. *Scr Mater*, 2006, 54: 363–367.

[60] Spaepen F. Microscopic mechanism for steady-state inhomogeneous flow in metallic glasses. *Acta Metall*, 1977, 25: 407–415.

[61] Argon AS. Plastic-deformation in metallic glasses. *Acta Metall*, 1979, 27: 47–58.

[62] Yavari AR, LeMoulec A, Inoue A, *et al.* Excess free volume in metallic glasses measured by X-ray diffraction. *Acta Mater*, 2005, 53: 1611–1619.

[63] Wright WJ, Hufnagel TC, Nix WD. Free volume coalescence and void formation in shear bands in metallic glass. *J Appl Phys*, 2003, 93: 1432–1437.

[64] Chen J, Shi J, Zhang M, *et al.* Effect of indentation speed on deformation behaviors of surface modified silicon: A molecular dynamics study. *Comput Mater Sci*, 2018, 155: 1–10.

[65] Bhowmick R, Raghavan R, Chattopadhyay K, *et al.* Plastic flow softening in a bulk metallic glass. *Acta Mater*, 2006, 54: 4221–4228.

[66] Gerbig YB, Michaels CA, Forster AM, *et al.* Indentation device for in situ Raman spectroscopic and optical studies. *Rev Sci Instrum*, 2012, 83: 125106.

[67] Gerbig YB, Michaels CA, Forster AM, *et al.* In situ observation of the indentation-induced phase transformation of silicon thin films. *Phys Rev B*, 2012, 85: 104102.

[68] Sanz-Navarro CF, Kenny SD, Smith R. Atomistic simulations of structural transformations of silicon surfaces under nanoindentation. *Nanotechnology*, 2004, 15: 692–697.

[69] Han J, Xu S, Sun JP, *et al.* Pressure-induced amorphization in the nanoindentation of single crystalline silicon. *RSC Adv*, 2017, 7: 1357–1362.

[70] Taylor TA, Barrett CR. Creep and recovery of silicon single-crystals. *Mater Sci Eng*, 1972, 10: 93.

[71] Gerbig YB, Michaels CA, Cook RF. In situ observations of Berkovich indentation induced phase transitions in crystalline silicon films. *Scr Mater*, 2016, 120: 19–22.

[72] Cao ZQ, Zhang X. Nanoindentation creep of plasma-enhanced chemical vapor deposited silicon oxide thin films. *Scr Mater*, 2007, 56: 249–252.

[73] Wang F, Li JM, Huang P, *et al.* Nanoscale creep deformation in Zr-based metallic glass. *Intermetallics*, 2013, 38: 156–160.

[74] Xue F, Wang F, Huang P, *et al.* Structural inhomogeneity and strain rate dependent indentation size effect in Zr-based metallic glass. *Mater Sci Eng A*, 2016, 655: 373–378.

[75] Cao ZH, Li PY, Meng XK. Nanoindentation creep behaviors of amorphous, tetragonal, and bcc Ta films. *Mater Sci Eng A*, 2009, 516: 253–258.

[76] Wang Z, Sun BA, Bai HY, *et al.* Evolution of hidden localized flow during glass-to-liquid transition in metallic glass. *Nat Commun*, 2014, 5: 5823.

[77] Liu ST, Wang Z, Peng HL, *et al.* The activation energy and volume of flow units of metallic glasses. *Scr Mater*, 2012, 67: 9–12.

[78] Pan D, Inoue A, Sakurai T, *et al.* Experimental characterization of shear transformation zones for plastic flow of bulk metallic glasses. *Proc Natl Acad Sci USA*, 2008, 105: 14769–14772.

[79] Sun J, Fang L, Han J, *et al.* Phase transformations of mono-crystal silicon induced by two-body and three-body abrasion in nanoscale. *Comput Mater Sci*, 2014, 82: 140–150.

[80] Fang L, Sun K, Shi J, *et al.* Movement patterns of ellipsoidal particles with different axial ratios in three-body abrasion of monocrystalline copper: A large scale molecular dynamics study. *RSC Adv*, 2017, 7: 26790–26800.

[81] Si L, Guo D, Luo J, *et al.* Planarization process of single crystalline silicon asperity under abrasive rolling effect studied by molecular dynamics simulation. *Appl Phys A*, 2012, 109: 119–126.

[82] Si LN, Guo D, Luo JB, *et al.* Monoatomic layer removal mechanism in chemical mechanical polishing process: A molecular dynamics study. *J Appl Phys*, 2010, 107: 064310.

[83] Chen R, Liang M, Luo J, *et al.* Comparison of surface damage under the dry and wet impact: Molecular dynamics simulation. *Appl Surf Sci*, 2011, 258: 1756–1761.

[84] Wang XD, Yu JX, Chen L, *et al.* Effects of water and oxygen on the tribochemical wear of monocrystalline Si(100) against SiO_2 sphere by simulating the contact conditions in MEMS. *Wear*, 2011, 271: 1681–1688.

[85] Barnette AL, Asay DB, Kim D, *et al.* Experimental and density functional theory study of the tribochemical wear behavior of SiO_2 in humid and alcohol vapor environments. *Langmuir*, 2009, 25: 13052–13061.

[86] Yu J, Kim SH, Yu B, *et al.* Role of tribochemistry in nanowear of single-crystalline silicon. *ACS Appl Mater Interfaces*, 2012, 4: 1585–1593.

[87] Wang X, Kim SH, Chen C, *et al.* Humidity dependence of tribochemical wear of monocrystalline silicon. *ACS Appl Mater Interfaces*, 2015, 7: 14785–14792.

[88] Ren J, Zhao J, Dong Z, *et al.* Molecular dynamics study on the mechanism of AFM-based nanoscratching process with water-layer lubrication. *Appl Surf Sci*, 2015, 346: 84–98.

[89] Boda D, Henderson D. The effects of deviations from Lorentz-Berthelot rules on the properties of a simple mixture. *Mol Phys*, 2008, 106: 2367–2370.

[90] Son SM, Lim HS, Ahn JH. Effects of the friction coefficient on the minimum cutting thickness in micro cutting. *Int J Mach Tools Manuf*, 2005, 45: 529–535.

[91] Zhang LC, Zarudi I. Towards a deeper understanding of plastic deformation in mono-crystalline silicon. *Int J Mech Sci*, 2001, 43: 1985–1996.

Chapter 7

Effect of Grain Boundaries on the Mechanical Properties of Nano-Polycrystalline Copper

7.1 General

Copper, one of the earliest non-ferrous metals discovered by mankind, boasts excellent ductility, electrical conductivity, abrasion resistance, corrosion resistance, and environmental friendliness. It is widely used in the electric power, electronics industry, machinery manufacturing, light industry construction, and various other fields [1–3]. Its characteristics make it commonly used in electric power, electronics, machinery manufacturing, light industry, and construction. Copper and its alloys are utilized in devices of varying scales, from nanoscale ultra-high precision devices to macro-scale components. With the recent development of China's semiconductor chip industry and micro-electromechanical systems (MEMS), copper has gained significant attention as a strategic advanced electronic material. Consequently, it has been included in China's key development plan for many years, covering the refining of copper, precision machining, and optimization of organizational properties through research.

Even though single-crystal copper has numerous advantages, its preparation is challenging and it is not easy to form and process, which limits its industrial application compared to polycrystalline copper. Polycrystalline copper is widely used and studied. As early as the beginning of the 20th century, research on polycrystalline materials began. It was recognized that grain refinement could significantly improve the mechanical properties of the material, such as strength and hardness. Grain boundaries are crucial in strengthening polycrystalline materials, and it is commonly believed that they impede the movement of dislocations, causing dislocation pile-up. In the 1950s, Hall and Petch proposed the famous Hall–Petch (H-P) formula to describe the relationship between strength and grain size [4, 5]. The mechanical properties of polycrystalline materials have been continuously improved under the guidance of this theory. However, when the grain size decreases to a critical value, the strength and hardness of the material decrease, known as the anti-H-P phenomenon. The main cause of this phenomenon has been attributed to the transformation of the deformation mechanism [6, 7]. When the grain size is small, the proportion of low-symmetry atoms in the grain boundaries increases, causing the deformation mechanism to change from dislocation motion to grain boundary activity, including grain boundary deformation, rotation, sliding, and creep.

In order to better understand and improve the conditions leading to the anti-H-P phenomenon, numerous experimental and simulation studies have been conducted on grain boundary activities. However, due to the complex structural characteristics of polycrystalline materials, a unified theory on the deformation mechanism of polycrystalline materials is still lacking. This is mainly due to the complexity of the structure of nano-polycrystalline materials, and the difficulty of observing and quantifying microscopic deformation using current experimental equipment. Nevertheless, experiments remain an important method for studying the deformation of polycrystalline metallic materials and provide a basis for simulation studies. For example, they can help determine the critical grain size of polycrystalline copper for the anti-H-P phenomenon, the release of dislocations, the formation of twins, and the strengthening mechanism, as well as the strain rate sensitivity (*SRS*). *SRS* is an important parameter reflecting the deformation mechanism, and its value is mainly influenced by the activity of grain boundaries.

The Molecular Dynamics (*MD*) method is widely used for simulating nano-polycrystalline systems due to its ability to model atomic interactions at the nanoscale and provide detailed insights into the deformation process. However, the method's limitations include the large strain rate of metal deformation due to time scale constraints. Despite this, valuable insights into deformation mechanisms can still be obtained from the simulation results. One drawback of the method is the oversimplified configuration of polycrystalline systems, which does not accurately represent real-world models and leads to reduced coupling between neighboring grains. As a result, certain phenomena, such as the initial deformation stage of grain boundaries, cannot be accurately replicated in the simulation. Current research on grain boundary activity primarily focuses on creep behavior at high temperatures and grain boundary deformation under shear strain [11, 12], with limited quantitative statistical analysis of grain boundary activity at room temperature.

In order to more accurately and comprehensively analyze the deformation mechanism of nano-polycrystalline copper, specifically the grain boundary activities, it is important to propose a method for constructing a more reasonable polycrystalline system for MD studies. To address this issue, this topic primarily combines the phase field theory to create a more natural polycrystalline system. It utilizes molecular dynamics to gain a deeper understanding of the deformation mechanism of polycrystalline copper film substrate materials at the nanoscale. This includes a comparative analysis of the new model and the traditional model, as well as a comparative study with experiments, the development of algorithms for quantitatively analyzing grain boundary activities, and a quantitative analysis of the effect of grain boundary activities on mechanical behaviors. Therefore, this study is of great significance for the in-depth understanding of the deformation mechanism and mechanical properties of polycrystalline copper.

7.2 Construction of Polycrystalline Systems Based on Multiphase Field Theory

7.2.1 *Introduction to multiphase field theory*

The multiphase field (*MPF*) model, initially proposed by Steinbach *et al.* [13] and enhanced by Takaki *et al.* [14, 15], was employed to simulate grain growth. The simulation results aligned well with theoretical values and those obtained through the meta-cellular automata method, indicating the model's high accuracy. Moreover, the improved model utilizes the Kim algorithm [16] to solve the phase field equations, thus ensuring good stability.

7.2.2 *Polycrystalline copper grain growth models*

The phase field equation is crucial for modeling grain growth in a polycrystalline system containing N grains. Depending on the structural differences within the crystal, the ordinal parameter Φ_i (where the value of i represents the specific grain number) is used to categorize the grains into three parts: the inside of grain i ($\Phi_i = 1$), the grain boundaries ($0 < \Phi_i < 1$), and other grains outside of grain $i(\Phi_{\text{other}} = 0)$. This is illustrated in Figure 7.1 [14].

All phase field ordinal coefficients in the n_{th} grain are not independent and must satisfy the following equation:

$$\sum_{i=1}^{N} \phi_i = 1 \tag{7.1}$$

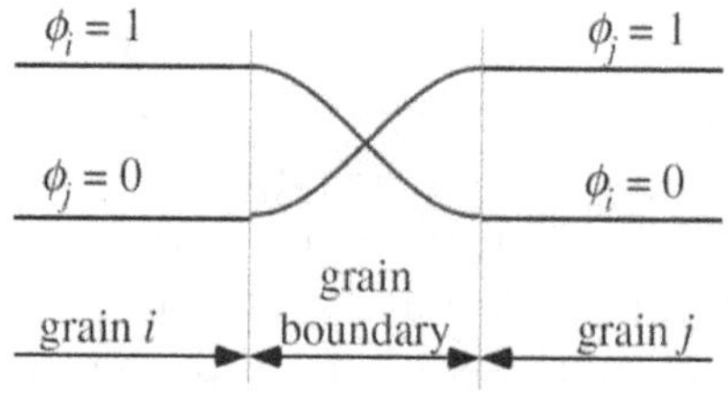

Figure 7.1. Sketch of the variation of phase field ordinal coefficients near grain boundaries [14].

The free energy equation in the phase field is defined as followed:

$$F = \int_v \left[\sum_{i=1}^{N} \sum_{j=i+1}^{N} \left(-\frac{a_{ij}^2}{2} \nabla\phi_i \nabla\phi_j + W_{ij}\phi_i\phi_j \right) + f_e \right] dV \tag{7.2}$$

where a_{ij} is the gradient coefficient; W_{ij} is the grain boundary energy barrier, and they are related to the grain boundary energy γ_{ij} and interface thickness δ_j, and f_e is the free energy density, which is a function of the ordinal parameter. N in Eqs. (7.1) and (7.2) can be replaced by $n = \sum_{i=1}^{N} \sigma_i$, and when $0 < \phi_i < 1$, $\sigma_i = 1$, and for other ϕ_i values, $\sigma_i = 0$, which can be obtained by substituting into Eq. (7.1):

$$\sum_{i=1}^{n} \phi_i = 1 \tag{7.3}$$

From the literature [14], we can obtain the initial phase field evolution equations:

$$\dot{\phi}_l = -\sum_{j=1}^{n} \frac{2M_{ij}^{\phi}}{n} \left(\frac{\delta F}{\delta\phi_i} - \frac{\delta F}{\delta\phi_j} \right) \tag{7.4}$$

where M_{ij}^{ϕ} is the grain boundary mobility/m in the phase field[4] K J^{-1} s^{-1}, $\frac{\delta F}{\delta\phi_i}$ can be expressed as follows:

$$\frac{\delta F}{\delta\phi_i} = \sum_{\substack{k=1 \\ k\neq i}}^{n} \left(W_{ik}\phi_k + \frac{a_{ik}^2}{2} \nabla^2\phi_k \right) + \frac{\partial f_e}{\partial\phi_i} \tag{7.5}$$

where ∇^2 is the Laplace operator. Substituting Eq. (7.5) into Eq. (7.4), we get

$$\dot{\phi}_l = -\sum_{j=1}^{n} \frac{2M_{ij}^{\phi}}{n} \left[\sum_{k=1}^{n} \left\{ (W_{ik} - W_{jk})\phi_k + \frac{1}{2}(a_{ik}^2 - a_{jk}^2)\nabla^2\phi_k \right\} + \left(\frac{\partial f_e}{\partial\phi_i} + \frac{\partial f_e}{\partial\phi_j} \right) \right] \tag{7.6}$$

The last term in the equation can be expressed as $\frac{\partial f_e}{\partial \phi_i} - \frac{\partial f_e}{\partial \phi_j} = \frac{8}{\pi}\sqrt{\phi_i\phi_j}\Delta E_{ij}$, where ΔE_{ij} is the driving force, i.e., the energy storage difference between grain i and grain j, which can be expressed as follows:

$$\Delta E_{ij} = \frac{1}{2}\mu b^2(\rho_i - \rho_j) \tag{7.7}$$

where μ is a constant, b is the Burgers vector, and ρ_i and ρ_j are the densities of dislocations for grain i and j, respectively. Bringing Eq. (7.7) into Eq. (7.6), we can obtain the final phase field evolution equation for grain growth:

$$\dot{\phi}_l = -\sum_{j=1}^{n} \frac{2M_{ij}^{\phi}}{n} \left[\sum_{k=1}^{n} \left\{ (W_{ik} - W_{jk})\phi_k + \frac{1}{2}(a_{ik}^2 - a_{jk}^2)\nabla^2\phi_k \right\} - \frac{8}{\pi}\sqrt{\phi_i\phi_j}\Delta E_{ij} \right] \tag{7.8}$$

In a_{ij} and $W_{ij}W_{ij}$, $M_{ij}^{\phi}M_{ij}^{\phi}$ can be expressed as follows:

$$a_{ij} = \frac{2}{\pi}\sqrt{2\delta\gamma_{ij}}, \quad W_{ij} = \frac{4\gamma_{ij}}{\delta}, \quad M_{ij}^{\phi} = \frac{\pi^2}{8\delta}M_{ij} \tag{7.9}$$

At this point, the phase field evolution model of grain growth has been established, and using the formula (7.8), we can simulate the grain growth process; in this case, ΔE_{ij} is a constant, but the formula is more complicated to deal with. To save computational time, we use a more efficient C^{++} language combined with parallel algorithms to solve the formula; the specific process of the formula is also used to solve the Kim algorithm [16] which is described in detail in the following.

7.2.3 *Parallel solution of three-dimensional multiphase fields*

In the 3D phase field model, the 3D space is divided into many lattice points along the x, y, and z directions. Since the phase field method is based on phenomenology, the length unit of the lattice points is not considered in the evolution equations. In this study, we define the unit length of each lattice point along the three directions as 1 Å, which provides a high resolution to fill the atoms for Cu with a lattice constant of 3.6149 Å. When the average grain size is about 14 nm and contains 10 grains, the number of lattice points in the space reaches 13,824,000. At each lattice point, we have to compute the evolution equation of each grain to get the ordinal parameter of each grain. This process consumes a lot of time. To improve computational efficiency, we have to use parallel computational processing.

In C^{++} algorithms, parallel program running is primarily implemented using the header file "*omp.h*" in the library. This is mainly used to parallelize the for loop in a program to speed up its execution. For a program without data dependencies, parallel running can be achieved by simply adding "*#paragmaomp parallel for*" to realize the for loop. However, in programs with data dependencies, resolving data association in parallel running is achieved by setting variable locks and assigning threads to the variables. In the phase field evolution equation, there are significant data correlations. Before performing parallel operations, it's important to understand some simple parallel process instructions.

(1) **Private variables (Firstprivate).** These variables are initialized in each thread. Although each thread has the same variable name and type, there is no conflict between threads.
(2) **Shared variables (Shared).** All threads can read and write to variables of this type. However, when using shared variables, it's important to remember the order of thread operations, especially for structural variables with multiple elements. When one thread reads the elements of a structure while another thread is writing to the structure, it may cause the pointer pointing to be unclear, resulting in an error. When operating with shared variables, it's essential to use locks to ensure proper access order. This can be achieved using locks of various types. The program uses "*#pragmaomp critical*" for this purpose.
(3) **The parallel threads setting (Num_Threads).** This sets the number of processor threads that will execute the statement block in parallel. If the number of threads is not specified, then the computer will maximize the use of idle threads.

In the evolutionary equations, when computer thread 1 calculates the value of the ordinal parameter for the current lattice point, it uses the ordinal parameter of the nearby lattice point. However, if another thread, like thread 2, is also calculating the ordinal parameter of a grain in that nearby lattice point and if the speed of thread 2 is greater than that of thread 1, it can change the ordinal parameter calculated by thread 1, which will affect the result. In a worst-case scenario, if the ordinal parameter of the grain in the neighboring lattice is less than 10^{-5} in thread 2, the thread will delete the grain information in the neighboring lattice. If thread 1 needs to access the ordinal parameter of a grain in the neighboring lattice that has been deleted by thread 2, it will cause an address access error. To prevent this, it's necessary to set variables and locks for this parallel segment. We can use private variables to temporarily store calculated grid information in each thread and shared variables for both the current and next grid information. During normalization, sorting, and deletion operations, each thread should only operate on its own temporary vector without destroying all the information of the current grid point, thus allowing normal access by each thread. After completing the thread, the data in the temporary grid point information variable should be written to all the vectors of grid point information for the next step.

Kim's algorithm [16] to solve the phase field evolution equations is as follows:

(1) In a 3D lattice, we initialize the ordinate parameter of the N_{th} grain and randomly select some regions (*N–1* ones) as the initial grains for the 1st to *(N–1)*$_{th}$ grains. In this study, the randomly chosen areas are circular, and the selection of the diameter depends on the number of grains in the lattice. If the number of grains is larger, then the diameter is smaller, ensuring that there will be no crossover of the individual grains before the evolution. The N_{th} grain has an ordinate of 1.0 in all lattice points, and the 1st to *(N–1)*$_{th}$ grains have an ordinal parameter of 1.0 in the selected region:

$$(\nabla^2 u_{i,j,k}) := \frac{u_{i-1,j,k} - 2u_{i,j,k} + u_{i+1,j,k}}{\Delta x^2} + \frac{u_{i,j-1,k} - 2u_{i,j,k} + u_{i,j+1,k}}{\Delta y^2} + \frac{u_{i,j,k-1} - 2u_{i,j,k} + u_{i,j,k+1}}{\Delta z^2} \tag{7.10}$$

where Δx is the spacing between grid points in the x direction, Δy is the spacing between lattice points in the y direction, Δz is the spacing between lattice points in the z direction. $u_{i-1,j,k}$, etc. are the variables of the Laplace operator, here denoted as ordinal parameters. Note that the periodic boundary conditions are taken into account when solving the operator. Subsequently, a judgment is made on the obtained ordinal parameters if Φ_i is greater than 1.0 or less than 10^{-5}, respectively, performs $\Phi_i = 1$ and $\Phi_i = 0$, respectively, and the grains with an ordinal parameter of 0 are deleted.

(1) Solving Eq. (7.8) yields the ordinal parameter of N grains at lattice point n, which is equal to the current value of the ordinal parameter plus the ordinal parameter increment. In three-dimensional space, the Laplace operator of Eq. (7.8) is as in Eq. (7.10).
(2) Repeat Step (2) until all grid points have been counted.
(3) Regularizing ordinal parameter and there is no information about a grain in a particular lattice point, but the ordinal parameter of that grain in the six adjacent lattice points is not zero, then the information about that grain needs to be stored in that lattice point.
(4) Repeat Steps (2)–(4) and then update the saved ordinal parameter values from the previous step with the currently calculated ordinal parameter values.
(5) Repeat Steps (2)–(5); when the proportion occupied by the N_{th} grain is less than 0.001, the grain boundaries between adjacent grains are formed, then grain growth is complete.

Subsequently, the grain boundary information was extracted and randomly filled with *FCC* copper atoms. Figure 7.2 contains the phase field model of 3D polycrystalline copper and the MD simulation model after relaxation.

7.2.4 *Introduction to MD simulation methods and parameters*

The software used for this study is called large-scale atomic/molecular massively parallel simulator (LAMMPS) [17]. It is an open-source program developed by Sandia National Laboratories in the

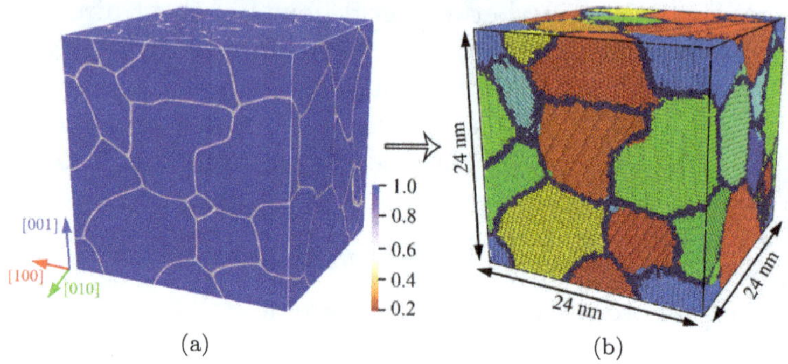

Figure 7.2. Phase field model of 3D polycrystalline copper and MD model of polycrystalline copper constructed from the phase field model [9] (in the diagrams, dark blue indicates grain boundary atoms). (a) Phase field model, (b) 3D polycrystalline copper.

USA. LAMMPS can be run on a single-processor computer as well as on a cluster of computers with multiple processors, which significantly improves the system and simulation efficiency.

The study focused on nano-polycrystalline copper films, 3D polycrystalline copper, and pre-set twinned copper. It aimed to investigate the effects of grain size, temperature, strain rate, and prepositioned twinning on the deformation mechanism and mechanical properties of polycrystalline copper. The EAM potential developed by Mishin *et al.* was adopted as the potential function for the simulations. The simulation process under different conditions and the parameters can be categorized as follows:

(1) In the simulation study of polycrystalline Cu film, we compared straight and curved grain boundaries and analyzed deformation behaviors under different grain sizes, temperatures, and strain rates. The polycrystalline copper film model (a) was created using the phase field method, with a 50 nm × 50 nm × 6 nm ($x \times y \times z$) size and 1,200,000 atoms, and a model (b) of 34.7 nm × 60 nm × 6 nm ($x \times y \times z$) size with 1,036,273 atoms, as depicted in Fig. 7.3. The straight-surface grain boundaries were formed using ortho-hexagonal grains with large-angle grain boundaries and the orientation approximation in model (a), maintaining the same average grain size. We conducted simulations using model (a) with variations in grain size ranging

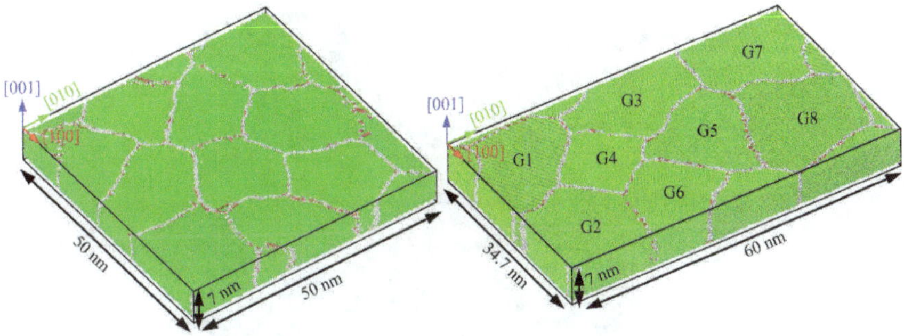

Figure 7.3. Two-dimensional polycrystalline copper films of different sizes constructed by the phase field method.

Note: The average grain size in both models is 17.84 nm, the atoms on the surface of the film (white) were removed to see the grain boundary distribution more clearly, and the green atoms are FCC copper atoms.

from 7.97 nm to 25.23 nm and a strain rate of 1×10^8/s. For the smaller model (b), we performed simulations with temperature variations from 0.1 K to 900 K and strain rates ranging from 7.5×10^7/s to 5×10^9/s. The energy minimization in MD utilized the conjugate gradient method, allowing 500 picoseconds of relaxation under the NPT ensemble for full relaxation of atoms at the grain boundaries. Uniaxial tensile was carried out under the NVT ensemble with the tensile direction along the y-axis.

(2) A study was conducted to simulate the behavior of three-dimensional polycrystalline copper. The study primarily focused on the impact of grain size, temperature, and strain rate on the material's mechanical properties and deformation behavior. The model used for simulation is illustrated in Figure 7.2(b), and its dimensions are 24 nm × 24 nm × 24 nm ($x \times y \times z$), containing approximately 1,100,000 atoms. The range of grain size variation was 6.5 nm to 18.8 nm, while the temperature variation ranged from 0.1 K to 900 K. The strain rate variation spanned from 7.5×10^7/s to 2.5×10^9/s. The energy minimization was done using the conjugate gradient method, and further relaxation was carried out under the NPT ensemble for 700 ps to achieve a stable structure. Subsequently, uniaxial tensile simulations were conducted under the same system, with the tensile direction along the z-axis.

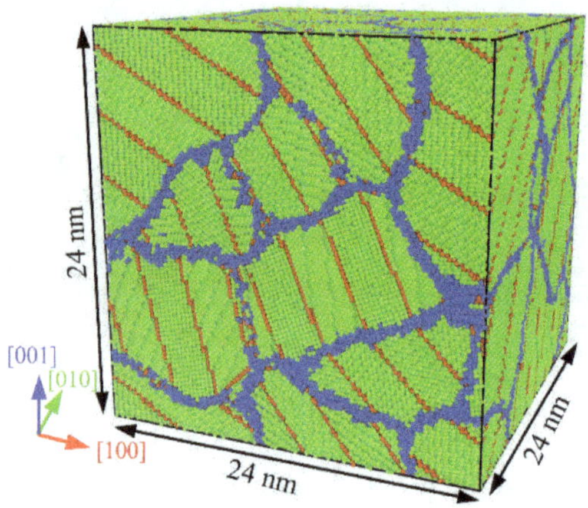

Figure 7.4. Preset twinned copper model.

Note: The twin boundary spacing in the figure is 2.5 nm, blue atoms are grain boundary atoms, and red atoms are twin boundary atoms. The green atoms are FCC copper atoms.

(3) In the simulation study of preset twinned copper, we primarily looked into the impact of preset twin spacing on deformation behavior and mechanical properties. The model is depicted in Figure 7.4, with an average grain size of 13.8 nm and a twin boundary spacing (twin spacing, l) of 2.5 nm. The model size is the same as in Section 2. The range of variation for the twin boundary spacing is 1.25 nm to 3.13 nm. The energy minimization and relaxation methods are identical to those in section (2), and the tensile strain rate is 1×10^8/s. The model construction involves two steps: first, the phase field model is built, and the conventional 3D polycrystalline copper is constructed using the phase field model. Subsequently, the twin structure is introduced into the copper, and this is done in the following steps:

(1) Each grain in the polycrystalline model is larger than the grain in the original phase field model. This is mainly done to prevent voids from appearing later when the atoms move according to the twin structure.
(2) The starting and ending points of the twin region in each grain are determined based on the distance between the twin

boundaries. We have set the total number of twin boundaries to 30 to ensure that the twin structure can cover each grain.

(3) To find the atoms belonging to the twin region, we use the start point, end point, and the normal vector of the (111) plane set in Step (2). The atoms are then numbered based on their perpendicular distances from the start point along the (111) plane ($n = 1, 2, 3, \ldots, n$).

(4) Move the numbered atoms a specific distance in the direction of the three coordinate axes: x direction $\frac{1}{6}(\sqrt{2} * n/la_{cu} * \cos\frac{\pi}{4})$, y direction $\frac{1}{6}(\sqrt{2} * n/la_{cu} * \cos\frac{\pi}{4})$, and z direction $\frac{1}{3}(n * la_{cu})$. In the formula, n is the number of the atom and la_{cu} is the lattice constant of copper.

(5) Steps (2)–(4) were repeated, and then each grain was introduced into the twin structure by, followed by modification of the grain boundary atoms as well as atoms collision between neighboring grains.

7.3 Comparison of Phase Field Models and Regular Polycrystalline

7.3.1 *Introduction*

In the discussion above, we focused on the current methods for constructing phase field models of polycrystalline materials and identified the drawbacks of existing methods. We proposed a new polycrystalline model based on the phase field method, where the growth of grains and the formation of grain boundaries are influenced by interfacial and free energies. This model is more consistent with real polycrystalline systems. To validate the superiority of the model, we conducted comparative simulations and created regular polycrystalline copper films using ortho-hexagonal shapes. In both models, the grain size was 17.84 nm. After relaxation, the grain boundaries in the regular polycrystalline model were all straight-sided, presenting a significant difference in grain boundary geometry compared to the new model. However, both models exhibited large-angle grain boundaries. We conducted a comparative analysis of the effect of grain boundary geometry on deformation under tensile stress. We quantitatively analyzed the stress concentration state of the grain boundaries and the changes in the HCP structure in the two models. Using this method, we then created polycrystalline copper films with varying grain sizes and investigated the deformation mechanism, the formation of the twin structure, and the effect of the twin structure on the deformation behavior in the copper films.

7.3.2 *Comparative analysis of mechanical properties and deformation behavior*

7.3.2.1 *Comparative analysis of mechanical properties*

The stress–strain curves obtained from uniaxial tensile are shown in Figure 7.5 with phase field model (*PFM*) and regular hexagonal model (*RHM*).

We can distinguish between the two materials by calculating the slope of the initial 0.5% strain in their mechanical curves. Young's modulus obtained for the *PFM* is 108 GPa, while it's 110 GPa for the pre-twinned copper film. This difference is attributed to the irregular grain boundaries in *PFM*, resulting in a larger degree of grain

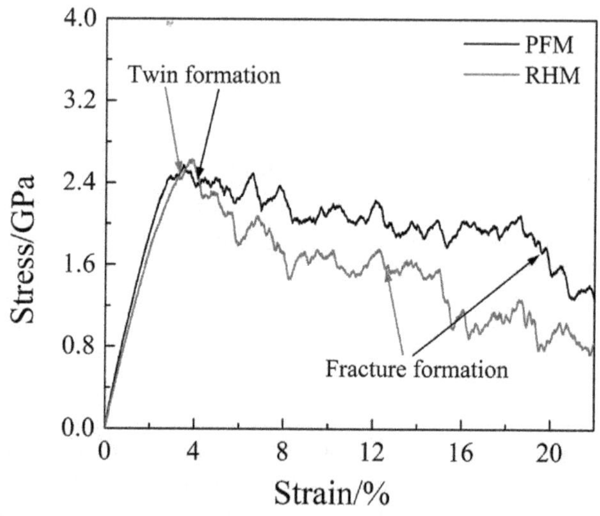

Figure 7.5. Stress–strain curves for uniaxial tensile of regular polycrystalline copper film (*RHM*) and phase field polycrystalline copper film (*PFM*) at a temperature of 300 K and a strain rate of 1×10^8/s.

coupling during deformation. Additionally, the yield stress of *RHM* is smaller than that of *PFM*. Moreover, the twin formation strain and fracture strain for *PFM* are both smaller than those of *RHM*. These differences are explained later in relation to the deformation of grain boundaries and the release of dislocations.

7.3.2.2 *Comparative analysis of grain boundary deformation*

The grain boundaries in both models were analyzed using *CNA* analysis before and after deformation. Figure 7.6 shows the extracted structures, with dashed lines indicating the contours of grain boundaries affected by deformation. A comparison between the initial state and a strain of 4.04% reveals significant grain boundary deformation in the *PHM* model. The straight-faced grain boundaries become more curved, matching experimental observations. The deformation behavior of the grain boundary aligns with the grain boundary deformation behavior observed through *in situ* TEM experiments conducted at 313 K [19]. In contrast, the *RHM* model did not show significant grain boundary deformation, except for disruption caused by dislocation release. Similarly, in the polycrystalline system created by Voronoi, the included grain boundaries were all straight-sided, and significant grain boundary deformation behavior was not observed

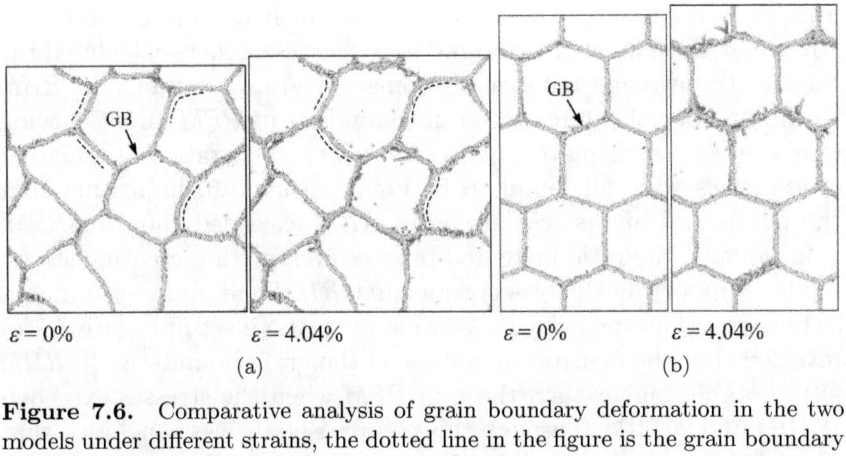

Figure 7.6. Comparative analysis of grain boundary deformation in the two models under different strains, the dotted line in the figure is the grain boundary contour with larger deformation. (a) Phase field polycrystalline model, (b) regular polycrystalline model.

during the initial deformation process [20]. It's important to note that grain boundaries not only transmit forces during deformation but also regulate the deformation of neighboring grains [9]. So, the deformation of grain boundaries can be influenced by the deformation of neighboring grains. For straight grain boundaries, the force exerted on the grain boundaries by neighboring grains is primarily along one direction, making it difficult to cause shear tangential force for deformation. However, for curved grain boundaries, the increased coupling between adjacent grains results in a non-uniform force on the adjacent grains, leading to tangential forces for grain boundary deformation during tensile deformation.

7.3.3 *Comparative analysis of grain boundary stress concentration*

In their research, Meyer *et al.* [21] highlighted that the deformation of grain boundaries can alleviate stress concentration on the grain boundaries, which has a significant impact on the release of dislocations. The study involved comparative analysis of the distribution of atomic stresses on grain boundaries in two models, as depicted in Figures 7.7(a) and 7.7(b). The analysis revealed that in the *RHM* model without grain boundary deformation, there was a notably higher number of highly stressed atoms on the grain boundaries compared to the *PFM*. To further illustrate the difference in stress distribution on grain boundaries, the researchers calculated the ratio of the average number of atoms per grain boundary in *RHM* to the number of atoms per grain boundary in *PFM* under varying stress levels, as displayed in Figure 7.7(c). The ratios at different stress levels were all found to be greater than 1.0, indicating that the number of high-stress atoms in *RHM* exceeded those in *PFM*. Additionally, the ratio increased progressively with increasing stress, further supporting the observation that *RHM* had a greater number of high-stress atoms than *PFM*. Moreover, the inset of Figure 7.7(c) revealed that the densities of atoms on the grain boundaries in *RHM* were notably higher than those in *PFM* when the stresses exceeded 18 GPa and 25 GPa. Through this comparison, it was concluded that the deformation of grain boundaries during the deformation process alleviates stress concentration on the grain boundaries, aligning with previous research findings.

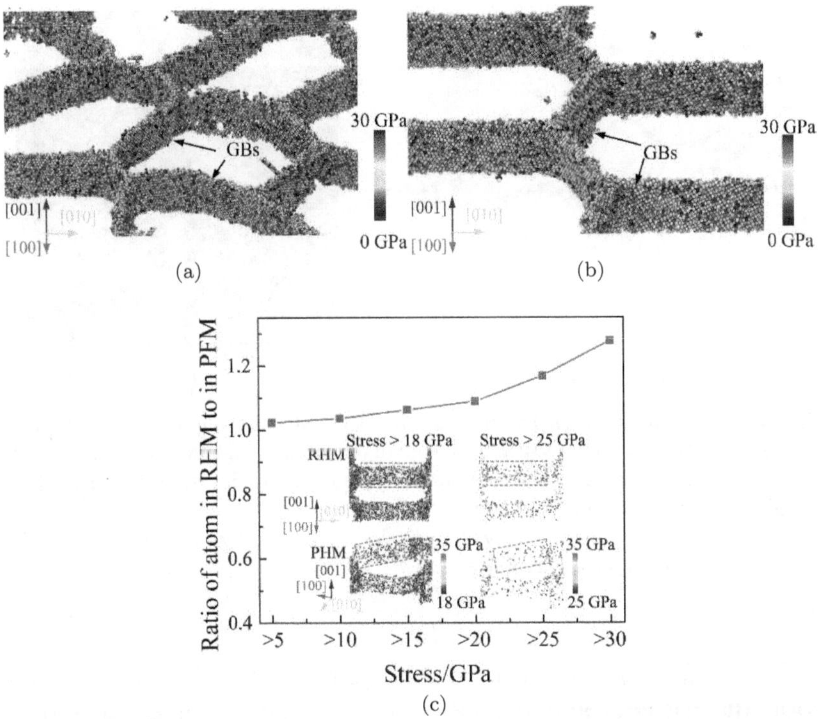

Figure 7.7. Comparison of the stress distribution of atoms on grain boundaries in the two models at a strain of 2.53% and the fraction of highly stressed atoms at grain boundaries in the two models averaged over each grain. (a) Stress distribution in *PFM* crystal boundaries, (b) stress distribution in *RHM* crystal boundaries, (c) the number fraction of highly stressed atoms at grain boundaries.

7.3.4 *Comparative analysis of dislocations and HCP structures (SFs and TBs)*

7.3.4.1 *Comparative analysis of dislocations*

The release of dislocations and the formation of twins in polycrystals largely depend on the stress concentration at the grain boundaries. In materials with a face-centered cubic (FCC) structure, dislocations are typically released in the form of $1/6\langle 112\rangle$ dislocations, consisting of leading and trailing partial dislocations. The release of leading partial dislocations is accompanied by the generation of stacking faults, while the release of trailing partial dislocations causes the stacking faults generated by the leading dislocations to disappear.

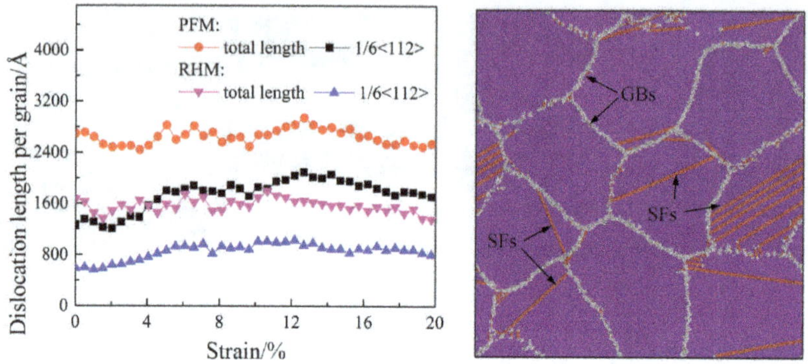

(a) (b)

Figure 7.8. Variation of the average total dislocation length and partial dislocation length per grain as a function of strain for the two models and the distribution of the different structures in the *PFM* model at a strain of 3.85% (purple, red, and white atoms in the figure denote FCCs, *HCPs* (*SFs* and *TBs*), and other structures, respectively). (a) Dislocation length per grain, (b) distribution of atoms for different structure.

Figures 7.8(a) and 7.8(b) display the average dislocation length (total dislocations and partial dislocations) in each grain and the distribution of stacking faults generated by leading dislocations in the material for the two models, respectively. The total dislocation length remains relatively constant, with minor fluctuations within a small range. The fluctuations in total dislocation length are primarily due to the generation of leading partial dislocations and the disappearance of trailing partial dislocations at grain boundaries.

RHM exhibits greater fluctuation compared to *PFM* due to the deformation regulation being mainly through the release and absorption of dislocations in the model, with no deformation regulation mechanism at grain boundaries. Additionally, higher stress concentration at grain boundaries facilitates the release of dislocations. Upon reaching a strain of 3–4%, dislocations gradually increase in both models, indicating that the material's deformation enters the yield stage. When comparing the *PFM* and *RHM* models, we observe that during the transition from the elastic stage to the plastic stage, dislocations are more released in the *PFM*. Even though the stress concentration on the grain boundaries is smaller in the *PFM* compared to the *RHM*, there are two main reasons for this:

(1) Both models have large-angle grain boundaries through which dislocation release mainly occurs via release and absorption, without a deformation regulation mechanism on the grain boundaries. Although both models have large-angle grain boundaries, there may be some differences in local misorientation.
(2) In the *PFM*, grain rotation is not limited to the x–y direction, unlike in the *RHM*. Thus, rotation can occur in the *PFM* model. Furthermore, irregular grain boundaries in the *PFM* can cause coupling, leading to the inevitable rotation of grain boundaries during the deformation process. This rotation creates more slip systems and consequently leads to the release of a large number of dislocations [22, 23].

7.3.5 *Comparative analysis of the structure of HCPs (SFs and TBs)*

The structure of hexagonal close-packed (*HCP*) materials, especially the structure of twin boundaries (*TBs*), has a significant impact on the mechanical properties of face-centered cubic (*FCC*) polycrystalline materials. The process of generating twin crystals in *FCC* materials can be divided into two steps:

(1) Intrinsic stacking faults are formed through leading partial dislocations, but the trailing partial dislocations are not released.
(2) With increased deformation, stress concentration rises, causing the generation and slip of new leading partial dislocations on the adjacent (111) surfaces of stacking faults (*SFs*). This causes the intrinsic stacking faults to become extrinsic stacking faults. As the dislocations continue to be released, the extrinsic stacking dislocations become micro-twins, as illustrated in the constitutive diagram in Figure 7.9(a) for a strain of 4.21%. The twin region gradually grows with the continuous release of dislocations in the neighboring planes, and this mechanism of twin formation and growth is consistent with previous findings [24]. The twin boundaries exhibit high mechanical stability in *FCC* polycrystalline materials, which can impede the movement of dislocations, thereby strengthening the material.

To better understand the differences in the mechanical properties of the two models, we analyzed and counted the twin structures

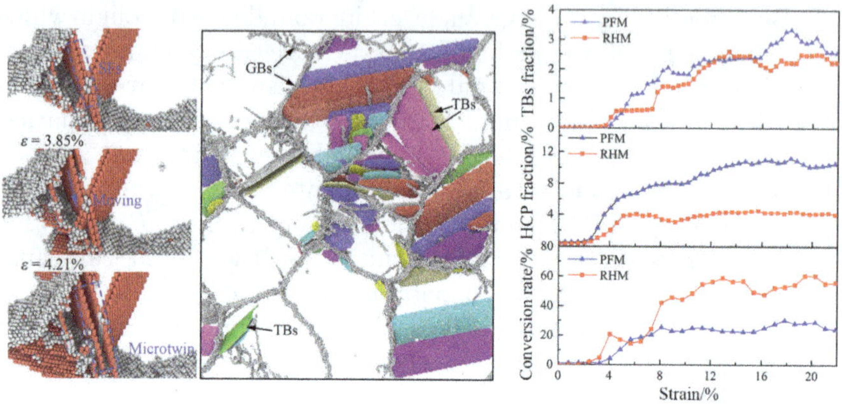

(a) (b) (c)

Figure 7.9. Twin formation process and statistics and analysis of the twin configuration by the quantitative twin analysis algorithm. (a) Twin formation process, (b) twin distribution in *PFM*, (c) percentage of *HCP* and *TB* conversion rate.

Note: The red and white colors in Fig. (a) are the *HCP* (*SFs* and *TBs*) structures and other structures, respectively, and the colors other than white in Fig. (b) represent the twin structures at different positions.

produced in the two models using a quantitative twin analysis algorithm. The twin configuration diagrams and statistical data are shown in Figures 7.9(b) and 7.9(c). These figures display the changes in the number of twins, the number of *HCPs*, and the conversion rate of *SFs* to *TBs* as a function of strain, as shown in Figure 7.9(c). From the figures, we observed that during the yield phase, both the *HCP* and *TB* fractions are larger in the *PFM* compared to the values in the *RHM*. This indicates that a higher number of *TBs* leads to higher yield stress. However, the lower conversion of *TBs* in *PFM* than in *RHM*, especially when the strain is greater than 8%, is mainly attributed to the deformation of grain boundaries. This leads to a decrease in stress concentration, whereas a larger stress concentration is required for the release and movement of partial dislocations in the neighboring (111) faces of stacked faults. As a result, the higher stress concentration in the *RHM* causes partial dislocations to be released on the neighboring faces of the *SFs*, leading to a higher twin conversion rate. Additionally, we observe that the strain formed by twins in the *PFM* is slightly larger than that formed in the *RHM*.

This is mainly due to the deformation of grain boundaries, resulting in smaller stress concentrations at grain boundaries in the *PFM*. Consequently, the release of dislocations on the proximal faces of *SFs* is delayed.

7.3.6 *Deformation mechanism of polycrystalline copper films at different grain sizes*

Afterward, we utilized the approach to form polycrystalline copper films with varying grain sizes ranging from 7.97 nm to 25.23 nm. We conducted uniaxial tensile simulations to analyze the changes in the mechanical properties of the different models and the impact of the twin structure on the deformation behavior of the grains. In nano-polycrystalline metals, as the grain size decreases, the number of grain boundary atoms gradually increases, significantly influencing the mechanical properties. The change in the number of grain boundary atoms is typically considered a key factor in the softening of the material. Furthermore, we calculated the number of grain boundary atoms for these polycrystalline copper films with different grain sizes, as illustrated in Figure 7.10, which aligns with the number of atoms in polycrystalline materials formed by the gas condensation cluster

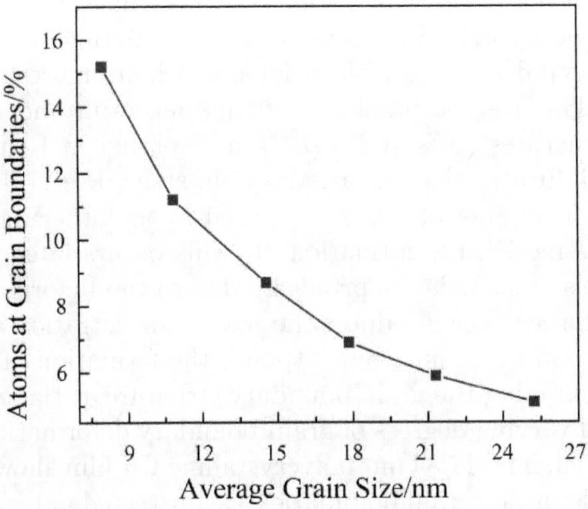

Figure 7.10. Number of grain boundary atoms in polycrystalline copper films with different grain sizes.

method [25, 26]. The number of atoms is consistent with the polycrystalline material created by the gas condensation cluster method. Due to the presence of irregular or curved grain boundaries, the number of grain boundaries in this model is slightly higher than in the model created by the simple geometry method.

7.3.7 *Mechanical properties and deformation behavior*

The stress–strain curves for the uniaxial tensile of polycrystalline copper films with different grain sizes, as well as the twin formation strain and fracture strain, are depicted in Figure 7.11(a). It is evident that Young's modulus of polycrystalline copper films is not significantly influenced by grain size. For example, Young's modulus is 104 GPa for a grain size of 7.97 nm and 115 GPa for a grain size of 25.23 nm. These values are quite similar to Young's modulus of 108 GPa obtained from uniaxial tensile experiments for copper films with a grain size of 26 nm [27]. Additionally, there is little variation in the yield stresses from the curves. However, there appears to be a relationship between the twin formation strain and the fracture strain, which increases with grain size, reaching a peak in the sample with a grain size of 17.84 nm before decreasing. This relationship suggests that the earlier the formation of twins, the earlier the material will crack, potentially due to prolonged interactions between twins and grain boundaries and dislocations, which are more likely to lead to cracks. The average number of twins per grain and the number of *HCP* structures (*SFs* and *TBs*) are displayed in Figure 7.11(b). The figure indicates that during the yield stage, the 17.84 nm model has a greater number of twins compared to the other models. However, in this model, the formation of twins occurs later than in the other models. This delay is primarily due to the deformation of the grain boundaries, which reduces the stress concentration at the grain boundaries and consequently postpones the formation of twins.

After examining the grain boundary structure at the initial stage, we observed varying degrees of grain boundary deformation behavior in each model. The 17.84 nm polycrystalline Cu film showed a larger number of deformed grain boundaries, as illustrated in Figure 7.11(c). The deformation of grain boundaries can help reduce stress concentration, as indicated by our comparative study. We also calculated

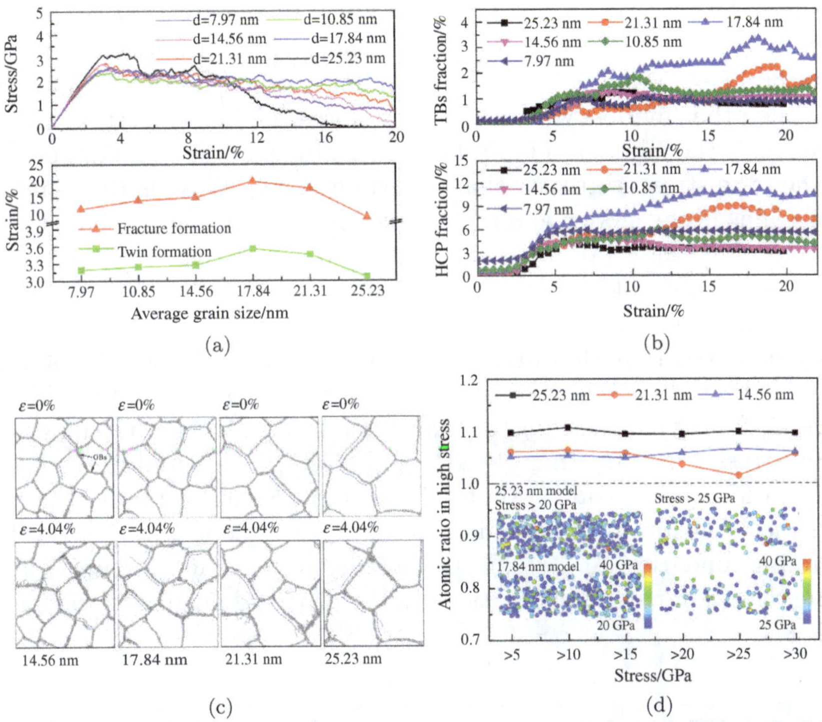

Figure 7.11. Stress–strain curves and twin formation strain and crack formation strain for different models of tensile stretching; variation of the average *HCP* (*SFs* and *TBs*) number and *TB* number with strain; grain boundary deformation in different models. (a) Stress, strain, and both correlated features, (b) *HCP* structure and their fractions, (c) deformed grain boundaries, (d) the ratio of atoms on the boundaries with high-stress to others.

Note: The contours of the grain boundaries that had larger deformation colored by blue dashed lines in the image; the ratio of the number of atoms on grain boundaries in the other models and the 17.84 nm model when the stress is larger than a certain value; and the stress distribution of atoms on the grain boundary showed in the inset image.

the ratio of high-stress atoms in the other models to those in the 17.84 nm model, considering the difference in the number of grain boundary atoms in different models. Consequently, we incorporated the number of grain boundary shares into the following equation:

$$f_{\text{atom}} = \frac{N_{\text{atom}} * 10}{N_{\text{atom10}} * N_{\text{grain}}} * \frac{N_{\text{GB}}}{N_{\text{GB10}}} \tag{7.11}$$

where N_{atom} is the total number of grain boundary atoms in the other models, N_{atom10} is the total number of grain boundary atoms in the 17.84 nm model, N_{GB} is the number of grain boundary atoms in the other models when the stress is greater than a certain value, N_{GB10} is the number of grain boundary atoms in the 17.84 nm model when the stress is greater than a certain value, N_{grain} is the number of grains in the model, and the number of grains in the 17.84 nm is 10.

In this study, we selected three models with grain numbers 5, 7, and 15 for calculation, corresponding to grain sizes of 7.97 nm, 10.85 nm, and 14.56 nm. The results can be seen in Figure 7.11(d). We observed that the ratio is greater than 1.0. Additionally, it was noted that the density of high-stress atoms at grain boundaries in the 25.23 nm model is higher than that in the 17.84 nm model when the stress is greater than 20 GPa and 25 GPa. This finding suggests that the number and ratio of deformed grain boundaries are higher in the 17.84 nm model, leading to lower stress concentration on the grain boundaries. As a result, the smaller stress concentration delays the release of partial dislocations in the proximal neighborhood faces of *SFs*.

7.3.8 *Effect of twinning on grain deformation*

Twin boundaries have a significant impact on the deformation of low-energy stacking fault copper, where they hinder the movement of dislocations and also affect grain deformation. However, most studies do not thoroughly discuss the role of deformation twins on the copper substrate, mainly because stacking faults are the primary deformation mechanism in polycrystalline copper. As a result, researchers have mainly used the pre-positioned twins method to study the twin strengthening mechanism and deformation behavior. Typically, twin growth occurs through partial dislocation sliding on adjacent (111) surfaces of the single-layer *TB* structure. However, when partial dislocation movement is impeded by defects such as clusters or grain boundaries, the twin region will not grow, as shown in Figure 7.12(a).

As deformation increases, partial dislocations form on neighboring faces of the twin boundary or twin region and move toward the hindered partial dislocations. Subsequently, the hindered partial dislocations continue to move forward under the impetus of the later-formed

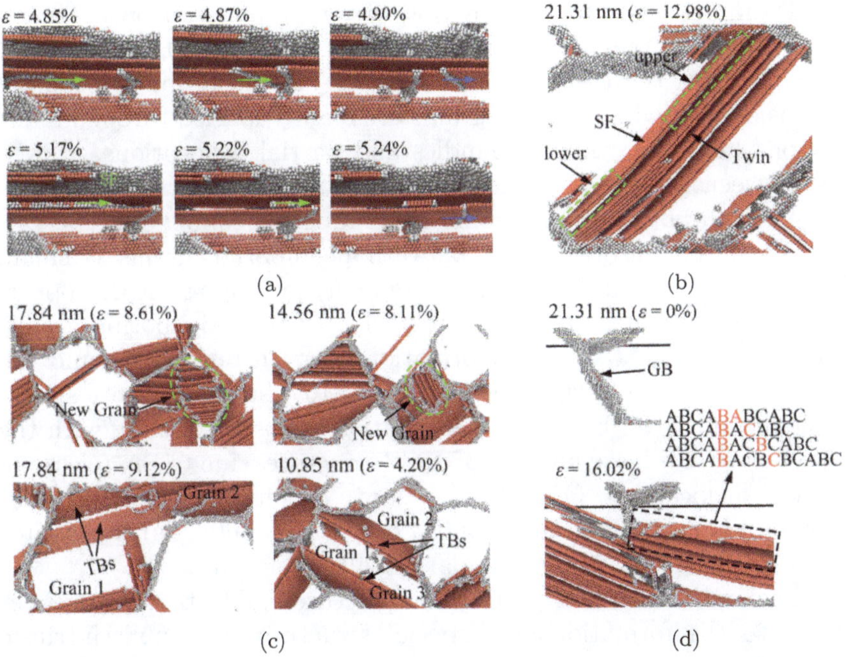

Figure 7.12. Interaction between twins and grain boundaries. (a) Growth of deformation twins driven by dislocations in the 10.85 nm model, (b) twins make grains appear twisted, (c) deformation twins observed in different models allow for the formation of new grains and fusion of neighboring grains, (d) formation of special structures: intrinsic stacking fault, extrinsic stacking fault, micro twin boundaries, and larger twin boundaries.

Note: The red atoms in the figure are *HCP* structures (*SFs* and *TBs*) and the white atoms are other structures.

partial dislocations, leading to gradual growth of the twin region. Figure 7.12(b) illustrates the twisting behavior of grains caused by twinning. From the atomic configuration diagram, it can be observed that during initial deformation, dislocations are released inside the grains, forming a flat stacking fault structure. With further dislocation release and twin structure formation, the grain boundaries bend, especially in the region of the *HCP* structure (stacking faults and twin boundaries), leading to twisted appearance. The bent grains change the slip system, causing gradual partial dislocation release from the lower part of the grains during the deformation process.

Twinning leads to the creation of new grains, fusion of neighboring grains, and formation of special structures. Figure 7.12(c) demonstrates the creation of new grains and fusion of neighboring grains due to twinning. New grains are mainly formed through interactions between twinning bundles and partial dislocations, as well as between grain boundaries. This process is similar to what was observed in both polycrystalline aluminum and copper [27, 28].

When the misorientation between neighboring grains is small, twins cross the grain boundaries, gradually leading to the disappearance of the grain boundaries and fusion of the neighboring grains. Figure 7.12(d) shows the deformation of grain boundaries and the creation of special structures at 0% and 16% strain. At 16% strain, a significant grain boundary deformation is observed along with the generation of a large number of *HCP* structures due to the release of partial dislocations. Our previous study concluded that the formation of *SFs* (stacking faults) and *TBs* (twin boundaries) accelerates the deformation of grain boundaries [29].

The interaction between these structures and grain boundaries leads to the formation of a special structure containing intrinsic stacking faults, extrinsic stacking faults, micro twins, and larger twins. This special structure is similar to what was obtained from simulations of polycrystalline aluminum [30].

7.3.9 *Interaction of twins with HCP structures (SFs and TBs)*

Compared with traditional grain boundaries, twin boundaries have higher stability and can effectively hinder the movement of dislocations, thereby strengthening the material. This mechanism is widely recognized [31, 32]. Therefore, there is a need for more studies on introducing pre-positioned twin structures in copper to enhance its strength. As depicted in Figure 7.13(a), it's evident that the motion of dislocations is impeded by the twin boundaries. Thus, in the 17.84 nm model, the greater number of twin fractions increases the possibility of interaction between partial dislocations and twin boundaries, leading to higher yield stresses. When twins with different orientations interact with each other inside the grain, they can form fivefold twins, as shown in Figure 7.13(b). This structure is more stable compared to conventional twin boundaries and is difficult

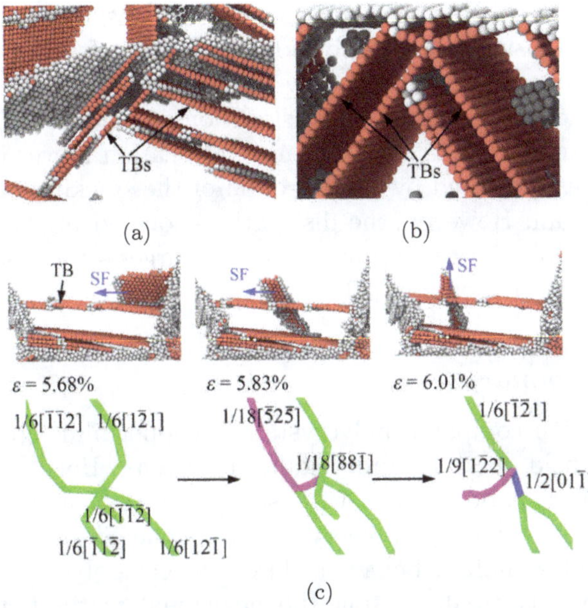

Figure 7.13. Twins and dislocation slide. (a) Twin crystals impede dislocation movement in the 17.84 nm model, (b) formation of fivefold twin structures in the 10.85 nm model, (c) 4.56 nm model in which twins change the direction of motion of dislocations.

Note: The red atoms in the figure are *HCP* structures (*SFs* and *TBs*) and the white atoms are other structures.

to destroy by moving partial dislocations. The phenomenon of fivefold twin formation and strengthening is similar to what has been observed in simulations [34] and experiments [35].

Twin boundaries not only impede dislocation motion but also alter the direction of dislocation slip, as seen in Figure 7.13(c). At a strain of 5.83%, a Burger's vector of $1/6[\bar{1}\ \bar{1}\ 2]$ of the leading partial dislocation and a Burger's vector of $1/6[1\ \bar{2}\ 1]$ of the trailing partial dislocation move along the twin boundary. In between the two dislocations, there is a stacking fault. When the leading partial dislocation meets defects on the twin boundary, its motion is hindered, and the partial dislocation decomposes with increased deformation, as shown in the following reaction equation of dislocations:

$$1/6[\bar{1}\ \bar{1}\ 2] \rightarrow 1/18[\bar{5}\ \bar{2}\ 5] + 1/18[\bar{8}\ 1\ \bar{1}] \tag{7.12}$$

The process results in the creation of two sessile dislocations, as shown in the dislocation distribution at a strain of 5.83% in 7.13(c). The stress concentration caused by these sessile dislocations leads to the formation of another partial dislocation at a strain of 6.01% with a Burger's vector $1/6[\bar{1}\ 2\ 1]$. This new dislocation moves in different directions, as indicated by the direction of the stacking faults in the 6.01% diagram. However, the dislocations located in the lower half of the twin boundary maintain their old Burger's vectors, a finding consistent with Lu's study [33].

7.3.10 *Summary*

The simulation compared polycrystalline copper films created using the phase field method and regular hexagons. Both models had large-angle grain boundaries with similar average grain sizes. The results showed that the polycrystals constructed using the phase field method exhibited behavior closer to real polycrystals. Unlike the polycrystals created using the hexagonal method, those constructed using the phase field method had irregular and curved grain boundaries. This led to significant differences in mechanical properties between the two models. Deformation behavior was observed at the initial stage in the phase field method model but not in the hexagonal method model produced by the Voronoi method using the reference recommended. Additionally, stress concentration was lower in irregular grain surfaces compared to straight grain boundaries. This lower stress concentration led to less formation of twin crystals in irregular polycrystals compared to those constructed using the hexagonal method. The higher stress concentration leads to the continuous release of partial dislocations on the adjacent (111) planes of the stacking faults, resulting in the formation of twins. Therefore, in regular polycrystals, the conversion rate from stacking faults (*SFs*) to twin boundaries (*TBs*) is higher than the conversion rate in the models constructed by the phase field method.

We used the method to create polycrystalline copper films with various grain sizes. We conducted tensile simulations and found that the mechanical properties and deformation behaviors matched both the experimental and simulated results. We observed significant grain boundary deformation during the initial stage of deformation in different models. The 17.84 nm model showed higher grain boundary deformation, resulting in a small stress concentration. As a result, twin crystals formed later in this model compared to the others. Twins significantly influenced the deformation of polycrystals, leading to the creation of new grains and the fusion of neighboring grains with smaller misorientation. Twin boundaries also contributed to the torsional behavior of grains during deformation, forming a stable fivefold structure that effectively impeded the motion of partial dislocations. Additionally, when partial dislocation motion was blocked at the twin boundary, partial dislocations decomposed and formed sessile dislocations, resulting in higher stress concentration and the release of new dislocations at the twin boundary. These findings suggest that the new method for constructing a polycrystalline system is more reasonable and contributes to our understanding of the deformation mechanism of polycrystalline copper.

7.4 Deformation Mechanisms of Polycrystalline Copper at Different Strain Rates and Temperatures

7.4.1 *Introduction*

Research on polycrystalline copper films has been extensive, focusing on their mechanical properties, fracture, twin release, pre-positioned twins, and ductility. It has been noted that significant grain boundary deformation exists in polycrystalline films, which influences stress concentration on the grain boundary and deformation behavior. This phenomenon is not typically found in conventional models due to the reduced coupling between grains. Grain boundary activities observed in simulations occur under specific conditions such as high temperatures, shear stresses, and special models where the activities are more obvious. However, dynamically observing grain boundary activities in space and time is challenging due to current experimental setups, leading to fewer investigations on grain boundary activities, deformation mechanisms, and factors affecting grain boundary deformation at low temperatures.

A polycrystalline copper film with an average grain size of 17.84 nm was created for MD simulation using the phase field method. The polycrystalline model is illustrated in Figure 7.3(b), showing the orientation differences between various grains, such as the orientation difference between grains is as follows: G1 and G4: 45.8° [001], G2 and G4: 34.4° [001], G2 and G6: 17.1° [03–2], G3 and G7: 40.1° [0–14], G3 and G5: 74.5° [01–6], G3 and G4: 17.1° [03–2], G5 and G6: 63° [001], G5 and G8: 17.1° [0–15], and G7 and G8: 97.4° [01–5]. Uniaxial tensile simulations of the copper film were conducted at different strain rates and temperatures. The simulation results revealed a connection between grain boundary activity, dislocation release, and mechanical properties. Additionally, an in-depth analysis of the mechanism of grain boundary deformation during the initial tensile deformation stage was carried out, along with an exploration of the factors influencing grain boundary deformation.

7.4.2 *Mechanical properties and deformation behavior at different strain rates*

7.4.2.1 *Mechanical properties and changes in HCP structure (SFs and TBs)*

The stress–strain curves for different strain rates during tensile at 300 K are presented in Figure 7.14(a), while the typical mechanical properties derived from the curves are displayed in Figure 7.14(b). Young's modulus can be determined as the slope of the curve prior to 0.5% strain. According to previous research, the average yield stress

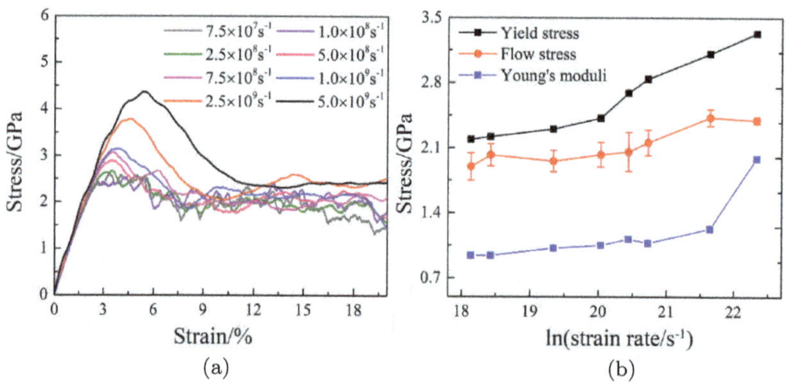

Figure 7.14. Stress–strain curves and related mechanical properties obtained by stretching at 300 K. (a) Stress–strain curves for uni-tensile, (b) related mechanical properties.

ought to be computed by identifying the point at which the stress–strain curve stabilizes [6, 7]. Consequently, for strain rates ranging from 7.5×10^7/s to 1.0×10^9/s, the stable phase strain falls between 6% and 15%. Conversely, for 2.5×10^9/s and 5.0×10^9/s strain rates, the stable phase is observed within the 12–18% range. Additionally, the yield stress is determined based on the stress level at which the first dislocation is released from the grain boundary [18]. Analysis of these mechanical property parameters reveals that the yield strength, average yield stress, and Young's modulus all increase as the natural logarithm of the strain rate (s^{-1}) increases. This trend aligns with results obtained from other simulations or experiments. Notably, the value of Young's modulus at a strain rate of 1×10^8/s is 111.3 GPa, a figure comparable to the results of 35 nm pre-twinned copper simulations (110 GPa) [18] and the values obtained in the approximate 26 nm polycrystalline copper film experiment (108 GPa) [27].

SRS is a parameter that describes the sensitivity of a material's mechanical properties (yield stress or hardness, etc.) to strain rate during deformation, i.e., the tendency of a material to strengthen as the strain rate increases [36]. The strain rate sensitivity S is defined as the rate at which a material is sensitive to strain rate:

$$S = \frac{k_B T}{V*} \tag{7.13}$$

where k_B is Boltzmann's constant/JK^{-1}, T is Kelvins temperature/K, and V^* is activation volume. The activation volume is the rate of decrease of the enthalpy of activation with respect to the yield stress or hardness, defined as follows:

$$V^* = \sqrt{3}kBT\left(\frac{\partial \ln \dot{\varepsilon}}{\partial \sigma}\right) \tag{7.14}$$

where $\dot{\varepsilon}$ is the strain rate and σ is the force/MPa. Combining Eqs. (7.13) and (7.14) yields a parameter for the strain rate sensitivity m [37], as shown in the following:

$$m = \frac{S}{\sigma} = \frac{k_B T}{\sigma V*} = \frac{\partial \ln \sigma}{\partial \ln \dot{\varepsilon}} \tag{7.15}$$

The strain rate sensitivity (*SRS*) can effectively indicate the deformation mechanism. A value of SRS (m) between 0.05 and 0.1 suggests that the deformation process is likely accompanied by grain boundary activities. For nanocrystalline materials, the value of m is higher compared to coarse crystals [38]. By using equation (7.15), we calculated the value of m for the tensile of a polycrystalline copper film, taking the stress as the average yield stress. The obtained m value is approximately 0.056, which is greater than 0.05. Huang [39] and colleagues conducted tensile experiments on polycrystalline copper films with an average grain size of 26 nm at room temperature and obtained an m value of about 0.059. Additionally, Lu [40] and colleagues performed tensile tests on polycrystalline copper films with an average grain size of 105 nm and obtained m values of about 0.06 ± 0.01 from indentation experiments. It is evident that under similar conditions, the results obtained from these simulations or experiments are relatively close, and these values are larger than those obtained for coarse-crystalline Cu [41]. As the grain size decreases to about 20 nm, the grain boundaries occupy a larger proportion of the grain, leading to increased grain boundary activity during deformation and larger m values. Based on the comparison of these results, it can be concluded that the value of the strain rate sensitivity of this model is in close agreement with the experimental observations and effectively reflects the deformation mechanism in the experiment.

In order to understand how strain rate affects the mechanical properties, we used *CNA* analysis and twin analysis algorithms to calculate the number of copies of *HCPs* (*SFs* and *TBs*) and *TB* copies for various strain rates, as illustrated in Figure 7.15. It was observed that as the strain rate increases, both types of copies also increase gradually. The increase in twin copies further restricts the movement of dislocations within the material matrix. As a result, the yield stress gradually rises with the increase in strain rate.

We have developed specific procedures for statistically analyzing the distribution of *HCP* (*SFs* and *TBs*) and *TB* structures within the matrix. Figures 7.16(a) and 7.16(b) display the distribution of *HCPs* and *TBs* respectively at a strain rate of 1×10^8/s during the yield stage. Different colors in the figure indicate the number of atoms in these structures, and the x and y axes represent the coordinates of the copper film along the two directions. The *HCP* structure is widely distributed in different grains, suggesting the occurrence of partial dislocation sliding in these grains, resulting in corresponding intrinsic and extrinsic stacking faults. The *HCP* structure is mainly

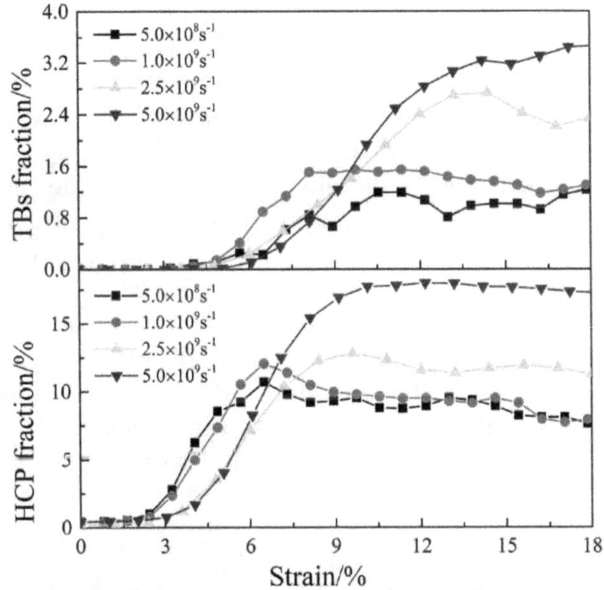

Figure 7.15. Variation of *HCP* structure (*SFs* and *TBs*) fraction and *TBs* fraction with strain in the model at different strain rates at 300 K.

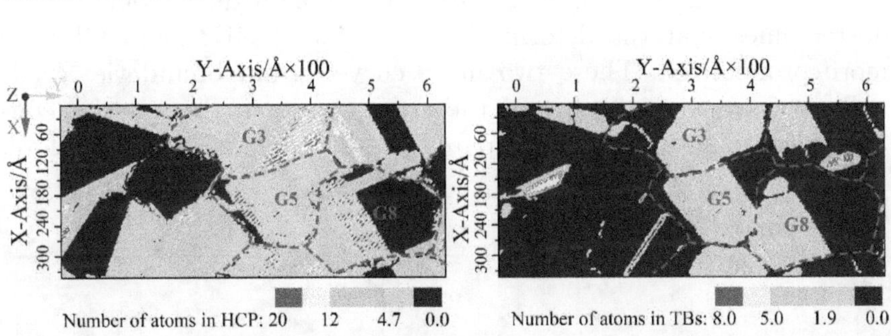

Figure 7.16. Strain rate at 300 K for 1×10^8/s, distribution of *HCP* (*SFs* and *TBs*) and *TB* structures in polycrystalline copper films (different colors represent the number of atoms contained in these structures). (a) Distribution of *HCP*, (b) distribution of *TBs*.

distributed in the three grains of G3, G5, and G8, and similarly, the distribution of the *TBs* also concentrates in these three grains. Later, we delve into a detailed discussion of the grain boundary activities in these grains.

For grains 2 and 4, the misorientation of the grain boundary is GB24 (34.4° [001]), which is close to the Σ5 (36.9° [100]) grain boundary. Zhang *et al.* [42] also conducted MD simulations using Σ5 bicrystalline copper and demonstrated that the release of dislocations from grain boundaries in this model depends on the energy of the grain boundaries. The energy gradually increases with the angle of the misorientation, reaches a maximum at around 26°, and then starts to decrease. As a result, the maximum tensile stresses of dislocation nuclei at misorientation angles of 0° and 45° are greater than those at other angles, primarily because these two grain boundaries have the lowest grain boundary energies. It is difficult for dislocations to be released from the grain boundary GB24 due to its lower grain boundary energy, ultimately leading to the difficulty of twinning in these two grains, especially at lower temperatures.

7.4.2.2 *Deformation behavior at grain boundaries*

The grain boundary structure was examined using *CNA* to compare the deformation behavior of the grain boundaries at the initial stage, as illustrated in Figure 7.17. It is evident from the red dotted lines that the deformation behavior of GB35 and GB58 is more pronounced. These two more curved grain boundaries gradually become straighter, and the straighter grain boundaries contain more chaotic atoms compared to the curved ones. Therefore,

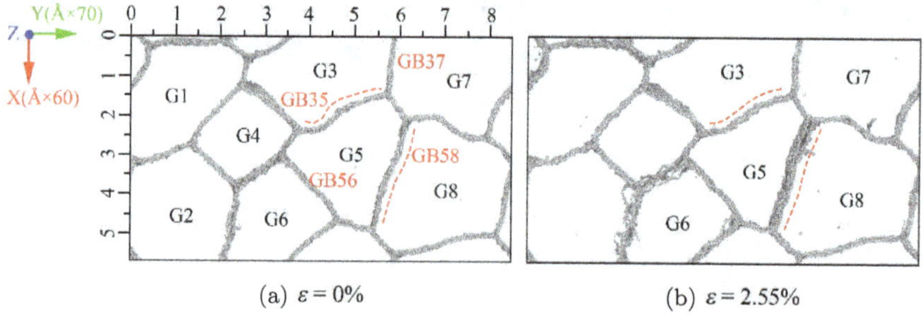

Figure 7.17. Distribution of grain boundaries under different strains at 300 K for a strain rate of 1×10^8/s, the distribution of grain boundaries at different strains.

Note: The red curves in the figure indicate grain boundaries with large deformations.

the curved grain boundaries are more inclined to deform toward the straight side when subjected to the corresponding shear stress. The deformation of other grain boundaries is not as apparent. An analysis of the misorientation reveals that the misorientation of GB14 (45.8° [001]) and GB24 (34.4° [001]) is closer to that of Σ29a (43.6° [100]) and Σ5 (36.9° [100]) of these two grain boundaries. Zhang *et al.* [43] found that large-angle [001] grain boundaries have higher stability compared to small-angle grain boundaries through simulation studies of bicrystalline copper. For Σ5 grain boundaries at different temperatures, the mobility of the grain boundaries decreases with the increase in the angle of the misorientation. Additionally, these two grain boundaries are also approximately flat grain surfaces, so the resulting coupling of neighboring grains during deformation is low, making it difficult for significant deformation behavior to occur.

From the above analysis, it is seen that the grain boundary activities are mainly concentrated in the grain boundaries of grains 3 and 5, and GB35 is more active. Therefore, here we mainly quantitatively calculated the deformation state of this grain boundary in the initial deformation stage, as shown in Figure 7.18, the cloud diagram of the movement of GB35 along the x-axis, where the x-axis indicates the position of GB35 and the y-axis indicates the change of strain.

It is calculated as $\frac{\sum X_\varepsilon}{N_\varepsilon} - \frac{\sum X_{\varepsilon 0}}{N_{\varepsilon 0}}$, where X_ε and $X_{\varepsilon 0}$ are the x-coordinate values of the atoms in the selected region in the current and initial state (0%), respectively. N_ε and $N_{\varepsilon 0}$ are the number of atoms in the currently and initially selected region, respectively, and the regions are selected sequentially along the y-axis at intervals of 5 Å for GB35.

In Figure 7.18, we can clearly observe that the movement of grain boundaries varies at different points on GB35, with different directions. For the part of the grain boundaries with y coordinates greater than 310 Å, the movement direction is along the positive x-axis and, conversely, along the negative x-axis. Comparing Figures 7.17 and 7.18, we note that at y-coordinate values of about 295 Å and 340 Å, the corresponding grain boundary sections are the most curved and exhibit the largest amount of movement. Comparing Figures 7.15 and 7.18, we can see that when dislocations are released from the grain boundaries, the amount of grain boundary movement increases

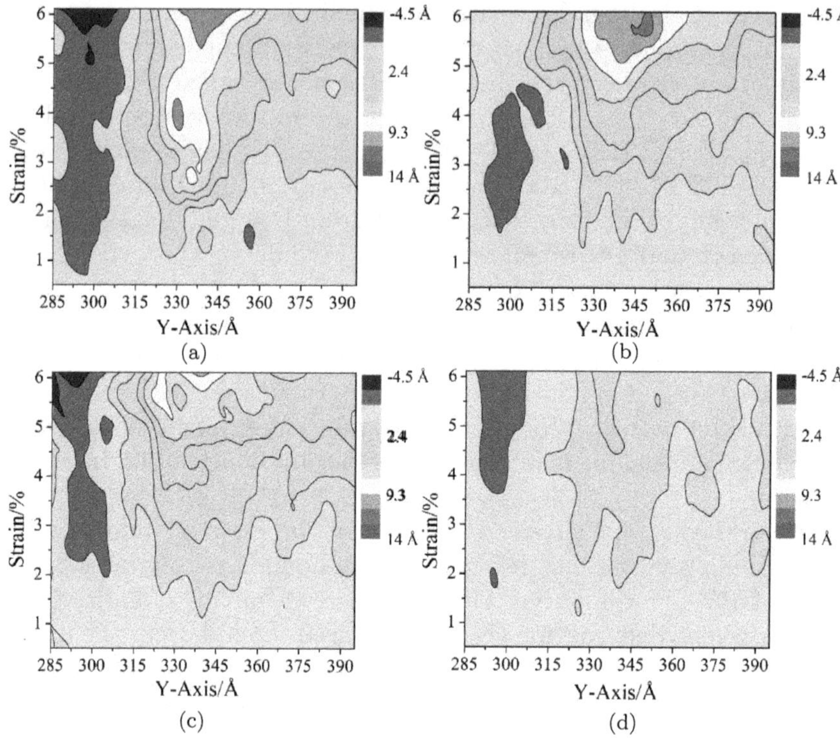

Figure 7.18. Displacement variation of GB35 distributed in y-axis along x-axis with strain for different strain rates at 300 K. (a) $\dot{\varepsilon} = 1 \times 10^8\,\mathrm{s}^{-1}$, (b) $\dot{\varepsilon} = 5 \times 10^8\,\mathrm{s}^{-1}$, (c) $\dot{\varepsilon} = 1 \times 10^9\,\mathrm{s}^{-1}$, (d) $\dot{\varepsilon} = 5 \times 10^9\,\mathrm{s}^{-1}$.

Note: The colors in the diagram indicate the amount of grain boundary movement.

rapidly. Additionally, as the strain rate increases, the amount of grain boundary movement decreases significantly. This is mainly because the time required to reach the same strain decreases as the strain rate increases, leading to shorter time for a grain boundary to deform at roughly equivalent shear stresses, resulting in smaller movement at higher strain rates. The movement of the grain boundary significantly affects stress concentration at the grain boundary. As seen in Figures 7.15 and 7.18, the stress concentration gradually increases with increasing strain rate, leading to the release of a large number of dislocations in the high-stress region. Consequently, the number of structural copies for *HCPs* (*SFs* and *TBs*) and *TBs* gradually increase.

7.4.3 *Mechanical properties and deformation behavior at different temperatures*

7.4.3.1 *Mechanical properties*

For the polycrystalline copper film under the strain rate of 1×10^8/s and tensile simulation at different temperatures, the stress–strain curves obtained and some typical mechanical properties extracted are shown in Figures 7.19(a) and 7.19(b). Comparing the stress–strain curves under different strain rates, it is difficult to see the difference in Young's modulus under different strain rates from Figure 7.14(a), which also confirms that the dependence of Young's modulus on strain rate is low; however, at different temperatures, both Young's modulus and the yield stresses appear to be significantly different, and late, we quantitatively analyze the effects of temperature and strain rate on Young's modulus. From Figure 7.19(b), it can be seen that the yield strength, yield stress and Young's modulus of the material gradually decrease as the temperature increases, and the simulated Young's modulus obtained at 300 K and 450 K for a grain size 26 nm and the experimental Young's modulus obtained at room temperature are in good agreement [27]. Young's modulus obtained at room temperature for polycrystalline copper with a grain size of 16 nm is also the same as those obtained in the experiments [44]. The temperature dependence of the calculated Young's modulus is about

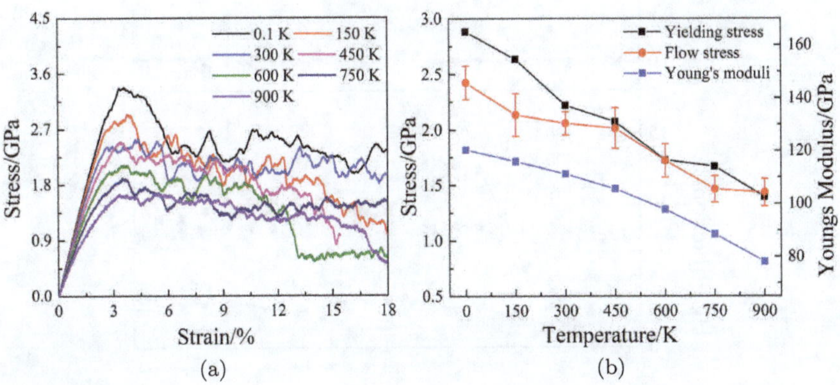

Figure 7.19. Stress–strain curves and related mechanical properties of tensile simulations at different temperatures for a strain rate of 1×10^8/s. (a) Stress–strain curves for tensile tests, (b) related mechanical properties in tensile tests.

-46.5 ± 3.5 MPa/K, which is slightly larger than the value obtained for polycrystalline copper with a grain size of 200 nm (-40 MPa/K) [45]. This is mainly due to the fact that the proportion of grain boundaries in the material gradually increases as the grain size decreases, and the temperature in deformation has a greater influence on the activity of the grain boundaries, which leads to an increase in the temperature dependence of Young's modulus.

7.4.3.2 *HCP structure (SFs and TBs) and grain boundary behavior*

The results of the simulations at different temperatures were analyzed using *CNA*, as well as quantitative analysis of the twin crystals. The changes in the *HCP* (*SFs* and *TBs*) and *TB* structures with strain at different temperatures are shown in Figure 7.20. It is observed that the number of copies of both structures decreases gradually with increasing temperature, especially at 900 K. It becomes difficult to

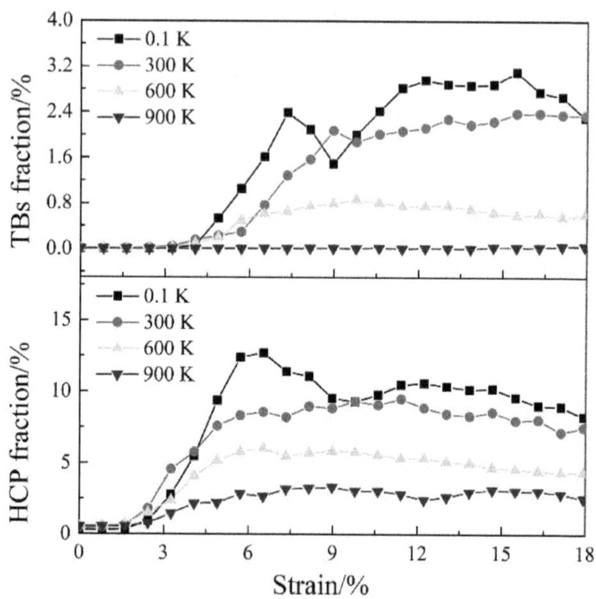

Figure 7.20. Variation of *HCP* (*SFs* and *TBs*) and *TB* structural fractions with strain at different temperatures for strain rates of 1×10^8/s.

identify the twin structure during the deformation, and at higher temperatures, the local vibration of atoms is intensified, particularly for the chaotic atoms on the grain boundaries. As a result, the grain boundary activity becomes more pronounced. Conversely, the vibration of atoms is weaker at temperatures below 0.1 K, making it difficult to carry out the grain boundary activity. Therefore, at low temperatures, the main mechanism of plastic deformation involves the release of dislocations from the grain boundaries and their absorption at the grain boundaries. As the temperature increases, the deformation mechanism shifts from the release of dislocations to the joint action of dislocations and grain boundary activity.

The highly active GB35 grain boundaries were quantitatively analyzed. The amount of grain boundary movement at different temperatures is depicted in Figure 7.21. The x-axis represents the

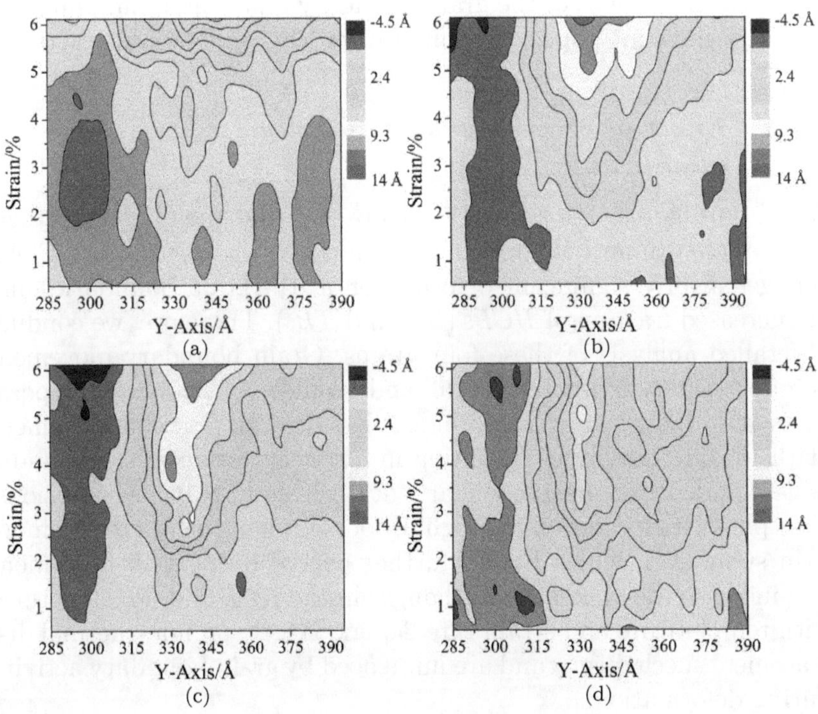

Figure 7.21. Strain rate of 1×10^8/s at different temperatures for GB35 (different colors in the figure represent the amount of grain boundary movement). (a) $T = 0.1\,\mathrm{K}$, (b) $T = 150\,\mathrm{K}$, (c) $T = 300\,\mathrm{K}$, (d) $T = 450\,\mathrm{K}$.

distribution of GB35 at different locations, while the y-axis indicates the strain. The figure shows that the range of GB35 deformation gradually increases with rising temperature. It is also evident that the strain decreases as GB35 starts to move at higher temperatures. According to equation (7.13), there is an exponential relationship between the inverse of grain boundary mobility and temperature, with higher mobility at higher temperatures. Consequently, at higher temperatures, GB35 requires less shear stress for movement, leading to preferential movement. At a location with a smaller radius of curvature (approximately 340 Å), the grain boundary movement amount is larger at 150 K, decreasing as temperature increases. Comparing with Figure 7.20, it is observed that the release of partial dislocations from the grain boundary to form an *HCP* structure (with a strain of 3–6%) accelerates the movement of this grain boundary. However, even with a large number of dislocations released, at a temperature of 0.1 K, the movement of the grain boundary is still minimal, illustrating the significant impact of temperature on grain boundary activity.

7.4.4 *Effect of grain boundary activity on mechanical properties*

Based on Figures 7.16 and 7.17, we observe that the microstructural changes are primarily happening in grains G3, G5, G7, and G8. These changes involve significant deformation at the grain boundaries and an increased fraction of *HCPs* (*SFs* and *TBs*). Therefore, we conduct a detailed analysis of these four grains. Grain boundary movement is influenced by temperature [46] and strain rate. At lower temperatures and higher strain rates, there is less deformation and movement at the grain boundaries, resulting in a concentration of stress and a release of dislocations, particularly at 0.1 K and 5×10^9/s conditions. This plastic deformation through dislocation movement creates more twin structures, which hinder further partial dislocation movement and increase the yield stress. Changes in the *HCP* and *TB* structures with temperature and strain rate depend largely on non-thermal displacement mechanisms and are influenced by grain boundary activity during deformation.

The strain at fracture of polycrystalline copper films is also affected by microstructural changes during initial deformation.

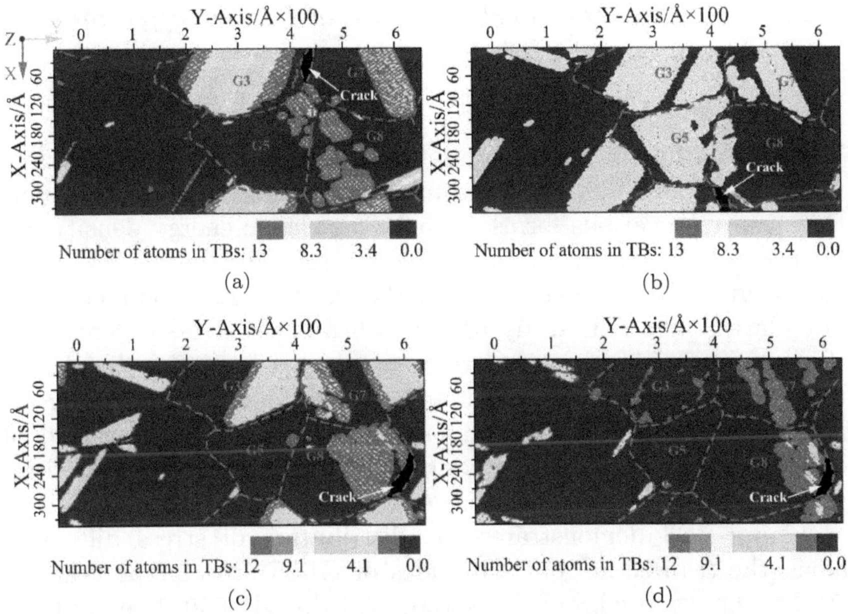

Figure 7.22. Distribution of *TB* structure and location of crack generation at different temperatures for a strain rate of 1×10^8/s, the distribution of *TB* structure, and the location of crack generation at different temperatures. (a) $\varepsilon = 20.17\%$ (0.1 K), (b) $\varepsilon = 14.47\%$ (150 K), (c) $\varepsilon = 14.26\%$ (450 K), (d) $\varepsilon = 12.83\%$ (600 K).

Note: Different colors in the graph indicate the number of atoms contained in *TBs*.

More twin boundaries at low temperatures and high strain rates hinder dislocation movement, leading to reduced ductility. Fracture locations differ significantly at low and high temperatures, as seen in Figure 7.22. At low temperatures (0 K and 150 K), the fracture location's y coordinate is around 430 Å, while at high temperatures (450 K and 600 K), it's about 600 Å. As temperature increases, the range of deformation and movement of GB35 expands, reducing stress concentration at grain boundaries. This lowers the fraction of special boundaries and twin boundaries in some grains (G3 and G5) at high temperatures, while increasing them in others (G7 and G8), leading to higher stress concentration and crack development at the boundary due to the interaction between the *HCP* structure (*SFs* and *TBs*) and grain boundaries.

7.4.5 *Deformation mechanisms and factors affecting grain boundary activity*

7.4.5.1 *Drivers of grain boundary shifts*

The deformation of grain boundaries is influenced by several factors including strain rate, temperature, and driving force. The driving force can take various forms, such as the elastic energy density difference across straight grain boundaries [47, 48], the net energy of curved grain boundaries [49], and the shear stress at curved grain boundaries [50]. This study has identified shear stress difference as the primary driving force for grain boundary deformation. Typically, the coupling of curved grain boundaries results in both movement and rotation of the grain boundaries. However, in two-dimensional materials, the rotational degrees of freedom are limited, so the driving force of rotation is not considered in this study.

Figure 7.23 demonstrates the distribution of stress difference along the x direction on both sides of GB35 at a temperature of 300 K. The x and y in the figure represent the GB35 at different locations and the strain changes, respectively. By using $\frac{\sum S_{x}+}{N_{px}} \pm \frac{\sum S_{x}-}{N_{nx}}$ for calculation, where S_x+ and S_x- are the stresses in the forward and reverse directions of the atoms along the x-axis and $N$is the number of atoms within a spacing of 5 Å along the y-axis.

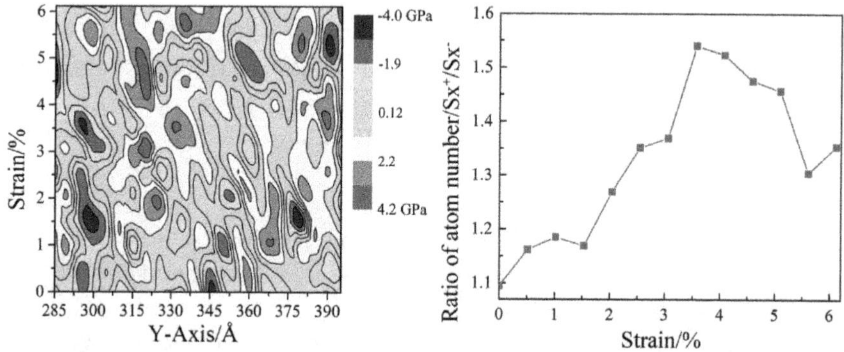

(a) (b)

Figure 7.23. GB35 grain boundary stress state at strain rate of 1×10^8/s and 300 K. (a) Stress difference along the x direction, (b) ratio of atom numbers for stress along the forward and reverse x direction.

Comparison of Figures 7.14(a) and 7.18(a) reveals that the two grain boundary shift cloud maps and stress difference cloud maps are in better agreement. At smaller radius positions ($y \approx 295\,\text{Å}$ and $y \approx 340\,\text{Å}$), there is a larger stress difference, likely due to the more curved and coupled grain boundaries. This larger stress difference results in a greater degree of grain boundary movement. Figure 7.23(b) also shows that the ratio of the stress difference along the two directions is close to 1.0 when the strain is 0%. This implies no notable stress difference, and the ratio gradually increases with deformation. This means that the stresses along the positive direction of the x-axis increase more than the stress along the negative direction, leading to a gradual movement of GB35 in the positive x-axis direction as a whole. When the strain is greater than 3.5%, this ratio gradually decreases due to the straightening of the grain boundary, which reduces the coupling of the neighboring grains. Additionally, the release of dislocations from the grain boundaries reduces the shear stress difference between the two sides.

7.4.5.2 *Grain boundary deformation mechanism*

In the previous section, we discussed the movement of grain boundaries due to differential shear stress. However, we did not delve into the exact mechanism of this movement. Previous research suggests that the main mechanism of grain boundary movement is the diffusion of atoms from one grain to another [51] or by the shuffling of atoms [52]. These mechanisms are typically observed under high temperature or plastic deformation conditions. However, the movement mechanism of curved grain boundaries under initial deformation conditions has not been reported.

In Figure 7.24(a), the strain rate is 1×10^8/s, and the temperature is 300 K. The atomic stress distribution on the grain boundaries of GB35, GB78, and GB56 and the movement process of GB35 is depicted in Figure 7.24(b) in the initial deformation stage. From the atomic stress distribution diagrams, we observe that the grain boundaries GB35 and GB78 present an obvious lamellar structure in the initial state, with more low-stress layers (denoted as L_y) and high-stress layers (denoted as H_y). On the other hand, the straight surface grain boundaries GB56 do not show an apparent lamellar structure in the initial state, and the distribution of atoms on the

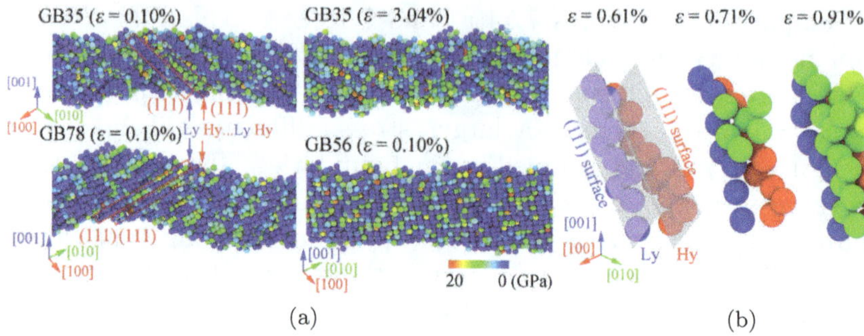

Figure 7.24. Atomic stress distribution at three grain boundaries and the movement of grain boundary atoms under GB35 initial strain. (a) Atom distribution in different crystal boundaries, (b) atom movement in different strains.

Note: The blue and red atoms in the diagram are the original low and high-stress atoms, respectively, and the green atoms are those that have moved from neighboring grains to grain boundaries.

straight grain boundary does not appear the lamellar structure and is more chaotic.

As the deformation increases, the lamellar structure of the curved grain boundary GB35 gradually disappears, and the atomic arrangement becomes chaotic. The movement of the curved grain boundary in the initial deformation stage is through the diffusion of atoms from the *FCC* structure of the neighboring grains along the (111) direction toward the grain boundary. The original atoms on the grain boundary gradually move toward the other neighboring grains and turn into the *FCC* structure. The blue and red atoms in Figure 7.24(b) represent the original grain boundaries, while the green atoms represent the new atoms added to the grain boundary. Thus, during the initial tensile stage, the curved grain boundaries gradually move along the (111) direction. As the grain boundary moves further, the atoms on the grain boundary become disordered, and the grain boundary gradually becomes flat, and then the deformation mechanism of the grain boundary becomes shuffling of the atoms. However, no movement of the grain boundary is found in the curved grain boundary GB78, mainly because the difference in shear stresses of the grain boundary is small in the grain, making it difficult to drive the movement of the curved grain boundary.

7.4.5.3 *Boundary orientation*

In Figure 7.25, both GB37 and GB58 are flat and distributed along the x-axis. However, their deformation behaviors are quite different, as explained below. When the strain reaches 2.0%, many partial dislocations are released from GB58 and then the grain boundary starts to cleave. With further deformation, the left portion of the cleaved grain boundary gradually moves toward grain G5 and disappears gradually. On the other hand, the grain boundary GB37 did not undergo cleavage. To deeply understand this phenomenon, we examined the two grain boundaries along the y-axis to analyze the distribution of the atomic stress difference and the movement of the grain boundary. The distribution of the atomic stress difference and the grain boundary movement is calculated and shown in Figure 7.25.

When comparing the distribution of the difference in stress for the two grain boundaries, it can be observed that the change in stress difference is not significantly different with the strain. The positive stress along the y-axis is greater than the negative direction, but the two grain boundaries move in opposite directions. Specifically, GB37 moves in the positive direction along the y-axis, while GB58 moves in the negative direction along the y-axis when the strain is greater than 2.0%.

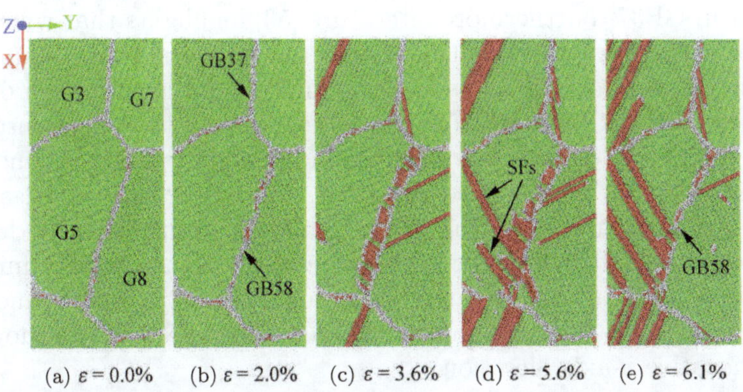

Figure 7.25. Atomic configurations of grain boundaries GB37 and GB58 at different strain rates of 1×10^8/s and temperature of 300 K. (a) $\varepsilon = 0.0\%$, (b) $\varepsilon = 2.0\%$, (c) $\varepsilon = 3.6\%$, (d) $\varepsilon = 6.1\%$.

Note: *FCC* and *HCP* structures (*SFs* and *TBs*) and other structures are shown in green, red, and white, respectively.

From the previous studies [53], we can conclude that the grain boundary cleavage and movement behavior depend on the shear stress and orientation difference, and the critical stress at which cleavage occurs can be calculated by the following equation:

$$\tau_c = 0.7D\omega \tag{7.16}$$

where $D = G/[2\pi(1-\gamma)]$, G is the shear modulus/GB, γ is Poisson's ratio, and ω is the orientation difference angle/rad.

Based on the given data, for copper with a shear modulus of 48 GPa and a Poisson's ratio of 0.34 [54], GB37 has an orientation difference angle $\omega = 0.7 \approx 40.1^\circ$ and GB58 has an orientation difference angle of $\omega = 0.3 \approx 17.1^\circ$. The critical shear stress for splitting the two grain boundaries is calculated using the above formulas, resulting in 5.7 GPa for GB37 and 2.4 GPa for GB58. Upon comparison of Figures 7.26(a) and 7.26(c), it is evident that the difference in shear stresses distributed on the grain boundaries is larger than the critical cleavage stresses for GB58 when the strain exceeds 2.0%. However, in the case of GB37, a very small area is found where the shear stress difference reaches the critical value only when the strain in GB37 reaches 3.0%. Moreover, most of the values are smaller than the critical value, indicating that it is challenging for cleavage behavior to occur in GB37. Furthermore, literature [53] highlights that when the misorientation of the grain boundary is small, the grain boundary where cleavage occurs tends to move in the opposite direction of the shear stress. In Figure 7.25, it is observed that the misorientation of the left half of the cleaved grain boundary is small, causing the stacking faults behind it to cross the grain boundary and leading to the disappearance of the grain boundary. Consequently, during cleavage, the left half of the grain boundary with the smaller misorientation moves in the negative y-axis direction by a larger amount. As a result, the grain boundary GB58 as a whole tends to move in the negative y-axis direction.

7.4.5.4 *Effect of dislocation release on grain boundary deformation*

The rate of movement of grain boundaries primarily depends on temperature and driving force. An increase in both temperature and

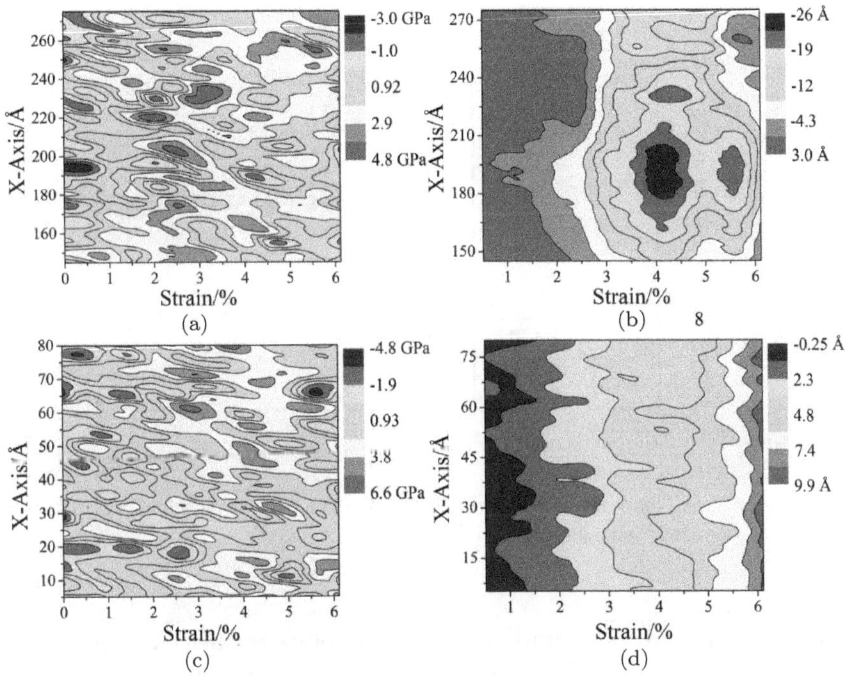

Figure 7.26. The distribution of shear stress difference on grain boundaries GB58 and GB37, and the distribution of the amount of movement of the two grain boundaries at strain rate of 1 × 108/s and temperature of 300 K. (a) Stress: GB58, (b) migration: GB58, (c) Stress: GB37, (d) migration: GB37.

driving force results in a higher rate of grain boundary movement. However, recent studies have revealed that the deformation behavior of grain boundaries is influenced not only by temperature and driving force, but also by defects near the grain boundaries, such as stacking faults and twinning structures [55, 56]. Figure 7.26 shows the distribution of grain boundary movement, indicating that the release of dislocations accelerates the movement of grain boundaries. As such, the average overall grain boundary movement with strain was calculated for GB35 at various temperatures.

Based on Figure 7.27, it is evident that the overall grain boundary movement increases with temperature. The movement of grain boundaries can be categorized into three stages: the initial movement stage, the *SF* acceleration stage, and the *TB* acceleration stage. In Stage I (strain between 0% and 2%), the grain boundaries'

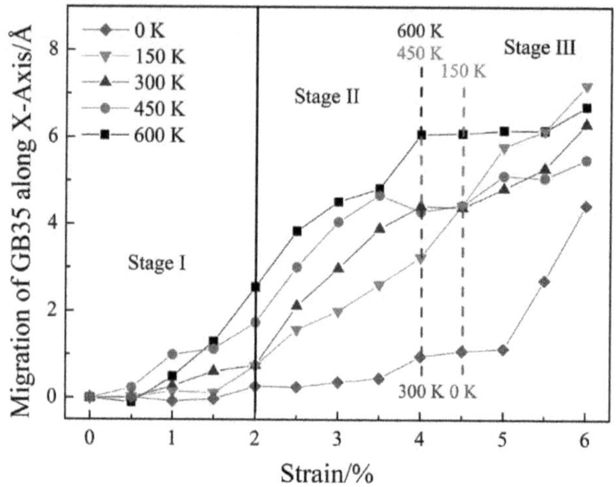

Figure 7.27. The average movement of GB35 grain boundaries at a strain rate of 1×10^8/s varied with strain at different temperatures.

movement is minimal and depends on temperature. As the strain surpasses 2%, the grain boundaries move into stage II, where partial dislocations are released from the grain boundaries, accelerating their movement. Stage II varies at different temperatures such as 2.0–4.5% for temperatures of 0.1 K and 150 K, and 2.0–4.0% for temperatures greater than or equal to 300 K. The movement of grain boundaries at this stage is still controlled by temperature. Therefore, the difference between the slopes of Stage I and Stage II is not very large when the temperatures are higher (450 K and 600 K). In Stage III, the number of twin boundaries (*TBs*) gradually increases as indicated in Figure 7.26, affecting the grain boundaries' movement. However, the velocity of grain boundary moving (the curve slope) slowly decreases with increasing temperature in this stage, mainly due to fewer twinned grain boundaries in grains 3 and 5 at high temperatures. The velocity of grain boundaries GB35 decreases at higher temperatures (450 K and 600 K) compared to Stage II, while at 0.1 K and 150 K, the velocity of grain boundaries in Stage III is greater than that in Stage II. Therefore, it can be concluded that in the plastic deformation stage, the velocity of grain boundaries is primarily controlled by temperature due to the influence of *SFs* and *TBs*.

7.4.6 *Summary*

We have studied a polycrystalline copper film with a grain size of 17.84 nm using phase field and MD uniaxial tensile simulations at different strain rates and temperatures. By analyzing the grain boundary activities, we found the main mechanism of grain boundary deformation under the initial state. We also established the relationship between the grain boundary activities and the mechanical properties of the polycrystalline copper film. Here are the main conclusions:

(1) The strain rate sensitivity (*SRS*) values obtained from tensile deformation closely match those obtained from experiments conducted under similar conditions. The *SRS* values are closely related to the deformation mechanism, indicating that the model effectively demonstrates the deformation mechanism of polycrystalline copper in the experiments.

(2) The movement of grain boundaries in the initial stage depends on the shear stress difference on the grain boundaries. Grain boundaries with smaller radii of curvature exhibit more pronounced movement and tend to deform toward the straight side to reduce energy.

(3) We found that the speed at which grain boundaries move is influenced by factors such as strain rate, stacking faults (*SFs*), and twin boundaries (*TBs*) structures, in addition to temperature and driving force. An increase in strain rate leads to faster grain boundary movement, and the presence of *SF* and *TB* structures also accelerates this movement.

(4) When it comes to curved grain boundaries, they tend to form a lamellar structure with more low and high-stress layers. On the other hand, the distribution of atoms on straightened grain boundaries is more chaotic due to the stress arrangement of the atoms. As a result, the initial deformation in curved grain boundaries occurs through the movement of *FCC* atoms of one grain along the (111) facing grain boundaries, gradually detaching from the grain boundaries and transforming into the *FCC* atoms of the other grain. This differs from the deformation mechanism of straightened grain boundaries, which involves the shuffling deformation of atoms. As the deformation increases, the curved grain boundary gradually transforms into a straighter

grain boundary with a more chaotic distribution of atoms, and the deformation mechanism also transforms into a shuffling of atoms.

(5) The mechanical properties of the material are influenced by grain boundary activity, which changes the number and distribution of hexagonal close-packed (*HCP*) *SF* and *TB* structures. As the strain rate increases and the temperature decreases, the grain boundary activity decreases gradually, leading to a larger stress concentration on the grain boundaries. This results in a large number of dislocations being released from the grain boundaries and the formation of a larger number of *TB* structures, which can impede the movement of dislocations and increase the yield stress of the material. The fracture position of the material differs at different temperatures. At higher temperatures, initial deformation of grain boundaries is greater, which reduces stress concentration. However, as deformation increases, stress concentration arises at other positions, leading to the release of a large number of dislocations from the *HCP* structure. The interaction between the *HCP* structure and the grain boundaries leads to the generation of cracks.

7.5 Deformation Mechanisms of 3D Polycrystalline Copper with Grain Sizes

7.5.1 *Introduction*

The text above discusses the quantitative analysis of grain boundary activity and its impact on the mechanical properties of a simple 2D polycrystalline copper model. It is then necessary to increase complexity in 3D polycrystalline models, where grain boundaries consist of *GB* plane, triple junction, and vertex so that the freedom degrees in grains cannot be limited like in 2D materials. It has also high challenges in experimentally investigating the microstructural evolution inside the grains in 3D materials. In mid-20th century, the literature further delves into the study of nanometallic materials with average grain sizes between 1 and 100 nm and emphasizes the softening behavior [57, 58] associated with a drastic increase in grain boundaries when the grain size drops below a critical value.

To understand the physical mechanisms involved, the grain boundary activities in nanometallic materials, including rotation, movement, and deformation, have been intensively investigated [59–61]. However, the construction of the models in most of the studies uses simple geometrical structures that limit the occurrence of grain boundary activities to specific conditions such as high temperatures or artificially constructed shear stresses, and for real polycrystalline systems, irregular grain boundaries leading to the occurrence of grain boundary activities are usually coupled. Therefore, it is necessary to quantitatively analyze grain boundary activity during deformation using more realistic models. Although those researches lay the foundation for understanding the mechanism of grain boundary activity and its influence on mechanical properties, there are still some shortcomings in the research as follows:

1. The Tyson polygon (Voronoi) method is commonly used to create 3D polycrystalline models. However, in this model, the grain faces are straight, reducing the interaction between neighboring grains and limiting grain boundary activity during deformation.
2. Typically, simulations of grain boundary activity focus on high temperatures to study the Arrhenius relationship for grain boundary movement. At high temperatures, grain boundary activity is more evident, and the driving force is thermodynamic rather than the pure shear stresses generated during deformation. As a result, it's challenging to establish a relationship between grain boundary activity and mechanical properties.
3. Research on deformation in the plastic phase of polycrystalline materials often involves the release of dislocations [62], localized stresses at grain boundaries [63], and grain boundary creep model [64]. Grain boundaries form a continuous network connecting the dispersed intragranulars. However, fewer studies have analyzed the interaction between grain boundary activities and intragranulars during the elastic phase due to the weakness of grain boundary activities in existing simple models.
4. Due to limitations in experimental equipment and techniques, there are few quantitative analyses of grain boundary activities through experiments [8], especially during the elastic phase. Consequently, for ideal polycrystalline metallic materials without

voids and defects, the relationship between grain boundary activity and mechanical properties has not yet been established.

The present authors utilized the phase field method to create a three-dimensional polycrystalline copper material with an average grain size of 6.5–18.8 nm. This material was then subjected to a tensile simulation using molecular dynamics at 300 K and a strain rate of 1×10^8/s. The authors also quantitatively analyzed the grain boundary activity at the initial stage of deformation using a previously developed grain boundary analysis algorithm. This was done to establish the relationship between grain boundary activity and the mechanical properties of the 3D polycrystalline material and to analyze the reasons for the emergence of the anti-H-P phenomenon. Additionally, deformation in the elastic phase of different models was simulated at smaller output steps to analyze the grain boundary activities and mechanical properties in the elastic phase. The results were then compared with previous theoretical studies to quantitatively analyze the reasons for the variability of Young's modulus in models with different grain sizes.

7.5.2 *Mechanical properties and grain boundary activity analysis*

The stress–strain curves for uniaxial tensile at 300 K and the mean strain, Young's modulus, and yield strength for different models are presented in Figures 7.28(a) and 7.28(b), respectively. Young's modulus is the slope of the strain curve at less than 0.5% strain, the mean yield stress is the average stress value at 5–10% strain, and the yield strength usually indicates the onset of plastic deformation and is defined as the stress when the strain deviates by 0.2% from the linear curve [6, 7]. At 300 K, Young's modulus varies in the range of 70–110 GPa, consistent with previous experimental [27] and simulation results [7]. In Figure 7.28(b), it is observed that the three mechanical properties initially increase and then gradually decrease as the grain size decreases. The three-dimensional polycrystalline copper exhibits an anti-H-P phenomenon, and the critical grain size for the H-P transition is approximately 14 nm, consistent with previous experimental, theoretical analysis, and simulation studies with critical grain size 10–15 nm [44, 65–67]. This supports further investigation into

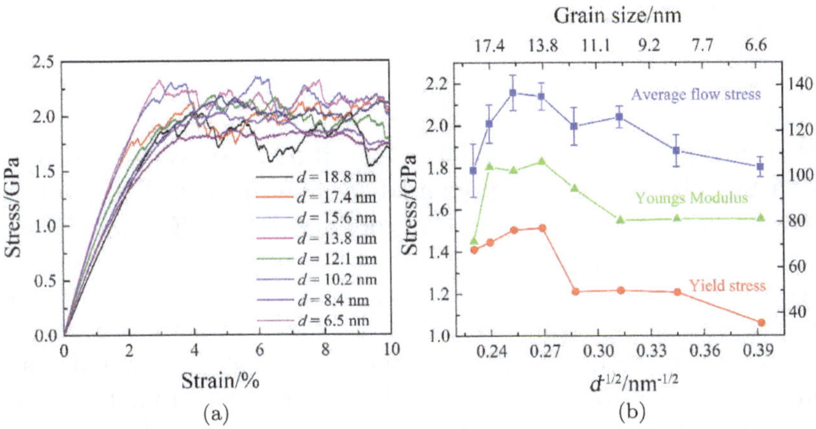

Figure 7.28. Stress–strain curves and typical mechanical properties of polycrystalline copper with different grain sizes when the strain rate is 1×10^8/s at 300 K in uniaxial tensile. (a) Stress–strain curve for uniaxial tensile, (b) related mechanical properties.

the deformation mechanism of 3D polycrystalline copper through a detailed analysis of the simulation results.

The study revealed that irregular grain boundaries are susceptible to deformation, altering the stress concentration on the grain boundaries. This deformation can lead to activities such as coupled slip processes, ultimately resulting in grain rotation. During the deformation process of the 3D grain boundary, we observed notable movement and rotation of the grain boundaries. Intriguingly, the initial deformation stage of the 3D grain boundary moves similarly to that of the 2D grain boundary. The stress distribution of atoms in the curved surface grain boundary contains numerous lamellar structures (e.g., $L_y, \ldots, H_y$) that are not found in straight surface grain boundaries. The movement of the curved grain boundary occurs in two steps: First, the *FCC* atoms of grain 1 move along (111) toward the grain boundary, causing the atoms on the opposite side of the grain boundary to gradually detach and transform into the *FCC* atoms of grain 2. Second, as the deformation increases, the curved grain boundary gradually transforms into a straightened grain boundary, causing chaos in the distribution of atoms on the boundary and resulting in the shuffling of the atoms. Unlike the 2D model, the 3D model does not restrict the rotational freedom of the grains, so the movement of the curved grain boundary may be accompanied by the rotation of the grains.

A comprehensive analysis was conducted to study the initial deformation of a grain boundary with a strain ranging from 0% to 6% using the Grain Boundary Quantitative Analysis Algorithm (GBAA). The results are presented in Figure 7.29. The analysis involved examining the change in goodness-of-fit (R-squared) and the rotation of the grain boundaries, which correspond to the change in the fitting coefficients of the fitted planes (a_0 and a_1). Moreover, the rotation of the grain boundaries correlates with the rotation of the grains. Figure 7.29(a) displays the contour of the original grain boundary as well as the contour of the deformed grain boundary after being fitted using the moving least squares method. This stage of initial deformation shows evident grain boundary activity, with the curved grain boundary gradually becoming flat when the strain exceeds 3.02%. Additionally, the deformation volume distribution

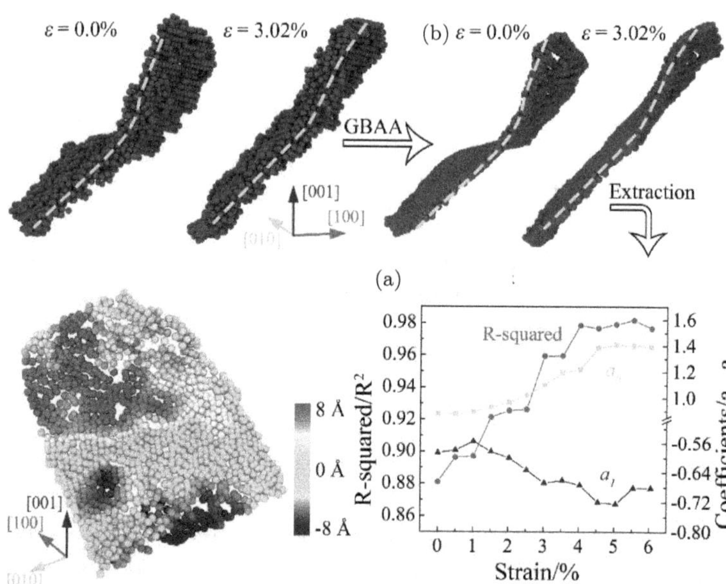

Figure 7.29. Deformation of curved grain boundaries in the 18.8 nm model and variation of the fitted R-squared parameters, a_0 and a_1 with strain. (a) Deformation of boundaries in different strains, (b) deformation distribution, (c) R-squares parameters.

graph in Figure 7.29(b) indicates a substantial amount of boundary deformation. The different colors in the graph represent varying degrees of deformation relative to the original position (strain 0%). In Figure 7.29(c), the graph shows how the goodness-of-fit and the fit coefficient change with strain, which measure the amount of grain boundary deformation. The R-squared value increases gradually with strain, indicating a better fit of the fitted plane and grain boundaries. This results in the grain plane becoming progressively flatter. Additionally, there is greater fluctuation in the fitting coefficient for the normal vector of the fitting plane. This indicates that the rotation of the grain boundary occurs during deformation, and the coupled activity of the grain boundary in the 3D model is similar to the findings of Upmanyu *et al.* [68] and Cahn [69].

7.5.3 *Elastic deformation and a grain boundary fluctuation model*

In contrast to grain intragranulars (GIs), grain boundaries contain loosely distributed atoms, impurities, and micropores. Therefore, the mechanical properties exhibited by grain boundaries are different from those of intragranulars. Additionally, grain boundaries are typically used as sites for dislocation release, leading to their plastic deformation behavior occurring before that of intragranulars. In polycrystalline materials, intragranulars can be thought of as discontinuous "islands," while grain boundaries form continuous networks. As a result, during the elastic deformation phase, the response of grain boundaries is faster than that of the intragranular of the grains. Research has indicated that the response of grain boundaries primarily involves an increase in dislocation density at grain boundaries and the occurrence of visco-plastic behavior reported by Meyers [23]. Furthermore, researchers have divided the deformation prior to yielding into three stages based on the incompatibility stress between grain boundaries and intragranulars [21, 70]. The material undergoes three stages of deformation:

First, neighboring grains exhibit different elastic responses, leading to incompatible stresses at grain boundary locations, resulting in higher stresses at the grain boundaries than within the grain ($\sigma_{GB} > \sigma_{GI}$). As a result, the grain boundaries experience yield plastic deformation first.

During the second stage, yield at the grain boundaries can release stress concentration by dislocation movement and a geometric adjustment. However, these dislocations still cannot be released from the grain boundaries to the intragranulars, mainly because stress concentration caused by the misorientation as well as the elastic incompatibility is not enough to make the dislocations slide toward the intragranular.

In the third stage, when the intercrystalline stress (σ_{GI}), grain boundary stress (σ_{GB}), and applied load (σ_{AP}) are the same, the incompatibility between the intragranulars and boundaries disappears, allowing dislocations to pass through the release of dislocations from the grain boundaries to the intragranulars. The material undergoes macroscopic yielding behavior.

Analyzing the grain boundary deformation in the previous subsection, it is evident that significant deformation occurs at the grain boundaries before yielding happens. Therefore, we analyze the incompatibility within the grain boundaries and intragranulars, based on Meyer's theory.

7.5.3.1 *Uneven deformation in intragranular and grain boundaries in the elastic stage*

To examine the lack of compatibility between grain boundaries and intragranulars before yielding, we conducted an analysis of stress variations at grain boundaries and intragranulars in models with different grain sizes. The results are presented in Figures 7.30(a–e). The insets in the figures show atomic configuration diagrams depicting the initial release of dislocations from the grain boundary into the intragranular, and Figure 7.30(f) provides the strain values corresponding to the first dislocation release when the stresses at the grain boundary and the intragranular are equal, and dislocations are released for the first time in the different models.

Although the stress variations do not visually demonstrate the generation process of incompatibility between the grain boundary and the intragranular, the difference in stress between the intragranular and the grain boundary before yielding is consistent with the first two stages of Meyer's model. The different stresses at the grain boundaries and the intragranular represent distinct deformation behaviors. From the atomic configuration diagrams, it is evident

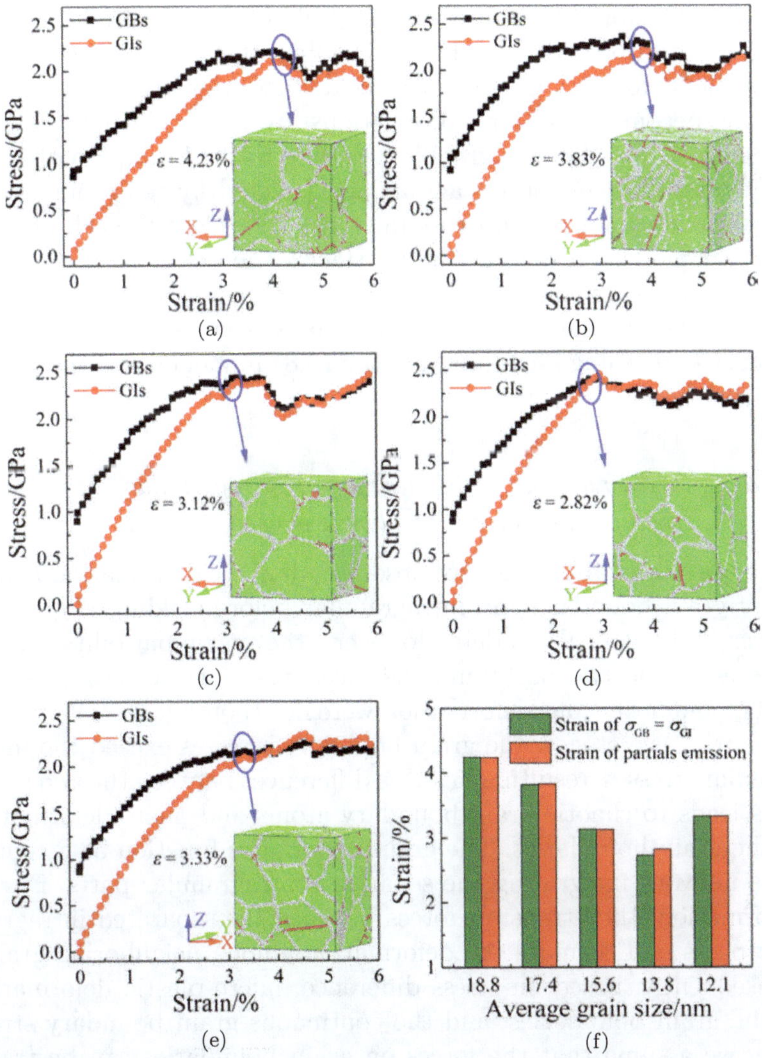

Figure 7.30. Variation of grain boundary stress and intragranular stress with strain in different models (the insets in the figure show the first release of dislocations from the grain boundary to the intragranular, where the white, green, and red atoms denote the other structures, the *FCC* structure and the *HCP* structure (*SFs* and *TBs*), respectively, as well as the statistical counts of the strain when the first release of partial dislocations from the grain boundary and intragranular and grain boundary stresses are equal in different models). (a) 18.8 nm, (b) 17.4 nm, (c) 15.6 nm, (d) 13.8 nm, (e) 12.1 nm, (f) statistical counts.

that micro-deformation occurs at the grain boundaries in the first two stages. Additionally, from Figure 7.30(d), it can be observed that when the deformation enters the third stage, the stresses in the two regions become equal, and dislocations are released from the grain boundaries into the intragranular, leading to the formation of stacking faults (*SFs*) in the intragranular (depicted by red atoms in the inset). The values of the two strains are very close to each other in different models in Figure 7.30(f). Hence, the deformation behavior in the third stage aligns with the theoretical model mentioned above. It is also noted that both strains gradually decrease with decreasing grain size, reaching a minimum in the 13.8 nm model, and then begin to increase.

7.5.3.2 *Modeling of grain boundary and intragranular interactions during the elastic phase*

In the previous discussion, we observed that the deformation behavior of grain boundary and intragranular before yielding aligns with Meyer's theoretical model. However, the incompatibility of the stresses in the two parts and the mechanism of the disappearance of the inhomogeneous deformation were not thoroughly explained. In the initial two stages, the grain boundary stresses exceed the intragranular stresses, resulting in stress differences between the two parts. This leads to chaotic grain boundary atoms and micro-deformation at the grain boundaries. The grain boundaries function as a continuous network integrating the separated intragranular parts. During deformation, they transmit forces between the separated intragranular parts and regulate the deformation among neighboring grains. Consequently, based on stress difference, micro-plastic deformation at the grain boundaries, and the continuous grain boundary structure, we assume that the forces on grain boundaries can be transferred to the intra-grain during the pre-yield stage, causing a gradual increase of forces within the grain. Although the perfect *FCC* structure inside the grain makes it difficult to deform, the interaction between the intragranular boundary atoms and the grain boundaries results in a lower symmetry of the intragranular lattice structure. We hypothesize that the intragranular and grain boundaries are connected by a rigidity matrix (k) and a damping matrix (c), as shown in Figure 7.31(a).

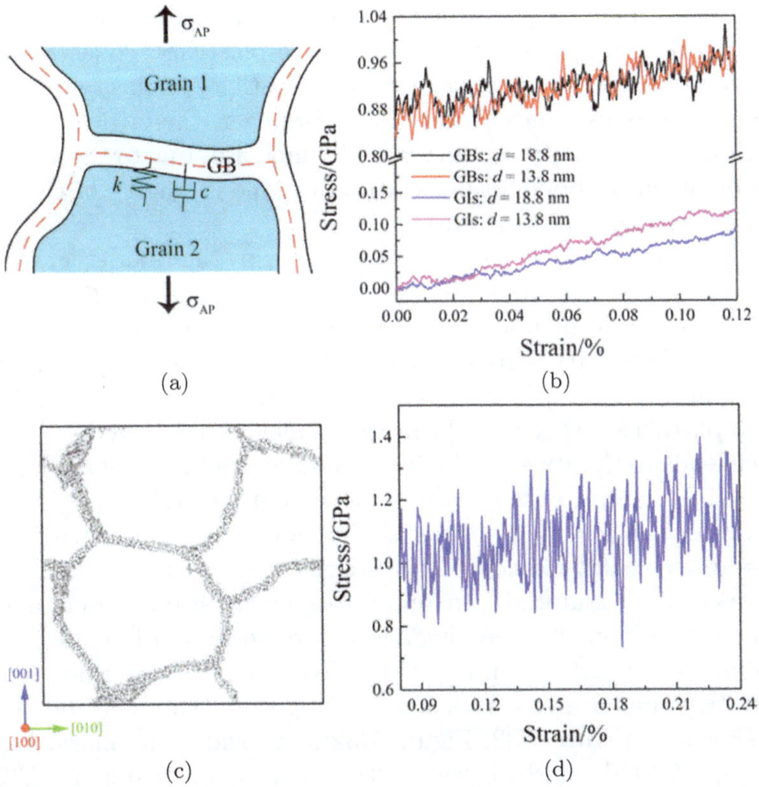

Figure 7.31. Modeling of grain boundary and intragranular interactions and changes in grain boundary and intragranular stresses. (a) Force transfer model, (b) stress variation of grain boundary and intragranular, (c) schematic of grain boundary, (d) stress of grain boundary in (c).

In order to validate this force transfer model, we simulated the deformation in the elastic phase for strains less than 0.3% using very small time steps. Figure 7.31(b) illustrates the variation of stress with strain for both the overall intragranular part and the grain boundary part in both models. It is evident that the stress at the grain boundary fluctuates widely, whereas the variation of the intragranular stress appears as a straight line without significant fluctuations.

To further investigate the change of stress at the grain boundary, a 1 nm thick segment of the grain boundary layer was extracted from the 15.6 nm model. The change of stress is calculated and depicted in Figures 7.31(c) and 7.31(d), revealing that the stress at the grain boundary exhibits nearly periodic fluctuation and contains more "micro-elasticity" and "micro-plasticity" phases. When the stress reaches a critical value, micro-plasticity begins to occur and releases the stress concentration [21]. However, the critical stress at which micro-plasticity occurs gradually increases due to the deformation of the grain boundaries, which alters the geometry of the grain boundaries and their orientation.

The stiffness and damping matrices in the model gradually increase the grain boundary stresses, but when micro-plasticity occurs at the grain boundaries, there is no significant decrease in the grain boundary stresses, primarily due to the high symmetry of the atoms in the intragranular. Consequently, it is difficult for micro-plasticity to occur in the intragranular during the elastic phase, which leads to the stress in the intragranular gradually increasing and exhibiting an approximately linear change in behavior.

Rey *et al.*'s [71] theoretical study demonstrates that grain boundaries are typically considered the source of stress generation within polycrystalline materials. In order to gain a better understanding of the interactions between intragranulars and grain boundaries, we conducted stress distribution calculations for both grain boundaries and intragranular areas, as depicted in Figure 7.31(c), and the results are shown in Figure 7.32. Figure 7.32(a) provides an enlarged view of Figure 7.31(d), and Figure 7.32(b) corresponds to Figure 7.32(a) with the micro-plasticity of the grain boundaries at points B and C. Figure 7.32(c) illustrates the grain boundary and intragranular stress distribution corresponding to the four points A–D in Figure 7.32(a). The stress distribution map reveals that the high-stress region is primarily concentrated near the grain boundaries, with a von Mises stress distribution similar to the Lebensohn *et al.*'s [72] study. A careful comparison of the stress distribution at different strains reveals that the high-stress region within the crystal expands as the strain increases from point A to point B or from point C to point D. When the strain reaches point B, micro-plasticity occurs in the grain boundary, leading to a gradual decrease in stress in the grain boundary to point C.

By comparing with the enlarged image of the white grain boundary in Figure 7.32(b), the deformation of the grain boundary can be observed. However, the strain from point B to point C does not show a significant change in stress within the grain. Therefore, the

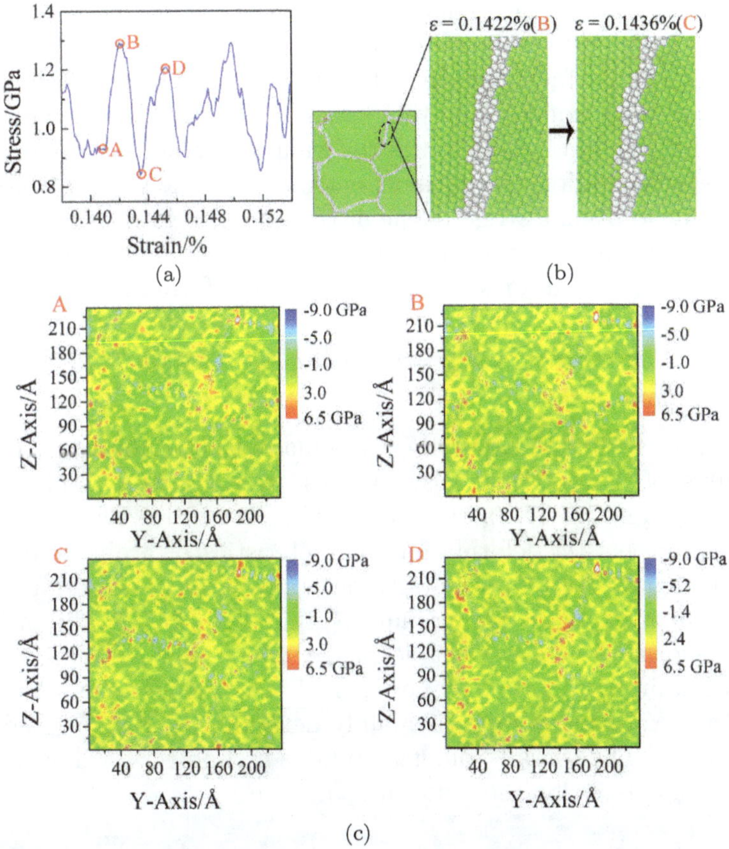

Figure 7.32. Stress variations, distribution, and micro-deformation at grain boundaries in Figure 7.31(d) at various stages. (a) Insert of Figure 7.31(d), (b) micro-deformation of grain boundary, (c) stress distribution in Figure 7.32(a).

Note: White, green, and red atoms in figure (b) indicate other structures, *FCC* structures, and *HCP* structures (*SFs* and *TBs*), respectively.

increased stress in the crystal is not released like the grain boundary stress but is preserved in the crystal. Additionally, by comparing the stress distributions at point A and point C, it can be found that the region of high stress within the crystal becomes larger when the strain is at point C. This means that the micro-elasticity of the grain boundary causes an increase in stress in intragranular within one cycle of the grain boundary stress, but the increase in stress within the crystal is not the same as that at the grain boundary, mainly due to the spacing damping matrix in the interaction between the two parts. Therefore, it is impossible for the stress on the grain boundary to be fully transferred to the intragranular, as stated in the theoretical model. From these phenomena, we can conclude that in the first two stages of initial deformation, grain boundaries and intragranular regions are interdependent, with the grain boundaries providing additional stress to the intragranular region, and the intragranular region providing space for micro-plastic deformation of the grain boundaries and the atoms needed for deformation.

7.5.3.3 *Quantitative analysis of Young's modulus at different grain sizes*

In addition to temperature, there are many factors affecting Young's modulus of nanopolycrystals, such as grain boundary shape [73], average grain size [7], and porosity [44]. Recently, Gao *et al.* [74] studied the elastic deformation characteristics of grain boundaries and intragranular of polycrystelline copper, and the results show that compared with intragranular, grain boundaries will produce local deformation in the elastic phase, which makes Young's modulus of grain boundaries smaller than that of the intragranular 30%, and then Young's modulus is mainly determined by the intragranular stresses. The overall Young's modulus in a polycrystalline system can be calculated by the following equation:

$$E = \frac{1}{\phi_{\mathrm{GI}}/E_{\mathrm{GI}} + (1 - \phi_{\mathrm{GI}})}, \quad \phi_{\mathrm{GI}} = \left(\frac{d}{d + 2d_{\mathrm{GB}}}\right)^3 \tag{7.17}$$

where ϕ_{GI} is the volume fraction occupied by the intragranular, E_{GI} is Young's modulus of the intragranular, d is the average grain size/nm, and d_{GB} is the thickness of the grain boundary/nm. Gao *et al.* [74] showed that the overall Young's modulus decreases with decreasing grain size.

The grain boundary stress in the elastic phase increases the intragranular stress during one fluctuation cycle. The slope of the grain boundary curve is significantly affected by the frequency of fluctuations in grain boundary stress. The fast Fourier transform (FFT) of the difference between the grain boundary stress and the mean stress value along the x and y axes is presented in Figures 7.33 and 7.34(a). In Figures 7.33(a) and 7.34(a), representing the FFT transform of a

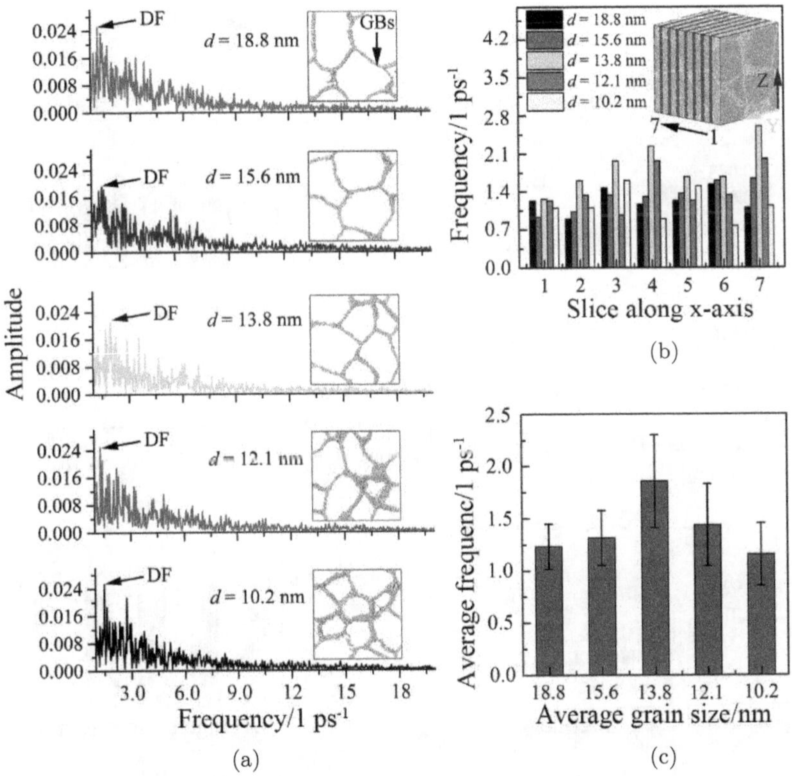

Figure 7.33. FFT calculations of grain boundary stresses distributed along the x direction in different models, and statistics of the frequency of grain boundary stress fluctuations. (a) FFT calculated in single layer boundary, (b) frequency statistics in x direction, (c) mean frequency value in x direction.

grain boundary layer in different grains selected along the two directions, the principal frequency of the fluctuation of the grain boundary stress is observable in the curves. The grain boundaries along the x and y axes for different models were selected based on a thickness of 1 nm, the FFT was calculated, and the obtained principal frequency of the fluctuation is depicted in Figures 7.33(a) and 7.34(b). Following this, the principal frequencies were averaged for different layer sheet grain boundaries in both directions, and the results are shown in Figures 7.33(c) and 7.34(c). These figures reveal that as the grain size decreases, the principal frequencies in both the x and y axes first increase and then decrease, reaching a maximum point in the model

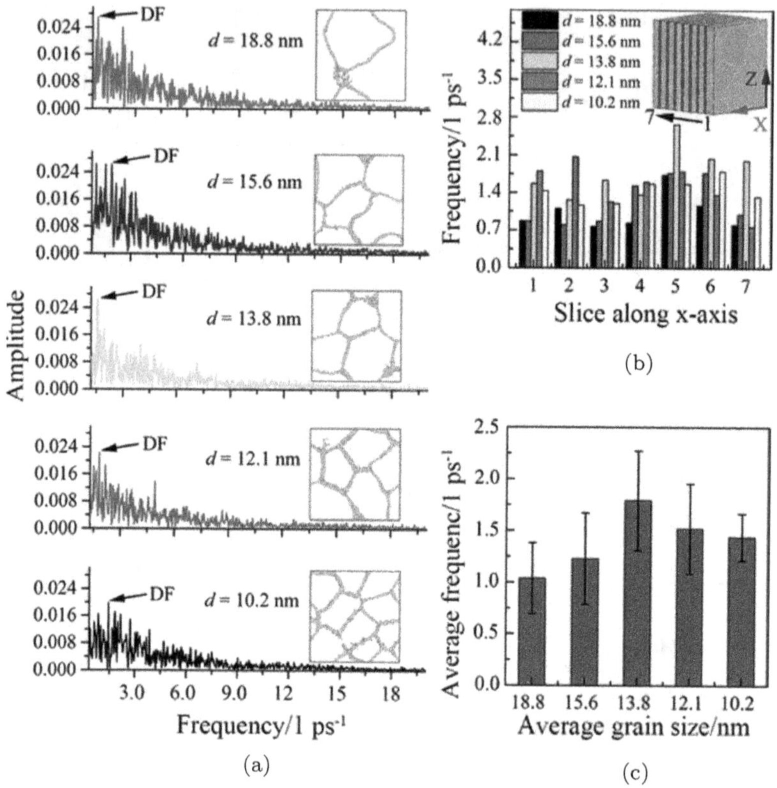

Figure 7.34. FFT calculations of grain boundary stresses distributed along the y direction in different models and statistics of the frequency of grain boundary stress fluctuations. (a) FFT calculated in single layer boundary, (b) frequency statistics in y direction, (c) mean frequency value in y direction.

with a grain size of 13.8 nm. In the model with a grain size of 13.8 nm, the slope of the stress increase inside the grain is larger compared to other models. As a result, Young's modulus value in this model is higher than in the other models. Additionally, the strain released from the grain boundary into the intragranular by dislocations is the smallest for the 13.8 nm model when the grain boundary stress and the intragranular stress are the same in Figure 7.30(f). This is mainly due to the fluctuation of the stress at the grain boundary in this model. The frequency of this fluctuation is large, leading to a faster increase of intragranular stress. As a result, the stresses in the two parts are the same within a shorter strain distance.

It can be observed that the principal frequency of grain boundary stress fluctuations in the 12.1 nm model is slightly higher than that in the 15.6 nm model. However, Young's modulus is lower than that in the 15.6 nm model. This is mainly due to the gradual increase in the number of grain boundary parts as the grain size decreases, leading to a decrease in the densification of the crystal. Moreover, the number of parts facing the grain boundary increases in smaller grains, resulting in a reduction in the coupling between grains.

7.5.4 *Deformation analyses in the plastic phase*

The relationship between grain boundary activity and the anti-H-P phenomenon is usually closely linked. However, the grain boundary activity during the initial deformation stage (0–6%) has not been quantitatively analyzed. To address this, the *GBAA* algorithm is primarily utilized for quantitative analysis of grain boundary activities in various models. As the grain size decreases, the number of flat faces in the polycrystals gradually increases. This limits grain boundary deformation and rotation. Conversely, smaller grains experience the emergence of grain boundary slip, greatly softening the material. For this reason, only models with average grain sizes ranging from 18.8 nm to 12.1 nm are chosen for the analysis of grain boundary deformation and rotation.

7.5.4.1 *Quantitative analysis of grain boundary activity*

The graphs in Figures 7.35(a–e) show how the grain boundary deformation parameter (R-squared) changes as strain increases for different models. These graphs only display grain boundaries with significant variations in this parameter. After that, the number of grain boundaries with a variation in ΔR^2 greater than 0.03 is calculated, and the final count is shown in Figure 7.35(f) for different models. From graphs (a–e), it's evident that in most cases, R^2 increases as strain increases, especially for the more curved grain boundaries. This indicates that most grain boundaries deform toward straight surfaces. This is primarily due to the larger number of disordered atoms in the curved grain boundaries compared to the straight grain boundaries, resulting in higher grain boundary energy for curved grain boundaries with the same misorientation. Consequently, this grain boundary

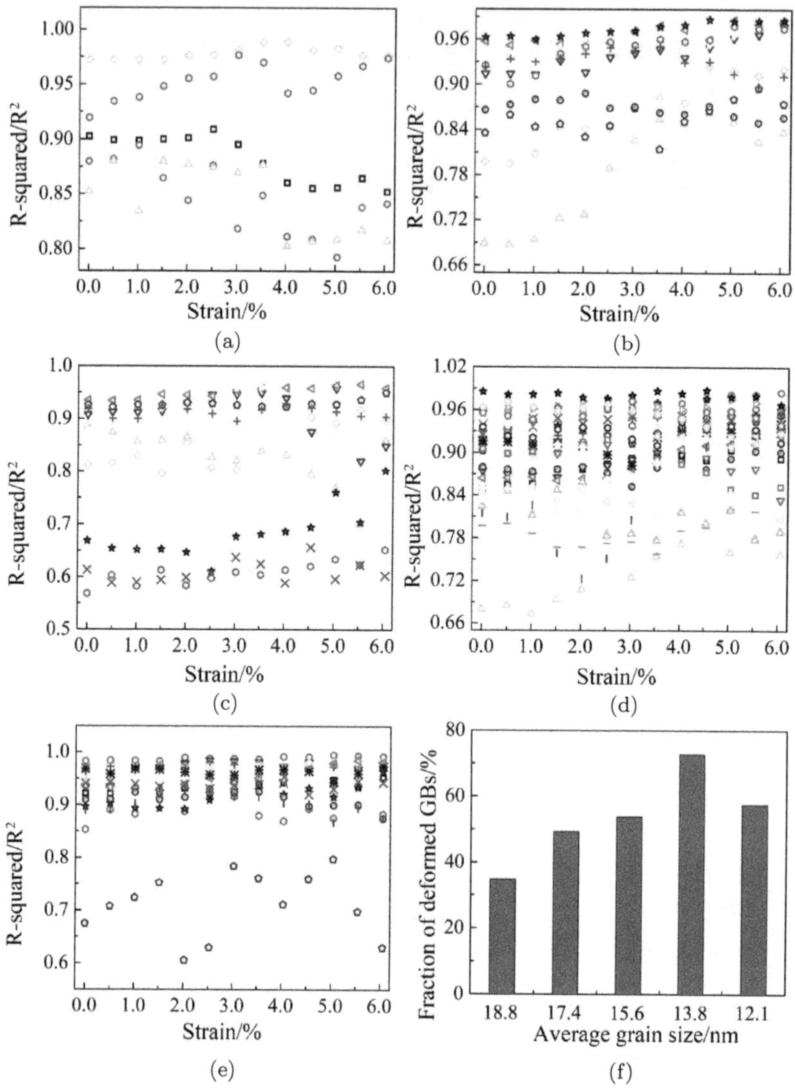

Figure 7.35. Variation of R-squared with strain for different grain boundaries in different models and larger grain boundary deformation in different models. (a) 18.8 nm, (b) 17.4 nm, (c) 15.6 nm, (d) 13.8 nm, (e) 12.1 nm, (f) deformation statistics.

Note: The change for grain boundary deformation in goodness-of-fit (R-squared) is larger than 0.03.

evolution reduces the overall grain boundary energy. As deformation increases, it becomes more challenging to deform the grain boundary, leading to an increase in the critical stress for grain boundary micro-plasticity, which was also mentioned in the previous section.

In Figure 7.35(f), it is apparent that as the grain size decreases, the number of deformed grain boundaries gradually increases, reaching a peak in the polycrystalline model with a grain size of 13.8 nm before starting to decrease. This can be attributed to the high frequency of grain boundary stresses in the 13.8 nm model prior to yielding. Consequently, this phase exhibits a substantial amount of micro-plastic behavior. The sum of the micro-plastic deformations before yielding and the deformations in the plastic phase results in the largest number of deformed grain boundaries in this model. Additionally, the release of dislocations in the plastic phase accelerates the deformation of the grain boundaries, leading to the largest number of partial dislocations in the yield phase in this model. Similarly, the rotation of grain boundaries was analyzed, and the variation of the rotation parameters of grain boundaries (a_0 and a_1) with strain in different models is depicted in Figures 7.36(a–e). Then, the grain boundaries with the variations Δa_0 and Δa_1 greater than 0.03 were counted. The final results of the number of rotations of grain boundaries in the different models are presented in Figure 7.36(f).

From Figure 7.36, it can be seen that as the grain size decreases, the number of rotating grain boundary gradually increases, again reaching a maximum in the 13.8 nm model, and then begins to decrease. A comparison of the grain boundary rotation coefficients and the grain boundary deformation coefficients reveals that the deformations are different, with a small range of fluctuations in the rotation parameter of the grain boundaries when the strain is less than 3.0%, but a larger fluctuation in this parameter when the strain is higher than this value. Therefore, it can be concluded that the rotation of grain boundaries usually occurs in the plastic phase, and this phenomenon is consistent with the grain boundary rotation behavior reported by Meyers [23]. However, for grain boundary deformation, it can be seen from the variation of the grain boundary deformation parameter that the grain boundary deformation occurs in both the elastic and plastic phases.

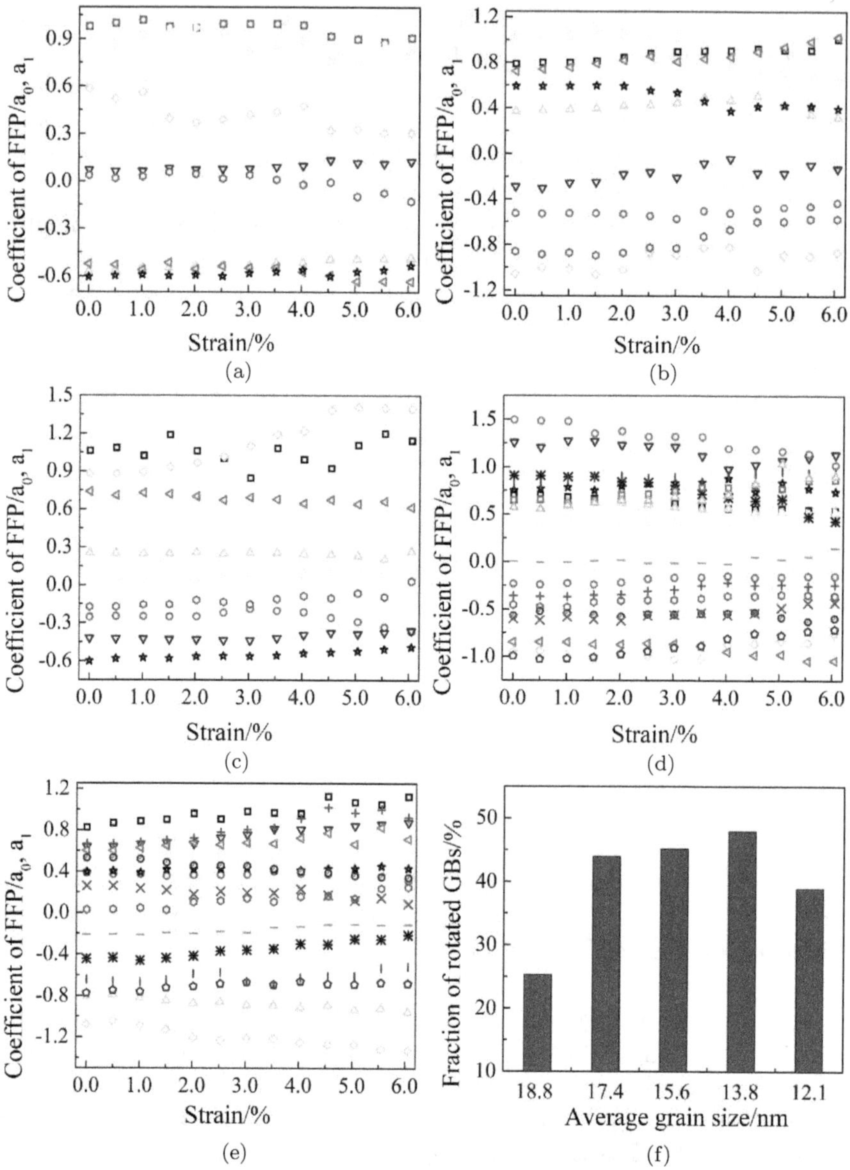

Figure 7.36. Variation of fitted facet coefficients and the number of larger grain boundary rotations (Δa_0 and $\Delta a_1 > 0.03$, where a_0 and a_1 are coefficients of the plane equation $a_0x+a_1y = z$ for the grain boundary) with strain for different grain boundaries in different models. (a) 18.8 nm, (b) 17.4 nm, (c) 15.6 nm, (d) 13.8 nm, (e) 12.1 nm, (f) deformation statistics.

7.5.4.2 *Changes in HCP structure (SFs and TBs) and the anti-H-P phenomenon*

In the study, it is observed that before reaching the yielding stage, the stresses at the grain boundaries cause a gradual increase in the stresses within the crystal, eventually equalizing the stress values in the two parts. This leads to the release of material stresses through dislocation into the yielding stage. Multiple grains undergo subsequent rotation of the grain boundaries, usually accompanied by rotation of the grains. This rotation changes the dislocation slip system and the number of stacked faults within the crystal. Figures 7.37(a) and 7.37(b) display the variation of the number of copies of *HCP* structures (*SFs* and *TBs*) and the average number of copies of *TBs* in each grain with strain in different models. The fluctuation range of *HCP* is found to be large, mainly because the motion of partial dislocations is the main deformation mechanism, and the motion of trailing partial dislocations leads to the disappearance of *SF* structures. Moreover, when the strain exceeds 3.0%, the *HCP* fraction shows a dramatic increase in the 13.8 nm model. This is attributed to the rotation of a larger fraction of grains in this model, resulting in a change in the Schmidt factor and dislocation slip system, ultimately causing the release of many dislocations from the grain boundaries to the intragranular under favorable slip systems.

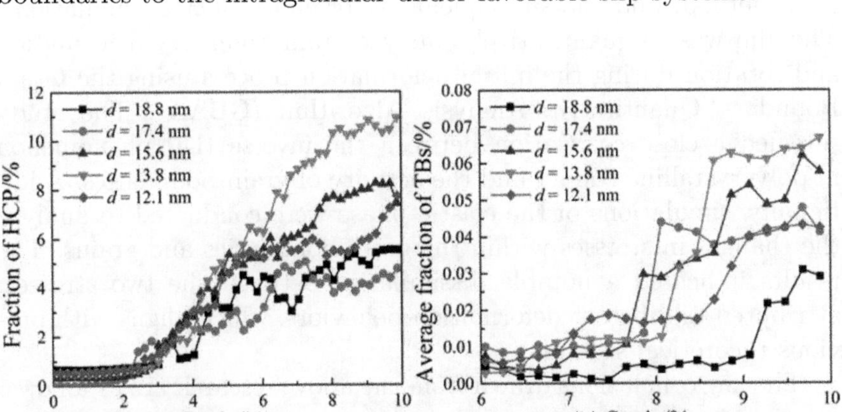

Figure 7.37. Variation of the number of *HCP* structures (*SFs* and *TBs*) and the fraction average *TBs* per grain with strain in different models. (a) *HCP* structure, (b) *TBs* structure.

In the 13.8 nm and 15.6 nm models, the presence of numerous dislocations can lead to the stacking of dislocations on grain boundaries or the formation of immovable dislocations through inter-dislocation reactions. This, in turn, causes stress concentration and results in the release of dislocations on the nearby interfaces, leading to the generation of a large number of twists structures, as shown in Figure 7.37(b). The stacking mechanism of dislocations is typically used to understand the relationship between yield strength and the material's H-P relations. However, the anti-H-P phenomenon for yield stress largely depends on grain boundary activities and twin structures. These structures are more stable and effectively hinder the movement of dislocations, thereby increasing the yield stress of the material. Therefore, we can conclude that in the case of *FCC* polycrystalline copper, the appearance of the anti-H-P phenomenon due to grain boundary activity is primarily achieved by indirectly modifying the number of stable microstructural components within the material.

7.5.5 *Summary*

This chapter focuses on the construction of polycrystalline copper models with varying grain sizes, primarily through the phase field method, and the subsequent performance of *MD* simulations. The aim was to quantitatively analyze grain boundary deformation and rotation during the initial deformation process using the Grain Boundary Quantitative Analysis Algorithm (GBAA). The study revealed a close correlation between the inverse H-P phenomenon in polycrystalline copper and the activity of grain boundaries. Additionally, simulations of the elastic phase were conducted to analyze the changes in stresses within the grain boundaries and grains. The results indicated a notable dissimilarity between the two stresses, attributed to different deformation behaviors, which aligns with previous theoretical studies.

The key conclusions drawn from the above research are as follows:

(1) The quantitative analysis algorithm of grain boundaries, built on the moving least squares (MLS) method, can accurately analyze the deformation and rotation of grain boundaries during the

initial deformation stage. This analysis sheds light on the mechanism of the anti-H-P phenomenon.

(2) The critical grain size for the emergence of anti-H-P aligns with previous studies, with significant grain boundary deformation and rotational behavior observed during uniaxial tensile. The deformation mechanism of the curved grain boundary at the initial stage mirrors that of the deformation of the curved grain boundary in the Cu film discussed in the previous chapter. Both 2D and 3D curved grain boundaries undergo two types of deformation behaviors: atoms moving along the (111) facing grain boundaries followed by deformation through atom shuffling.

(3) In light of prior theories and the simulation results presented in this study, we propose a model to elucidate the interaction between grains and grain boundaries during the elastic phase. This model effectively captures the stress variations within grain boundaries and grains, as well as the mutual influence between grain boundaries and intragranular areas before yielding. Specifically, the model posits that grain boundaries impart added stress to internal grains, while intragranular regions provide the space and atoms necessary for microplastic deformation of the grain boundaries.

(4) Furthermore, we observed periodic fluctuations in grain boundary stresses during the elastic phase, with a cycle in these stresses leading to increased intragranular stresses within a cycle. Consequently, the fluctuation frequency of grain boundary stress is shown to affect Young's modulus of polycrystals. Notably, in the 13.8 nm model, the higher frequency of grain boundary stress fluctuations results in a larger Young's modulus compared to other models.

(5) The phenomenon of anti-H-P yield stress is intricately linked to grain boundary activity. In the 13.8 nm model, the rotation of a larger fraction of grains instigates the release of a significant number of dislocations, leading to their accumulation at grain boundaries and subsequent stress concentration. This, in turn, gives rise to a higher occurrence of twins in this model. The twin structure impedes the movement of dislocations, leading to higher yield stress.

7.6 Mechanisms of Three-Dimensional Polycrystalline Copper Deformation at Different Temperatures and Strain Rates

7.6.1 *Introduction*

The influence of temperature and strain rate on the mechanical properties is primarily attributed to the evolution of microstructure, particularly the grain boundary structure in polycrystals. Factors such as grain boundary energy, creep rate, and grain boundary mobility are directly linked to temperature. In a 2D copper film simulation, it is evident that strain rate and temperature significantly impact grain boundary deformation. The degree of grain boundary deformation alters stress concentration, subsequently affecting dislocation release and twin crystal formation. This is reflected in the stress–strain curves, where varying mechanical properties such as Young's modulus, yield stress, and fracture location are observed under different deformation conditions. Notably, the dependence of Young's modulus on strain rate is not apparent. Recent studies suggest that Young's modulus is less sensitive to strain rate [7, 74], with the underlying mechanism attributed to the limitations of the polycrystalline model and experimental conditions. These aspects hinder a comprehensive statistical analysis of grain boundary activity, especially the minimal deformation of grain boundaries in the elastic phase.

Hence, it is essential to conduct MD simulations on 3D polycrystalline copper with an average grain size of 13.8 nm at varying temperatures and strain rates. Furthermore, a comprehensive analysis of the stresses within the grain boundaries during the elastic phase is performed to quantify the influence of temperature and strain rate on Young's modulus. This analysis is carried out in conjunction with the grain boundaries and intragranular interactions model proposed in the preceding section, which explores the fluctuations of the grain boundary stresses under different conditions. The quantitative GBAA and the quantitative analysis of twins and statistical algorithms are also employed to scrutinize the grain boundary activity of 3D polycrystalline copper under diverse conditions, establishing the relationship between grain boundary activity and the mechanical properties of the plastic phase at different temperatures and strain rates.

7.6.2 *Mechanical properties and grain boundary activity*

7.6.2.1 *Mechanical properties*

In Figure 7.38, stress–strain curves and typical mechanical properties of uniaxial tensile of 3D polycrystalline copper with a grain size of 13.8 nm at different temperatures and strain rates are presented. The mechanical properties include average yield stress, yield strength, and Young's modulus. The average yield stress is determined as the

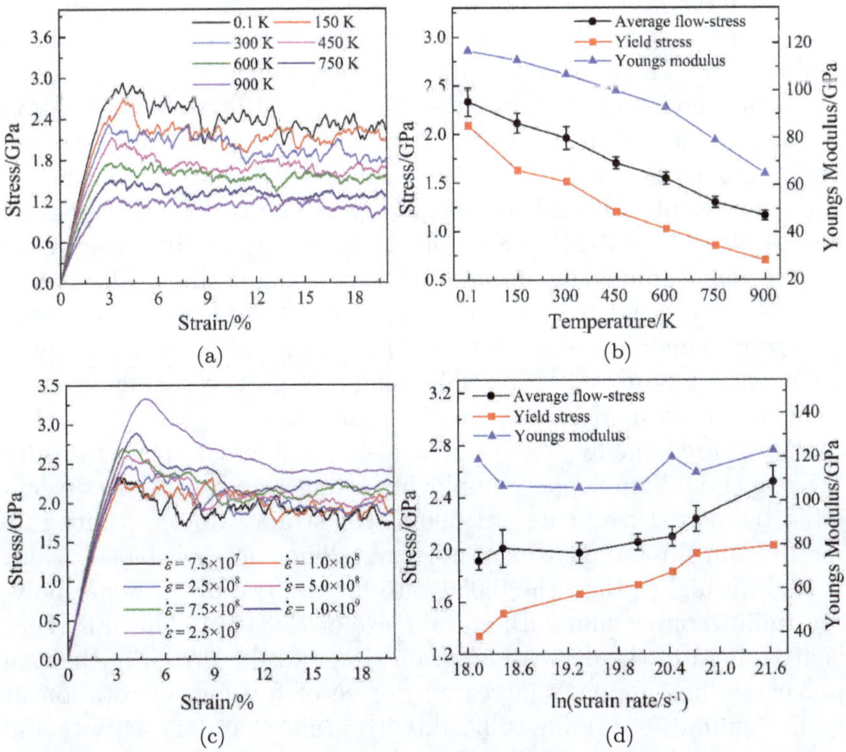

Figure 7.38. Stress–strain curves and some typical mechanical properties obtained by uniaxial tensile of polycrystalline copper with an average grain size of 13.8 nm at different temperatures and strain rates. (a) Stress–strain curves at strain rate 1×10^8/s in different temperatures, (b) mechanical properties at strain rate 1×10^8/s in different temperatures, (c) stress–strain curves at temperature 300 K in different strain rates, (d) mechanical properties at temperature 300 K in different strain rates.

average of the stresses in the strain range of 8–16%. The calculation of Young's modulus is adjusted for strains less than 0.3% due to the material yielding smaller strains at high temperatures, causing the stress–strain curve to curve. Similarly, yield strength, which signifies the onset of plastic deformation, is defined as the stress at which the strain deviates from the linear curve by 0.2%.

In Figures 7.38(b) and 7.38(d), it is apparent that both the yield stress and yield strength gradually decrease with increasing temperature and decreasing strain rate. The behavior of Young's modulus, however, varies, and this can be directly observed from the stress–strain curves. Young's modulus is more influenced by temperature and less affected by the strain rate, which aligns with recent research findings [74]. The obtained temperature dependence of Young's modulus from the fitting is approximately −56.67 MPa/K, which is lower than the value obtained from the simulation of polycrystalline copper with a grain size of 5.2 nm (−60±18 MPa/K) [7]. This value is higher than the result obtained for coarse crystalline copper with a grain size of 200 nm (−40 MPa/K) [45]. Additionally, the increase in the proportion of grain boundaries with decreasing grain size leads to a heightened dependency of grain boundary activity on temperature. The results highlight the strong influence of grain boundary activity on Young's modulus [74, 75], indicating a heightened dependence on temperature in models with smaller grain sizes.

We utilized the average yield stresses at different strain rates for 3D polycrystalline copper, employing strain rate sensitivities calculated by $m = \partial \ln \sigma / \partial \ln \dot{\varepsilon}$. At 300 K, the strain rate sensitivity rate for 13.8 nm three-dimensional polycrystalline copper is 0.066, which is slightly higher than the 0.056 rate for the two-dimensional polycrystalline copper film with a grain size of 17.84 nm. This difference is attributed to the decrease in grain size, and the three-dimensional polycrystalline copper's increased degree of freedom of rotation at grain boundaries, leading to heightened grain boundary activity and larger strain rate sensitivity. Experimental results from grain sizes of 105 nm and 9 nm polycrystalline copper showed sensitivities of 0.0601 [40] and 0.063 [39], respectively. These values are very close to the simulated results and slightly larger than those obtained for polycrystalline copper with a grain size of 26 nm (0.059 [39]) and for coarse crystalline copper [41]. The greater strain rate sensitivity is associated with increased grain boundary activity, as evidenced by the observation of significant grain boundary deformation, discussed in detail for both elastic and plastic phases as follows.

7.6.2.2 *Crystal boundary activity*

The GBAA algorithm was used to analyze the grain boundary activity in this model at different temperatures and strain rates. Figures 7.39(a) and 7.39(b) show the results for two conditions: a temperature of 450 K and a strain rate of 1.0×10^8/s, and a temperature of 300 K and a strain rate of 5.0×10^8/s. The R-squared and fitted plane coefficients with strain are graphed, displaying only the parameters with a variation greater than 0.03. The deformation, rotation, and unaltered shares of these boundaries are counted, and the results for different conditions are shown in Figure 7.39(c).

At high temperatures, the thermal vibration of atoms within polycrystalline copper intensifies, leading to the appearance of other types of grain boundary-like structures within the intragranular. This makes it difficult to separate the grain boundaries from the chaotic atoms, affecting the accuracy of the GBAA calculations. Only the activity of grain boundaries at temperatures below 450 K is considered here. Additionally, Trautt *et al.* [50] showed that the coupling motion of surface grain boundaries will inevitably lead to rotation. However, at temperatures higher than 0.8 times the melting point of copper, the rotational behavior of the grain boundaries will be sharply reduced or may even not occur, leading to the disappearance of the coupling motion. Therefore, at high temperatures, the grain boundary activity is mainly the slip of the grain boundary.

From Figure 7.39, it is evident that the percentage of deformation and rotation of grain boundaries is higher at high temperatures compared to low temperatures. The behavior of rotation and deformation of grain boundaries aligns with the results of the previous results. When the strain is less than 3%, there is no significant rotation of grain boundaries. However, when the strain exceeds 3%, a large number of dislocations contribute to the rotation of grain boundaries, while grain boundary deformation consistently occurs. The statistical results demonstrate that as the temperature increases and the strain rate decreases, the proportion of deformed and rotated grain boundaries gradually increases, and the proportion of unaltered grain boundaries decreases. Therefore, the deformation behavior of

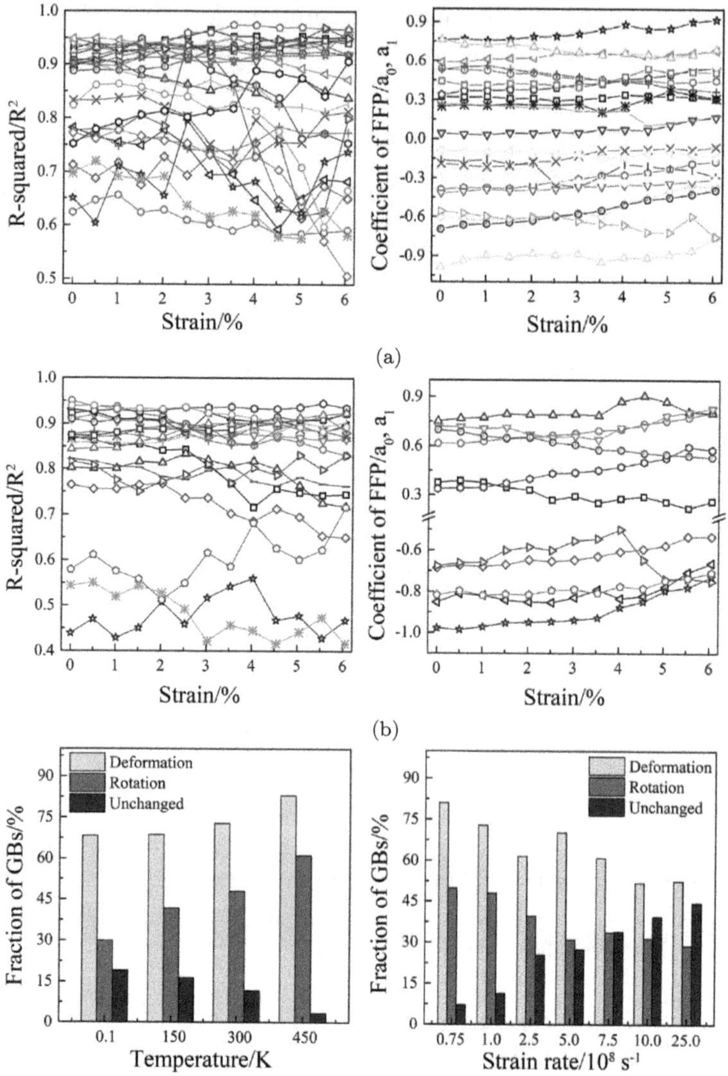

(c)

Figure 7.39. Variation of R-squared and fitted plane coefficients (a_0 and a_1) as a function of strain for different conditions, and the corresponding statistical analyses of grain boundaries deformations and rotations, with the statistics in figure (c) for the percentage of grain boundaries with values of variation of the fitted parameters R^2 and Δa_1 greater than 0.03. (a) Fitted parameters in strain rate 1×10^8/s at 450 K, (b) fitted parameter in strain rate 5×10^8/s at 450 K, (c) boundary deformation and rotation statistics in different strain rates and at different temperatures.

grain boundaries at different strain rates and temperatures aligns with the results observed in polycrystalline copper films. At high temperatures and low strain rates, the deformation of larger portions of grain boundaries in the matrix can reduce stress concentration on the grain boundaries, ultimately reducing the release of dislocations. However, the larger rotation of the grain boundaries provides more slip systems for the release of dislocations, ultimately resulting in polycrystalline materials exhibiting different mechanical properties under the combined effect of grain boundary deformation and rotation. These properties are discussed in detail later.

7.6.3 *Quantitative analysis of Young's modulus*

In the previous section, it was observed that during the elastic phase, the stress on the grain boundary fluctuates approximately periodically, while the stress inside the grain increases approximately linearly and gradually. It was found that the increase in the stress at the grain boundary in one cycle leads to the increase in the stress inside the grain through the stress distribution diagram. This led to the construction of a model for the interaction between grain boundaries and intragranular in the elastic phase. It was concluded that grain boundaries and intragranulars are interdependent before yielding: Grain boundaries provide extra stress to intragranular, and intragranular provides space and atoms for microplastic deformation of grain boundaries.

Additionally, recent studies have shown that local deformation occurs at the grain boundaries during the elastic phase, resulting in the value of Young's modulus at the grain boundaries being less than 30% of Young's modulus within the grain. Therefore, Young's modulus of the whole polycrystalline system depends strongly on the rate of stress growth within the crystal. In this subsection, MD simulations are carried out for an elastic phase with a strain of less than 0.3% at a very small output step, followed by an analysis of the grain boundary and intercrystallite stresses. The model is then used for a quantitative analysis of Young's modulus at different temperatures and strain rates.

The stress at grain boundaries and within the grains at different temperatures is illustrated in Figure 7.40(a). It shows that the stress fluctuation at grain boundaries increases gradually as the

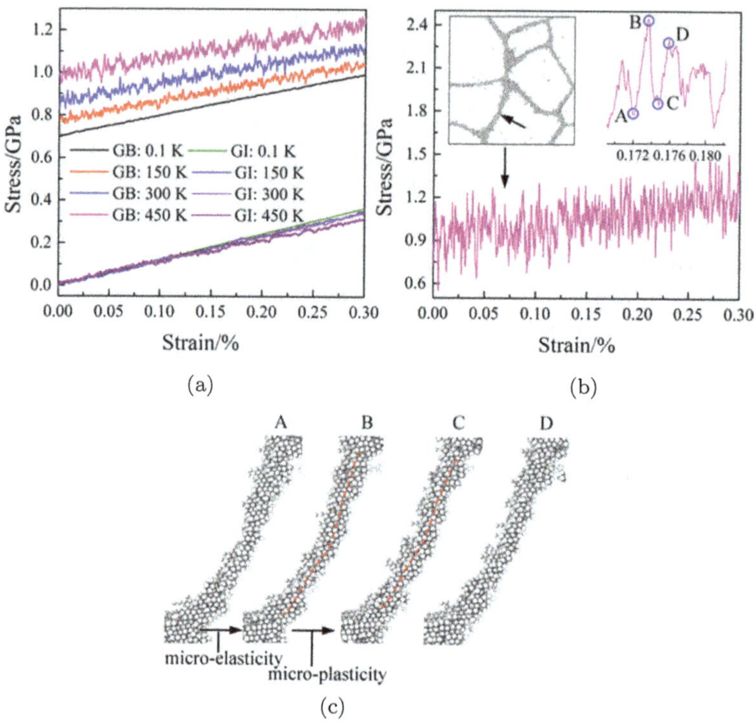

Figure 7.40. Analysis of grain boundary and intragranular stresses at different temperatures and the configuration of micro-variation of grain boundaries. (a) Stress in grain boundary and intragranular in different temperatures, (b) stress at grain boundary at 450 K in the insert, (c) micro displacements for points A-D in figure (b).

temperature rises, while the stress within the grains shows an almost linear increase with temperature. At 0.1 K, there is no stress fluctuation at the grain boundaries. This is because the microplastic deformation of grain boundaries occurs through the diffusion of atoms. Therefore, when there is no thermal vibration or minimal microplastic deformation, it is challenging for the atoms to diffuse, resulting in mostly elastic deformation of the grain boundary at 0.1 K. This causes the intercrystalline stresses to increase, leading to a higher Young's modulus at 0.1 K.

Figure 7.40(b) displays a cross-section of crystals with a 1 nm thickness (inset) for stress calculations at 450 K. The stresses appear to fluctuate significantly with an approximate periodicity. A close examination of the grain boundaries corresponding to the four points A–D in the inset reveals micro-elastic and micro-plastic deformation behaviors of the grain boundaries, as shown in Figure 7.40(c).

7.6.3.1 *Young's modulus at different temperatures*

Based on the analysis above, it can be concluded that Young's modulus is primarily influenced by the growth rate of intragranular stress, which in turn is associated with the frequency of fluctuation of grain boundary stress. To analyze these fluctuations, Fast Fourier Transformations (FFT) of stresses in different grain boundary layers with a 1 nm thickness at various temperatures were performed. The principal frequency of the grain boundary stress fluctuation and intragranular stress was obtained from these calculations. Notably, at a temperature of 0.1 K, the grain boundary stress exhibited an approximately linear growth trend. Therefore, the stress fluctuations at 150 K, 300 K, and 450 K were analyzed. Figure 7.41(a) presents the FFT calculation results for a grain boundary layer at 150 K, illustrating distinct fluctuating principal frequencies. Additionally, Figure 7.41(b) demonstrates the principal frequencies of stress fluctuations in different grain boundary layers along the x and y axes. The averaged principal frequencies in different directions at various temperatures are shown in Figure 7.41(c), indicating a gradual decrease with increasing temperature.

Burker and Turnbull [76], as well as Smoluchowski [77], have applied reaction theory to grain boundary slip caused by atom hopping across grain boundaries. They showed that the rate of slip at a grain boundary can be described as the product of mobility and driving force:

$$v = M\Delta F, \quad M = M_0 \exp\left(\frac{-Q_{gb}}{k_B T}\right) \tag{7.18}$$

where M is the mobility/m^2 s^{-1}, M_0 is a constant, ΔF is the driving force/GPa, $-Q_{gb}$ is the activation energy (enthalpy) of grain boundary movement/KJ mol^{-1}, k_b is the Boltzmann's constant/J-K^{-1}, and T is the temperature/K.

In Eq. (7.18), it is evident that the mobility of grain boundaries increases with temperature due to the temperature-dependent atomic diffusion activity in microplastic deformation. This increased

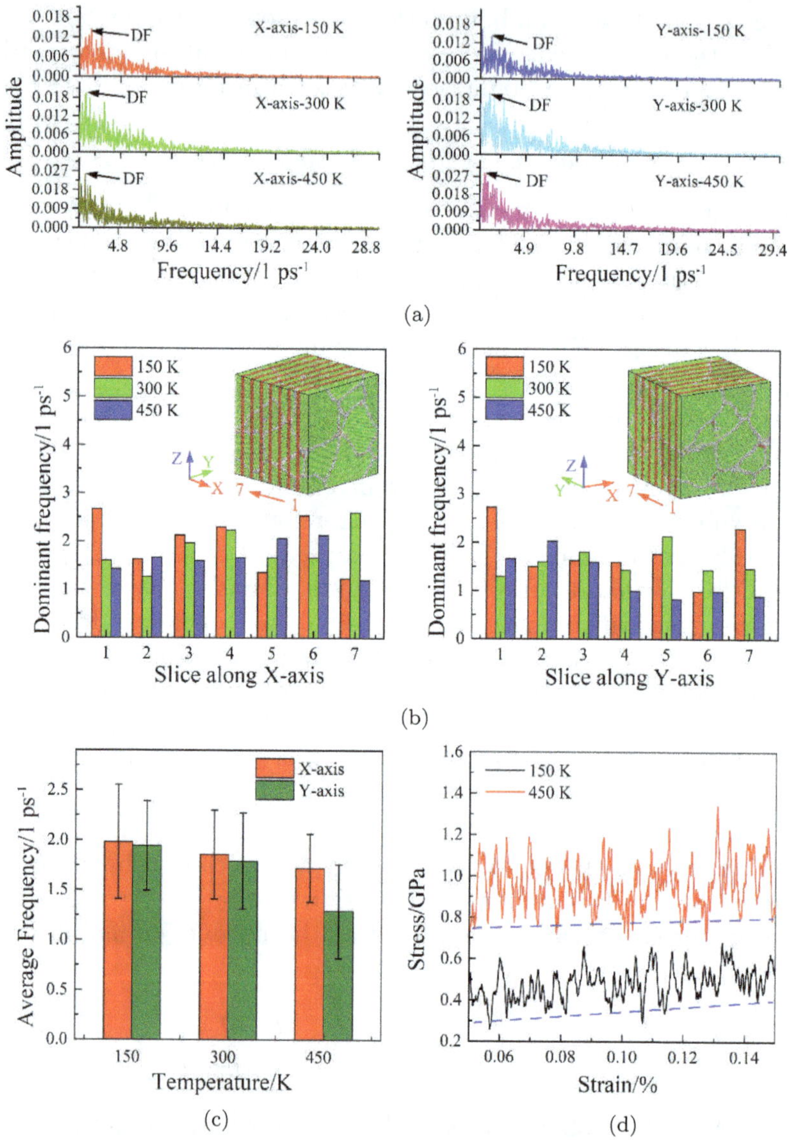

Figure 7.41. FFT analysis and statistics of grain boundary stress fluctuations, and fluctuations of grain boundary stress at 150 K and 450 K for a thickness of 1 nm grain boundary in different temperatures. (a) FFT calculation of stress along x and y axes in single-layer grain boundary, (b) frequency statistics of stress fluctuation along x and y axes in grain boundary, (c) average of stress fluctuation along x and y axes, (d) stress changes of grain boundary at 150 K and 450 K.

mobility results in easier and larger deformation of the grain boundaries. Consequently, the release of grain boundary stresses is more complete in the microplastic phase, leading to a longer duration of the microplastic phase and a reduction in overall stress fluctuations, as illustrated by the red and black curves in Figure 7.41(d). Comparing the stresses at two temperatures in the figure, it is also evident that the stress fluctuation range at grain boundaries is greater at higher temperatures. Therefore, the initial stress in the microelasticity phase at high temperatures is smaller, as indicated by the dashed line in Figure 7.41(d), resulting in a smaller contribution to the stresses within the grain. Hence, it can be concluded that Young's modulus depends not only on the material's softening within the crystal at high temperatures but also on the fluctuation frequency of grain boundary stress and the initial stress of microelasticity at the grain boundary.

7.6.3.2 *Young's modulus at different strain rates*

Based on previous studies, it has been observed that temperature [74], grain boundary shape [73], average grain size [7], and porosity [44] significantly influence Young's modulus. However, the strain rate does not have a significant impact on Young's modulus. To further explore this phenomenon, we conducted simulations of elastic phase deformation under different strain rates. We analyzed the change in grain boundary stress over time and strain at various strain rates. The results are depicted in Figure 7.42.

In each subplot of Figure 7.42, the lower horizontal coordinate represents time, while the upper horizontal coordinate represents strain. Upon comparing these plots, it becomes apparent that the stress trend under different strain rates is consistent, and the fluctuation period of different grain boundary stresses at the same time is also consistent. Therefore, it can be inferred that the fluctuation frequency of grain boundary stresses remains the same in different strain rates. Within the 0–0.3% strain range, there are more cycles of grain boundary stress fluctuation at a lower strain rate. Consequently, the number of times the grain boundary acts on the grain is higher at low strain rates, leading to a greater increase in the stress within this strain range. However, Figure 7.38(d) does not show any variation in Young's modulus. Upon comparing the curves in the figure, it is

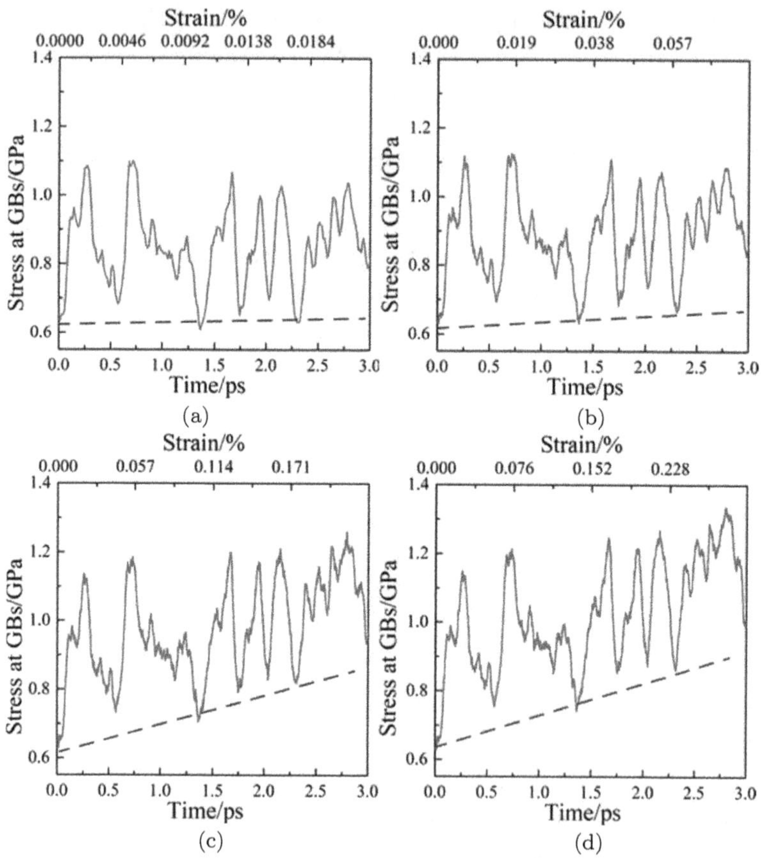

Figure 7.42. Variation of grain boundary stress with time and stress at different strain rates. (a) $\dot{\varepsilon} = 7.5 \times 10^7 \text{s}^{-1}$, (b) $\dot{\varepsilon} = 2.5 \times 10^8 \text{s}^{-1}$, (c) $\dot{\varepsilon} = 7.5 \times 10^8 \text{s}^{-1}$, (d) $\dot{\varepsilon} = 1 \times 10^9 \text{s}^{-1}$.

evident that the initial stress in the microplastic phase of the grain boundary increases with the rise in strain rate. Thus, at high strain rates, despite the lower number of grain boundary fluctuations, the initial stress provided by the grain boundary to the intracrystalline structure is greater, resulting in less dependence on Young's modulus on strain rate.

In polycrystalline materials, increasing Young's modulus by reducing the void ratio is difficult. However, we can enhance Young's modulus by increasing the frequency of grain boundary stress fluctuation or the initial stress of grain boundary micro-elasticity.

For instance, we can introduce stable structures such as twins or nano-clusters, or polarize other atoms near the grain boundaries to hinder micro-plastic deformation of the grain boundaries, thereby reducing the micro evolution period and increasing the frequency of stress fluctuations at grain boundaries.

7.6.4 *Quantitative analysis of yield stresses*

Based on the research on grain boundary deformation and rotational fractions at different temperatures and strain rates, it is observed that both fractions gradually increase with rising temperatures or decreasing strain rates. This leads to a decrease in the stress concentration on the grain boundaries, reducing the presence of Hexagonal Close-Packed (*HCP*) structures such as stacking faults (*SFs*) and twin boundaries (*TBs*). The changes in *HCP* (*SFs* and *TBs*) and *TB* structures with strain at different temperatures and strain rates are depicted in Figure 7.43. The figure illustrates that there are no significant differences between the two fractions, mainly because the rotational copies of the grain boundaries are more pronounced at higher temperatures and lower strain rates. This provides more slip systems for the dislocations, resulting in more dislocations being released and leading to an increase in the *HCP* structure despite the lower stress concentration at the grain boundaries under these conditions.

Comparison with 2D polycrystalline Cu shows that the distribution of *HCPs* and TBs in polycrystalline Cu films varies significantly at different temperatures and strain rates. This is primarily due to the limited rotational freedom of the grain boundaries in 2D polycrystalline materials, which impacts the structural distribution of *HCPs* and *TBs* based on the deformation of the grain boundaries. The studies conclude that it is not feasible to distinguish the differences in the yield stress of 3D polycrystalline copper under different conditions based on the *HCP* and *TB* structures, which are not significantly different.

In order to better understand the variations in yield stress in the plastic phase, we examined the size of twins (measured by the number of atoms) in the plastic phase (9.13–12.18%) at different temperatures and strain rates using a quantitative twin analysis algorithm. The results are displayed in Figure 7.44(a). The figure indicates that

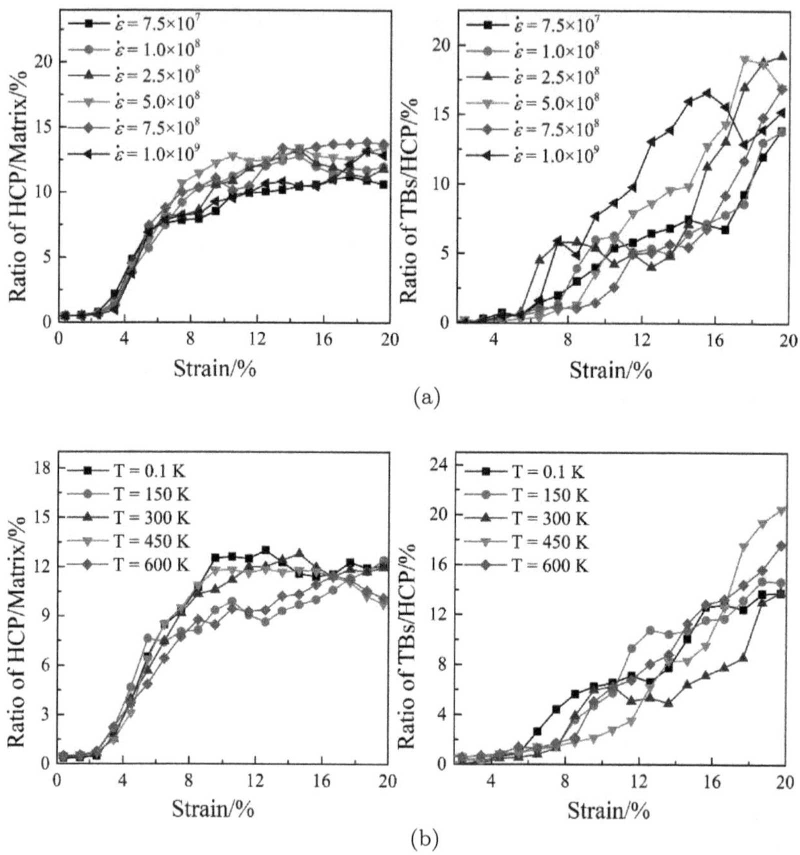

Figure 7.43. Variation of the number fractions (%) of *HCP* (*SFs* and *TBs*) and *TB* structures with strain at different strain rates and temperatures. (a) Number fractions of HCPs and TBs in different strain rates, (b) number fractions of *HCPs* and *TBs* in different temperatures.

as the temperature increases and the strain rate decreases, the size of the twin boundaries decreases gradually. Additionally, comparing the distribution of twin boundaries at different temperatures in Figure 7.44(c), we observed that the number of twins at the same strain at 600 K is greater than at 0.1 K. However, at 0.1 K, the twins are more uniform and smaller, with twin boundaries containing more atoms and being larger. The larger size can effectively increase the possibility of partial dislocation being obstructed by the twin crystals, thereby increasing yield stress.

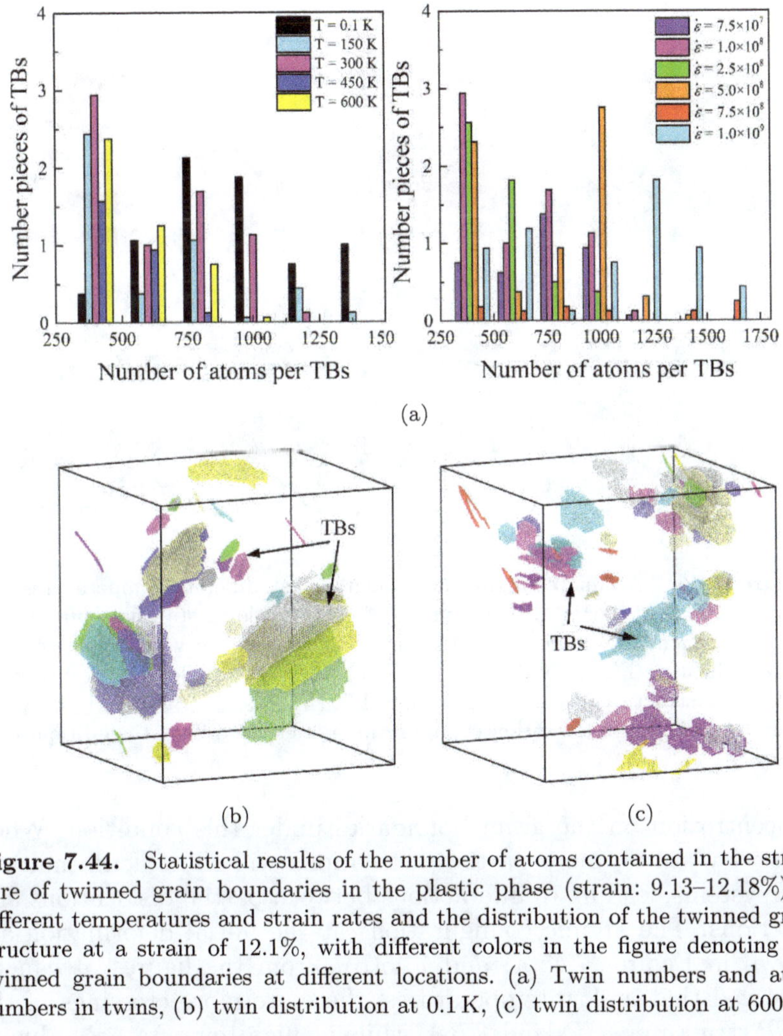

Figure 7.44. Statistical results of the number of atoms contained in the structure of twinned grain boundaries in the plastic phase (strain: 9.13–12.18%) at different temperatures and strain rates and the distribution of the twinned grain structure at a strain of 12.1%, with different colors in the figure denoting the twinned grain boundaries at different locations. (a) Twin numbers and atom numbers in twins, (b) twin distribution at 0.1 K, (c) twin distribution at 600 K.

To further analyze the variability in the size of the twin grain boundaries under different conditions such as high and low temperatures, as well as high and low strain rates, the processes of twin formation and growth under these opposing conditions were carefully observed. The results indicate that at low temperature and high strain rate conditions, the deformation of the grain boundaries is minimal in the initial stages and the rotation of the grain boundaries is less at strains less than 3%. As a result, there is a greater stress

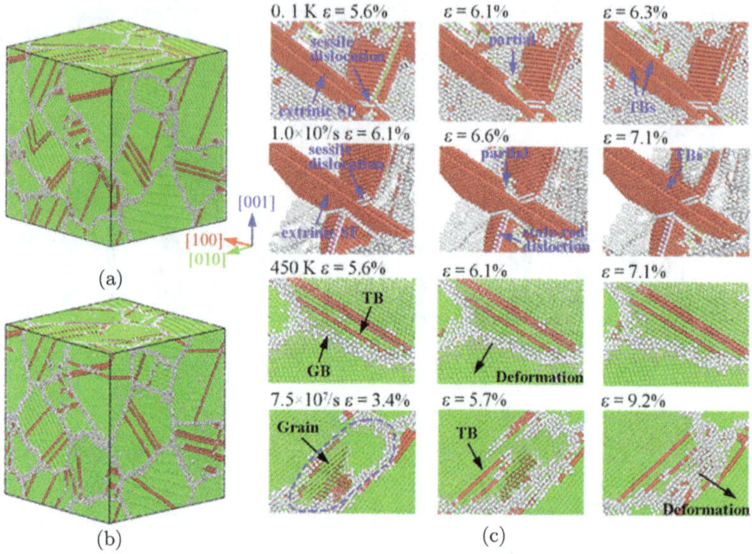

Figure 7.45. Atomic configurations diagrams at different temperatures and the formations of partial dislocations, sessile dislocations, stacking faults, twins and micrograins at different temperatures and strain rate with green, red and white indicating FCC, HCP (SFs and TBs) and other structures, respectively. (a) Crystal structure at 0.1 K and strain 6.1%, (b) crystal structure at 450 K and strain 6.1%, (c) twin formation configuration under the different conditions.

concentration on the grain boundaries under this condition. When plasticity starts, more partial dislocations will slide along multiple slip systems, leading to interactions between partial dislocations and reactions. The atomic configurations at the initial deformation are shown in Figures 7.45(a) and 7.45(b). From the figures, it can be observed that at low temperatures, there are more crossings of the *HCP* structures (*SFs* and *TBs*), while at high temperatures, due to the larger number of deformed copies of the grain boundaries, there are smaller stress concentrations at the grain boundaries, resulting in most of the *HCP* structures being parallel.

Diverse mechanisms of twin formation are observed under contrasting conditions. Figure 7.45(c) presents two mechanisms of twin formation. The more common mechanism involves the sliding formation of partial dislocations on the outer stacking faults, which primarily occurs at low temperatures and high strain rates, as depicted in the configuration diagrams at 0.1 K and 1.0×10^9/s. Due to dislocation interactions at low temperatures or high strain rates, stable dislocations such as sessile dislocations and stair-rod dislocations are generated, which are more stable and less prone to destruction. Consequently, stress concentration increases, leading to the release of partial dislocations from grain boundaries or the vicinity of these stable dislocations, which then slide on the proximity (111) surfaces of the outer stacked faults, ultimately resulting in the formation and gradual growth of micro-twin crystals.

However, the mechanisms of twin formation and growth differ significantly at high temperatures and low strain rates, as illustrated in the configurations corresponding to 450 K and 7.5×10^7/s in Figure 7.45(c), where stress concentration on the grain boundaries is minimal under these conditions. Hence, the formation of a stable dislocation structure through dislocation reactions is challenging. Dislocations also encounter increased resistance to movement on the adjacent surfaces of *SFs*. However, significant deformation of grain boundaries in this scenario results in the creation of micro-twins. This is demonstrated by the atomic configurations exhibiting a 5.6% strain at 450 K and a 3.4% strain at 7.5×10^7/s. When intrinsic stacking faults (*SFs*) are generated near the grain faces, the movement of the grain boundary causes disappearance of a layer of *HCP* structures (*SFs* and *TBs*) of intrinsic stacking faults, leading to the formation of a single twin boundary. At this stage, the twin boundary is small, but with the movement of the grain boundary in the opposite direction of the twin, as depicted in Figure 7.45 with 6.1% atomic configuration at 450 K and 5.7% at 7.5×10^7/s, the twin boundary gradually grows. The motion of the grain boundary also causes the grains to shrink, reducing the grain size at a rate of 7.5×10^7/s with strains of 3.4% and 9.2% configuration. The amount of grain boundary deformation remains constrained despite high temperatures or low strain rates, making twin growth through this mechanism difficult. Analyzing these findings, it can be concluded that differences in grain boundary activity under opposing conditions lead to variations in the mechanism of twin formation, ultimately resulting in differences in the size of the twin boundaries.

7.6.5 *Summary*

The study used a phase field method to create 3D polycrystalline copper with a grain size of 13.8 nm. Then, traditional MD tensile simulations were conducted at various strain rates and temperatures. To understand how Young's modulus varies under different conditions, the deformation in the elastic phase was also simulated. The study quantitatively analyzed the dependence of Young's modulus on temperature and strain rate using a grain-boundary and intragranular interaction model. The analysis of grain boundary activity at the start of deformation using the GBAA algorithm, along with statistical values of twin size obtained using the quantitative twin analysis algorithm, helped explain the differences in yield stresses in the plastic phase under different conditions. The main findings can be summarized as follows:

(1) The temperature dependence of Young's modulus and strain rate sensitivity of 3D polycrystalline copper align with previous experiments and simulations. Additionally, stress variations at grain boundaries and within the grain are consistent with previous chapters and theoretical studies.
(2) As temperature increases, the fluctuation frequency of grain boundary stress gradually decreases, leading to a decrease in the rate of increase of intragranular stress, thus causing a decrease in Young's modulus. With an increase in strain rate, the fluctuation frequency of grain boundary stress does not change significantly. Therefore, the dependence of Young's modulus on strain rate is low, which is consistent with other findings.
(3) As temperature increases and strain rate decreases, the number of deformed and rotated grain boundaries in three-dimensional polycrystalline copper steadily increases, while the number of unaltered grain boundaries gradually decreases.
(4) Under varying conditions, there is no significant difference between the structural shares of *HCPs* and *TBs* in the plastic phase. This is primarily due to the combined effect of grain boundary deformation, which reduces stress concentration, and the rotation of grain boundaries, which increases the slip system of dislocation sliding. However, there is a notable difference in the size of twin boundaries under different conditions.

Larger twin boundaries at low temperatures and high strain rates can effectively hinder the movement of dislocations, increasing yield stress. Conversely, at high temperatures and low strain rates, the size of the twin boundaries is smaller.

(5) At low temperatures and high strain rates, the initial stage of grain boundary activity is diminished, leading to increased stress concentration on the grain boundary. As the material begins to yield, multiple slip system dislocations are released, and these dislocations interact with each other to form a stable dislocation structure. As deformation increases, the dislocations slide on the near-neighboring surfaces of the stacking faults, resulting in the formation and growth of twin crystals. In contrast, under opposing conditions, the formation and growth of twin crystals primarily depend on grain boundary deformation, ultimately leading to smaller twin boundaries due to limited grain boundary deformation.

7.7 Three-Dimensional Polycrystalline Copper Deformation Mechanisms with Prefabricated Twins

7.7.1 *Introduction*

The improvement of mechanical properties of materials during deformation has been a subject of extensive research in recent years. Compared to traditional methods such as solute atoms, precipitates, nanoclusters, and grain boundaries, twins can hinder the motion of dislocations and act as nucleation points for dislocations, thereby increasing the yield stress of the material without losing its plasticity [78, 79]. These studies have been conducted in both 2D and 3D, using polycrystalline copper simulations. In copper with low stacking fault energy, dislocation motion is typically the primary deformation mechanism in coarse-grained copper. Early studies have indicated that twinning in coarse-grained copper usually occurs at low temperatures or high strain rates [80, 81]. However, when the grain size is reduced to the nanoscale, deformation twinning can also occur in polycrystalline copper at room temperature and low strain rates [24, 81, 82]. For pre-twinned materials, there are various methods to introduce twins, including mechanical processing or recrystallization.

These two types of twins are referred to as growth twins and annealing twins, respectively [40].

In recent years, there has been an increase in research on the introduction of twin structures in polycrystalline copper. Studies have focused on exploring the conditions that lead to the maximum yield strength in this material [33, 83], the deformation mechanism [84–86], the Schmidt factor [79], and the interaction between twins and cavities [87]. However, there have been few reports on the effect of prepositioned twin boundaries on the grain boundary activity during the initial deformation phase. This is mainly because the previous models used regular geometries, which may limit the grain boundary activity.

To address this, models of polycrystalline copper with an average grain size of 13.8 nm were created using the phase field method. These models introduced a twin structure with varying twin spacing from 1.25 nm to 3.13 nm. Subsequently, MD uniaxial tensile simulations were conducted on these constructed models, as well as on a model without the twin structure. These simulations, in conjunction with the polycrystalline analysis algorithm (GBAA), aimed to analyze the grain boundary activity during the initial stages of deformation. The goal was to quantify the effect of the pre-twin boundaries on the grain boundary activity.

7.7.2 *Mechanical properties and grain boundary activity*

7.7.2.1 *Mechanical properties*

The stress–strain curves of different models at a temperature of 300 K and a strain rate of 1×10^8/s are presented in Figure 7.46. It is observed that the stress changes of all models in the elastic phase are relatively similar, indicating that the presence of twin boundaries has little effect on the micro-plastic deformation of grain boundaries in this phase. Therefore, Young's modulus of polycrystalline copper cannot be increased by the pre-twin structure. However, in the plastic phase, the model with preset twins shows significantly higher yield stress compared to the model without twin structures. The yield stress tends to increase with decreasing twinning distance, albeit with a small increase. Previous studies [83] have shown that

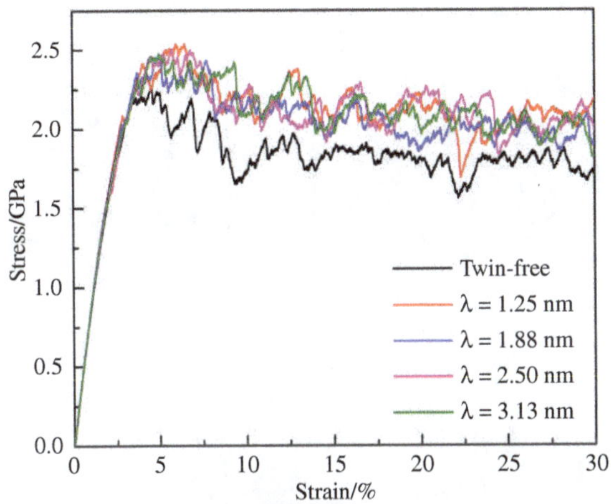

Figure 7.46. Stress–strain curves of uniaxial tensile for different models at 300 K and strain rate of 1×10^8/s.

smaller grain size and twin boundary spacing result in increased metal strength. However, the increase in yield stress in this study is not as significant, mainly due to differences in the model and the method used. Irregular grain boundaries are more prone to grain boundary activity, resulting in different stress concentrations compared to studies [83]. Further analysis is conducted to examine the activity of grain boundaries and their effect on deformation in detail.

7.7.2.2 *Grain boundary activity and grain boundary stress concentration*

Twinned grain boundaries exhibit high mechanical stability, so the presence of twinned grain boundaries may influence grain boundary activity during the initial deformation phase. We conducted an analysis of grain boundary activities in various models using the quantitative grain boundary analysis algorithm (GBAA). The variation of the coefficients of the fitted R-squared and fitted planes with strain in the model without twins are depicted in Figure 7.47(a). Figure 7.47(b) presents the statistical values of grain boundary deformation, rotation, and unaltered copies in different models, satisfying the conditions where the amount of deformation and rotation satisfies ΔR^2, Δa_0, and $\Delta a_1 \geq 0.03$. From Figure 7.47(b), it is evident that preset

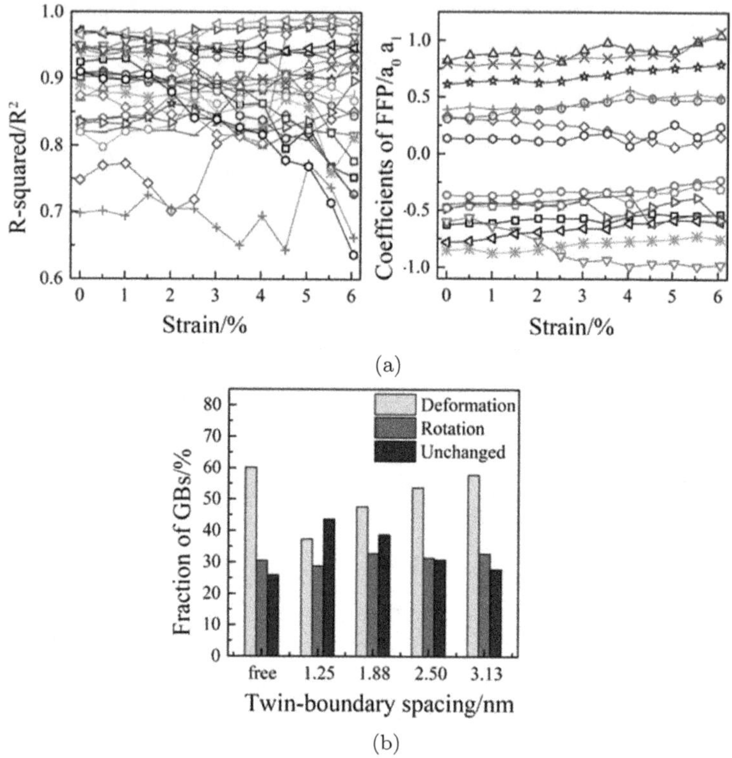

Figure 7.47. Changes in the goodness-of-fit R-squared and the fit plane coefficients (a_0 and a_1) with strain and the number of grain boundary deformation, rotation, and unaltered copies in different models, with grain boundary deformation and rotation parameter variations greater than or equal to 0.03 being counted. (a) Fitted parameter variations in twin-free model, (b) number statistics of grain boundary activity for all models.

twins significantly impede the deformation of the grain boundaries. The deformation of the grain boundaries decreases as the distance of the twin boundaries decreases, and the amount of unaltered grain boundaries in the polycrystals also gradually decreases. However, the impact of the twin boundary on obstructing grain boundary rotation is not substantial. The rotational behavior of the grain boundary is relatively consistent across all models, including those without a twin structure. This is mainly because the rotation of the grains accompanies the rotation of the grain boundary as a whole, and the driving force of the rotation is the force of the grains along the tangential direction of the grain boundary. Therefore, the rotation of the grains also impels the twin grain boundaries to rotate together. On the contrary, the driving force for deformation is the normal tangential force along the grain boundary, so the twin grain boundary hinders grain boundary deformation.

Deformation of grain boundaries reduces stress concentration on the grain boundaries and thereby affects the release of dislocations. In order to comprehend the mechanism of plastic deformation in preset twins, we also evaluated the stress concentration at grain boundaries in these models. The graph in Figure 7.48 illustrates the change in the ratio of the number of atoms on the grain boundaries in the preset twin model to the number of atoms on the grain boundaries in the model without twins as a function of strain for stresses exceeding 5.0 GPa. It is evident that for strains less than 3.0%, most ratios are greater than 1.0, implying a higher number of high-stress atoms in the preset twin model. Additionally, it can be observed that the ratio gradually increases as the size of the twin boundary decreases, mainly because smaller twin boundaries impede grain boundary deformation, leading to increased stress concentration. The ratio decreases rapidly at around 3.0% strain, especially

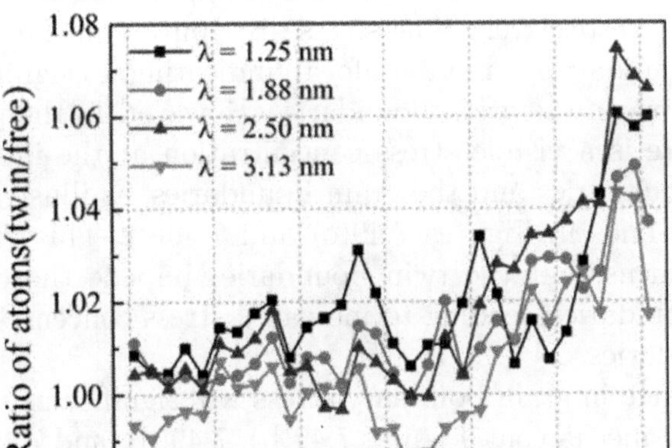

Figure 7.48. Ratio of the number of atoms on the grain boundaries in the preset twin model to the number of atoms on the grain boundaries in the twin-free model for stresses greater than 5.0 GPa.

for the model with a twin spacing of 3.13 nm, due to the release of dislocations, which reduces stress concentration at the grain boundaries [10, 29]. As strain exceeds 3.0%, the ratio for all models is greater than 1.0 and continues to increase gradually. This is primarily because, in polycrystals without twins, the release of dislocations accelerates grain boundary deformation, and the formation of deformation twins further accelerates this process [29], leading to additional stress release on the grain boundary. However, in the pre-twinned model, the inhibitory effect of the twins on grain boundary deformation results in less effective stress release, causing the ratio to gradually increase.

7.7.3 *Deformation mechanisms of grain boundary*

In the above analysis, we observed that the presence of twin grain boundaries significantly hinders the deformation of grain boundaries, leading to a change in stress concentration at the grain boundaries. To examine the local properties of grain boundaries more effectively, we selected grain boundary lamellae with a thickness of 1 nm in different models and calculated the stress distribution at a strain of 3.04%. Figure 7.49 depicts the stress distributions for models without twin boundaries and pre-twin models with 1.88 nm and 3.13 nm in thickness, respectively. The stress distribution in the model without twin boundaries is more uniform, indicating no significant stress concentration on the grain boundaries. However, in the pre-twinned model, there is a visible stress concentration at the intersection of the twin boundaries and the grain boundaries, as illustrated by the red dashed lines in Figures 7.49(c) and 7.49(e). This visual observation confirms that the twin boundaries impede the deformation of grain boundaries, leading to increased stress concentration at the grain boundaries.

The change in grain boundary stress will significantly impact the deformation mechanism. Figures 7.49(b), 7.49(d), and 7.49(f) display the distribution of *HCP* structures (*SFs* and *TBs*) on crystal laminae with a thickness of 1 nm in the plastic phase in three different models, respectively. A comparison of Figures 7.49(a) and 7.49(b) reveals that in the model without a twin structure, the uniform stress distribution on the grain boundaries results in the release of dislocations along different slip systems in different grains, leaving behind a relatively

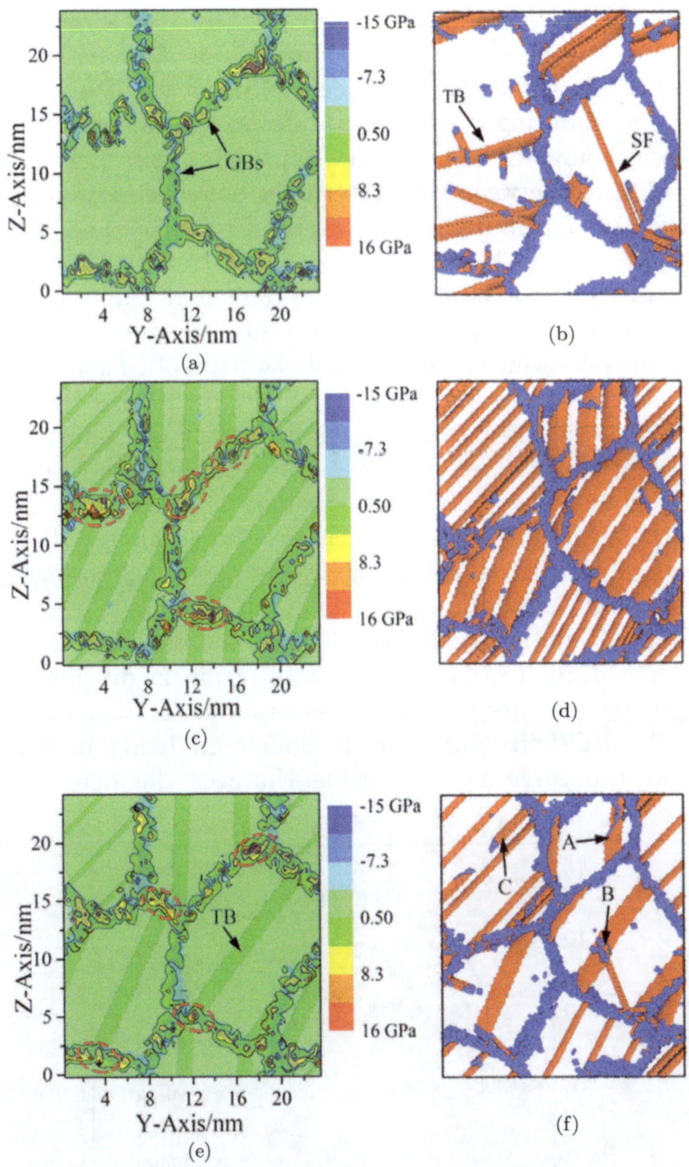

Figure 7.49. Stress distribution on grain boundaries with 1 nm thickness in different models at 3.04% strain. (a) Stress distribution of grain boundary for twin-free model, (b) *HCP* distribution for twin-free model, (c) stress distribution of grain boundary with twin spacing 1.88 nm, (d) *HCP* distribution for twin spacing 3.13 nm, (e) stress distribution of grain boundary with twin spacing 3.13 nm, (f) *HCP* distribution for twin spacing 3.13 nm.

heterogeneous *SF* structure. In Figure 7.49(d), the twinning interval is 1.25 nm due to high stress at the intersection of grain and twinning boundaries. Consequently, the release of dislocations always occurs along the neighboring faces of the preset twins, causing the original preset twin boundaries to start moving, thus increasing or decreasing the distance between the preset twin boundaries. Figure 7.49(f) showcases the distribution of *HCP* structures (*SFs* and *TBs*) with a twin spacing of 3.13 nm. The stress concentration in this model is similar to that of the model without preset twins, allowing dislocations to slip along different slip systems. Additionally, the interaction between the pre-twin boundaries and the pairs of *SFs* occurs in this model, as depicted by points A, B, and C in Figure 7.49(f). This interaction is the most common reinforcement mechanism for pre-twinned materials. These results lead to the conclusion that as the distance between the preset twin boundaries decreases, the deformation mechanism shifts from dislocations sliding along different slip systems to dislocations sliding along the near-neighboring faces of the preset twin boundaries.

In Figure 7.50, the graph demonstrates the change in the number of *HCP* structures (*SFs* and *TBs*) with strain in different models. It is observed that at the onset of plastic deformation (around 3% strain), the *HCP* structures in all models gradually increase with strain and stabilize at a certain value. However, the increase in *HCP*

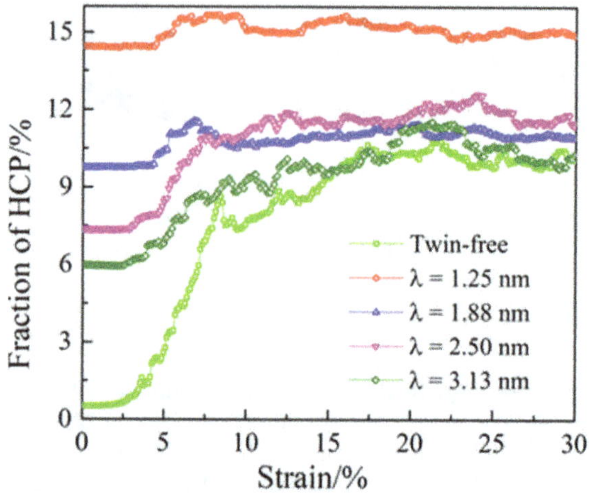

Figure 7.50. Variation of the number (%) of *HCP* structures with strain in different models.

in the pre-twin model is smaller compared to the non-pre-twin model. Additionally, as the twin spacing decreases, the increase in *HCP* also decreases. This is mainly attributed to the increase in stress concentration at the intersection of twins and grain boundaries at smaller twin distances, leading to the release of dislocations. This indicates that as the distance between twins decreases in the pre-twin model, the strengthening mechanism of the material shifts from the interaction between dislocations and twin boundaries to a reduction in the amount of dislocation release.

7.7.4 *Summary*

The fabrication of three-dimensional polycrystalline copper using the phase field method was followed by the application of a correlation algorithm to introduce twin grain boundaries with varying spacings in the models. Subsequently, these models underwent MD tensile simulations, and the activities of grain boundaries during the initial deformation phase were analyzed using GBAA. The main findings can be summarized as follows:

(1) Preset twins have minimal impact on the elastic phase but significantly increase yield stress in the plastic phase.
(2) Pre-positioned twins can impede grain boundary activity in the initial deformation stage, leading to increased stress concentration on the grain boundaries. The number of copies of deformed grain boundaries gradually decreases as the spacing of pre-positioned twins decreases. However, the twin boundaries have less effect on the rotation of grain boundaries in different models.
(3) Decreasing preset twin spacing results in larger stress concentration at the intersection location of the twin and grain boundaries. Subsequently, dislocations slide on the neighbor faces of the preset twins during the plastic deformation phase. In larger twin spacing models, dislocations can slip along different surfaces, leading to interaction between the dislocations and the preset twin boundaries.
(4) As the twin spacing decreases, the twin strengthening mechanism gradually changes from the interaction of dislocations and twins to a reduction in partial dislocation release.

References

[1] Wang Juncai. Development and strategies of China's Copper Processing Industry in the New Era. *World Nonferrous Met*, 2003, 0: 8–12.
[2] Li Junmeng. Current status and development trends of China's Copper Processing Industry. *World Nonferrous Met*, 2005, 0: 24–30.
[3] Davis JR. *Copper and Copper Alloys*. University of California, Davis: ASM Specialty Handbook, 2001.
[4] Hall EO. The deformation and ageing of mild steel .3. discussion of results. *P Phys Soc Lond B*, 1951, 64: 747–753.
[5] Petch NJ. The Cleavage strength of polycrystals. *J Iron Steel I*, 1953, 174: 25–28.
[6] Schiotz J, Di Tolla FD, Jacobsen KW. Softening of nanocrystalline metals at very small grain sizes. *Nature*, 1998, 391: 561–563.
[7] Schiotz J, Vegge T, Di Tolla FD, *et al.* Atomic-scale simulations of the mechanical deformation of nanocrystalline metals. *Phys Rev B*, 1999, 60: 11971–11983.
[8] Panzarino JF, Pan ZL, Rupert TJ. Plasticity-induced restructuring of a nanocrystalline grain boundary network. *Acta Mater*, 2016, 120: 1–13.
[9] Zhang M, Sun K, Fang L. Influence of grain boundary activities on elastic and plastic deformation of nanocrystalline Cu as studied by phase filed and atomistic simulation. *Int J Mech Sci*, 2020, 187: 105911.
[10] Zhang M, Xu T, Fang L. Grain boundaries dependence of plastic deformation in nanocrystalline Cu film investigated by phase field and molecular dynamics methods. *Mater Chem Phys*, 2020, 254: 123506.
[11] Olmsted DL, Holm EA, Foiles SM. Survey of computed grain boundary properties in face-centered cubic metals-II: Grain boundary mobility. *Acta Mater*, 2009, 57: 3704–3713.
[12] Homer ER, Foiles SM, Holm EA, *et al.* Phenomenology of shear-coupled grain boundary motion in symmetric tilt and general grain boundaries. *Acta Mater*, 2013, 61: 1048–1060.
[13] Steinbach I, Pezzolla F. A generalized field method for multiphase transformations using interface fields. *Physica D*, 1999, 134: 385–393.
[14] Takaki T, Hirouchi T, Hisakuni Y, *et al.* Multi-phase-field model to simulate microstructure evolutions during dynamic recrystallization. *Mater Trans*, 2008, 49: 2559–2565.
[15] Takaki T, Hisakuni Y, Hirouchi T, *et al.* Multi-phase-field simulations for dynamic recrystallization. *Comp Mater Sci*, 2009, 45: 881–888.

[16] Kim SG, Kim DI, Kim WT, *et al.* Computer simulations of two-dimensional and three-dimensional ideal grain growth. *Phys Rev E*, 2006, 74: 061605.

[17] Plimpton S. Computational limits of classical molecular-dynamics simulations. *Comp Mater Sci*, 1995, 4: 361–364.

[18] Cao AJ, Wei YG. Molecular dynamics simulation of plastic deformation of nanotwinned copper. *J Appl Phys*, 2007, 102: 083511.

[19] Wang LH, Xin TJ, Kong DL, *et al. In situ* observation of stress induced grain boundary migration in nanocrystalline gold. *Scripta Mater*, 2017, 134: 95–99.

[20] Smith L, Zimmerman JA, Hale LM, *et al.* Molecular dynamics study of deformation and fracture in a tantalum nano-crystalline thin film. *Model Simul Mater Sc*, 2014, 22: 045010.

[21] Meyers MA, Ashworth E. A Model for the effect of grain-size on the yield stress of metals. *Philos Mag A*, 1982, 46: 737–759.

[22] Wang YM, Ma E, Chen MW. Enhanced tensile ductility and toughness in nanostructured Cu. *Appl Phys Lett*, 2002, 80: 2395–2397.

[23] Meyers MA, Mishra A, Benson DJ. Mechanical properties of nanocrystalline materials. *Prog Mater Sci*, 2006, 51: 427–556.

[24] Huang CX, Wang K, Wu SD, *et al.* Deformation twinning in polycrystalline copper at room temperature and low strain rate. *Acta Mater*, 2006, 54: 655–665.

[25] Siegel RW. What do we really know about the atomic-scale structures of nanophase materials. *J Phys Chem Solids*, 1994, 55: 1097–1106.

[26] Siegel RW. Cluster-assembled nanophase materials. *Annu Rev Mater Sci*, 1991, 21: 559–578.

[27] Sanders PG, Weertman JR, Eastman JA. Tensile behavior of nanocrystalline copper. In: Suryanarayana, C., Singh, J., Froes, F.H. (Eds.), *Processing and Properties of Nanocrystalline Materials.* The Minerals, Metals & Materials Society, 1996: 379–386.

[28] Li YS, Tao NR, Lu K. Microstructural evolution and nanostructure formation in copper during dynamic plastic deformation at cryogenic temperatures. *Acta Mater*, 2008, 56: 230–241.

[29] Zhang M, Chen J, Xu T, *et al.* Effect of grain boundary deformation on mechanical properties in nanocrystalline Cu film investigated by using phase field and molecular dynamics simulation methods. *J Appl Phys*, 2020, 127: 125303.

[30] Yamakov V, Wolf D, Phillpot SR, *et al.* Dislocation processes in the deformation of nanocrystalline aluminium by molecular-dynamics simulation. *Nat Mater*, 2002, 1: 45–48.

[31] Shabib I, Miller RE. A molecular dynamics study of twin width, grain size and temperature effects on the toughness of 2D-columnar nanotwinned copper. *Model Simul Mater Sc*, 2009, 17: 055009.

[32] Tucker GJ, Foiles SM. Quantifying the influence of twin boundaries on the deformation of nanocrystalline copper using atomistic simulations. *Int J Plasticity*, 2015, 65: 191–205.

[33] Lu K, Lu L, Suresh S. Strengthening materials by engineering coherent internal boundaries at the nanoscale. *Science*, 2009, 324: 349–352.

[34] Cao AJ, Wei YG. Formation of fivefold deformation twins in nanocrystalline face-centered-cubic copper based on molecular dynamics simulations. *Appl Phys Lett*, 2006, 89: 041919.

[35] Zhu YT, Liao XZ. Formation mechanism of fivefold deformation twins in nanocrystalline face-centered-cubic metals. *Appl Phys Lett*, 2005, 86: 103112.

[36] Khan AS, Liu HW. Variable strain rate sensitivity in an aluminum alloy: Response and constitutive modeling. *Int J Plasticity*, 2012, 36: 1–14.

[37] Wei Q, Cheng S, Ramesh KT, *et al.* Effect of nanocrystalline and ultrafine grain sizes on the strain rate sensitivity and activation volume: fcc versus bcc metals. *Mat Sci Eng A*, 2004, 381: 71–79.

[38] Mohamed FA, Li Y. Creep and superplasticity in nanocrystalline materials: current understanding and future prospects. *Mat Sci Eng A*, 2001, 298: 1–15.

[39] Huang P, Wang F, Xu M, *et al.* Dependence of strain rate sensitivity upon deformed microstructures in nanocrystalline Cu. *Acta Mater*, 2010, 58: 5196–5205.

[40] Chen J, Lu L, Lu K. Hardness and strain rate sensitivity of nanocrystalline Cu. *Scripta Mater*, 2006, 54: 1913–1918.

[41] Guduru RK, Darling KA, Scattergood RO, *et al.* Mechanical properties of electrodeposited nanocrystalline copper using tensile and shear punch tests. *J Mater Sci*, 2007, 42: 5581–5588.

[42] Zhang L, Lu C, Tieu K. Atomistic simulation of tensile deformation behavior of sigma 5 tilt grain boundaries in copper bicrystal. *Sci Rep*, 2014, 4: 5919.

[43] Zhang L, Lu C, Michel G, *et al.* Molecular dynamics study on the atomic mechanisms of coupling motion of [001] symmetric tilt grain boundaries in copper bicrystal. *Mater Res Express*, 2014, 1: 015019.

[44] Sanders PG, Eastman JA, Weertman JR. Elastic and tensile behavior of nanocrystalline copper and palladium. *Acta Mater*, 1997, 45: 4019–4025.

[45] Lebedev AB, Burenkov YA, Romanov AE, *et al.* Softening of the elastic modulus in submicrocrystalline copper. *Mat Sci Eng A*, 1995, 203: 165–170.

[46] Yang XS, Wang YJ, Wang GY, *et al.* Time, stress, and temperature-dependent deformation in nanostructured copper: Stress relaxation tests and simulations. *Acta Mater*, 2016, 108: 252–263.

[47] Zhang H, Mendelev MI, Srolovitz DJ. Computer simulation of the elastically driven migration of a flat grain boundary. *Acta Mater*, 2004, 52: 2569–2576.

[48] Janssens KGF, Olmsted D, Holm EA, *et al.* Computing the mobility of grain boundaries. *Nat Mater*, 2006, 5: 124–127.

[49] Zhang H, Upmanyu N, Srolovitz DJ. Curvature driven grain boundary migration in aluminum: molecular dynamics simulations. *Acta Mater*, 2005, 53: 79–86.

[50] Trautt ZT, Mishin Y. Grain boundary migration and grain rotation studied by molecular dynamics. *Acta Mater*, 2012, 60: 2407–2424.

[51] Gorkaya T, Molodov DA, Gottstein G. Stress-driven migration of symmetrical $\langle 1\ 0\ 0 \rangle$ tilt grain boundaries in Al bicrystals. *Acta Mater*, 2009, 57: 5396–5405.

[52] Cahn JW, Mishin Y, Suzuki A. Coupling grain boundary motion to shear deformation. *Acta Mater*, 2006, 54: 4953–4975.

[53] Bobylev SV, Ovid'ko IA. Stress-driven migration, convergence and splitting transformations of grain boundaries in nanomaterials. *Acta Mater*, 2017, 124: 333–342.

[54] Niu RM, Han K. Strain hardening and softening in nanotwinned Cu. *Scripta Mater*, 2013, 68: 960–963.

[55] Zhang YW, Wang TC. Molecular dynamics simulation of interaction of a dislocation array from a crack tip with grain boundaries. *Model Simul Mater Sc*, 1996, 4: 231–244.

[56] Wang W, Lartigue-Korinek S, Brisset F, *et al.* Formation of annealing twins during primary recrystallization of two low stacking fault energy Ni-based alloys. *J Mater Sci*, 2015, 50: 2167–2177.

[57] Khan AS, Zhang HY. Mechanically alloyed nanocrystalline iron and copper mixture: behavior and constitutive modeling over a wide range of strain rates. *Int J Plasticity*, 2000, 16: 1477–1492.

[58] Tang YZ, Bringa EM, Meyers MA. Inverse Hall-Petch relationship in nanocrystalline tantalum. *Mat Sci Eng A*, 2013, 580: 414–426.

[59] Rahman MJ, Zurob HS, Hoyt JJ. A comprehensive molecular dynamics study of low-angle grain boundary mobility in a pure aluminum system. *Acta Mater*, 2014, 74: 39–48.

[60] Wu ZX, Zhang YW, John MH, *et al.* Anatomy of nanomaterial deformation: Grain boundary sliding, plasticity and cavitation in nanocrystalline Ni. *Acta Mater*, 2013, 61: 5807–5820.

[61] Trautt ZT, Mishin Y. Capillary-driven grain boundary motion and grain rotation in a tricrystal: A molecular dynamics study. *Acta Mater*, 2014, 65: 19–31.

[62] Zhu B, Asaro RJ, Krysl P, *et al.* Transition of deformation mechanisms and its connection to grain size distribution in nanocrystalline metals. *Acta Mater*, 2005, 53: 4825–4838.

[63] Kamaya M, Kawamura Y, Kitamura T. Three-dimensional local stress analysis on grain boundaries in polycrystalline material. *Int J Solids Struct*, 2007, 44: 3267–3277.

[64] Kim HS, Estrin Y, Bush MB. Plastic deformation behaviour of fine-grained materials. *Acta Mater*, 2000, 48: 493–504.

[65] Wei YJ, Gao HJ. An elastic-viscoplastic model of deformation in nanocrystalline metals based on coupled mechanisms in grain boundaries and grain interiors. *Mat Sci Eng A*, 2008, 478: 16–25.

[66] Hahn EN, Meyers MA. Grain-size dependent mechanical behavior of nanocrystalline metals. *Mat Sci Eng A*, 2015, 646: 101–134.

[67] Schiotz J, Jacobsen KW. A maximum in the strength of nanocrystalline copper. *Science*, 2003, 301: 1357–1359.

[68] Upmanyu M, Srolovitz DJ, Lobkovsky AE, *et al.* Simultaneous grain boundary migration and grain rotation. *Acta Mater*, 2006, 54: 1707–1719.

[69] Cahn JW, Taylor JE. A unified approach to motion of grain boundaries, relative tangential translation along grain boundaries, and grain rotation. *Acta Mater*, 2004, 52: 4887–4898.

[70] Benson DJ, Fu HH, Meyers MA. On the effect of grain size on yield stress: extension into nanocrystalline domain. *Mat Sci Eng A*, 2001, 319: 854–861.

[71] Rey C. Effects of grain-boundaries on the mechanical-behavior of grains in polycrystals. *Rev Phys Appl*, 1988, 23: 491–500.

[72] Lebensohn RA, Bringa EM, Caro A. A viscoplastic micromechanical model for the yield strength of nanocrystalline materials. *Acta Mater*, 2007, 55: 261–271.

[73] Hill R. The elastic behaviour of a crystalline aggregate. *P Phys Soc Lond A*, 1952, 65: 349–355.

[74] Gao GJJ, Wang YJ, Ogata S. Studying the elastic properties of nanocrystalline copper using a model of randomly packed uniform grains. *Comp Mater Sci*, 2013, 79: 56–62.

[75] Zhang M, Rao ZX, Xu T, *et al.* Quantifying the influence of grain boundary activities on Hall-Petch relation in nanocrystalline Cu by using phase field and atomistic simulations. *Int J Plasticity*, 2020, 135: 102846.

[76] Burke JE, Turnbull D. Recrystallization and grain growth. *Prog Met Phys*, 1952, 3: 220–292.

[77] Smoluchowski R. Theory of grain boundary motion. *Phys Rev*, 1951, 83: 69–70.

[78] Lu L, Chen X, Huang X, *et al.* Revealing the maximum strength in nanotwinned copper. *Science*, 2009, 323: 607–610.

[79] Stukowski A, Albe K, Farkas D. Nanotwinned fcc metals: Strengthening versus softening mechanisms. *Phys Rev B*, 2010, 82: 224103.

[80] Blewitt TH, Coltman RR, Redman JK. Low-temperature deformation of copper single crystals. *J Appl Phys*, 1957, 28: 651–660.

[81] Johari O, Thomas G. Substructures in explosively deformed Cu and Cu-Al alloys. *Acta Metall Mater*, 1964, 12: 1153–1159.

[82] Liao XZ, Zhao YH, Srinivasan SG, *et al.* Deformation twinning in nanocrystalline copper at room temperature and low strain rate. *Appl Phys Lett*, 2004, 84: 592–594.

[83] Li XY, Wei YJ, Lu L, *et al.* Dislocation nucleation governed softening and maximum strength in nano-twinned metals. *Nature*, 2010, 464: 877–880.

[84] Zhao X, Lu C, Tieu AK, *et al.* Deformation mechanisms in nanotwinned copper by molecular dynamics simulation. *Mat Sci Eng A*, 2017, 687: 343–351.

[85] Cao AJ, Wei YG, Mao SX. Deformation mechanisms of face-centered-cubic metal nanowires with twin boundaries. *Appl Phys Lett*, 2007, 90: 151909.

[86] Zhang HW, Fu YF, Zheng YG, *et al.* Molecular dynamics investigation of plastic deformation mechanism in bulk nanotwinned copper with embedded cracks. *Phys Lett A*, 2014, 378: 736–740.

[87] Wang L, Sun J, Li Q, *et al.* Coupling effect of twin boundary and void on the mechanical properties of bulk nanotwinned copper: An atomistic simulation. *J Phys D Appl Phys*, 2019, 52: 055303.

Index

C

D

E

N

O

P

Q

R

www.ingramcontent.com/pod-product-compliance
Lightning Source LLC
LaVergne TN
LVHW020752231125
825833LV00002B/100